Catalysis: Science and Engineering

Catalysis: Science and Engineering

Edited by Ross Beckett

CLANRYE INTERNATIONAL
www.clanryeinternational.com

Clanrye International,
750 Third Avenue, 9th Floor,
New York, NY 10017, USA

ISBN: 978-1-63240-629-3

Cataloging-in-Publication Data

Catalysis : science and engineering / edited by Ross Beckett.
p. cm.
Includes bibliographical references and index.
ISBN 978-1-63240-629-3
1. Catalysis. 2.Catalysts. I. Beckett, Ross.
QD505 .C38 2017
541.395--dc23

For information on all Clanrye International publications
visit our website at www.clanryeinternational.com

CLANRYE INTERNATIONAL

Printed in the United States of America.

Contents

Preface

The rate of increase caused to a chemical reaction through the addition of another substance is known as catalysis. This book discusses the principles of the process of catalysis. It is a significant process in many industries such as the petroleum industry, chemical manufacturing and the food processing industries. This book unravels the recent studies in this field. Chapters in this book delve into the scientific processes as well as industrial applications of catalysis. It includes contributions of experts and scientists which will provide innovative insights into this field. The readers would gain knowledge that would broaden their perspective about catalysis.

The researches compiled throughout the book are authentic and of high quality, combining several disciplines and from very diverse regions from around the world. Drawing on the contributions of many researchers from diverse countries, the book's objective is to provide the readers with the latest achievements in the area of research. This book will surely be a source of knowledge to all interested and researching the field.

In the end, I would like to express my deep sense of gratitude to all the authors for meeting the set deadlines in completing and submitting their research chapters. I would also like to thank the publisher for the support offered to us throughout the course of the book. Finally, I extend my sincere thanks to my family for being a constant source of inspiration and encouragement.

Editor

1

A combined approach for deposition and characterization of atomically engineered catalyst nanoparticles

Q. Yang[1], D. E. Joyce[2], S. Saranu[2], G. M. Hughes[1], A. Varambhia[1], M. P. Moody[1] and P. A. J. Bagot[1]

[1]Department of Materials, University of Oxford, Parks Road, Oxford OX1 3PH, UK
[2]Mantis Deposition Ltd., Thame OX9 3RR, UK

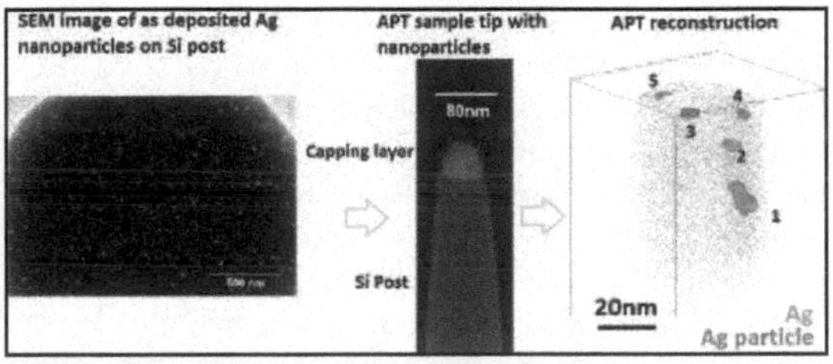

Abstract The structure and composition of catalytic silver nanoparticles (Ag-NPs) fabricated through a novel gas condensation process has been characterized by Scanning Electron Microscopy (SEM) and Atom Probe Tomography (APT). SEM was used to confirm the number density and spatial distribution of Ag-NPs deposited directly onto standard silicon microposts used for APT experiments. Depositing nanoparticles (NPs) directly by this method eliminates the requirement for focussed ion beam (FIB) liftout, significantly decreasing APT specimen preparation time and enabling far more NPs to be examined. Furthermore, by encapsulating deposited particles before final FIB sharpening, the APT reconstruction methodologies have been improved over prior attempts, as demonstrated by comparison to the SEM data. Progress in these areas is vital to enable large-scale catalyst research efforts using APT, a technique, which offers significant potential to examine the detailed atomic-scale chemistry in a wide variety of catalytic NPs.

Keywords Nanoparticles, Atom probe tomography, Heterogeneous catalysis

Introduction

Heterogeneous catalysts facilitate the production of a wide range chemicals that are critical to a broad spectrum of industries. In addition to their importance financially, they also often play a significant role for the environment, for example, by removing atmospheric pollutants from internal combustion engine exhausts.[1] Current research in the development of heterogeneous catalyst must balance two major, and at times, conflicting goals. These are to limit the total loading of expensive transition metals, while simultaneously improving catalyst performance in terms of selectivity and endurance. To achieve these demanding targets, a far more detailed understanding of the fundamental structure and chemistry of catalyst materials is required, and importantly how these may alter during reaction conditions. Such catalysts are most commonly employed in the form of nanoparticles (NPs), therefore advanced characterization methods are necessary to study them in detail. A variety of experimental techniques have to date been applied to examine the structure and performance of numerous NPs, including transmission electron microscopy (TEM), X-ray diffraction (XRD), X-ray absorption spectroscopy (XAS), and Brunauer, Emmett and Teller (BET) isotherm tests, all of which supply valuable, complementary information. TEM is extensively used for determining the morphology and size distribution of particles, see, for example, Refs.[2–7]. More recently, the application of Scanning TEM (STEM) has also enabled chemical analysis of individual NPs,[8–10] while XRD can also be employed to investigate crystallinity and particle size.[11,12] The local atomic structure and in particular chemical state of NPs are becoming increasingly accessible using XAS,[13,14] while finally BET isotherm tests are used to detect another vitally important parameter for catalysts, namely overall active surface area.[15,16]

The industrial production of NPs typically utilize wet chemical synthesis methods. A pressing issue is ensuring that all of the synthesis NPs are all of a similar size, shape, and with the desired structure, e.g. core-shell. Without this

*Corresponding author, email qifeng.yang@materials.ox.ac.uk

fine level of control, the efforts to design atomically engineered catalyst materials, to address the challenges outlined above, will be negated. A further difficulty is the risk that the high-resolution characterization tools and associated sample preparation methods used to characterize these NPs may yield unrepresentative data owing to the small numbers of NPs that can be realistically examined in the course of an experimental campaign.

In this study, a novel method to synthesize NPs of well-defined sizes by means of a cluster beam deposition method is described, together with the development of an approach to prepare these NPs in a form suitable for characterization by Atom Probe Tomography (APT). Atom probe tomography offers a unique combination of 3D atomic-scale spatial and chemical resolution. Indeed number of recent studies using this approach to examine catalytic NPs have been attempted.[17–19] These have highlighted the clear benefits of this method, particularly in the case of nano-engineered alloyed structures such as core-shell NPs; even the most advanced STEM instruments can struggle to resolve the atomic-scale chemical structure of these, particularly for elements of similar masses (and hence electron scattering factors). Despite the benefits, a current disadvantage for APT studies is that they require highly controlled field evaporation of ions from a smooth very sharp needle-shaped specimen to ensure maximum spatial resolution. Thus examining individual catalyst NPs, often dispersed over a porous support material remains a considerable challenge, and by no means routine. Therefore, a second goal of the current study is to demonstrate further developments in the approach to prepare specimen suitable for APT, using Focussed Ion Beam (FIB) methods to encapsulate a small number of NPs prepared by cluster beam deposition within a suitable sample geometry.

Experimental

Silver (Ag) NPs were prepared by means of terminated gas condensation, using a Mantis Nanogen 50 source. A schematic of the equipment and method is shown in Fig. 1. During sample preparation, a metallic vapor was generated in the condensation zone by magnetron sputtering of an Ag target under an argon environment. The number density of atoms, ions, and clusters in this vapor was controlled by setting the appropriate magnetron power and gas flowrate conditions. A constriction in the outlet of the aggregation zone creates an elevated stagnation pressure inside the chamber on the order of a few tens of Pascals, equivalent to a mean free path of a few tens of micrometers. The large number of collisions in this region thermalizes the vapor, promoting the nucleation and subsequent coalescence of NPs. The magnetron is mounted on a linear translation arm, which provides a means to increase the dwell time of the NPs in the aggregation zone. An expansion zone (circled in red in Fig. 1) creates a rapid pressure drop where excess carrier gas is pumped away through the differential pumping port and a NP beam emerges. The sputtering power, aggregation length, inert gas flowrate, stagnation pressure, and chamber wall temperature can thus all strongly influence NP size and structure.

One of the features of this method is that a large fraction of the NPs become charged as a consequence by passing through a plasma in the aggregation zone. A quadrupole can therefore be used to filter NPs by their mass-to-charge ratio. The resulting charge state is a function of the experimental conditions, but generally a large proportion of the NPs (>70% for Ag) have a single negative charge, meaning the total flux remains quite high even after mass filtering. Because of the associated charge, the NPs can also be accelerated toward a biased substrate to ensure good adhesion. The quadrupole mass filter is attached after the expansion zone (shown as a green rectangular in Fig. 1), and a mass spectrum can be generated by scanning the frequency of the voltage bias on the quadrupole rods. In order to mass-select nanoclusters, AC and DC voltage is applied to four straight metal rods inside the quadrupole. Any ionized NP passing through the rods will be forced into an oscillating path. For a given AC frequency and amplitude, only one mass (within the resolution of the instrument) will continue on a stable oscillating path and other masses will be rejected. In this way, particles having the mass-to-charge-ratio match with predetermined value will make their way through the quadrupole and on to the substrate. In this experiment, the threshold was set as 10 nm for the diameter of particles as an upper limit.

For APT experiments, the substrate onto which the Ag-NPs were deposited took the form of Si multi-tip coupon. These

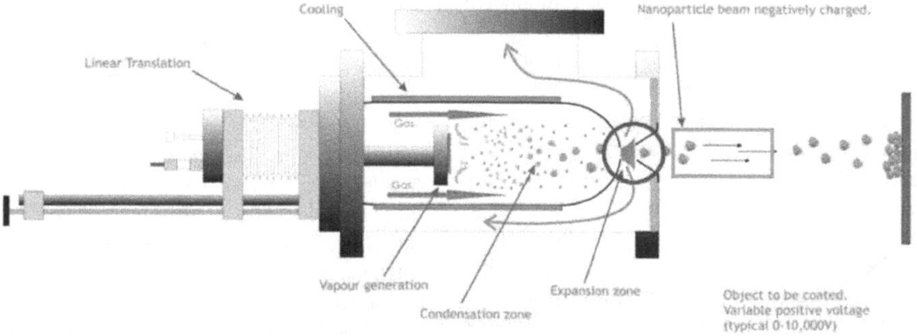

Figure 1 Schematic configuration of the Mantis Nanogen 50 source. The circled red area is the zone where pressure drop happens and particles are expelled. The rectangular green zoom indicates the mass spectrometer, used to filter charged particles (Courtesy Mantis Deposition Ltd. (Thame, UK))

coupons as used extensively to prepare conventional FIB liftout samples for APT studies (however FIB liftout was not required in this study). The APT sample preparation procedures in the current work are schematically depicted in Fig. 2. In Table 1, the operational parameters used in the Nanogen deposition are listed. The aim was to deposit a uniform layer of NPs across the array of flat-topped post-s on the Si coupon, within which the NPs were clearly separated yet still with relatively high number density on the surface. After particle deposition, the tops of the Si posts were covered in a platinum layer, using the e $^-$ beam of a Zeiss Auriga Dual-Beam FIB microscope. The organic precursor used for this coating is $C_9H_{16}Pt$ (methylcyclopentadienly(trimethyl)platinum). The purpose of this step is to completely encapsulate the particles in a matrix, after which Ga ion beam milling can be used to mill the samples into their final needle shape suitable for APT. Overall, the goal of this sample preparation route serves to embed a small number of NPs near the apex of the APT needle. This should offer more controlled field evaporation process as the tip is analysed and hence improved reconstructions in comparison to previous electrophoresis-based specimen preparation methods that provided a layer of free-standing NPs completely exposed at the surface of the needle.[17,18] Prepared samples were then analyzed using a LEAP 3000 HR^{TM}, running with 200 kHz laser pulses at 0.2 nJ and a stage temperature of 50 K.

Results

Deposition

For the deposition process, the assumptions are that particles are in a single (1 +) charge state, spherical, and of known density then allows determination of their diameters, which are shown as a size distribution in Fig. 3. In this experiment, an upper limit of the NP size selection was set to 10 nm, so particles larger than this should be eliminated. However, inevitably some fraction of the produced particles are either neutral or doubly charged. Therefore, a small proportion of all particles deposited will lie out with the size range shown. Overall however, a narrow distribution of NPs is indeed observed, with the majority of particles deposited having a diameter ~6 nm.

Scanning Electron Microscopy

The quality of the deposition was examined after preparation by SEM. Figure 4 shows SEM images of as-deposited Ag-NPs on the apex of a single Si post. Bright spots on the image indicate individual NPs. The size of the NPs, as measured by SEM, is approximately 10 nm and these particles have a uniform size distribution. However, some particles larger than 20 nm can be observed, which could be the result of coalescence or non-singly charged ions. Overall, the SEM-images are in relatively close agreement with the results from the quadrupole mass filter (Fig. 3), indicating a good control of particle size. In addition, the higher magnification image of Fig. 4*b* shows an even coverage of well separated NPs, which is promising for the described APT specimen preparation route. To obtain further spatial information on the surface distribution of the NPs, a nearest neighbour (NN) distribution was calculated using an in-house developed image analysis algorithm based on MATLAB. The first stage in this analysis is to determine the centers of individual NPs and record the 2D coordinates (Fig. 5*a*). Following this, a NN distribution can be calculated (based on the 1-NN). The values obtained are plotted in Fig. 5*b*, which shows that most NPs lie within 10 nm of each other.

Atom probe tomography

Figure 6*a* shows an Ag-NP-coated post-along with a series of tests aimed at burying the NPs below a region of Pt deposited using the FIB electron beam (which appear as the bright protrusions in the picture). One of these Pt-based deposition protrusion was then subject to annular milling using the Ga + beam to produce the final sample ready for

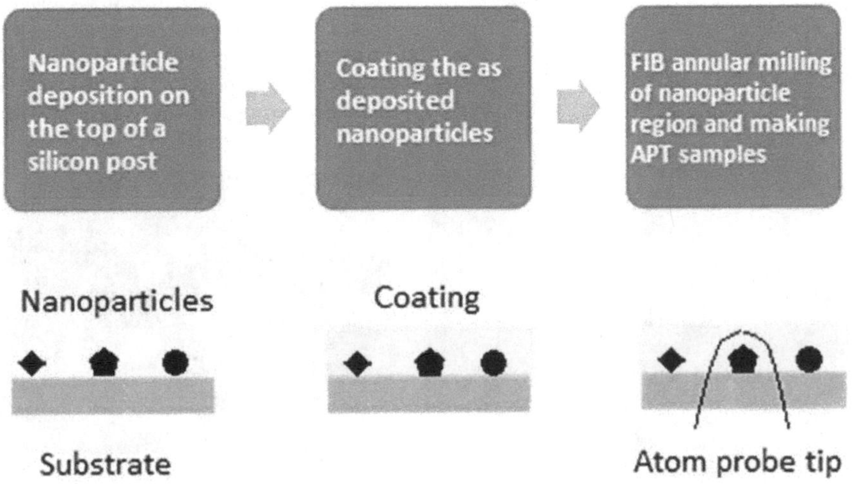

Figure 2 Simple schematic for producing atom probe specimens that incorporate the deposited NPs. Pink substrate represents the flat top of a silicon post. Black NPs are the products of cluster beam deposition. The yellow region is the Pt coating as deposited in the focussed ion beam (FIB)

Table 1 Operation parameters for the Nanogen 50

Running conditions	Parameters
Base pressure	4×10^{-6} mbar
Argon flowrate	100 sccm
Aggregation length	50 mm
Pulsed DC power	20 kHz, 1 μs, 100 W (0.34 A, 298 V)
MesoQ filter	10 nm
Substrate bias	0.5 kV
Deposition rate	60 ngm s^{-1}
Deposition time	45 s

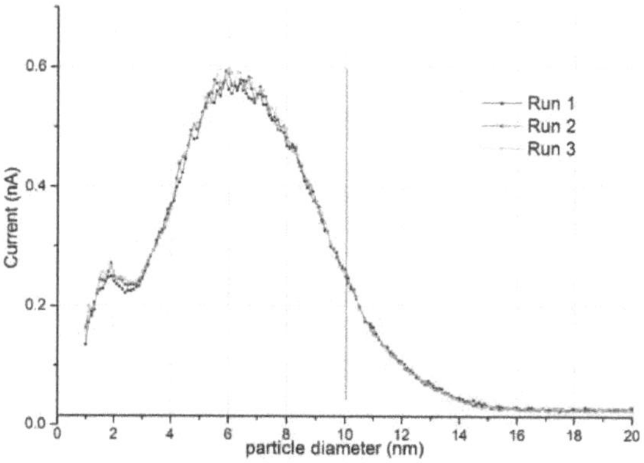

Figure 3 Size distribution of the produced NPs. The data is obtained from the mass-to-charge ratio detected from the quadrupole mass filter. The legends represent different scans during experiment, showing little change in the size distribution between different measurements

APT analysis as shown in Fig. 6*b*. In this, the Pt capping layer and silicon post have good contrast, which aids final sharpening; the Ag-NPs lie at the interface between the bright and dark regions. The prepared needle samples were then analyzed via APT, and the resulting atom map is shown in Fig. 6c.

In the reconstruction, the individual Ag-NPs, coating layer and supporting post-are clearly revealed upon application of the iso-concentration surface. The data set thus closely correlates with the SEM image of Fig. 6*b*. In the upper region, aside from Ag the major detected ions are O, Pt, and Ga. Oxygen is abundant on the outer surface of the dataset, which is usual for samples exposed to the atmosphere. However, the oxygen layer is thin, indicating that the deposited Pt layer has preserved the embedded NPs. The precursor species used to form the Pt layer contains substantial amounts of C, which is also detected, along with Ga^+ from ion implantation during the FIB milling process.[20] Efforts to refine the quality of the deposited layers, utilizing higher purity dedicated coating instrumentation are actively being pursued for future refinements of this method.

To examine the distribution of the Ag atoms separate atom maps were generated, which are shown in Fig. 7. The first of these, Fig. 7*a*, shows that a significant quantity of the Ag atoms are distributed throughout the analysis volume, rather than all being contained within distinct NPs. The origin of this Ag distribution may be owing to a combination of effects. One cause may be trajectory aberrations in the evaporation of atoms during APT analysis, owing to the significant difference between the electric field strength required to evaporate Ag and Pt atoms respectively. Furthermore, the known high surface mobility of Ag under intense electric fields could further contribute to any trajectory aberration.[21] Another explanation is the possible presence of very small NPs/individual Ag atoms on the microtip coupon following deposition. Evidence for the latter is shown in the NN distribution of Fig. 5*b*, which indicates a substantial fraction of very small species (<1 nm) present. Refining deposition conditions (including substrate bias which may also be causing NP breakup on impact) and those of the FIB-capping layer would help to identify the origin of the spread in Ag distribution. Nevertheless, the atom maps demonstrate a considerable improvement on previous attempts using electrophoresis-based methods, with distinct intact Ag-NPs clearly resolvable in the data by means of an Ag iso-concentration surface (set at two Ag atoms/nm^3). This is also shown alone in Fig. 7*a*, highlighting five isolated Ag-NPs detected, which are labeled from '1–5'.

The volume and calculated equivalent diameters of the five isolated particles are listed in Fig. 7*b*. The volume data is directly derived from isosurfaces, from which the equivalent diameters are then calculated assuming all particles are spherical. The calculated equivalent diameters are all below 10 nm, in agreement with the mass filtering and SEM results in Figs. 3 and 4. The calculated NN distances for the five NPs are also listed in Fig. 7*b*, and these, with the exception of NP, 5,

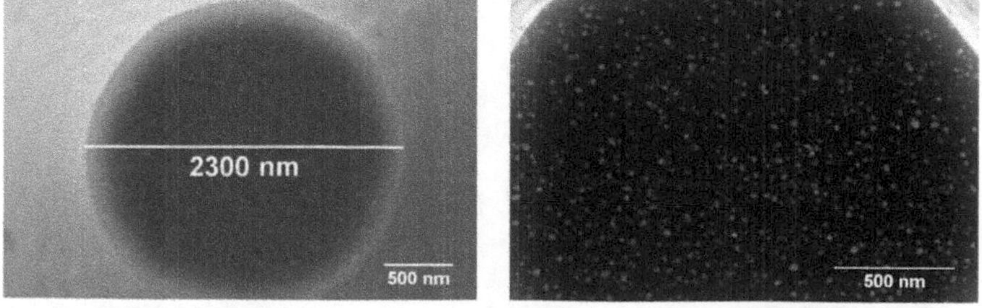

a b

Figure 4 SEM images of the deposited NPs on the top of a silicon microtip post. *a* A full picture of the flat end of a silicon post, the diameter of which is about 2300 nm. *b* A magnified picture of the silicon post, bright dots on which are individual Ag-NPs

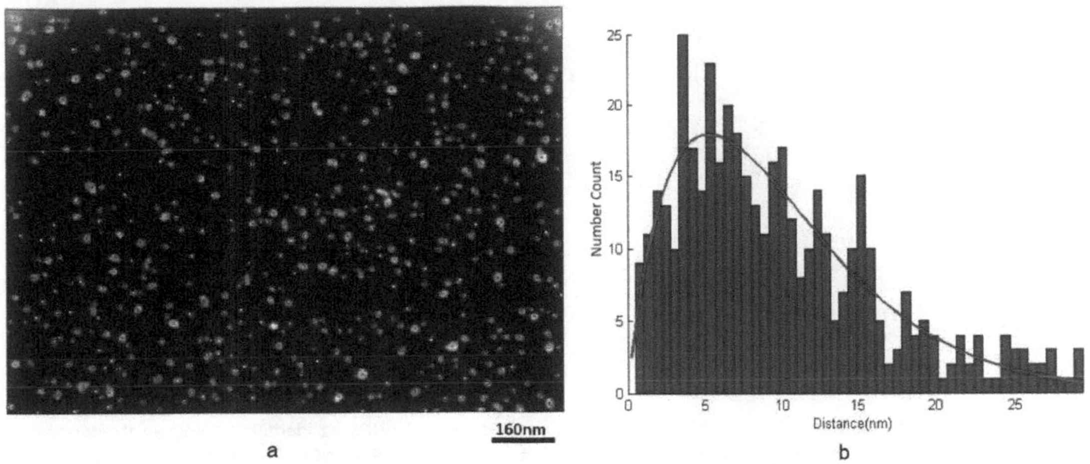

Figure 5 *a* **SEM image with green dots shows the detected centers of deposited particles. Totally, 464 coordinates of those dots have been recorded.** *b* **Resulting nearest neighbour (NN) distibution of all detected particles. The distribution indicates that most particles are within 10 nm of each other**

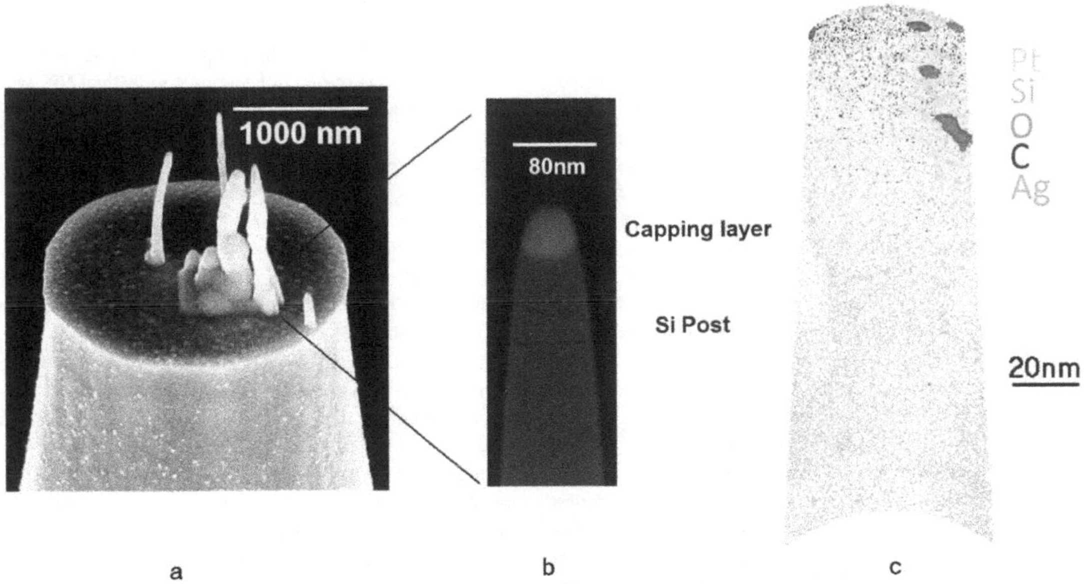

Figure 6 Atom probe sample preparation under high resolution SEM. *a* **A single Si post-following Ag-NP deposition and subsequent capping by Pt using the electron beam.** *b* **Final sample produced by annular milling. The radius of the tip is around 50 nm.** *c* **Atom map reconstruction tip shown in** *b***. The data set shows distinct regions: an upper layer consisting of Pt and C species, then a layer of Ag-NPs before the Si atoms in the post-are detected. (Ag particles are enclosed in blue color)**

all lie within the broad distribution based on the SEM observations in Fig. 5*b*. Nanoparticle five resides just within the APT analysis field of view, and it may well be that further nearer NPs are located just outside the analysis volume.

A particle composition analysis shows there are a range of other elements enclosed within the isolated Ag particles. These are mostly Si, Pt, and Ga, which are most likely introduced through trajectory aberrations/Ag surface mobility along with sample preparation issues in the FIB. Alternatively, the existence of Si within identified Ag-NPs could be the result of the particle deposition process, whereby a fraction of Ag particles impact on the surface and mix with Si atoms from the surface.[22] However, the accuracy of compositional analyses through this method could be easily enhanced by improvements in deposition/sample preparation conditions.

Discussion

In this study, we have demonstrated two important steps for the synthesis and accurate characterization of NPs. First, a novel method allowing the production of tightly controlled

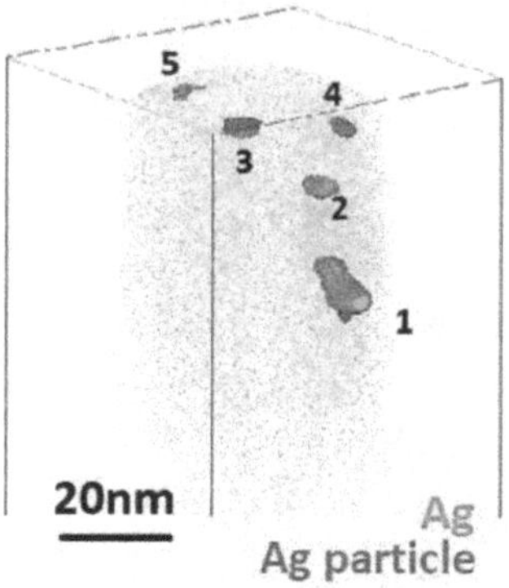

a

No.	Volume (nm^3)	Equivalent diameter (nm)	Nearest Neighbour distance (nm)
1	332	8.6	18.4
2	80.7	5.4	15.0
3	66.0	5.0	18.4
4	30.4	3.9	15.0
5	14.7	3.0	33.6

b

Figure 7 ***a*** **Ag atoms detected during APT analysis, along with distinct Ag-NPs as identified using an iso-concentration surface (2 atoms/nm^3). Isolated NPs labeled '1–5'.** ***b*** **Dimensions of the isolated Ag-NPs. The volumes are derived from the isosurfaces and the equivalent diameters are based on a spherical volume. Nearest neighbor (NN) distances are calculated based on the central coordinates of those particles within the atom data reconstruction**

distributions of NPs was presented. Importantly, this approach is highly suitable for the controlled deposition of deposition of NPs onto a substrate and hence can underpin a new specimen preparation route in providing much higher quality APT analyses.

The terminated gas condensation method yields reproducible batches of NPs with uniform sizes, but unlike conventional wet chemistry approaches, it does not result in the co-deposition of extensive carbon support materials/surfactants that can particularly complicate APT analysis.[23] Furthermore, terminated gas condensation also offers the capability of subsequent direct deposition onto either characterization substrates as shown here or onto support materials for catalysis applications, such as electrode assemblies for fuel cells, under the same deposition conditions. This presents an opportunity to assess using advanced electron microscopy, APT and other methods (e.g. AFM) whether the NPs engineered have been intended, if they have been deposited at the optimum number density to ensure maximum catalytic performance, and potentially even post-service analyses to assess particle stability during operation. An important additional benefit to APT is that entire Si microtip arrays can be coated, rather than just single tips as used previously.[18,19] This approach also avoids the traditionally used FIB liftout approach for APT sample preparation, which significantly reduces preparation time and hence increases sample throughput.[24] The latter of these benefits is particularly important to ensure representative data on catalyst particles can be obtained. Table 2 summarizes the pros and cons of the currently developed methods for preparing viable APT NP-containing samples. Note that in some cases sample form will also play a significant factor; for example the electrophoresis method requires NPs to be suspended in a suitable solvent, while particle deposition methods require suitable target sources to be available.

The most promising finding in the characterization results obtained here is the identification of individual NPs within the APT analysis volume, with a size a spatial separation in good agreement with the Nanogen 50/SEM data. This clearly establishes the suitability of this approach, while also indicating areas where further attention needs to be paid to improve the method. For such small NPs, the use of TEM would provide higher resolution shape and size information that would aid the reconstructions, although the SEM examination is also valuable for providing accurate statistical data on number density and particle separation.

In terms of the chemical analysis of the particles, the current data shows them to contain high levels of Si as well as expected Ag. This indicates that the particles have perhaps penetrated the Si support. The mixing of these two elements within the reconstructed NP cores may be a result of either the known high surface mobility of Ag atoms under the intense electric field applied by the APT analysis,[21] intermixing of elements during NP deposition, or else trajectory aberrations in the field evaporation process resulting in a blurring of the spatial positions. This latter effect is a well known phenomenon, seen in other material systems, where there are significant differences in evaporation fields between analyzed species, for example Fe matrix atoms apparently located inside Y_2O_3 particles in oxide-dispersion strengthened steels.[26,27] Furthermore, a similar recent APT analysis on PtFe NPs revealed the same issue of matrix penetration into the particles.[22] However a lower substrate

Table 2 **Comparison of different sample preparation methods for nano-particle analysis. There are four methods listed and the first three has been reported in various literatures. The fourth one is proposed in this work**

Methods used for NP sample preparation	Electrophoresis method[17,18,23]	Liftout method[19,22,24]	*In-situ* growth method[25]	Particle deposition method
Reconstruction accuracy	Low	High	High	High
Cost	Low	High	Medium	Medium
Time for preparation	Fast	Slow	Slow	Medium

bias and/or the use of different techniques (e.g. sputter coaters) to embed the NPs in more closely matched matrix materials, in terms of the field required to evaporate, would greatly improve if not entirely eradicate these issues.

Preparing a closely evaporation field-matched matrix material that fully surrounds the NPs would unlock the full power of the APT technique.[22,28] Further to the model NP materials examined here, recent studies have already given strong indications of this potential, for example it has been possible using APT to correlate the thickness of Pd shell layers in Ag@Pd core-shell NPs to their efficacy as formic-acid cracking catalysts.[18] Modern catalysts are becoming increasingly complex in terms of their compositions, and APT can play a significant role in their characterization and hence further design, as a previous study of AuAg catalysts prepared for APT by FIB liftout has demonstrated.[19] Both these investigations have, however, been limited by the numbers of NPs that were able to be successfully analyzed, yet the current work suggests a clear approach to resolving many of these issues making routine analysis of catalytic NPs by APT a desirable and achievable goal.

Conclusions

A novel approach has been developed for preparing NPs using a terminated gas condensation method, which are tightly controlled in terms of size, shape, and number density as deposited on substrate materials. Following deposition, the clear benefits of using a multi-technique advanced characterization approach based on SEM/APT to examine deposited NPs in detail have been demonstrated. For APT analysis, deposition of NPs directly onto Si microtip coupons, followed by subsequent encapsulation/shaping using a FIB provides a new method for greatly increasing the yield and quality (for example in comparison to electrophoresis) of resulting APT data. This combined preparation/characterization approach offers great potential for producing nano-engineered catalysts with far improved understanding of the links between atomic-scale structure and catalytic performance.

Conflicts of interest

The authors declare that there are no conflicts of interest.

Acknowledgements

The authors want to thank M.D. Green from Mantis Deposition Ltd for the initial discussion of the results.

References

1. J. Kašpar, P. Fornasiero and N. Hickey: *Catal. Today*, 2003, **77**, (4), 419–449.
2. S. B. Simonsen, I. Chorkendorff, S. Dahl, M. Skoglundh, J. Sehested and S. Helveg: *J. Am. Chem. Soc.*, 2010, **132**, (23), 7968–7975.
3. R. He, Y.-C. Wang, X. Wang, Z. Wang, G. Liu, W. Zhou, L. Wen, Q. Li, X. Wang, X. Chen, J. Zeng and J. G. Hou: *Nat. Commun.*, 2014, **5**, 4327.
4. S. Helveg and P. L. Hansen: *Catal. Today*, 2006, **111**, (1-2), 68–73.
5. S.-J. Cho, J.-C. Idrobo, J. Olamit, K. Liu, N. D. Browning and S. M. Kauzlarich: *Chem. Mater.*, 2005, **17**, (12), 3181–3186.
6. S. Trasobares, M. L. ópez-Haro, M. Kociak, K. March and F. de: *Angew. Chem. Int. Ed. Engl*, 2011, **50**, (4), 868–872.
7. M. Lopez-Haro, L. Dubau, L. Guétaz, P. Bayle-Guillemaud, M. Chatenet, J. André, N. Caqué, E. Rossinot and F. Maillard: *Appl. Catal. B Environ.*, 2014, **152-153**, 300–308.
8. P. D. Nellist and S. J. Pennycook: *Science*, 1996, **274**, (5286), 413–415.
9. H. E. P. D. Nellist, S. Lozano-Perez and D. Ozkaya: *J. Phys. Conf. Ser.*, 2010, **241**, (1), 012067.
10. P. D. Nellist, S. Lozano-Perez and D. Ozkaya: *J. Phys. Conf. Ser.*, 2012, **371**, (1), 012027.
11. M. Min, J. Cho, K. Cho and H. Kim: *Electrochim. Acta*, 2000, **45**, (25-26), 4211–4217.
12. K. Uchino, E. Sadanaga and T. Hirose: *J. Am. Ceram. Soc.*, 1989, **72**, (8), 1555–1558.
13. R. Kaegi, A. Voegelin, B. Sinnet, S. Zuleeg, H. Hagendorfer, M. Burkhardt and H. Siegrist: *Environ. Sci. Technol.*, 2011, **45**, (9), 3902–3908.
14. J.-I. Park, M. G. Kim, Y. Jun, J. S. Lee, W. Lee and J. Cheon: *J. Am. Chem. Soc.*, 2004, **126**, (29), 9072–9078.
15. R. Mueller, L. Mädler and S. E. Pratsinis: *Chem. Eng. Sci.*, 2003, **58**, (10), 1969–1976.
16. S. Nakade, Y. Saito, W. Kubo, T. Kitamura, Y. Wada and S. Yanagida: *J. Phys. Chem. B*, 2003, **107**, (33), 8607–8611.
17. C. Eley, T. Li, F. Liao, S. M. Fairclough, J. M. Smith, G. Smith and S. C. E. Tsang: *Angew. Chem. Int. Ed. Engl.*, 2014, **53**, (30), 7838–7842.
18. K. Tedsree, T. Li, S. Jones, C. W. Chan, K. M. Yu, P. A. Bagot, E. A. Marquis, G. D. Smith and S. C. Tsang: *Nat. Nanotechnol.*, 2011, **6**, (5), 302–307.
19. P. Felfer, P. Benndorf, A. Masters, T. Maschmeyer and J. M. Cairney: *Angew. Chemie Int. Ed.*, 2014, **126**, (42), 11372–11375.
20. M. K. Miller, K. F. Russell, K. Thompson, R. Alvis and D. J. Larson: *Microsc. Microanal.*, 2007, **13**, (6), 428–436.
21. K. Moazed: *AIME*, 1964, **230**, 234.
22. E. Folcke, R. Lardé, J. M. Le Breton, M. Gruber, F. Vurpillot, J. E. Shield, X. Rui and M. M. Patterson: *J. Alloys Compd.*, 2012, **517**, 40–44.
23. T. Li, P. A. J. Bagot, E. Christian, B. R. C. Theobald, J. D. B. Sharman, D. Ozkaya, M. P. Moody, S. C. E. Tsang, G. D. W. Smith and A. C. S. Catal: *ACS Catal.*, 2014, **4**, (2), 695–702.
24. P. Felfer, T. Li, K. Eder, H. Galinski, A. Magyar, D. Bell, G. D. W. Smith, N. Kruse, S. P. Ringer and J. M. Cairney: *Ultramicroscopy*, 2015. doi:10.1016/j.ultramic.2015.04.014
25. O. Moutanabbir, D. Isheim, H. Blumtritt, S. Senz, E. Pippel and D. N. Seidman: *Nature*, 2013, **496**, (7443), 78–82.
26. D. Larson, P. Maziasz, I. -S. Kim and K. Miyahara: *Scr. Mater.*, 2001, **44**, (2), 359–364.
27. A. J. London, S. Lozano-Perez, M. P. Moody, S. Amirthapandian, B. K. Panigrahi, C. S. Sundar and C. R. M. Grovenor: *Ultramicroscopy*, 2015. doi:10.1016/j.ultramic.2015.02.013
28. D. J. Larson, A. D. Giddings, Y. Wu, M. A. Verheijen, T. J. Prosa, F. Roozeboom, K. P. Rice, W. M. M. Kessels, B. P. Geiser and T. F. Kelly: *Ultramicroscopy*, 2015. doi:10.1016/j.ultramic.2015.02.014

2

Determining surface structure and stability of ε-Fe_2C, χ-Fe_5C_2, θ-Fe_3C and Fe_4C phases under carburization environment from combined DFT and atomistic thermodynamic studies

Shu Zhao[1,2,3], Xing-Wu Liu[1,2,3], Chun-Fang Huo[1,2], Yong-Wang Li[1,2], Jianguo Wang[1] and Haijun Jiao*[1,4]

[1]State Key Laboratory of Coal Conversion, Institute of Coal Chemistry, Chinese Academy of Sciences, Taiyuan 030001, China
[2]National Energy Center for Coal to Liquids, Synfuels China Co., Ltd, Huairou District, Beijing 101400, China
[3]University of Chinese Academy of Sciences, No. 19A Yuquan Road, Beijing 100049, China
[4]Leibniz-Institut für Katalyse e.V. an der Universität Rostock, Albert-Einstein Strasse 29a, 18059 Rostock, Germany

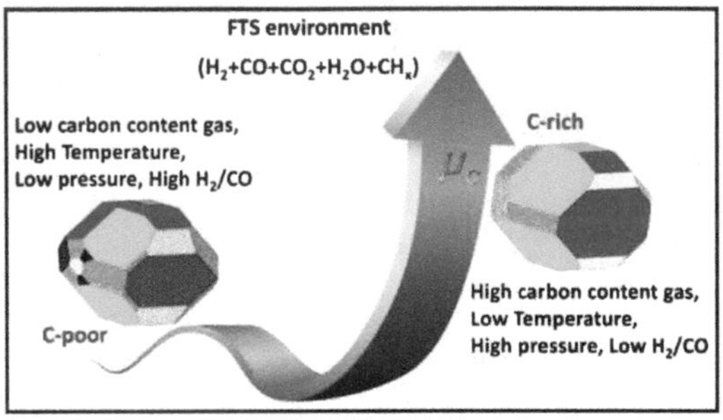

Abstract The chemical–physical environment around iron based FTS catalysts under working conditions is used to estimate the influences of carbon containing gases on the surface structures and stability of ε-Fe_2C, χ-Fe_5C_2, θ-Fe_3C and Fe_4C from combined density functional theory and atomistic–thermodynamic studies. Higher carbon content gas has higher carburization ability; while higher temperature and lower pressure as well as higher H_2/CO ratio can suppress carburization ability. Under wide ranging gas environment, ε-Fe_2C, χ-Fe_5C_2 and θ-Fe_3C have different morphologies, and the most stable non-stoichiometric termination changes from carbon-poor to carbon-rich (varying surface Fe/C ratio) upon the increase in $\Delta\mu_C$. The most stable surfaces of these carbides have similar surface bonding pattern, and their surface properties are related to some common phenomena of iron based catalysts. For these facets, χ-Fe_5C_2-(100)-2.25 is most favored for CO adsorption and CH_4 formation, followed by θ-Fe_3C-(010)-2.33, ε-Fe_2C-(1$\bar{2}$1)-2.00 and Fe_4C-(100)-3.00, in line with surface work function and the charge of the surface carbon atoms.

Keywords Iron carbides, Carburization, Morphology, Fischer–Tropsch synthesis, DFT

Introduction

Fischer–Tropsch synthesis (FTS) is an important technology in converting synthesis gas generated from coal, natural gas and biomass into oil and value added chemicals.[1] Almost a hundred years after its discovery, FTS has been attracting increasing interest worldwide due to the increasing oil prices. Despite of the large scale industrial applications and extensive studies on this important technology in the past decades, the detailed FTS mechanisms are still not fully understood and many explanations to the experimental observations are premature and lack of scientific rationalization. One of these uncertainties is the surface structures, the corresponding active sites and their roles under FTS conditions.

Suitable FTS catalysts for industrial applications are iron or cobalt based, and iron based catalysts may become more dominant along with the expanding of FTS capacity due to the higher availability and lower cost of iron compared to cobalt. Freshly prepared iron based FTS catalysts are generally iron oxides [mainly hematite (α-Fe_2O_3) and also small amount of maghemite (γ-Fe_2O_3) or even ferrihydrate] and they have to be reduced before becoming FTS active. During the reduction, α-Fe_2O_3 is firstly reduced to magnetite

*Corresponding author, email haijun.jiao@catalysis.de

(Fe_3O_4) by using H_2, synthesis gas or CO and then partially transferred to metallic iron and iron carbides in varying proportions depending on the operating conditions.[2,3] Such multiple phases with very fine crystalline dimensions (several to tens of nanometers) make the characterization and identification of the active phases very difficult, and often such phases will change once the environment around changes as found in most *ex situ* analyses. This leads to the misunderstanding and misinterpretation of the experimentally observed phenomena from real FTS reaction tests.[4]

In Fe based FTS, ε-Fe_2C, χ-Fe_5C_2 and θ-Fe_3C phases have been detected experimentally.[3,5] Both ε-Fe_2C and χ-Fe_5C_2 phases have a hexagonally close packed structure[6,7] but differ in interstitial carbon sites. In ε-Fe_2C, the carbon atom is in the iron octahedral center, while in θ-Fe_3C and χ-Fe_5C_2, the carbon atom is in the iron trigonal prismatic center. Hexagonal carbide (ε-Fe_2C) has been identified as the carburization product of H_2 reduced iron and CO at low temperature,[8] and is the sole component up to 520 K and stable up to 600 K.[8,9] Not formed during FTS at low temperature (<575 K),[10] θ-Fe_3C is only found after carburization above 720 K.[11,12] Under both CO and synthesis gas, ε-Fe_2C is considered as χ-Fe_5C_2 precursor, which is subsequently transformed into θ-Fe_3C at high temperature.[8,13] Königer *et al.*[6] observed that ε-Fe_2C can be converted to χ-Fe_5C_2 after annealing at 423 K, and the χ-Fe_5C_2 phase starts to transform to θ-Fe_3C at 573 K, which is the dominant phase after annealing at 723 K.[6] The exact transformation temperature for the formation of the specific carbide phases depends on many factors, such as crystallite size, morphology, surface texture and promoters or inhibitors as well as the other environment conditions (pressure and gas composition). By using *ab initio* atomistic thermodynamics to investigate the stability of bulk carbide phases, de Smit *et al.*,[3] found that the stable carbide phases depend highly on carbon chemical potential (μ_C) imposed by gas phase surroundings and emphasized the importance of the controlling chemical–physical environment around the catalyst for forming an efficient FTS system.

Despite the fact that pretreatments can affect the catalytic performance of iron based catalysts, the corresponding studies of the surface properties along with the change of the gas environment (or chemical potential) are rare and most have focused on the defined stoichiometric terminations and the non-stoichiometric terminations of carbides have not been considered.[14,15] The stability and structure as well as electronic and magnetic properties of ε-Fe_2C[16–18] and θ-Fe_3C[14,19–22] have been investigated intensively. In addition, the adsorption and activation of CO and H_2 as well as C_xH_y formation on the (100), (001) and (010) surfaces of Fe_3C have been computed.[23–26] Although not directly detected under FTS conditions, Fe_4C can be formed by incorporating carbon atoms into the face centered cubic γ-iron lattices[27] and we included Fe_4C in our study for comparison. The properties of the (100), (110), (111) surfaces of Fe_4C[28] and the CO adsorption properties on these surfaces also have been investigated in our previous work.[29] Recently we found that pretreating conditions, such as temperature, pressure and H_2/CO ratios of an idealized and closed equilibrium system, have significant impact on the relative stability of the χ-Fe_5C_2 facets in different Fe/C ratios.[30] However, the effects of non-idealized and wide varying operating environments on surface composition and stability of other iron carbides (ε-Fe_2C, θ-Fe_3C and Fe_4C) have not ever been considered. In fact, the real FTS chemical–physical environment may result in wide varying non-equilibrium nature for the catalytic system and the change trends of the catalyst phases will be driven by carbon chemical potential (μ_C) under conditions with complicated mechanisms. Although the gas environment may result in non-equilibrium, the catalysts can be considered to reach the steady state for a continuous flow of reactants and products at defined conditions. For a fundamental understanding into the FTS mechanisms, systematic studies of the relationship between catalyst surface structures and the thermodynamic parameters for pushing surface structure evolution on the basis of the FTS environment are highly desired.

In this work, the surface structure and stability of the ε-Fe_2C, θ-Fe_3C and Fe_4C as well as χ-Fe_5C_2 phases have been investigated on the basis of density functional theory (DFT) calculations and atomistic thermodynamics by considering the influence of temperature, pressure and H_2/CO ratio under simplified and wide varying non-equilibrium environment. The CO activation and the reactivity analysis on the obtained stable surfaces are also conducted aiming at approaching the overall landscape of Fe based FTS catalysts under real operating conditions.

Methodology

Structure calculation

The catalyst structures were calculated at the level of DFT with Vienna *ab initio* simulation package.[31,32] Electron exchange and correlation energy was treated within the generalized gradient approximation and the Perdew–Burke–Ernzerhof scheme (PBE).[33] Electron ion interaction was described by the projector augmented wave method.[34,35] Spin polarization was included in all calculations on the ferromagnetic iron carbide systems (ε-Fe_2C, θ-Fe_3C and Fe_4C) and this is essential for an accurate description of the magnetic properties. Iterative solutions of the Kohn–Sham equations were done using a plane wave basis with energy cutoff of 400 eV, and the samplings of the Brillouin zone were generated from the Monkhorst–Pack scheme. A second order Methfessel–Paxton[36] electron smearing with σ=0.2 eV was used to ensure accurate energies with errors due to smearing of less than 1 meV per unit cell. The convergence criteria for the force and electronic self-consistent iteration were set to 0.03 eV $Å^{-1}$ and 10^{-4} eV, respectively.

Catalyst models

The bulk structures and the corresponding Monkhorst–Pack grid of k points of the ε-Fe_2C, θ-Fe_3C and Fe_4C phases are listed in Table 1. The optimized lattice parameters agree well with those of the experiments and other calculations.[23,28,37–41] In calculating the ε-Fe_2C bulk structure, a $2 \times 2 \times 1$ supercell was used, and the detailed information is given in the Appendix. For all surface calculations, symmetrical slab surface models were chosen. Each surface was represented by a slab in 10–15 Å thickness, enough to avoid significant influence on the surface energies from our benchmarks.[30] A

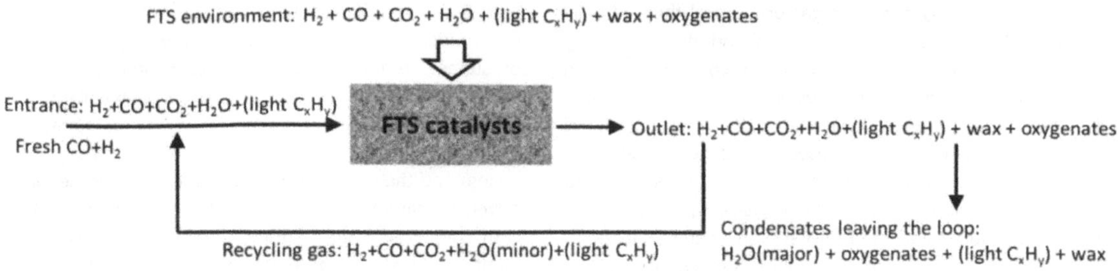

Figure 1 Overall scheme of complicated chemical environment of FTS reactions

vacuum layer of 15 Å was set to exclude the interactions among the periodic slabs, and all atoms were fully relaxed during the calculations. The Monkhorst–Pack grid of k points for each of the corresponding slab models is included in the Supplementary Material (Table S1).

Atomistic thermodynamics

The surface stability influenced by temperature, pressure and gas composition was investigated by using *ab initio* atomistic thermodynamics.[42,43] Since this is the same procedure used in our previous study on the surface composition and morphology of the χ-Fe_5C_2 phase,[30] detailed information can be found either in the Supplementary Material or in our previous work. Using the total energy of an isolated carbon atom (E_C) as reference for the variable μ_C, $\Delta\mu_C=\mu_C-E_C$, the minimum $\Delta\mu_C$ for the ε-Fe_2C, χ-Fe_5C_2,[30] θ-Fe_3C and Fe_4C phases is −7.83, −7.80, −7.80 and −8.01 eV, respectively. The total energies of gas phase molecules and carbon atom were calculated using a single k point (gamma point), where the periodic molecules were separated with 15 Å vacuum distances. These critical $\Delta\mu_C$ values indicate the lowest μ_C for the formation ofstable carbides. Since the vibrational contribution to the Gibbs free energy of the χ-Fe_5C_2 slab is negligible[30] and this is also true for most solid matter, we used only the total energy (E_{slab}) as the predominant term obtained directly from DFT calculations.

CO adsorption

For systematic comparison of the surface properties, we computed CO adsorption on the most stable facets of these carbides. The adsorption energy per CO molecule (E_{ads}) is defined as $E_{ads}=[E_{CO/slab}-(nE_{CO}+E_{slab})]/n$; where $E_{CO/slab}$, E_{CO} and E_{slab} are the total energies of the slab with adsorbed CO on the surface, an isolated CO molecule and the slab of the clean surface, respectively, and n is the number of adsorbed CO molecules. The coverage (θ) is defined as the number of CO molecules over the number of the exposed layer iron atoms. The surface C atoms (C_s) of iron carbides can be considered as adatoms on the defective surfaces, and the binding energy of C_s can be obtained as $E_{ads}(C_s)=E_{slab}-E_{slab/defect}-E_C$, where E_{slab}, $E_{slab/defect}$ and E_C are the total energies of the slab, the defective slab and an isolated carbon atom, respectively. Since PBE functional can give reasonable optimized geometries but overestimates the adsorption energies,[44] we used PBE for structure optimization and RPBE single point energy for estimating the adsorption energy. The Bader charges are used for discussing the effects of charge transfer.[45–47]

Results and discussion

Carbon chemical potential (μ_C) models for real FTS environment

The chemical–physical environment of real FTS catalysts should be properly defined in terms of the operating conditions. Our focus is on the trend of the phase transition of iron carbides in FTS reaction, for which the iron based catalysts have been treated as iron carbides obtained from reduction steps.[3,8,11] For this purpose, the driving force of the phase transition of iron carbides is μ_C, which can be defined using *ab initio* atomistic thermodynamics. Due to the complicated phenomena of the carburization processes,[3,48]

Table 1 Bulk properties of iron carbides (experimental values in parentheses) and used Monkhorst–Pack grid of k points

Crystal	Cell parameters	μ_B (Fe)
ε-Fe_2C	a=5.472 Å (2×2.794 Å)*	1.66 (1.70–1.72)§
Hexagonal	b=5.639 Å (2×2.794 Å)*	
P63/mmc	c=4.280 Å (4.360 Å)*	
(5×5×6)	β=121.0° (120.0°)*	
θ-Fe_3C	a=5.025 Å (5.080 Å)†	1.91 (1.72–1.79)§
Orthorhombic		
Pnma		
(7×5×9)		
Fe_4C	b=6.726 Å (6.730 Å)†	
Cubic	c=4.471 Å (4.510 Å)†	
$Pm\bar{3}m$	β=90.0° (90.0°)†	
(7×7×7)	a=3.761 Å (3.750 Å)‡	Fe(I): 3.04, Fe(II): 1.75

*Ref. 37.
†Ref. 38.
‡Ref. 39.
§Ref. 41.

many species in FTS system can either increase or decrease μ_C imposed over the iron based catalyst surfaces, as discussed in previous studies.[49–55] A comprehensive modeling of the chemical–physical environment of real FTS catalysts should consider most of the key factors, which will have normally non-linear type of contribution to the phase transition phenomena to be investigated. With this in mind, it is necessary to recall the schemes of a real FTS process as Fig. 1.

As given in Fig. 1, the chemical species involved in real FTS system include hydrogen, carbon monoxide, carbon dioxide, water and hydrocarbons within a wide carbon number distribution as well as small portions of oxygenates (mainly alcohols). The composition of these chemical species can easily be determined from industrial operation measurements and/or detailed material balance calculations for a real FTS system. This defines the rather accurate boundaries of the chemical–physical environment around working FTS catalysts, and the catalyst evolution trend may be predicted by using *ab initio* atomistic thermodynamics. For this goal, the key issue is to systematically develop the model to describe μ_C over the surfaces imposed by environment. However, this model is not very straight forward because of the wide varying and non-equilibrium nature of real FTS systems, namely the theoretical trends of the change in catalyst structures will largely depend on the kinetic factors of all the events related to the chemical species in the catalyst phases, and these events cover both carburization and decarburization reactions. Table 2 lists an overall summary of the events relating to the possible reactions between the catalyst phases (C_s) and the chemical species.

It should be noted that the reactions in Table 2 may occur thermodynamically under typical FTS conditions and affect μ_C. In fact the exact behaviors of μ_C will be also highly related to the rates of these reactions under FTS conditions. However, one can always study the thermodynamic trends by using energetics on the basis of the data from *ab initio* atomistic thermodynamics.

Obviously, CO is the most potential carburization agent in FTS and can easily deposit carbon atoms on the iron surface from the Boudouard reaction ($2CO \rightarrow C + CO_2$, reaction (2)). Gas phase molecular hydrogen and the adsorbed hydrogen which plays important roles in the transition of catalyst phases are in equilibrium.[56] Oxygen removal from the surface is rate limiting for carbide formation in pure CO, but this step becomes rapid in the presence of hydrogen, therefore addition of H_2 to CO can accelerate carbon deposition ($CO + H_2 \rightarrow C_s + H_2O$, reaction (3)).[57,58] However, the most important role of H_2 is the hydrogenation of surface carbon atoms ($C_s + H_2 \rightarrow -CH_2-$, reaction (1)) resulting in hydrocarbons as the primary products of FTS.[50,56] On the other hand, the light C_xH_y can also be transferred into surface carbons ($-CH_2- \rightarrow C_s + H_2$, reaction (7),[57,59] and $2(-CH_2-) + CO_2 \rightarrow 3C_s + 2H_2O$, reaction (8)). In addition, CO_2 can also consume hydrogen ($CO_2 + 2H_2 \rightarrow C_s + 2H_2O$, reaction (5)), and the reaction extent is limited by water content and temperature.[50,58] Otherwise, CO_2 and H_2O as byproducts can act as decarburizing agents ($C_s + CO_2 \rightarrow 2CO$, reaction (4); and $C_s + H_2O \rightarrow CO + H_2$, reaction (6)). The presence of CO_2 even in small quantities requires high CO concentration to balance this decarburizing reaction at elevated temperature.[48]

It is suggested that reaction (3) has the fastest kinetics on the basis of the high metal dusting rates in CO/H_2 environment,[48,55,57,59] while reaction (2) is also rapid and the rate of carbon deposition decreases with the increasing CO_2 content.[53] Olsson and Turkdogan[53] showed that in CO–H_2 mixtures reaction (2) is most important for H_2 content less than 50%, while the contribution of reaction (3) to the total rate is dominant for more than 50% H_2. In their study, H_2O has great influence on the rate of carbon deposition. When H_2O is added into CO–H_2 mixture, the rate of carbon deposition decreases with the increasing water vapor content, and this is due to reaction (6) (the reverse of reaction (3)). On the other hand, under CO condition, the rate of reaction (2) increases with the increasing H_2O content.[53] Koeken *et al.*,[60] found that increasing the total pressure can increase carbon deposition rate for a H_2/CO ratio of 1 : 1, but when H_2/CO ratio is higher than 4 : 1, higher total pressure can suppress the carbon deposition, and increasing H_2/CO ratio can also decrease the rate of carbon deposition. Ando and Kimura[61] also found that the amount of deposited carbon on iron apparently increases by adding small amount of H_2 to pure CO, while an excessive H_2 retards carbon deposition. These results imply that the carbon deposition rate is sensitive to the operating conditions (temperature, pressure and gas composition).

Apart from reactions (2) and (3), we also considered the carbon transfer from light hydrocarbons (C_xH_y) in reaction (7) ($C_xH_y \rightarrow xC_{(Fe)} + y/2H_2$) to estimate their carburization ability. In this case, $\mu_C\left[\mu_C = 1/x\left(\mu_{CxHy} - y/2\mu_{H_2}\right)\right]$ is determined by decomposition of light hydrocarbon. The influences of temperature, pressure and H_2/C_xH_y ratio on $\Delta\mu_C$ are given in Fig. 2. As temperature increases from 450 to 650 K at 30 atm with a 15% molar percentage of C_xH_y (Fig. 2*a*), $\Delta\mu_C$ changes hardly under C_2H_4 and C_2H_6 as gas reservoirs, while slightly decreases under C_2H_2 and increases under

Table 2 **List of possible reactions between catalyst phases (C_s) and FTS species**

Species	Events		Effect§
H_2	$C_s + H_2 \rightarrow -CH_2-$	(1)*	−
CO	$2CO \rightarrow C_s + CO_2$	(2)†	+
	$CO + H_2 \rightarrow C_s + H_2O$	(3)‡	+
CO_2	$C_s + CO_2 \rightarrow 2CO$	(4)†	−
	$CO_2 + 2H_2 \rightarrow C_s + 2H_2O$	(5)	+
H_2O	$C_s + H_2O \rightarrow CO + H_2$	(6)‡	−
$-CH_2-$	$-CH_2- \rightarrow C_s + H_2$	(7)	+
	$2(-CH_2-) + CO_2 \rightarrow 3C_s + 2H_2O$	(8)	+

*Molecular hydrogen may undergo rapid decomposition ($H_2 \rightarrow 2H_s$) on catalyst surfaces and surface hydrogen atoms may hydrogenate surface carbon atoms (FTS key steps).
†Under CO–CO_2 mixture, the carburizing mechanism consists of two elementary reactions: $CO \rightarrow C_s + O_s$ and $CO + O_s \rightarrow CO_2$, the later one was found to be rate limiting.
‡Hydrogen has been shown to be an accelerator of CO decomposition over iron based catalysts, while H_2O has been found to both accelerate and retard CO decomposition.[75] The rates of reactions (2) and (3) depend on the rates of the reactions of CO and H_2 with an oxygen atom to produce CO_2 and H_2O, respectively.[76]
§Positive effect (+) for carburization, and negative effect (–) for decarburization.

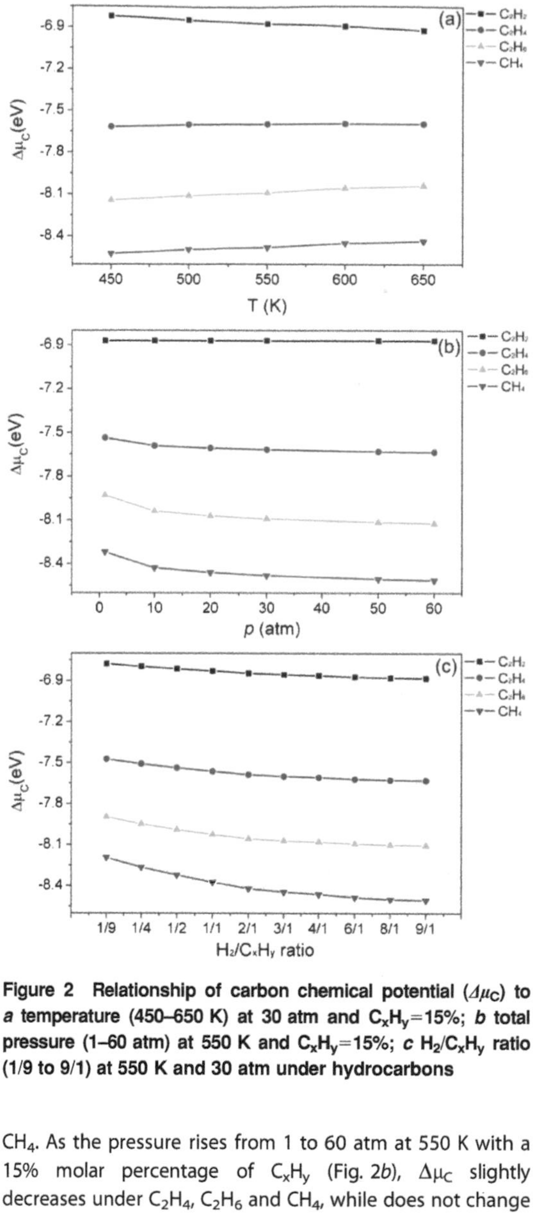

Figure 2 Relationship of carbon chemical potential ($\Delta\mu_C$) to *a* temperature (450–650 K) at 30 atm and C_xH_y=15%; *b* total pressure (1–60 atm) at 550 K and C_xH_y=15%; *c* H_2/C_xH_y ratio (1/9 to 9/1) at 550 K and 30 atm under hydrocarbons

CH_4. As the pressure rises from 1 to 60 atm at 550 K with a 15% molar percentage of C_xH_y (Fig. 2*b*), $\Delta\mu_C$ slightly decreases under C_2H_4, C_2H_6 and CH_4, while does not change under C_2H_2. As expected, increasing the H_2/C_xH_y ratio from 1/9 to 9/1 at 550 K and 30 atm lowers $\Delta\mu_C$ in some extent (Fig. 2*c*).

Figure 2 shows that the carburization ability of light hydrocarbons decreases with the decrease in carbon content from acetylene to saturated hydrocarbons, i.e. $C_2H_2 > C_2H_4 > C_2H_6 > CH_4$. It is noted that $\Delta\mu_C$ under saturated hydrocarbons (C_2H_6 and CH_4) becomes lower than the critical values for stable iron carbide phases, i.e. −7.83 eV for ε-Fe_2C, −7.80 eV for χ-Fe_5C_2[30] and θ-Fe_3C as well as −8.01 eV for Fe_4C, and consequently the carbide phases will transform to metallic iron phase. Therefore, we used ethylene as light hydrocarbon model for our discussion and comparison.

The μ_C to real FTS catalysts can be estimated under different conditions relating to the possible modes, e.g. hydrogen rich modes for starting-up and shutdown of the process, the normal modes typically for oil production, and CO rich mode due to some unexpected reasons in the whole process. In this work, we try to understand the tendency of the change of the iron carbide phases in the above major situations. As discussed above, in CO/H_2 mixture, only reactions (2), (3) and (7) can contribute to carbon deposition at different levels (Table 2) and their reversible reactions can give overall description of the major physical and chemical relations to carbon deposition on real FTS catalysts. By considering only reaction (3) ($CO+H_2 \rightarrow C+H_2O$), the $\Delta\mu_C$ is much higher than the minimum of carbides (−7.80 eV). It is also true by raising the H_2/CO ratio from 1/1 to 100/1, the $\Delta\mu_C$ (from −6.76 to −6.91 eV) is still far away from the minimum. This would mean that the H_2/CO ratio could not affect the $\Delta\mu_C$, and the carbon-rich facets would remain stable at very high H_2/CO ratio. Obviously, this disagrees with the experimental results, because high H_2 partial pressure will retard carbon deposition and even reduce iron carbide into metallic iron. Instead of using only single reaction to estimate the changes of the $\Delta\mu_C$, we combined different reactions. However, it should be noted that these independent reactions impose chemical force to the phase transition of catalysts and the extent of each reaction to carbon balance of the catalyst has not been well defined. Therefore, we supposed three different extents of each reaction and discuss these situations respectively

Scheme A: (2)+(3)+(7)

$$1/4C_2H_4+3/4CO \rightarrow C_s+1/4H_2+1/4CO_2+1/4H_2O$$

$$\mu_C=3/4\mu_{CO}+1/4\mu_{C_2H_4}-1/4\mu_{H_2}-1/4\mu_{H_2O}-1/4\mu_{CO_2}$$

Scheme B: (2)×2+(3)+(7)

$$1/5C_2H_4+CO \rightarrow C_s+1/5H_2+2/5CO_2+1/5H_2O$$

$$\mu_C=\mu_{CO}+1/5\mu_{C_2H_4}-1/5\mu_{H_2}-1/5\mu_{H_2O}-2/5\mu_{CO_2}$$

Scheme C: (2)+(3)+(7)×2

$$1/3C_2H_4+1/2CO \rightarrow C_s+1/6H_2+1/6CO_2+1/6H_2O$$

$$\mu_C=1/2\mu_{CO}-1/3\mu_{C_2H_4}-1/6\mu_{H_2}-1/6\mu_{H_2O}-1/6\mu_{CO_2}$$

In a typical Fe based FTS process, the most important parameters influencing the catalyst performance are the chemical composition of fluids surrounding the catalyst, temperature and pressure. In this study, the gas environment is designed in composition with $CO+H_2$ (varying H_2/CO ratio), $-CH_2-$ (light hydrocarbons), CO_2 and H_2O of 75, 10, 12 and 3%, respectively, representing the typical industrial conditions. Since we did not consider the contribution of the condensed heavy hydrocarbons and oxygenates, the ratios of carbon, oxygen and hydrogen are not stoichiometric. For comparison, we also included a gas phase free from CO with H_2 and hydrocarbons, which simulates the weaker carburization environment for the iron catalyst as tested in fundamental studies in FTS.[62]

Under real CO involved environment, the influences of temperature and pressure on $\Delta\mu_C$ are evaluated for H_2/CO ratio of 2, 4 and 8 with other gas compositions presented in Table 3, and the main results are shown in Fig. 3. At a total pressure fixed at 30 atm (Fig. 3*a*), it is found that higher temperature leads to lower (more negative) $\Delta\mu_C$ for all three schemes with different H_2/CO ratios, in consistent with the results of de Smit *et al.*[63] At a given temperature the μ_C

determined in Scheme B is the highest, followed by those in Scheme A and Scheme C. This trend implies that CO has stronger carburization ability than C_2H_4.

Figure 3*b* shows the results with varying pressure at 550 K. For Scheme B and Scheme A, higher pressure leads to higher (less negative) $\Delta\mu_C$, which indicates lower pressure can retard carbon deposition. In the whole pressure range (1 to 60 atm), the $\Delta\mu_C$ determined by Scheme C keeps constant.

Figure 3*c* presents the H_2/CO ratio influence on $\Delta\mu_C$ at 550 K and 30 atm, with the gas composition presented in Table 4. As H_2/CO ratio increases from 1/1 to 20/1, the $\Delta\mu_C$ determined by all three schemes decreases and this implies that excess hydrogen would retard carbon deposition. However, it should be noted that even at extremely high H_2/CO ratio (20/1) $\Delta\mu_C$ does not become lower than the critical value for stable iron carbide phases, i.e. −7.83 eV for ε-Fe_2C, −7.80 eV for χ-Fe_5C_2 and θ-Fe_3C as well as −8.01 eV for Fe_4C. This reveals the thermodynamic possibility for carbon deposition at very high H_2/CO ratios beyond stable iron carbide phases under extended FTS operation conditions. Such carbon deposition destroys the mechanical structure of catalysts as observed in industrial practices and deactivates the catalysts.[57] It is noted that the same trends in Schemes A–C have been found for using CH_4 as light hydrocarbon model (Supplementary Material).

How to keep the activity of the catalysts by adjusting the chemical and physical parameters remains to be a headache problem and challenging. Rising temperature can retard carbon deposition thermodynamically, but accelerate carburization kinetically. Lowering pressure is thermodynamically and kinetically promising, but reduces significantly the process productivity. Increasing H_2/CO ratio is therefore a reasonable choice with thermodynamic and kinetic advantages as well as controllable process productivity. It should be noticed that our current calculations depend very strongly on three different carburization schemes that hopefully can cover all possibilities in real FTS systems. The exact schemes should be determined by experimental studies, which show the chemical extents of different carbon formations steps, namely (2), (3) and (7) in Table 2. This should be done by experiments with well defined techniques and operation conditions. Such studies provide a precise thermodynamic basis for the fundamental investigation along with *ab initio* atomistic thermodynamics developed in this work.

In FTS reaction system, the influence of CO_2 and H_2O contents on the $\Delta\mu_C$ is presented in Fig. 4, with the gas composition listed in Tables 5 and 6. It shows clearly that increasing the content of CO_2 (5 to 25%) and H_2O (1 to 15%) lowers $\Delta\mu_C$ at very low degree. The results indicate that CO_2

Table 3 Gas composition under H_2/CO ratios of 2, 4 and 8

H_2/CO	2	4	8
H_2 (%)	50.00	60.00	66.67
CO (%)	25.00	15.00	8.33
CH_4 (%)	10.00	10.00	10.00
CO_2 (%)	12.00	12.00	12.00
H_2O (%)	3.00	3.00	3.00

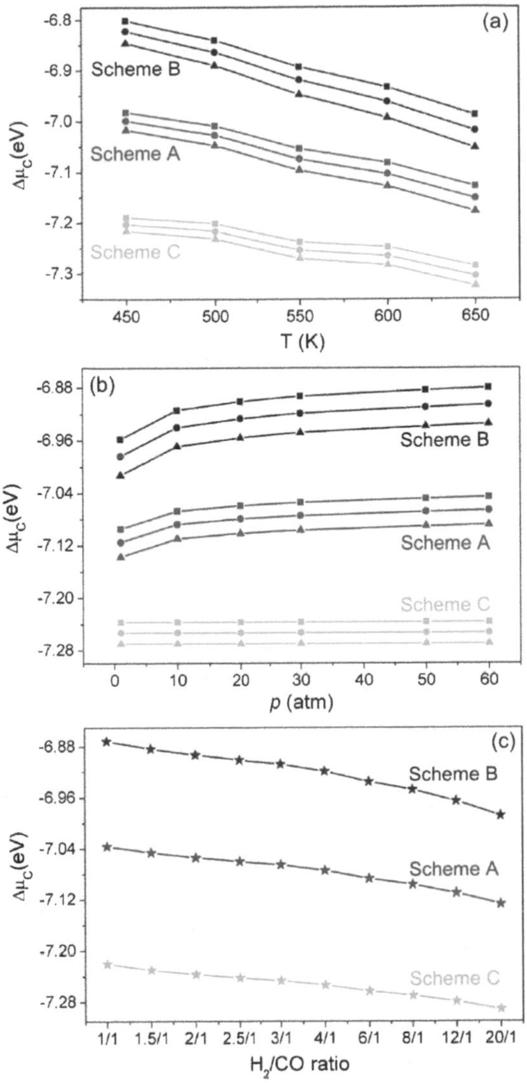

Figure 3 Relationship of carbon chemical potential ($\Delta\mu_C$) to *a* temperature (450–650 K) at 30 atm and H_2/CO ratio of 2, 4 and 8; *b* pressure (1–60 atm) at 550 K and H_2/CO ratio of 2, 4 and 8; *c* H_2/CO ratio (1 to 20) at 550 K and 30 atm (■ for H_2/CO=2; • for H_2/CO=4 and ▲ for H_2/CO=8)

Table 4 Gas composition under different H_2/CO ratios at 550 K and 30 atm

H_2/CO	H_2/%	CO/%	CH_4/%	CO_2/%	H_2O/%
1.0	37.50	37.50	10.00	12.00	3.00
1.5	45.00	30.00	10.00	12.00	3.00
2.0	50.00	25.00	10.00	12.00	3.00
2.5	53.57	21.43	10.00	12.00	3.00
3.0	56.25	18.75	10.00	12.00	3.00
4.0	60.00	15.00	10.00	12.00	3.00
6.0	64.29	10.71	10.00	12.00	3.00
8.0	66.67	8.33	10.00	12.00	3.00
12.0	69.23	5.77	10.00	12.00	3.00
20.0	71.43	3.57	10.00	12.00	3.00

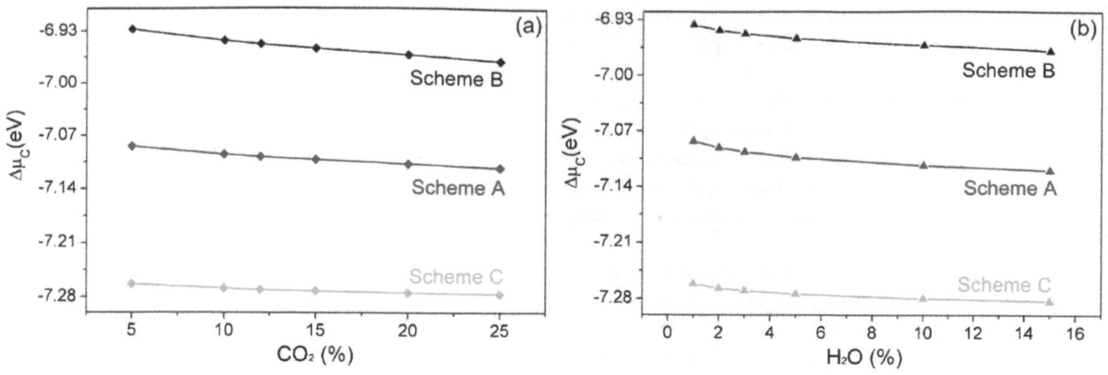

Figure 4 Relationship of carbon chemical potential ($\Delta\mu_C$) to *a* CO_2 content and *b* H_2O content at 550 K, 30 atm and H_2/CO=8

and H_2O, the byproduct in FTS, play only subordinate role in controlling phase transition process and should be removed from the process as usual (Fig. 1) for achieving other process benefits.

These compared results in Schemes A–C imply that we may use different unsaturated hydrocarbons and CO to optimize the environment for getting stable catalyst phases, especially for initializing the FTS process. It has been proved experimentally that unsaturated light hydrocarbons and H_2, instead of CO and H_2, can conduct chain growth reactions over iron based FTS catalysts.[62] In order to conduct efficient FTS reactions it is necessary to optimize the carburization ability of the chemical–physical environments (temperature, pressure and H_2/CO ratio) in terms of μ_C. The insight behind μ_C is the change of the stable iron carbide phases as well as the surface structure and composition.

Surface stability

To get the equilibrium shapes of ε-Fe_2C, θ-Fe_3C and Fe_4C under different operation conditions, we studied both low and high Miller index facets of these carbides, which contain all low Miller index surfaces and the characteristic peaks in X-ray diffraction.[37,64,65] All calculated surfaces and the equivalent Miller index are listed in the Supplementary Material (Table S3). Because of their complex bulk structures, each surface has several terminations (including both stoichiometric and non-stoichiometric terminations), e.g. five terminations for each of the (101), (102) and (103) surfaces of ε-Fe_2C; 16 terminations for each of the (111), (113), (133) and (131) surfaces of θ-Fe_3C; and each facet of Fe_4C has two terminations. Here we used the surface Fe/C ratio ($\alpha=n_{Fe}/n_C$) to distinguish these terminations as discussed previously.[30] In the following discussion, the number following the Miller index indicates the surface Fe/C ratio. The surface free energies of these terminations within the $\Delta\mu_C$ range from -8.50 to -6.00 eV are given in the Supplementary Material (Fig. S3) for comparison, and only the results of the most stable termination of each facet are used for discussion.

Figures 5–7 show the relationship between surface free energy ($\gamma(T,p)$) of the most stable facets of ε-Fe_2C, θ-Fe_3C and Fe_4C and $\Delta\mu_C$. Similar to Fe_5C_2,[30] carbon-rich termination with lower α value becomes more stable at higher (less negative) $\Delta\mu_C$ for all iron carbides, while the carbon-poor terminations are more favorable at lower (more negative) $\Delta\mu_C$. With the increasing $\Delta\mu_C$, the most stable termination changes from carbon-poor (higher Fe/C ratio) to carbon-rich (lower Fe/C ratio), and the turn points represent the change of the stable termination and they differ from facet to facet. By combining theory and *in situ* XPS studies de Smit *et al.*[63] also found that body centered cubic Fe and surface/subsurface carbon are more stable at high temperature (low μ_C), while the carbon-rich χ-Fe_5C_2 (100) surface becomes thermodynamically more stable upon lowering the temperature (high μ_C). Since excessive carbon deposition will deactivate the catalysts and lower the catalytic performance,[57] avoiding carbon deposition can be improved by using proper temperature, pressure and gas environment.

For ε-Fe_2C (Fig. 5), $(1\bar{2}1)$-2.00 and (101)-1.50 are the most stable facets, followed by $(2\bar{2}1)$-2.67/1.33, and $(0\bar{1}1)$-2.00/1.33. The least stable surfaces are $(\bar{2}01)$-2.67/1.33, and (103)-2.50/1.00. The other surfaces, $(0\bar{1}3)$-2.67/1.50, (001)-4.00/1.00, (110)-2.00/1.33, (112)-2.00/1.00, (100)-2.00/1.00, $(0\bar{1}2)$-4.00/1.00, (102)-4.00/2.00, (111)-3.00/1.33, have intermediate stability. It is also noted that for the $(1\bar{2}0)$, (010), $(1\bar{2}1)$ and $(1\bar{2}2)$ facets, the stoichiometric terminations are most stable (Fig. S3).

Table 5 Gas composition at 550 K and 30 atm with H_2/CO ratio of 8 for different CO_2 contents

CO_2/%	H_2/%	CO/%	CH_4/%	H_2O/%
5.00	72.89	9.11	10.00	3.00
10.00	68.44	8.56	10.00	3.00
12.00	66.67	8.33	10.00	3.00
15.00	64.00	8.00	10.00	3.00
20.00	59.56	7.44	10.00	3.00
25.00	55.11	6.89	10.00	3.00

Table 6 Gas composition at 550 K and 30 atm with H_2/CO ratio of for different H_2O contents

H_2O/%	H_2/%	CO/%	CH_4/%	CO_2/%
1.00	68.44	8.56	10.00	12.00
2.00	67.56	8.44	10.00	12.00
3.00	66.67	8.33	10.00	12.00
5.00	64.89	8.11	10.00	12.00
10.00	60.44	7.56	10.00	12.00
15.00	56.00	7.00	10.00	12.00

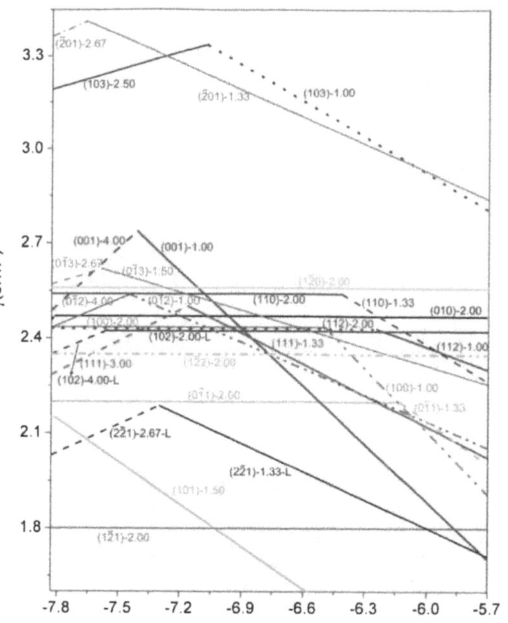

Figure 5 Relationship of surface free energies of most stable facets of ε-Fe_2C to $\Delta\mu_C$ (indices given in parentheses indicates corresponding Miller index, and second term of indices provides corresponding surface α=Fe/C ratio)

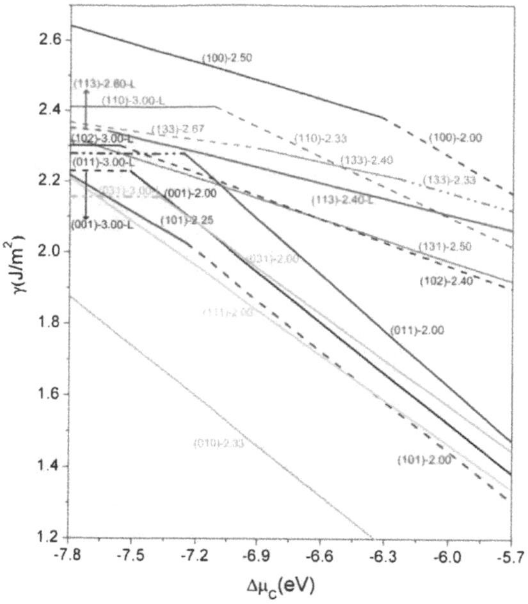

Figure 6 Relationship of surface free energies of most stable facets of θ-Fe_3C to $\Delta\mu_C$ (indices given in parentheses indicates corresponding Miller index, and second term of indices provides corresponding surface α=Fe/C ratio)

For θ-Fe_3C (Fig. 6), (010)-2.33 is the most stable one. The least stable surface is (100) with α=2.50 or 2.00. The other surfaces, (110)-3.00/2.33, (133)-2.67/2.40/2.33, (113)-2.80/2.40, (131)-2.50, (102)-3.00/2.40, (011)-3.00/2.00, (001)-3.00/2.00, (101)-2.25/2.00, (111)-2.00, (031)-3.00/2.00, have intermediate stability.

For Fe_4C (Fig. 7), (100) with α=3.00 is the most stable facet. The surface free energies of other facets are significantly higher than (100), and the least stable surface is (133)-6.00/3.00, followed by (110)-3.00, (111)-4.00/2.00, (131)-6.00/2.67 and (210)-5.00/3.33, have intermediate stability.

For Fe_5C_2[30] as reported previously, the (100) termination is most stable with α=2.25, followed by (111)-2.17/1.75, (510)-2.50, and (110)-2.40/2.00. The least stable surfaces are (10$\bar{1}$)-2.75/2.25, (001)-2.50; (11$\bar{3}$)-2.50, (113)-2.00 and (101)-1.50. In addition, (110)-2.40/2.00, (010)-2.50; (133)-1.75; (11$\bar{1}$)-2.50, (511)-2.25, (221)-3.00, ($\bar{4}$11)-2.50, (011)-2.40/2.20 have stability in between.

Crystallite morphology

In order to estimate the crystallite morphology of these iron carbides, it is necessary to determine the equilibrium crystal shape by using the standard Wulff construction.[66] In the standard Wulff construction, the surface free energy for a given closed volume is minimized and the exposure of a facet depends not only on surface free energy but also on orientation in crystal.[67] Since the surface free energy of each facet is a function of μ_C, the crystal shape should also be a function of the $\Delta\mu_C$ that corresponds to different experimental conditions of temperature, pressure and atmosphere. Figure 8 presents the morphology of the ε-Fe_2C, θ-Fe_3C and Fe_4C crystals at different $\Delta\mu_C$, corresponding to different gas compositions at 550 K and 30 atm, respectively, along with that of χ-Fe_5C_2 (slightly modified form compared to our previous report, where an incorrect default crystal parameter was used; however, this does not affect our conclusion). The proportions of exposed terminations of ε-Fe_2C, χ-Fe_5C_2 and θ-Fe_3C are listed in Tables 7, 8 and 9.

At lower $\Delta\mu_C$ (−7.60 eV), the crystallite of ε-Fe_2C has 11 exposed surface terminations in different Fe/C ratios, (100), (010), (001), (101), (0$\bar{1}$1), (110), (111), (1$\bar{2}$1), (102), (0$\bar{1}$2) and (2$\bar{2}$1). The (1$\bar{2}$1) termination has the largest portion (35.3%) of the surface area, followed by the (101) and (2$\bar{2}$1) terminations (27.7 and 16.7%, respectively), and they cover about 80% of the total surface area of the crystal. As the $\Delta\mu_C$ increases to −7.10 eV, (0$\bar{1}$1), (001) and (102) are

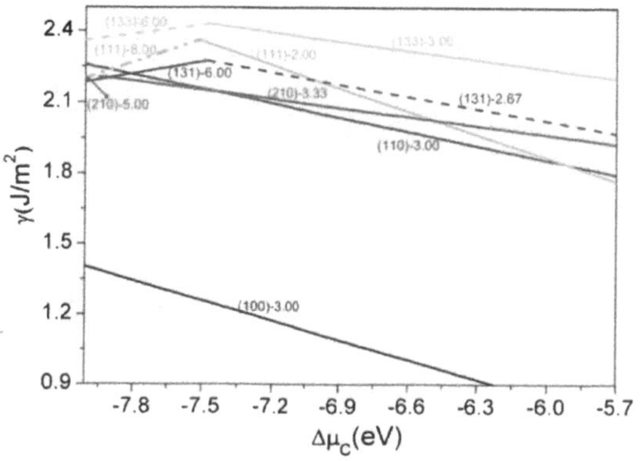

Figure 7 Relationship of surface free energies of most stable facets of Fe_4C to $\Delta\mu_C$ (indices given in parentheses indicates corresponding Miller index, and second term of indices provides corresponding surface α=Fe/C ratio)

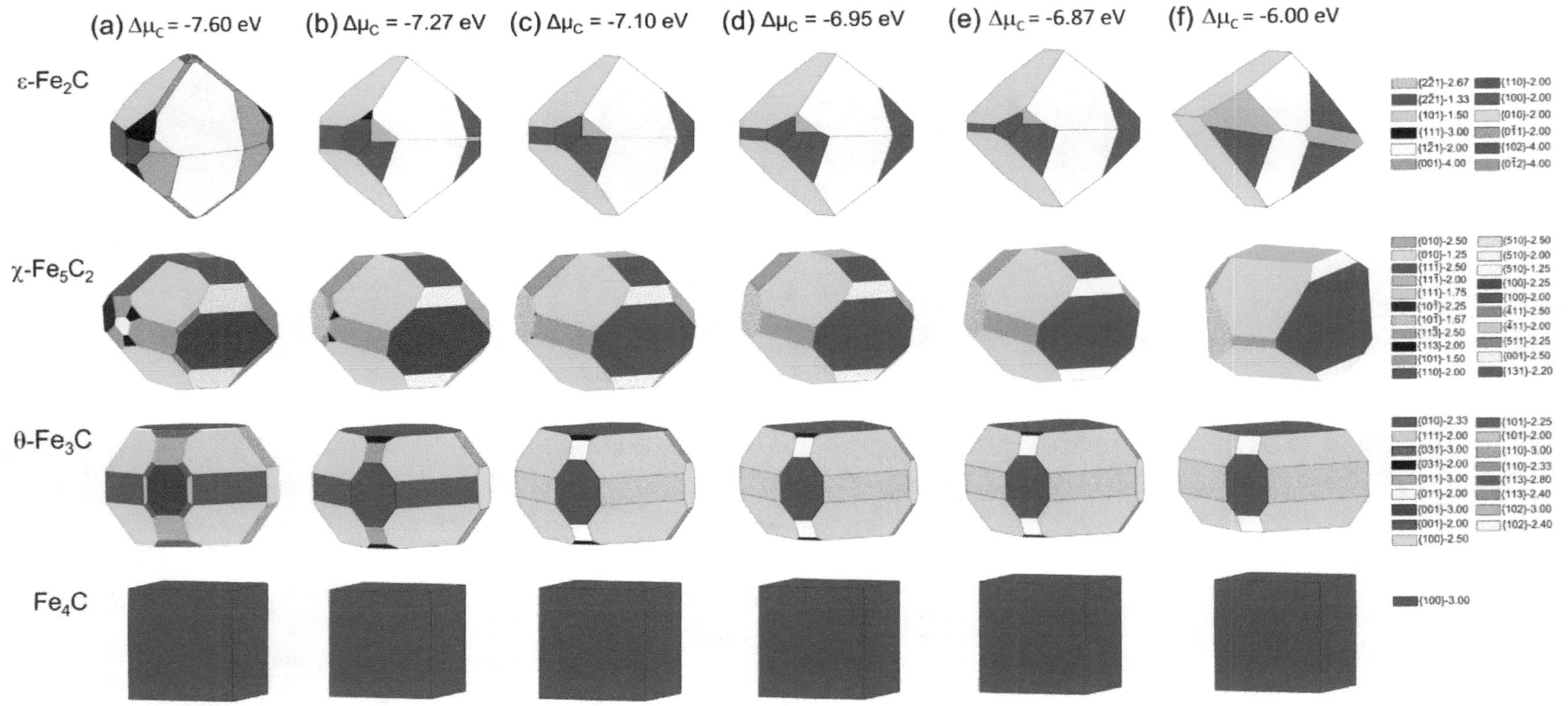

Figure 8 Morphologies of ε-Fe_2C, χ-Fe_5C_2, θ-Fe_3C and Fe_4C at different $\Delta\mu_C$ at 550 K and 30 atm: *a* for C_2H_4 (15%) condition; *b* for Scheme C with $H_2/CO=8.0$; *c* for Scheme A with $H_2/CO=8.0$; *d* for Scheme B with $H_2/CO=8.0$; *e* for C_2H_2 (15%) condition. Indices given in legend indicates corresponding Miller index, and second term of indices provides corresponding surface α=Fe/C ratio)

disappeared, and the carbon-rich $(2\bar{2}1)$-1.33 termination becomes more stable than the carbon-poor $(2\bar{2}1)$-2.67 termination. As the $\Delta\mu_C$ increases, the proportion of (100), (111), $(1\bar{2}1)$ and $(0\bar{1}2)$ decreases, while the area of (101) and (010) increases. When the $\Delta\mu_C$ reaches to −6.0 eV, the crystallite of ε-Fe_2C has only four exposed surface terminations and they are (101), $(1\bar{2}1)$, (010) and $(2\bar{2}1)$. The (101) becomes the largest exposed surface (59.5%), and the facets $(1\bar{2}1)$, 101) and $(2\bar{2}1)$ still cover the most surface area (92%) of the crystal.

Table 7 **Facets contributions (%) to total surface area in Wulff construction of ε-Fe_2C presented in Fig. 8**

Facet	(a) C_2H_4	(b) Scheme C	(c) Scheme A	(d) Scheme B	(e) C_2H_2	(f) $\Delta\mu_C$=−6.0 eV
(100)-2.00	5.52	3.74	2.75	1.85	1.36	0.00
(010)-2.00	0.55	1.45	0.86	0.64	0.53	7.99
(001)-1.00	0.18	0.00	0.00	0.00	0.00	0.00
(101)-1.50	27.70	40.81	44.13	46.96	48.52	59.52
$(0\bar{1}1)$-2.00	4.14	0.00	0.00	0.00	0.00	0.00
(110)-2.00	3.43	4.96	3.96	3.01	2.51	0.00
(111)-1.33	...	...	0.03	0.06	0.07	0.00
(111)-3.00	4.91	0.28	...	...	...	...
$(1\bar{2}1)$-2.00	35.25	39.02	36.41	34.15	32.93	17.32
(102)-2.00	0.97	0.00	0.00	0.00	0.00	0.00
$(0\bar{1}2)$-1.00	0.65	0.03	0.00	0.00	0.00	0.00
$(2\bar{2}1)$-1.33	...	9.70	11.85	13.34	14.09	15.18
$(2\bar{2}1)$-2.67	16.71	...	...	...	...	...

Table 8 **Facets contributions (%) to total surface area in Wulff construction of χ-Fe_5C_2 presented in Fig. 8**

Facet	(a) C_2H_4	(b) Scheme C	(c) Scheme A	(d) Scheme B	(e) C_2H_2	(f) $\Delta\mu_C$=−6.0 eV
(010)-2.50	1.20	...	...	...	...	...
(010)-1.25	...	0.92	2.06	2.83	3.33	10.44
$(11\bar{1})$-2.50	11.72	...	...	...	...	...
$(11\bar{1})$-2.00	...	9.76	9.42	9.06	8.91	2.25
(001)-2.50	0.92	0.00	0.00	0.00	0.00	0.00
(510)-2.50	9.58	...	...	...	...	...
(510)-2.00	...	6.38	5.79	6.05	...	...
(510)-1.25	...	...	...	...	6.25	5.37
(113)-2.00	0.91	0.35	0.03	0.00	0.00	0.00
$(\bar{4}11)$-2.50	9.14	...	...	...	...	...
$(\bar{4}11)$-2.00	...	6.35	4.56	2.74	1.87	0.00
$(11\bar{3})$-2.50	3.33	1.02	0.00	0.00	0.00	0.00
(101)-1.50	4.53	4.92	4.65	4.24	4.00	1.42
(100)-2.25	16.63	20.24	...	...	...	...
(100)-2.00	...	...	21.67	23.07	23.76	31.84
(110)-2.00	10.49	12.90	11.87	10.54	9.64	0.00
(111)-1.75	23.81	30.41	32.60	33.65	34.22	39.99
(511)-2.25	1.19	0.00	0.00	0.00	0.00	0.00
(131)-2.20	3.34	0.68	0.20	0.00	0.00	0.00
$(10\bar{1})$-2.25	3.21	...	...	...	...	...
$(10\bar{1})$-1.67	...	6.07	7.35	7.82	8.02	8.69

Table 9 **Facets contributions (%) to total surface area in Wulff construction of θ-Fe_3C presented in Fig. 8**

	(a) C_2H_4	(b) Scheme C	(c) Scheme A	(d) Scheme B	(e) C_2H_2	(f) $\Delta\mu_C$=−6.0 eV
(001)-3.00	5.03	...	...	...	...	...
(001)-2.00	...	6.80	6.82	6.82	6.79	6.81
(010)-2.33	20.31	23.94	24.74	25.44	25.77	29.99
(100)-2.50	3.46	3.36	3.06	2.53	2.24	0.00
(101)-2.25	13.25	12.60	...	...	...	...
(101)-2.00	...	...	13.06	13.89	14.49	21.39
(110)-3.00	9.28	4.96	...	...	...	...
(110)-2.33	...	...	3. 04	2.38	2.04	0.00
(011)-3.00	5.60	2.36	...	...	...	...
(011)-2.00	...	...	2.28	2.42	2.50	2.92
(111)-2.00	35.88	44.38	45.75	45.63	45.45	38.88
(113)-2.80	1.21	...	...	...	...	...
(113)-2.40	...	0.00	0.00	0.00	0.00	0.00
(031)-3.00	3.92	...	...	...	...	...
(031)-2.00	...	1.61	1.25	0.89	0.71	0.00
(102)-3.00	0.85	...	...	...	...	...
(102)-2.40	...	0.00	0.00	0.00	0.00	0.00

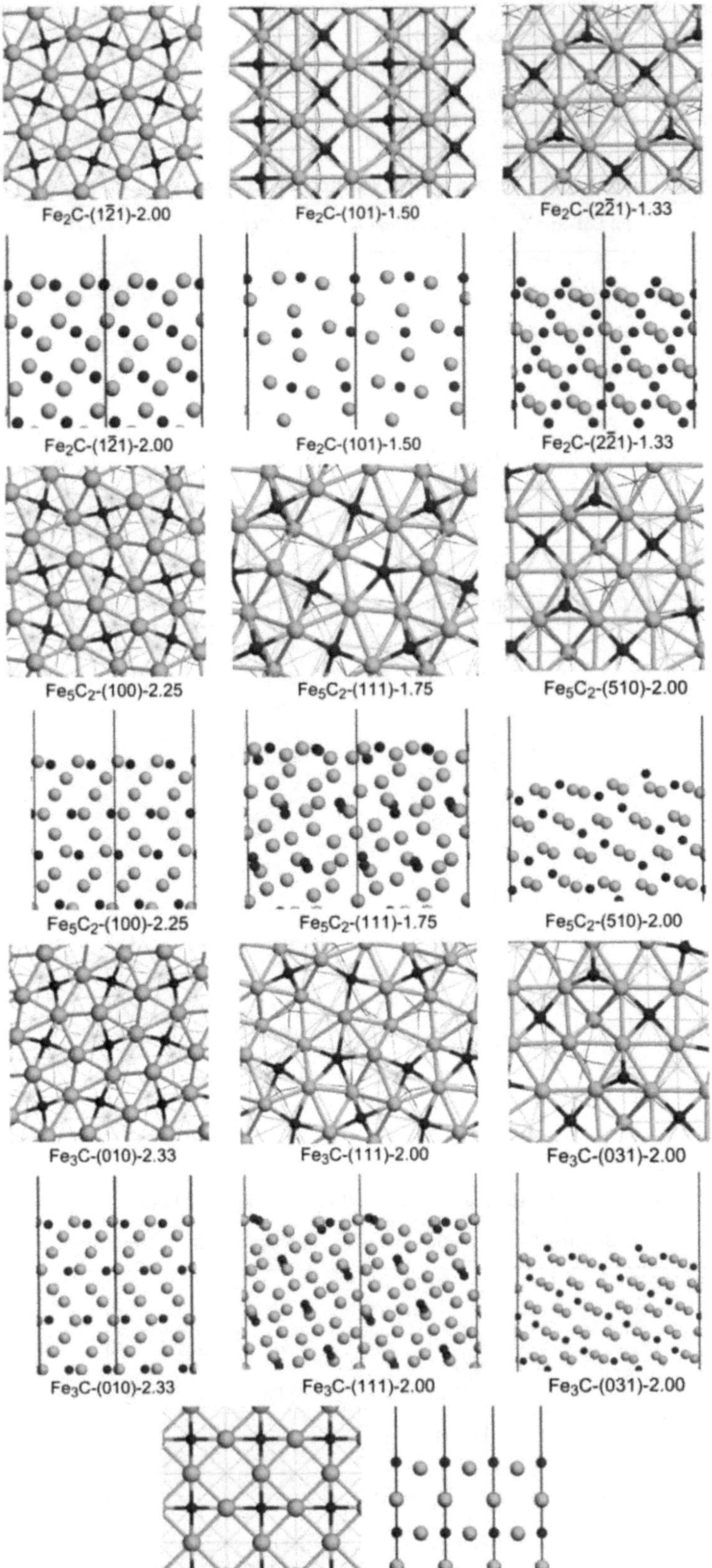

Fe_2C-(1$\bar{2}$1)-2.00
Fe_2C-(101)-1.50
Fe_2C-(2$\bar{2}$1)-1.33
Fe_2C-(1$\bar{2}$1)-2.00
Fe_2C-(101)-1.50
Fe_2C-(2$\bar{2}$1)-1.33
Fe_5C_2-(100)-2.25
Fe_5C_2-(111)-1.75
Fe_5C_2-(510)-2.00
Fe_5C_2-(100)-2.25
Fe_5C_2-(111)-1.75
Fe_5C_2-(510)-2.00
Fe_3C-(010)-2.33
Fe_3C-(111)-2.00
Fe_3C-(031)-2.00
Fe_3C-(010)-2.33
Fe_3C-(111)-2.00
Fe_3C-(031)-2.00
Fe_4C-(100)-3.00
Fe_4C-(100)-3.00

Figure 9 Surface structures of most stable surfaces and surfaces that have largest exposed surface area in Wulff construction of ε-Fe_2C, χ-Fe_5C_2, θ-Fe_3C and Fe_4C (indices given in parentheses indicates corresponding Miller index, and the third term of the indices provides the corresponding surface α=Fe/C ratio, Fe atoms are shown by blue balls, C atoms are shown by black balls)

At lower $\Delta\mu_C$ (−7.60 eV), the crystallite of χ-Fe_5C_2 has 14 exposed surface terminations in different Fe/C ratios, (010), (11$\bar{1}$), (001), (510), (113), ($\bar{4}$11), (11$\bar{3}$), (101), (100), (110), (111), (511), (131) and (10$\bar{1}$). The (111), (100) and (11$\bar{1}$) surfaces cover 52.1% of the total surface area of the crystal (23.8, 16.6 and 11.7%, respectively). When the $\Delta\mu_C$ increases to −6.95 eV, the facets (001), (511), (113), (11$\bar{3}$), and (131) are disappeared, and the carbon-rich termination of (010), (11$\bar{1}$), (10$\bar{1}$), (510), ($\bar{4}$11) and (100) are exposed. As the $\Delta\mu_C$ increases, the proportion of (100), (010), (111) and (10$\bar{1}$) increases, while the exposed areas of all the other surfaces decrease. At higher $\Delta\mu_C$ (−6.00 eV), only seven facets are still exposed, among which the (111) and (100) terminations cover as much as 71.8% of the surface area of the crystal.

When the $\Delta\mu_C$ is −7.60 eV, the crystallite of θ-Fe_3C has 11 exposed surface terminations, (001), (100), (010), (101), (110), (011), (111), (113), (131), (102) and (031). The (111) facet has the largest portion of the total surface area (35.9%), followed by the (010) (20.3%) and (101) (13.3%) facets. As the $\Delta\mu_C$ increases to −6.95 eV, the proportions of (111) and (010) increase to 45.6 and 25.4%, respectively. The carbon-rich (031), (011), (001), (101) and (110) terminations are exposed under higher $\Delta\mu_C$ compared with lower μ_C. There are five facets still exposed at higher $\Delta\mu_C$ (−6.00 eV), among them the (111), (010) and (101) terminations cover 90.3% of the total surface area of the crystal (38.9, 30.0 and 21.4%, respectively).

In the whole range of $\Delta\mu_C$ that we considered, the crystallite of Fe_4C only exposes the (100)-3.00 termination.

Surface property

In the most stable terminations of each iron carbides (Fig. 9), ε-Fe_2C-(1$\bar{2}$1)-2.00, χ-Fe_5C_2-(100)-2.25, θ-Fe_3C-(010)-2.33 and Fe_4C-(100)-3.00, each surface carbon atom coordinates with four surface iron atoms, and each surface iron atom coordinates with two surface carbon atoms. In addition, the most exposed surfaces, χ-Fe_5C_2-(111)-1.75 and θ-Fe_3C-(111)-2.00, have similar atom arrangement on partial surface structures. The third exposed facet of ε-Fe_2C, (2$\bar{2}$1)-1.33, has also some similar surface structure with χ-Fe_5C_2-(510)-2.00 as well as θ-Fe_3C-(031)-2.00. The computed density of states of the surface layer atoms (Fig. 10) also revealed the similarity of these surface structures.

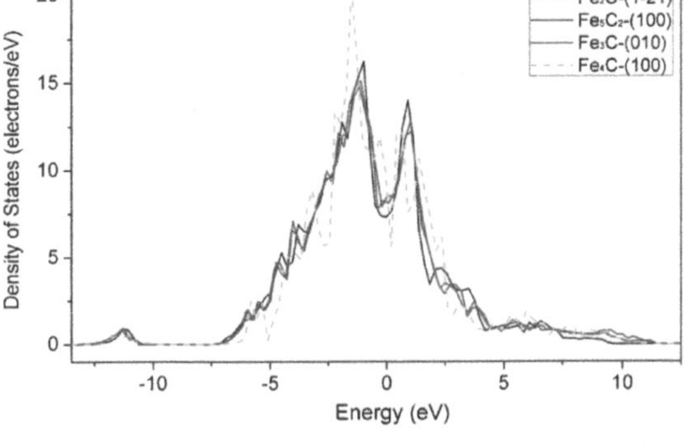

Figure 10 Density of states of surface layer atoms of most stable facets

Since the pattern and density of carbonaceous deposit on surface can significantly influence the catalytic performance,[68,69] similar and unique catalytic activities of the carbides facets with the same atom arrangement on surface layer should be expected. At first we analyzed the surface properties, e.g. the charge and binding energy of the surface carbon atoms, as well as the surface work function (difference between the electrostatic potential energy in the vacuum region and the Fermi energy of the slab), which is an important electronic indicator of a surface, i.e. lower work function indicates the higher electron donating ability of the surface. In addition, we also computed the adsorption structure and energy of CO on these surfaces. As shown in Fig. 11, the most stable CO adsorption site is the Fe-top site

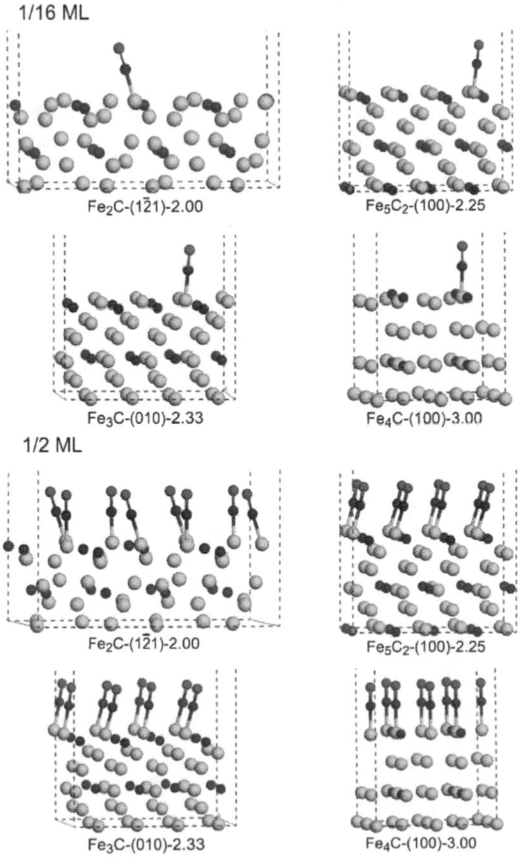

Figure 11 Adsorption of CO on most stable facet of ε-Fe_2C, χ-Fe_5C_2, θ-Fe_3C and Fe_4C with different coverage (indices given in parentheses indicates corresponding Miller index, and third term of indices provides corresponding surface α=Fe/C ratio, Fe atoms in blue, C atoms in black and O atoms in red)

on the ε-Fe_2C-(1$\bar{2}$1)-2.00, χ-Fe_5C_2-(100)-2.25 and θ-Fe_3C-(010)-2.33 surfaces. The Fe_4C-(100)-3.00 facet has hollow site on the surface, the most stable CO adsorption site is the 4-fold site. In order to compare with other three carbides, the less stable Fe-top site of CO adsorption is taken into account (adsorption energy of CO on the Fe-top site is only 0.054 eV higher than that of the 4-fold site).

At 1/16 and 1/2 ML (Table 10), as the surface work function increases in the order of χ-Fe_5C_2-(100)-2.25<θ-Fe_3C-(010)-2.33<ε-Fe_2C-(1$\bar{2}$1)-2.00<Fe_4C-(100)-3.00, both CO adsorption energies and C-O bond elongation as well as the net negative charge of the adsorbed CO molecules decrease, implying that CO favors to adsorb on the surface with lower work function. Consequently, the χ-Fe_5C_2-(100)-2.25 has the largest CO adsorption energy, followed by θ-Fe_3C-(010)-2.33, ε-Fe_2C-(1$\bar{2}$1)-2.00 and Fe_4C-(100)-3.00, respectively. The C–O bond activation degree is almost the same on χ-Fe_5C_2-(100)-2.25, θ-Fe_3C-(010)-2.33 and ε-Fe_2C-(1$\bar{2}$1)-2.00, while Fe_4C-(100)-3.00 has the weakest ability to activate the C–O bond. For each surface, when the coverage of CO increases from 1/16 to 1/2 ML, both CO adsorption energies and C–O bond elongation decrease.

On the basis of the computed binding energy of surface carbon atoms, it is evident that less negatively charged surface carbon atoms have weaker bonding to the surface. Since CH_4 formation energy exhibits a linear relationship with the charge of surface carbon atom,[24] one can expect that CH_4 formation is most favored thermodynamically on χ-Fe_5C_2-(100)-2.25, followed by θ-Fe_3C-(010)-2.33, ε-Fe_2C-(1$\bar{2}$1)-2.00 and Fe_4C-(100)-3.00.

Conclusion

In this work, we employed DFT calculations and *ab initio* atomistic thermodynamics to investigate the surface structure and stability of the low and high Miller index surfaces of the ε-Fe_2C, χ-Fe_5C_2, θ-Fe_3C and Fe_4C phases as well as their crystal shapes. The goal is to understand the effects of the FTS conditions on the structure and stability of iron carbides as FTS catalysts as well as their differences in surface properties.

The chemical–physical environment around iron based FTS catalysts under working conditions is described from thermodynamic aspect. With different carbon containing gas environments under real FTS operating conditions, it is found that the carburization ability depends mainly on the carbon content of the gas environments, i.e. the higher carbon content of C containing gases, the higher the carburization ability. It is also found that higher temperature, lower pressure and higher H_2/CO ratio can suppress carburization ability and retard carbon deposition.

The crystal shapes of ε-Fe_2C, χ-Fe_5C_2, θ-Fe_3C and Fe_4C have been determined by using the standard Wulff construction on the basis of the calculated surface free energies. Under different pretreatment conditions, the surface morphologies of ε-Fe_2C, χ-Fe_5C_2 and θ-Fe_3C are different in termination and proportion of each facet area, and the most stable non-stoichiometric termination changes from carbon-poor to carbon-rich (varying surface Fe/C ratio) upon the increase in $\Delta\mu_C$. The surface structure and composition of the most stable terminations have similar atom arrangement on the surface layer and the catalytic activities of these facets have been investigated. It is found that lower work function of the surface leads to larger adsorption energy of CO. Less negatively charged surface carbon atoms have weaker binding energy on the surface. Among these four carbides, χ-Fe_5C_2-(100)-2.25 is most favored for CO adsorption and CH_4 formation, followed by θ-Fe_3C-(010)-2.33, ε-Fe_2C-(1$\bar{2}$1)-2.00 and Fe_4C-(100)-3.00, respectively. The activation degree of C–O bonds are almost same on χ-Fe_5C_2-(100)-2.25, θ-Fe_3C-(010)-2.33 and ε-Fe_2C-(1$\bar{2}$1)-2.00, while Fe_4C-(100)-3.00 has the weakest ability for activating the C–O bond.

Appendix

Configuration modeling of fractional site occupancy in ε-Fe_2C

Figure A1 shows the unit cell structure of ε-Fe_2C, and this unit cell has the occupancy of carbon atoms of only 0.5

Table 10 Calculated adsorption energies (E_{ads}, eV) per CO, bond lengths (*d*, Å), net charges (*q*, e) and CO stretching frequencies (ν, cm^{-1}) on carbide surfaces, as well as surface properties of each facet

	χ-Fe_5C_2	θ-Fe_3C	ε-Fe_2C	Fe_4C
Surface	**(100)-2.25**	**(010)-2.33**	**(1$\bar{2}$1)-2.00**	**(100)-3.00**
Work function/eV	3.847	3.991	4.009	4.713
$q^*(C_s)$	−1.062	−1.090	−1.098	−1.119
$E_{ads}(C_s)$	−8.442	−8.459	−8.580	−9.126
E_{ads}(1/16 ML)†	−1.464	−1.442	−1.345	−0.946
E_{ads}(1/16 ML)‡	−1.741	−1.719	−1.667	−1.263
d(C–O) (1/16 ML)	1.171	1.172	1.170	1.164
q (CO) (1/16 ML)*	−0.273	−0.277	−0.254	−0.226
ν_{CO} (1/16 ML)	1937	1933	1939	1990
E_{ads}(1/2 ML)†	−1.142	−1.135	−1.133	−0.317
E_{ads}(1/2 ML)‡	−1.482	−1.483	−1.489	−0.758
d(C–O) (1/2 ML)	1.162	1.163	1.163	1.159
*q**(CO) (1/2 ML)	−0.179	−0.176	−0.178	−0.091
ν_{CO} (1/2 ML)	1976	1971	1973	2017

*From Bader charge analysis.
†RPBE energies
‡PBE energies.

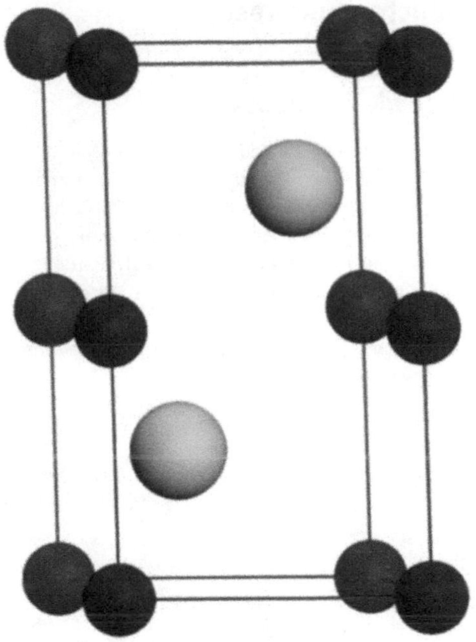

Figure A1 Unit cell structure of ε-Fe_2C (Fe atoms in blue balls, C atoms in black balls)

instead 1.0. This means that Fe atoms form a hexagonal close packed array and half of the octahedral interstitial sites are occupied by carbon atoms in a random way. Similar site-occupancy disorder structures are also found in β-Mo_2C[70] and Fe_2N.[71] According to this bulk structure, the Fe/C ratio is 1 to 1. In order to keep the 2 to 1 stoichiometry, the usually used practice is to delete half of carbon atoms from the bulk structure. However, it will generate several configurations and the number of the possible configurations increases dramatically with the supercell size. Since there are no systematic investigations into the bulk structure of ε-Fe_2C known, several structures of ε-Fe_2C were used in previous studies. Jack reported a ε-Fe_2C structure by deleting carbon atoms on the vertices of the unit cell (Fig. A1).[72] Jang *et al.*[17] and Fang *et al.*[18] calculated the ε-Fe_2C bulk structure with the space group *P*$6_3$22, which is different from the experimental data (*P63/mmc*).[37] In this work, we used the unit cell of ε-Fe_2C with the space group *P63/mmc* to generate the supercell by deleting half of the carbon atoms from the supercell. For supercell with different size, we calculated all possible structures in order to give a reasonable configuration of ε-Fe_2C.

Firstly, we generated the 2×2×1 and 2×2×2 supercells by deleting half of the carbon atoms from these

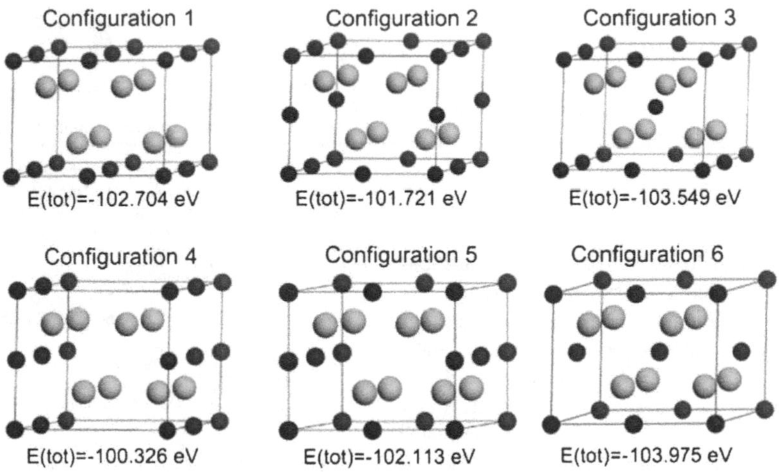

Figure A2 Optimized structures of 2×2×1 supercell of ε-Fe_2C (Fe atoms in blue balls, C atoms in black balls)

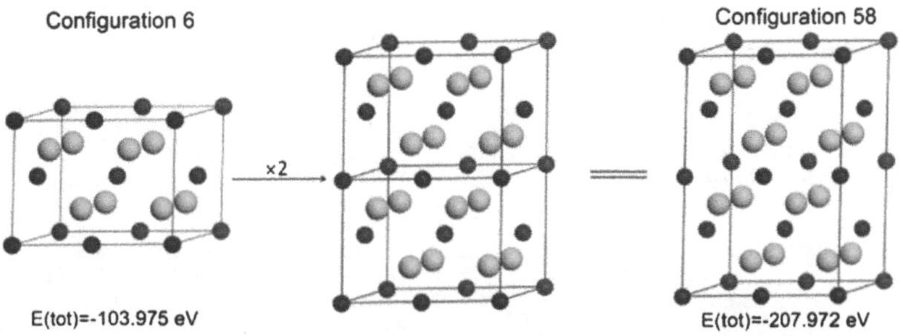

Figure A3 Optimized structures of 2×2×2 supercell of ε-Fe_2C (Fe atoms are shown by blue balls, C atoms are shown by black balls)

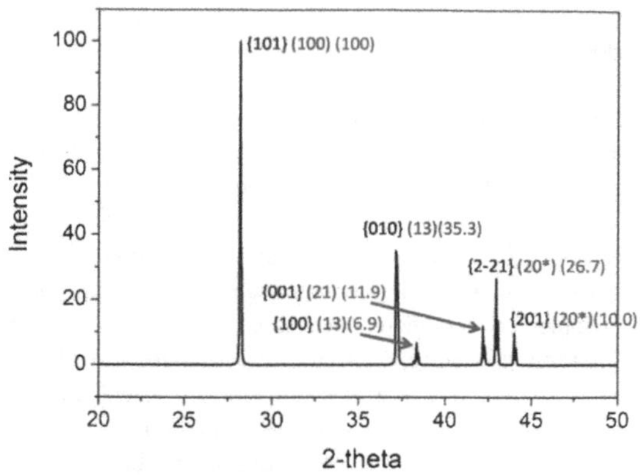

Figure A4 X-ray diffraction of ε-Fe_2C (detected intensities in red and calculated intensities in blue)

two structures. During this process, the site-occupancy disorder program was used to obtain all the supercell structures.[73] By taking the advantage of isometric transformation, site-occupancy disorder excludes the equivalent configurations and reduces the configurations only to the independent ones (Table A1). All the energies were obtained using the method described in the section on 'Structure calculation'. The occurrence probability of each configuration at temperature T can be obtained from equation (9)

$$P_n = \frac{1}{Z}\exp(-E_n/k_BT) \tag{9}$$

where k_B=8.6173 × 10^{-5} eV K^{-1}, E_n is the energy of that configuration and

$$Z = \sum_{n=1}^{N}\exp(-E_n/k_BT) \tag{10}$$

The 2 × 2 × 1 supercell has six independent configurations; the optimized structures and the corresponding total energies are shown in Fig. A2. Among these six structures, configuration 6 is most stable and its occurrence probability is 100% from 0 to 1000 K. After optimizing all 128 configurations of the 2 × 2 × 2 supercell, we found configuration 58 to be most stable (Fig. A3). Its occurrence probability is also 100% from 0 to 1000 K. In addition, configuration 58 is two times of configuration 6, therefore they are the same structure. The simulated XRD spectrum is presented in Fig. A4, the relative intensity of the characteristic peaks agrees well with the experimental data.[37,74] This rationalizes our calculated structure of ε-Fe_2C.

Table A1 Number of independent configurations for a series of supercell of ε-Fe_2C

Supercell	N_0*	N†	M‡
2 × 2 × 1	96	70	6
2 × 2 × 2	192	12,870	122

*Number of symmetry operations.
†Total number of configurations.
‡Number of independent configurations.

Conflicts of interest

The authors declare no conflicts of interest.

Acknowledgements

This work was supported by National Natural Science Foundation of China (Grant No. 21273261), National Basic Research Program of China (Grant No. 2011CB201406), and Chinese Academy of Science and Synfuels China Co., Ltd.

Supporting Information Available: The atomic thermodynamic method used for calculating surface free energies and the calculated surface free energies are available. This material is available free of charge via the Internet at http://www.publicationethics.org.

References

1. F. Fischer and H. Tropsch: *Brennstoff-Chem.*, 1923, **4**, 276–285.
2. F. H. Herbstein, J. Smuts and J. N. van Niekerk: *Anal. Chem.*, 1960, **32**, 20–24.
3. E. de Smit, F. Xinquini, A. M. Beale, O. V. Safonova, W. van Beek, P. Sautet and B. M. Weckhuysen: *J. Am. Chem. Soc.*, 2010, **132**, 14928–14941.
4. E. de Smit, I. Swart, J. F. Creemer, G. H. Hoveling, M. K. Gilles, T. Tyliszczak, P. J. Kooyman, H. W. Zandbergen, C. Morin, B. M. Weckhuysen and F. M. F. de Groot: *Nature*, 2008, **456**, 222–226.
5. A. K. Datye, Y. Jin, L. Mansker, R. T. Motjope, T. H. Dlamini and N. J. Coville: *Stud. Surf. Sci. Catal.*, 2000, **130**, 1139–1144.
6. A. Königer, C. Hammerl, M. Zeitler and B. Rauschenbach: *Phys. Rev. B*, 1997, **55B**, 8143–8147.
7. D. L. Williamson, K. Nakazawa and G. A. Krauss: *Metall. Trans. A*, 1979, **10A**, 1351–1363.
8. M. Manes, A. D. Damick, M. Mentser, E. M. Cohn and L. J. E. Hofer: *J. Am. Chem. Soc.*, 1952, **74**, 6207–6209.
9. E. M. Cohn and L. J. E. Hofer: *J. Chem. Phys.*, 1953, **21**, 354–359.
10. J. F. Shultz, W. K. Hall, T. A. Dubs and R. B. Anderson: *J. Am. Chem. Soc.*, 1955, **78**, 282–285.
11. L. J. E. Hofer and E. M. Cohn: *Nature* 1951, **167**, 977–978.
12. S. Nagakura: *J. Phys. Soc. Jpn*, 1959, **14**, 186–195.
13. H. Merkel and F. Weinrotter: *Brennstoff-Chem.*, 1951, **32**, 289–297.
14. W. C. Chiou Jr and E. A. Carter: *Surf. Sci.*, 2003, **530**, 88–100.
15. D. C. Sorescu: *J. Phys. Chem. C*, 2009, **113C**, 9256–9274.
16. Z. Q. Lv, S. H. Sun, P. Jiang, B. Z. Wang and W. T. Fu: *Comput. Mater. Sci.*, 2008, **42**, 692–697.
17. J. H. Jang, I. G. Kim and H. K. D. H. Bhadeshia: *Scr. Mater.*, 2010, **63**, 121–123.
18. C. M. Fang, M. A. van Huis and H. W. Zandbergen: *Scr. Mater.*, 2010, **63**, 418–421.
19. H. I. Faraoun, Y. D. Zhang, C. Esling and H. Aourag: *J. Appl. Phys.*, 2006, **99**, 093508–093515.
20. C. Jiang, S. G. Srinivasan, A. Caro and S. A. Maloy: *J. Appl. Phys.*, 2008, **103**, 043502–043509.
21. B. Hallstedt, D. Djurovic, J. von Appen, R. Dronskowski, A. Dick, F. Körmann, T. Hickel and J. Neugebauer: *CALPHAD: Comput. Coupling Phase Diagr. Thermochem.*, 2010, **34**, 129–133.
22. C. K. Ande and M. H. F. Sluiter: *Metall. Mater. Trans. A*, 2012, **43A**, 4436–4444.
23. L. J. Deng, C. F. Huo, X. W. Liu, X. H. Zhao, Y. W. Li, J. G. Wang and H. J. Jiao: *J. Phys. Chem. C*, 2010, **114C**, 21585–21592.
24. C. F. Huo, Y. W. Li, J. G. Wang and H. J. Jiao: *J. Am. Chem. Soc.*, 2009, **131**, 14713–14721.
25. X. Y. Liao, D. B. Cao, S. G. Wang, Z. Y. Ma, Y. W. Li, J. G. Wang and H. J. Jiao: *J. Mol. Catal. A: Chem.*, 2007, **269**, 169–178.
26. X. Y. Liao, S. G. Wang, Z. Y. Ma, Y. W. Li, J. G. Wang and H. J. Jiao: *J. Mol. Catal. A: Chem.*, 2008, **292**, 14–20.
27. B. Q.Wei, M. Shima, R. Pati, S. K. Nayak, D. J. Singh, R. Ma, Y. Li, Y. Bando, S. Nasu and P. M. Ajayan: *Small*, 2006, **6**, 804–809.
28. C. M. Deng, C. F. Huo, L. L. Bao, X. R. Shi, Y. W. Li, J. G. Wang and H. J. Jiao: *Chem. Phys. Lett.*, 2007, **448**, 83–87.
29. C. M. Deng, C. F. Huo, L. L. Bao, G. Feng, Y. W. Li, J. G. Wang and H. J. Jiao: *J. Phys. Chem. C*, 2008, **112C**, 19018–19029.
30. S. Zhao, X. W. Liu, C. F. Huo, Y. W. Li, J. G. Wang and H. J. Jiao: *J. Catal.*, 2011, **294**, 47–53.
31. G. Kresse and J. Furthmüller: *Comput. Mater. Sci.* 1996, **6**, 15–50.
32. G. Kresse and J. Furthmüller: *J. Phys. Rev. B*, 1996, **54B**, 11169–11186.

where Res$[k]$ and Res$[-k]$ are the residues in the poles k and $-k$ given, respectively, by

$$\mathrm{Res}[k] = \lim_{\alpha \to k} [(\alpha - k) f(\alpha)] = -\frac{1}{2k} \quad \mathrm{Res}[-k] = \lim_{\alpha \to -k} [(\alpha + k) f(\alpha)] = \frac{1}{2k}$$

Of course the problem of decomposition yields many solutions that differ from one another by the possible presence of entire functions. We define as standard decomposition the one in which the decomposed functions $f_+(\alpha)$ and $f_-(\alpha)$ either vanish or have the lowest growth order as α tends to infinity.

2.3 Multiplicative decomposition or factorization

Given an analytic function $f(\alpha)$, the problem of multiplicative decomposition (or factorization) consists in finding two functions $f_+(\alpha)$ and $f_-(\alpha)$ satisfying the equation

$$f(\alpha) = f_-(\alpha) f_+(\alpha) \tag{4}$$

and such that they, and their inverses, are regular in the half-planes $\mathrm{Im}[\alpha] \geq 0$ and $\mathrm{Im}[\alpha] \leq 0$, respectively.

To avoid confusion with the additive decomposition defined in the previous section 2.2, occasionally we will use the following symbols to indicate the factorized functions $f_-(\alpha)$, $f_+(\alpha)$ of $f(\alpha)$:

$$f_-(\alpha) = [f(\alpha)]_-, \quad f_+(\alpha) = [f(\alpha)]_+ \tag{5}$$

The Weirstrass factorization of entire functions (Spiegel, 1964) often provides the factorizations involved in many practical situations (section 3.2.4). However, in general we are faced with many-valued functions, and the fundamental idea to factorize a given function $f(\alpha)$ is based on the concept of logarithmic decomposition. By introducing

$$f(\alpha) = e^{\log[f(\alpha)]} = e^{\psi(\alpha)} \tag{6}$$

the decomposition of

$$\psi(\alpha) = \mathrm{Log}[f(\alpha)] = \psi_-(\alpha) + \psi_+(\alpha) \tag{7}$$

yields

$$\begin{aligned} f(\alpha) &= \exp[\psi(\alpha)] = \exp\left[\psi_-(\alpha) + \psi_+(\alpha)\right] \\ &= \exp[\psi_-(\alpha)]\exp\left[\psi_+(\alpha)\right] \end{aligned} \tag{8}$$

This approach involves a commutative algebra. For instance, it works well for scalar functions. Conversely, in the matrix case a not-commutative algebra is involved, and this approach may be ineffective. The matrix factorization is possible in closed form only in some particular cases that will be considered in chapter 4. The general explicit matrix factorization problem is presently still an open problem that despite many efforts does not yet have a general solution.

Example 2

Factorize $f(\alpha) = \sqrt{k^2 - \alpha^2}$ in plus and minus functions that are regular in $\mathrm{Im}[\alpha] \geq 0$ and $\mathrm{Im}[\alpha] \leq 0$, respectively.

The simplicity of the function to be factorized makes it possible to find the explicit solution by inspection:

$$f_+(\alpha) = \sqrt{k - \alpha}, \quad f_-(\alpha) = \sqrt{k + \alpha} \tag{9}$$

Of course, the factorization problem yields many solutions that differ for the possible presence of entire functions as long as their inverses are entire as well. We call standard factorization the factorization that involves factorizing functions having algebraic behavior as $\alpha \to \infty$.

2.4 Solution of the W-H equation

2.4.1 Solution of the nonhomogeneous equation

Let us consider the case where the unknowns $F_+(\alpha)$ and $F_-^s(\alpha)$ are conventional (section 1.1) plus and minus functions. The standard factorization of $G(\alpha) = G_-(\alpha)G_+(\alpha)$ yields

$$G_+(\alpha)F_+(\alpha) = G_-^{-1}(\alpha)F_-^s(\alpha) + G_-^{-1}(\alpha)F_{o+}(\alpha) \tag{10}$$

By decomposing the second term of the second member in a standard way:

$$G_-^{-1}(\alpha)F_{o+}(\alpha) = S_-(\alpha) + S_+(\alpha) \tag{11}$$

and separating the plus and minus functions yields

$$G_+(\alpha)F_+(\alpha) - S_+(\alpha) = G_-^{-1}(\alpha)F_-^s(\alpha) + S_-(\alpha) = w(\alpha) \tag{12}$$

Considering now (12) in the whole complex plane α, the analytic function $w(\alpha)$ is regular both in the upper half-plane $\mathrm{Im}[\alpha] \geq 0$ (considering the first member) and in the lower half-plane $\mathrm{Im}[\alpha] \leq 0$ (considering the second member). Since the two half-planes are partially overlapping, this means that $w(\alpha)$ is an entire function.

The functions $F_+(\alpha)$, $F_-(\alpha)$ and $F_{o+}(\alpha)$ are generally Laplace transforms. Thus, they are vanishing as $\alpha \to \infty$. This means that $G(\alpha)$ has to show a suitable behavior for $\alpha \to \infty$. For instance, the factorization $G(\alpha) = G_-(\alpha)G_+(\alpha)$ must involve factorized function $G_-(\alpha)$ and $G_+(\alpha)$ such that $\frac{G_\pm(\alpha)}{\alpha}$ and $\frac{G_\pm^{-1}(\alpha)}{\alpha}$ are vanishing as $\alpha \to \infty$. By looking at the behavior of the standard decomposed and factorized functions, it is possible to ascertain that the first two members of (12) are Laplace transforms. Furthermore, since they are vanishing as $\alpha \to \infty$, it means that also the entire function $w(\alpha)$ is vanishing as $\alpha \to \infty$. Consequently, using Liouville's theorem (Spiegel, 1964), $w(\alpha)$ is zero everywhere, which yields the solutions

$$F_+(\alpha) = G_+^{-1}(\alpha)S_+(\alpha), \quad F_-^s(\alpha) = -G_-(\alpha)S_-(\alpha) \tag{13}$$

To extend the possibilities of this ingenious technique, it is also possible to deal with generic plus and minus functions in (12) that are not vanishing for $\alpha \to \infty$, and consequently the entire function $w(\alpha)$ is not vanishing. In this case the following procedure applies (Noble, 1958, p. 37). Let us suppose that it could be shown that

$$\begin{aligned} |G_+(\alpha)F_+(\alpha) - S_+(\alpha)| &< |\alpha|^p \quad \text{as } \alpha \to \infty, \ \operatorname{Im}[\alpha] \geq 0 \\ |G_-^{-1}(\alpha)F_-^s(\alpha) + S_-(\alpha)| &< |\alpha|^q \quad \text{as } \alpha \to \infty, \ \operatorname{Im}[\alpha] \leq 0 \end{aligned} \tag{14}$$

By applying the extended form of Liouville's theorem, it follows that $w(\alpha)$ is a polynomial $P(\alpha)$ of degree less than or equal to the integral part of $\min(p, q)$, that is,

$$G_+(\alpha)F_+(\alpha) - S_+(\alpha) = P(\alpha) \tag{15}$$

$$-G_-^{-1}(\alpha)F_-^s(\alpha) + S_-(\alpha) = P(\alpha) \tag{16}$$

These equations determine $F_+(\alpha)$ and $F_-(\alpha)$ to within the arbitrary polynomial $P(\alpha)$, that is, to within a finite number of arbitrary constants that have yet to be determined.

Example 3

Let us consider the W-H equation:

$$f(x) - \int_0^\infty \frac{4}{3}e^{-3|x-x'|}f(x')dx' = \frac{1}{4}e^{-2x}, \quad 0 \leq x < \infty \tag{17}$$

Taking into account that

$$g(x) = \delta(x) - \frac{4}{3}e^{-3|x|} \tag{18}$$

yields the kernel

$$G(\alpha) = 1 - \int_{-\infty}^{\infty} \frac{4}{3}e^{-3|x|}e^{j\alpha x}dx = \frac{\alpha^2 + 1}{\alpha^2 + 9} \tag{19}$$

In the spectral domain we obtain

$$G(\alpha)F_+(\alpha) = F_-^s(\alpha) + F_o(\alpha) \quad \text{or} \tag{20}$$

$$\frac{\alpha^2 + 1}{\alpha^2 + 9}F_+(\alpha) = F_-^s(\alpha) + \frac{1}{\alpha^2 + 4} \tag{21}$$

where

$$F_+(\alpha) = \int_{-\infty}^{\infty} f(x)e^{j\alpha x}dx \tag{22}$$

$$F_o(\alpha) = \int_{-\infty}^{\infty} \frac{1}{4}e^{-2|x|}e^{j\alpha x}dx = \frac{1}{\alpha^2 + 4} \tag{23}$$

and $F_-^s(\alpha)$ is an unknown minus function.

Factorization of $G(\alpha) = \frac{\alpha^2+1}{\alpha^2+9} = \frac{\alpha-j}{\alpha-3j}\frac{\alpha+j}{\alpha+3j}$ yields

$$G_-(\alpha) = \frac{\alpha - j}{\alpha - 3j}, \quad G_+(\alpha) = \frac{\alpha + j}{\alpha + 3j}$$

Decomposition of $S(\alpha) = G_-^{-1}(\alpha)\, F_o(\alpha) = S_-(\alpha) + S_+(\alpha)$ gives

$$S_+(\alpha) = \frac{\text{Res}\left\{\frac{\alpha - 3j}{\alpha - j}\frac{1}{\alpha^2 + 4}\right\}\Big|_{\alpha=-2j}}{\alpha + 2j} = \frac{5j}{12(\alpha + 2j)} \tag{24}$$

thus yielding the solution

$$F_+(\alpha) = G_+^{-1}(\alpha) S_+(\alpha) = \frac{5j(\alpha + 3j)}{12(\alpha + j)(\alpha + 2j)} \tag{25}$$

The inverse Laplace transform provides

$$f(x) = \frac{1}{2\pi}\int_B F_+(\alpha) e^{-j\alpha x} d\alpha = \frac{1}{2\pi}\int_{-\infty}^{\infty} F_+(\alpha) e^{-j\alpha x} d\alpha \tag{26}$$

Since in this case the Bromwich contour B coincides with the real axis, we have the solution of the W-H equation:

$$f(x) = \frac{1}{2\pi}\int_{-\infty}^{\infty} F_+(\alpha) e^{-j\alpha x} d\alpha = -\frac{5}{12}e^{-2x} + \frac{5}{6}e^{-x}, \quad x \geq 0 \tag{27}$$

Example 4

The solution of the W-H equation relevant to the half-plane problem (1.1.1):

$$-\frac{k\, Z_o}{4}\frac{2}{\sqrt{k^2 - \alpha^2}}\tilde{J}_+(\alpha) = V_-^s(\alpha) - \frac{j}{\alpha - \alpha_o}E_o = V_-(\alpha) \quad \text{Im}[\alpha_o] < 0$$

can be accomplished readily by observing that the factorization of the kernel

$$G(\alpha) = \frac{k\, Z_o}{2\sqrt{k^2 - \alpha^2}} = G_-(\alpha) G_+(\alpha)$$

is immediate:

$$G_-(\alpha) = \frac{k\, Z_o}{2\sqrt{k + \alpha}}, \quad G_+(\alpha) = \frac{1}{\sqrt{k - \alpha}}$$

We find

$$\tilde{J}_+(\alpha) = 2\frac{\sqrt{k - \alpha}\sqrt{k + \alpha_o}}{kZ_o}\frac{j}{\alpha - \alpha_o}E_o, \quad V_-^s(\alpha) = -\frac{\sqrt{k + \alpha_o}}{\sqrt{k + \alpha}}\frac{j}{\alpha - \alpha_o}E_o + \frac{j}{\alpha - \alpha_o}E_o$$

The obtained solution refers to $\text{Im}[\alpha_o] < 0$. An analytic continuation renders this solution valid also for $\text{Im}[\alpha_o] > 0$.

2.4.2 Remote source

When the source of the problem is very far (e.g., in the case of plane wave excitation of an open structure or in the case of incident mode excitation of a closed structures), the W-H equations assume the following form:

$$G(\alpha)F_+(\alpha) = F_-(\alpha) \tag{28}$$

where $F_+(\alpha)$ and $F_-(\alpha)$ are Laplace transforms related to the total fields.

Remarkably, in the presence of a remote source the W-H equations are homogeneous. This does not produce vanishing solutions for $F_+(\alpha)$ and $F_-(\alpha)$, since some Laplace transforms become nonconventional (e.g., a plus function could have the pole relevant to the source in the upper half plane $\text{Im}[\alpha] > 0$) and the procedure indicated in the previous section cannot be applied.

The standard factorization of

$$G(\alpha) = G_-(\alpha)G_+(\alpha) \tag{29}$$

yields

$$G_+(\alpha)F_+(\alpha) = [G_-(\alpha)]^{-1}F_-(\alpha) \tag{30}$$

Without loss of generality, let us suppose that $F_+(\alpha)$ possesses a first-order nonconventional pole at $-\beta_o$, and $F_-(\alpha)$ possesses a first-order nonconventional pole at α_o.

Equation (30) can be rewritten as

$$\begin{aligned} &G_+(\alpha)F_+(\alpha) - G_+(-\beta_o)\frac{T}{\alpha+\beta_o} - [G_-(\alpha_o)]^{-1}\frac{R}{\alpha-\alpha_o} \\ &= G_-^{-1}(\alpha)F_-(\alpha) - G_+(-\beta_o)\frac{T}{\alpha+\beta_o} - [G_-(\alpha_o)]^{-1}\frac{R}{\alpha-\alpha_o} = w(\alpha) \end{aligned} \tag{31}$$

where T and R are the residues of $F_+(\alpha)$ and $F_-(\alpha)$, respectively, at $-\beta_o$ and α_o. The entire function $w(\alpha)$ is vanishing, provided that $F_+(\alpha)$ and $F_-(\alpha)$ are Laplace transforms and the factorization (29) is standard. From (31) we get the solutions:

$$\begin{aligned} F_+(\alpha) &= [G_+(\alpha)]^{-1}\cdot G_+(-\beta_o)\cdot\frac{T}{\alpha+\beta_o} + [G_+(\alpha)]^{-1}\cdot[G_-(\alpha_o)]^{-1}\cdot\frac{R}{\alpha-\alpha_o} \\ F_-(\alpha) &= G_-(\alpha)\cdot G_+(-\beta_o)\cdot\frac{T}{\alpha+\beta_o} + G_-(\alpha)\cdot[G_-(\alpha_o)]^{-1}\cdot\frac{R}{\alpha-\alpha_o} \end{aligned} \tag{32}$$

To apply (32), we must know the vectors T and R that define the nonconventional part of $F_+(\alpha)$ and $F_-(\alpha)$. We observe that the contribution of the pole at α_o (or $-\beta_o$) in the spectral domain α, represents a plane wave or a mode. Therefore T and R can be obtained by physical considerations. Unfortunately, in some instances such as in planar stratifications or wedge problems the same nonconventional pole occurs not only for the incident field but also for reflected fields. In these cases the evaluation of the nonconventional contribution must be performed very carefully. Alternatively, this contribution may be assumed initially known and determined a posteriori by taking into account physical considerations.

Example 5

Let us examine again the solution of the W-H equation related to the half-plane problem (1.1.1). In this case (30) becomes

$$\frac{1}{\sqrt{k-\alpha}}\tilde{J}_+(\alpha) = -\frac{2\sqrt{k+\alpha}}{k\,Z_o}V_-(\alpha) \tag{33}$$

Let us consider $\text{Im}[\alpha_o] > 0$. It means that this pole is nonconventional for $J_+(\alpha)$. We obtain

$$\frac{1}{\sqrt{k-\alpha}}\tilde{J}_+(\alpha) - \frac{1}{\sqrt{k-\alpha_o}}\frac{T}{\alpha-\alpha_o} = -\frac{2\sqrt{k+\alpha}}{k\,Z_o}V_-(\alpha) - \frac{1}{\sqrt{k-\alpha_o}}\frac{T}{\alpha-\alpha_o}$$

which yields the solutions

$$\tilde{J}_+(\alpha) = \frac{\sqrt{k-\alpha}}{\sqrt{k-\alpha_o}}\frac{T}{\alpha-\alpha_o}$$

$$V_-(\alpha) = -\frac{k\,Z_o}{2\sqrt{k+\alpha}\sqrt{k-\alpha_o}}\frac{T}{\alpha-\alpha_o}$$

We may employ several methods to determine the constant T.

(a) The residue of the pole α_o present in $V_-(\alpha)$ is related only to the incident plane wave present on the aperture $-\infty < x < 0,\ y = 0$:

$$V_-(\alpha) = V_-^s(\alpha) - \frac{j}{\alpha-\alpha_o}E_o$$

From (33) the residue T of $J_+(\alpha)$ is given by

$$T = -\frac{2\sqrt{k^2-\alpha_o^2}}{k\,Z_o}(-jE_o)$$

and this yields the same result previously obtained in example 4:

$$\tilde{J}_+(\alpha) = j\frac{\sqrt{k-\alpha}}{\sqrt{k-\alpha_o}}\frac{2\sqrt{k+\alpha_o}}{k\,Z_o}\frac{E_o}{\alpha-\alpha_o}$$

(b) Since the incident plane wave and the reflected plane wave have the same pole α_o, the characteristic part $\frac{T}{\alpha-\alpha_o}$ must include both these waves. Geometric optics consideration yields

$$J^g(x) = -\left[H_x^i(x,0_+) + H_x^r(x,0_+)\right] - 2H_x^i(x,0_+) = 2\frac{\sin\varphi_o}{Z_o}e^{jk\,x\cos\varphi_o}E_o$$

and therefore the residue

$$T = 2\frac{\sin\varphi_o}{Z_o}jE_o$$

which is the same value obtained in (a).

(c) The residue of $V_-(\alpha) = -\frac{k\,Z_o}{2\sqrt{k+\alpha}\sqrt{k-\alpha_o}}\frac{T}{\alpha-\alpha_o}$ is $-jE_o$. This yields the same value obtained with the procedures in (a) and (b).

The obtained solution remains the same if $\text{Im}[\alpha_o] < 0$. This validates the analytic continuations used in example 2.

2.5 Unbounded plus and minus unknowns

Sometimes the formulation of a specific problem in terms of Wiener-Hopf equations yields the presence of a plus function, call it $\hat{F}_+(\alpha)$, that even though it is regular in the upper half-plane, is not a classical Laplace transform, in the sense that it does not show the vanishing asymptotic behavior for $\alpha \to \infty$ required by a proper Laplace transform.

For instance $\hat{F}_+(\alpha)$ may be defined as

$$\hat{\mathbf{F}}_+(\alpha) = \mathbf{p}(\alpha) \cdot \mathbf{F}_+(\alpha), \tag{34}$$

where the vector $\mathbf{F}_+(\alpha)$ is a classical Laplace transform and $\mathbf{p}(\alpha)$ is a polynomial matrix or, more generally, an entire matrix. Following the Wiener-Hopf procedure that yields (12), a non-vanishing entire vector $\mathbf{w}(\alpha)$ could be involved. If $\mathbf{p}(\alpha)$ is a polynomial factor, the asymptotic behavior as $\alpha \to \infty$ shows that $\mathbf{w}(\alpha)$ must also be a polynomial. The coefficients of the polynomial $w(\alpha)$ must be determined by imposing that $F_+(\alpha) = [p(\alpha)]^{-1}\hat{F}_+(\alpha)$ be a plus function. Some examples will be considered later, in dealing with some important specific diffraction problems (sections 7.5 and 9.5).

Similar considerations apply in the presence of unbounded minus functions:

$$\hat{\mathbf{F}}_-(\alpha) = \mathbf{p}(\alpha) \cdot \mathbf{F}_-(\alpha) \tag{35}$$

2.6 Factorized matrices as solutions of the homogeneous Wiener-Hopf problem

We recall that the solutions of the W-H equation of order n are given by

$$\begin{aligned} F_+(\alpha) &= G_+^{-1}(\alpha)G_-^{-1}(\alpha_o)\frac{R}{\alpha-\alpha_o} \\ F_-(\alpha) &= G_-(\alpha)G_-^{-1}(\alpha_o)\frac{R}{\alpha-\alpha_o} \end{aligned} \tag{36}$$

Since for R known the factorized matrices $G_-(\alpha)$ and $G_+(\alpha)$ provide immediately the solution for every value of α and/or α_o, it is important to evaluate the factorized matrices once and for all, through suitable solutions of the W-H equations. To this end, consider first the homogeneous Wiener Hopf problem:

$$G(\alpha)X_+(\alpha) = X_-(\alpha) \tag{37}$$

where $G(\alpha)$ is the kernel matrix. This problem is a particular case of the Riemann-Hilbert problem that has been studied by many authors (Vekua, 1967). It is easy to show that n

independent solutions $\{X_{i+}(\alpha), X_{i-}(\alpha)\}$, $\{i = 1, 2, \ldots, n\}$ of (37) provide the factorized matrices of $G(\alpha) = G_-(\alpha)G_+(\alpha)$ through

$$G_-(\alpha) = |X_{1-}(\alpha), X_{2-}(\alpha), \ldots, X_{n-}(\alpha)|, \quad G_+(\alpha) = |X_{1+}(\alpha), X_{2+}(\alpha), \ldots, X_{n+}(\alpha)|^{-1} \tag{38}$$

Of course the existence and uniqueness of n independent solutions $\{X_{i+}(\alpha), X_{i-}(\alpha)\}$, of (37) follows from the application of sophisticated theorems to the given matrix $G(\alpha)$. However, in all the cases of practical interest we can be sure, on physical considerations, that the problem is mathematically well defined, so to factorize $G(\alpha)$ the only problem to face is to determine the set $\{X_{i+}(\alpha), X_{i-}(\alpha)\}$. To this purpose, let us introduce the functions:

$$F_{i+}(\alpha) = \frac{X_{i+}(\alpha)}{\alpha - \alpha_p} \tag{39}$$

where α_p is not a structural singularity with $\text{Im}[\alpha_p] < 0$.[1] The homogeneous equation can be rewritten as

$$G(\alpha) \cdot F_{i+}(\alpha) = \frac{X_{i-}(\alpha) - X_{i-}(\alpha_p)}{\alpha - \alpha_p} + \frac{X_{i-}(\alpha_p)}{\alpha - \alpha_p} = F_{i-}(\alpha), \quad i = 1, 2, \ldots, n \tag{40}$$

or:

$$G(\alpha) \cdot F_{i+}(\alpha) = F_{i-}^s(\alpha) + \frac{E_i}{\alpha - \alpha_p}, \quad i = 1, 2, \ldots, n \tag{41}$$

where $E_i = X_{i-}(\alpha_p)$ and the plus $F_{i+}(\alpha)$ and the minus $F_{i-}^s(\alpha) = \frac{X_{i-}(\alpha) - X_{i-}(\alpha_p)}{\alpha - \alpha_p}$ are standard Laplace transforms. In the following we will set

$$E_1 = [1, 0, 0, 0, \ldots]^t, \quad E_2 = [0, 1, 0, 0, \ldots]^t, \ldots, E_n = [0, 0, 0, 0, \ldots, 1]^t \tag{42}$$

Consequently, the solution of the n W-H equations:

$$G(\alpha) \cdot F_{i+}(\alpha) = F_{i-}^s(\alpha) + \frac{E_i}{\alpha - \alpha_p} = F_{i-}(\alpha), \quad i = 1, 2, \ldots, n \tag{43}$$

provides the determinations of the factorized matrices $G_-(\alpha)$ and $G_+(\alpha)$ through the eqs. (38), (36) and (39). It can be shown that by changing α_p, the factorized matrices are modified by a multiplicative constant matrix such that the equation $G(\alpha) = G_-(\alpha)G_+(\alpha)$ is always satisfied.

In fact, let us introduce the factorized matrices $G_{-,+}(\alpha)$, $G_+(\alpha, \alpha_1)$, $G_+(\alpha, \alpha_2)$ where $G_{-,+}(\alpha)$ are arbitrary standard factorized matrix of $G(\alpha)$ and $G_+(\alpha, \alpha_1)$ and $G_+(\alpha, \alpha_2)$ have been determined with the previous technique by assuming two different values of α_p: $\alpha_p = \alpha_1$ and $\alpha_p = \alpha_2$, respectively. The Wiener-Hopf solution of (43) yields the following expressions for the vectors X_{i+}^1 and X_{i+}^1:

$$X_{i+}^{1,2} = \frac{G_+^{-1}(\alpha) \cdot G_-^{-1}(\alpha_{1,2})}{\alpha - \alpha_{1,2}} R_i \tag{44}$$

---

[1] For instance notice from (39) that if $\det[G(\alpha_p)] = 0$ (α_p is a pole of $G^{-1}(\alpha)$), it must be $X_{i-}(\alpha_p) = 0$ and $X_{i+}(\alpha_p) = \text{NullSpace}[G(\alpha_p)]$ otherwise $X_{i+}(\alpha_p) = \infty$. Conversely if $\det[G^{-1}(\alpha_p)] = 0$ (α_p is a pole of $G(\alpha)$), it must be $X_{i+}(\alpha_p) = 0$ and $X_{i-}(\alpha_p) = \text{NullSpace}[G^{-1}(\alpha_p)]$ otherwise $X_{i-}(\alpha_p) = \infty$.

By taking into account eq. (38), which define $G_+(\alpha,\alpha_2)$ and $G_+(\alpha,\alpha_1)$, algebraic manipulations yield

$$G_+(\alpha,\alpha_2) = K\, G_+(\alpha,\alpha_1) \tag{45}$$

where the constant matrix K is given by

$$\begin{aligned} K = {} & |G_-^{-1}(\alpha_2)\cdot R_1, G_-^{-1}(\alpha_2)\cdot R_2, G_-^{-1}(\alpha_2)\cdot R_3, G_-^{-1}(\alpha_2)\cdot R_4|^{-1}\cdot \\ & |G_-^{-1}(\alpha_1)\cdot R_1, G_-^{-1}(\alpha_1)\cdot R_2, G_-^{-1}(\alpha_1)\cdot R_3, G_-^{-1}(\alpha_1)\cdot R_4| \end{aligned} \tag{46}$$

In the following, we do not examine the problem of the better location of α_p in the half-plane $\mathrm{Im}[\alpha] < 0$. This problem has relevance in numerical solutions: a not suitable choice of α_p could lead to unsatisfactory numerical expressions for factorized matrices. In fact, α_p introduces an apparent singularity; it does not produce effects on closed-form solutions but in numerical solutions does increases the possibility of having significant errors in a region of the α − plane near α_p. To avoid this problem, we suggest defining α_p in the context of a physical problem related to the factorization of the matrix. For instance, in open structures a pole α_p relevant to a physical source is constituted by a plane wave.

2.7 Nonstandard factorizations

We are often dealing with the factorization of matrices having the form

$$G(\alpha) = R(\alpha)\cdot M(\alpha)\cdot T(\alpha) \tag{47}$$

where $R(\alpha)$ and $T(\alpha)$ are rational matrices and the factorized matrices $M_-(\alpha)$, and $M_+(\alpha)$ of the matrix $M(\alpha) = M_-(\alpha)M_+(\alpha)$ are explicitly known. We can formally write the factorization

$$G(\alpha) = \tilde{G}_-(\alpha)\tilde{G}_+(\alpha) \tag{48}$$

where

$$\tilde{G}_-(\alpha) = R(\alpha)\cdot M_-(\alpha), \quad \tilde{G}_+(\alpha) = M_+(\alpha)\cdot T(\alpha) \tag{49}$$

$\tilde{G}_-(\alpha)$, $\tilde{G}_+(\alpha)$, and their inverses present algebraic behavior at infinity without being true minus and plus functions because of the presence of a finite number of offending poles. These poles are all the poles of $R(\alpha)$ and $T(\alpha)$, located in $\mathrm{Im}[\alpha] < 0$ and $\mathrm{Im}[\alpha] > 0$, respectively. This type of factorization has been referred to in the literature as weak factorization (Büyükaksoy & Serbest, 1993; Idemen, 1977, 1979; Abrahams & Wickham, 1990). It is possible to eliminate the offending poles by using many different techniques. To illustrate one of them, it can be observed that it is not restrictive to let $R(\alpha) = 1$ in (47). In fact, if we had to factorize $G(\alpha) = R(\alpha)\cdot M(\alpha)$, we could use the procedure that we are about to outline by reversing the roles of the plus and minus functions. Hence, we consider only the factorization of

$$G(\alpha) = M(\alpha)\cdot T(\alpha) \tag{50}$$

Letting $T(\alpha) = \frac{P(\alpha)}{d(\alpha)}$ where $P(\alpha)$ and $d(\alpha)$ are matrix and scalar polynomials, respectively, in α, eq. (37) becomes

$$M_+(\alpha)\frac{P(\alpha)}{d(\alpha)}X_+(\alpha) = M_-^{-1}(\alpha)X_-(\alpha) \tag{51}$$

The vector $M_+(\alpha)P(\alpha)X_+(\alpha) = \hat{X}_+(\alpha)$ is a plus function. The factorization of $d(\alpha)$ yields

$$\frac{\hat{X}_+(\alpha)}{d_+(\alpha)} = d_-(\alpha)M_-^{-1}(\alpha)X_-(\alpha) = w(\alpha) \tag{52}$$

where $w(\alpha)$ is an unknown polynomial vector. We observe that (52) provides the plus and minus functions within a finite number of constants (the coefficients of $w(\alpha)$). From here on, these representations will be called *weak representations*.

The polynomial $w(\alpha)$ can be determined by considering the first and third members of (52):

$$\begin{aligned} X_+(\alpha) &= P^{-1}(\alpha)M_+^{-1}(\alpha)\hat{X}_+(\alpha) = d_+(\alpha)P^{-1}(\alpha)M_+^{-1}(\alpha)w(\alpha) \\ &= \frac{d_+(\alpha)}{\Delta(\alpha)}P_a(\alpha)M_+^{-1}(\alpha)w(\alpha) \end{aligned} \tag{53}$$

where

$$P^{-1}(\alpha) = \frac{P_a(\alpha)}{\Delta(\alpha)} \tag{54}$$

has been introduced, with $\Delta(\alpha)$ and $P_a(\alpha)$ the determinant and the adjoint matrix of $P(\alpha)$, respectively. Let us suppose that all the zeroes of $\Delta(\alpha)$ are simple, and let us indicate with α_{-i} $(i = 1, 2, \dots)$ the zeroes located in the upper half-plane $\mathrm{Im}(\alpha) > 0$. Since $X_+(\alpha)$ is a plus vector, it follows from (53) that

$$P_a(\alpha_{-i})M_+^{-1}(\alpha_{-i})w(\alpha_{-i}) = 0 \quad (i = 1, 2, \dots) \tag{55}$$

and therefore $w(\alpha_{-i})$ must satisfy the equation

$$w(\alpha_{-i}) = c_i M_+(\alpha_{-i})u_i \quad (i = 1, 2, \dots) \tag{56}$$

where u_i is in the null space of $P_a(\alpha_{-i})$ and c_i is an arbitrary constant. Equation (56) allows the evaluation of the matrix polynomial coefficients in terms of the constants c_i, $(i = 1, 2, \dots)$. By changing the values of these constants, in well posed problems there are n and only n independent $w(\alpha)$: $\{w_1(\alpha), w_2(\alpha), \dots, w_n(\alpha)\}$ satisfying (52); hence, we have obtained n independent $X_+(\alpha)$ and $X_-(\alpha)$ given by

$$X_+(\alpha) = \frac{d_+(\alpha)}{\Delta(\alpha)}P_a(\alpha)M_+^{-1}(\alpha)w(\alpha) \tag{57}$$

$$X_-(\alpha) = M_-(\alpha)\frac{w(\alpha)}{d_-(\alpha)} \tag{58}$$

According to the results of section 2.6 these functions provide the factorization of $G(\alpha) = G_-(\alpha)G_+(\alpha)$.

By assuming $M(\alpha) = 1$, the previous technique can also be used to factorize rational matrices.

Example: Factorize the rational matrix $G(\alpha) = \begin{vmatrix} 1 & jq\frac{\alpha^2+A^2}{\alpha^2+B^2} \\ jq & 1 \end{vmatrix}$ *where q, A and B are real.*

Let us assume:

$$P(\alpha) = \begin{vmatrix} B^2+\alpha^2 & jq(A^2+\alpha^2) \\ jq(B^2+\alpha^2) & B^2+\alpha^2 \end{vmatrix} \quad P_a(\alpha) = \begin{vmatrix} B^2+\alpha^2 & -jq(A^2+\alpha^2) \\ -jq(B^2+\alpha^2) & B^2+\alpha^2 \end{vmatrix}$$

$$d(\alpha) = \alpha^2 + B^2, \quad \Lambda(\alpha) = (B^2+\alpha^2)(B^2 + A^2q^2 + \alpha^2 + q^2\alpha^2)$$

$$\alpha_{-1} = jB, \quad \alpha_{-2} = j\frac{\sqrt{B^2+A^2q^2}}{\sqrt{1+q^2}}$$

Equation (52) implies that the degree of $w(\alpha)$ is 1, so it is possible to rewrite $w(\alpha)$ in the following form:

$$w(\alpha) = \begin{vmatrix} a\alpha + b \\ c\alpha + d \end{vmatrix}$$

The null spaces of $P_a(\alpha_{-1})$ and $P_a(\alpha_{-2})$ are, respectively, the vectors $\begin{vmatrix} 1 \\ 0 \end{vmatrix}$ and $\begin{vmatrix} -j/q \\ 1 \end{vmatrix}$. Because of (56), this means that: $w(\alpha_{-1}) = C_1\begin{vmatrix} 1 \\ 0 \end{vmatrix}$, $w(\alpha_{-2}) = C_2\begin{vmatrix} -j/q \\ 1 \end{vmatrix}$ where C_1 and C_2 are arbitrary constants, thus leading to the following four equations in the a, b, c, d unknowns:

$$a\alpha_{-1} + b = C_1, \quad c\alpha_{-1} + d = 0, \quad a\alpha_{-2} + b = -C_2\frac{j}{q}, \quad c\alpha_{-2} + d = C_2$$

$$a = -\frac{\left(C_1 + \frac{jC_2}{q}\right)}{-jB + \frac{j\sqrt{B^2+A^2q^2}}{\sqrt{1+q^2}}}, \quad b = C_1 + \frac{jB\left(C_1 + \frac{jC_2}{q}\right)}{-jB + \frac{j\sqrt{B^2+A^2q^2}}{\sqrt{1+q^2}}}$$

$$c = \frac{jC_2\sqrt{1+q^2}}{B\sqrt{1+q^2} - \sqrt{B^2+A^2q^2}}, \quad d = \frac{BC_2\sqrt{1+q^2}}{B\sqrt{1+q^2} - \sqrt{B^2+A^2q^2}}$$

By letting $\{C_1 = 1, \quad C_2 = 0\}$ or $\{C_1 = 0, \quad C_2 = 1\}$ in these expressions, it is possible to obtain two independent solutions of (56):

$$W_1(\alpha) = \begin{vmatrix} 1 + \dfrac{B}{-B + \frac{\sqrt{B^2+A^2q^2}}{\sqrt{1+q^2}}} - \dfrac{\alpha}{-jB + \frac{\sqrt{B^2+A^2q^2}}{\sqrt{1+q^2}}} \\ 0 \end{vmatrix}$$

$$W_2(\alpha) = \begin{vmatrix} \dfrac{jB}{-Bq + \frac{q\sqrt{B^2+A^2q^2}}{\sqrt{1+q^2}}} - \dfrac{\alpha}{-Bq + \frac{q\sqrt{B^2+A^2q^2}}{\sqrt{1+q^2}}} \\ \dfrac{B\sqrt{1+q^2}}{B\sqrt{1+q^2} - \sqrt{B^2+A^2q^2}} + \dfrac{j\sqrt{1+q^2}\alpha}{B\sqrt{1+q^2} - \sqrt{B^2+A^2q^2}} \end{vmatrix}$$

By substitution in (52) one obtains the factorized matrices:

$$G_-(\alpha) = \begin{vmatrix} \dfrac{j\left(\sqrt{B^2+A^2q^2}+j\sqrt{1+q^2}\alpha\right)}{\left(-B\sqrt{1+q^2}+\sqrt{B^2+A^2q^2}\right)(B+j\alpha)} & \dfrac{1}{Bq-\frac{q\sqrt{B^2+A^2q^2}}{\sqrt{1+q^2}}} \\ 0 & -\dfrac{j}{-B+\frac{\sqrt{B^2+A^2q^2}}{\sqrt{1+q^2}}} \end{vmatrix}$$

$$G_+(\alpha) = \begin{vmatrix} 0 & Bq-\dfrac{q\sqrt{B^2+A^2q^2}}{\sqrt{1+q^2}} \\ \dfrac{\left(-B\sqrt{1+q^2}+\sqrt{B^2+A^2q^2}\right)\left(\sqrt{B^2+A^2q^2}-j\sqrt{1+q^2}\alpha\right)}{q(B-j\alpha)} & -\dfrac{\left(Bq-\frac{q\sqrt{B^2+A^2q^2}}{\sqrt{1+q^2}}\right)(jB+\alpha)}{q(B-j\alpha)} \end{vmatrix}$$

The inverse matrices $G_-^{-1}(\alpha)$ and $G_+^{-1}(\alpha)$, not reported explicitly, are minus and plus functions, respectively.

Another nonstandard factorization of the kernel $G(\alpha)$ yields

$$G(\alpha) = \hat{G}_-(\alpha)\hat{G}_+(\alpha) \tag{59}$$

where $\hat{G}_-(\alpha)$ and $\hat{G}_+(\alpha)$ are (together with their inverses) conventional plus and minus functions showing nonalgebraic behavior for $\alpha \to \infty$. In these cases the W-H procedure leads to an entire function $w(\alpha)$ that is nonvanishing for $\alpha \to \infty$. The determination of this function is generally very difficult to obtain, even if only a discrete spectrum is present (Daniele, 1986; Jones, 1986). Hence, in general it seems preferable to avoid the presence of the entire function $w(\alpha)$ and to obtain standard factorized matrices $G_-(\alpha)$ and $G_+(\alpha)$ working on the nonstandard matrices $\hat{G}_-(\alpha)$ and $\hat{G}_+(\alpha)$.

If $G(\alpha)$ is a scalar, factorized functions $\hat{G}_\pm(\alpha)$ having exponential behavior $e^{\pm q\,\alpha}$ as $\alpha \to \infty$ arises from factorization equations that use the Weierstrass factorization of entire functions (section 3.2.4). In this case, the offending behavior is easily eliminated by setting

$$G_\pm(\alpha) = \hat{G}_\pm(\alpha)e^{\mp q\,\alpha}$$

2.8 Extension of the W-H technique to the GWHE

Most of the results obtained with the W-H technique can be extended to obtain the solution of the generalized W-H equations (Daniele, 2001, 2003b, 2004a, 2010). For the sake of simplicity this extension is omitted from this book.

2.9 Important mappings for dealing with W-H equations

2.9.1 The $\chi = \sqrt{\tau_o^2 - \alpha^2}$ mapping

In many propagation problems the functions involved in the Wiener-Hopf equations depend on the $\chi = \sqrt{\tau_o^2 - \alpha^2}$ function. This happens when in the considered physical problems, geometrical discontinuities are present in isotropic media with wave number τ_o. It is not restrictive to assume that τ_o is the free space propagation constant k, so in the following we occasionally set $\tau_o = k$.

The function $\chi = \sqrt{\tau_o^2 - \alpha^2}$ presents two branch points $\alpha = \pm\tau_o$, and, after having considered two arbitrary branch lines, we define the principal branch that yields $\chi = \tau_o$ as $\alpha = 0$. This is a compulsory choice, due to the physical existence of the Green function. To understand the properties of the mapping $\chi = \sqrt{\tau_o^2 - \alpha^2}$, it is convenient to trace on the $\alpha -$plane the lines where $\mathrm{Im}[\chi] = \mathrm{Im}\left[\sqrt{\tau_o^2 - \alpha^2}\right] = 0$ and $\mathrm{Re}[\chi] = \mathrm{Re}\left[\sqrt{\tau_o^2 - \alpha^2}\right] = 0$.

In Fig. 1, these lines consist of arcs of equilateral hyperbolas. If the branch lines are the vertical lines as indicated in Fig. 1, the $\alpha -$ plane is divided into subregions A, B, C, D that allow to identify the principal branch for every point considered in this plane. For instance, starting from $\alpha = 0$ we can reach one point of the region B using different paths; in particular, if we choose a path that crosses the arc where $\mathrm{Im}[\chi] = \mathrm{Im}\left[\sqrt{\tau_o^2 - \alpha^2}\right] = 0$, for reasons of continuity we must assume $\mathrm{Re}[\chi] > 0$, $\mathrm{Im}[\chi] > 0$. Choosing another path that links the

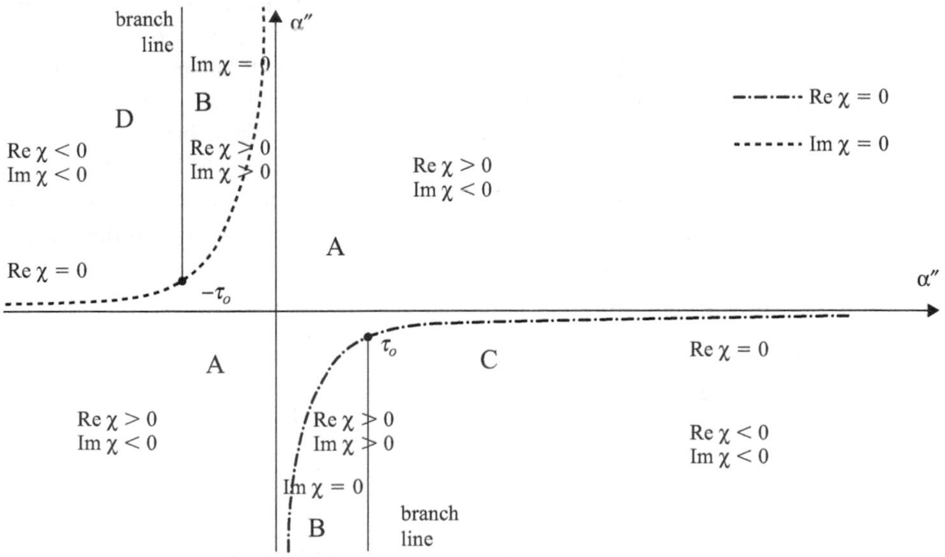

Fig. 1: Mapping $\chi = \sqrt{\tau_o^2 - \alpha^2}$

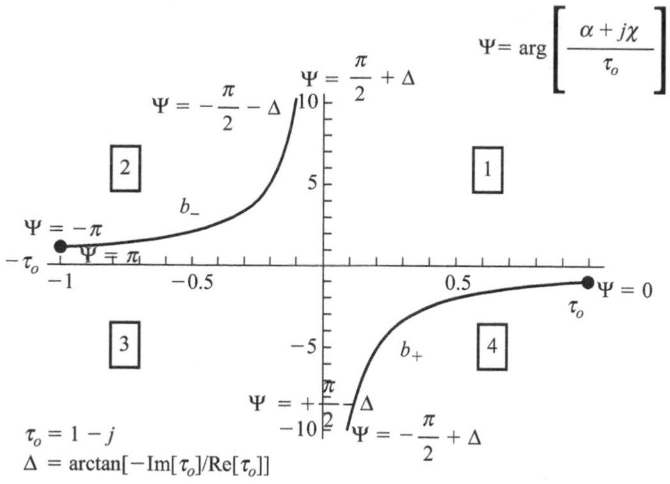

Fig. 2: Standard branch lines in the α – plane.

point $\alpha = 0$ to a point of the region B yields the same conclusion. In fact, if we select the path that crosses the arc where $\text{Re}[\chi] = \text{Re}\left[\sqrt{\tau_o^2 - \alpha^2}\right] = 0$, we first reach region D where $\text{Re}[\chi] < 0$, $\text{Im}[\chi] < 0$. Successively we can reach one point of region B by crossing the assumed vertical branch line where there is a discontinuity of the principal branch. This procedure again yields $\text{Re}[\chi] > 0$, $\text{Im}[\chi] > 0$.

One must observe that the use of other branch lines could modify the sign of $\text{Im}[\chi] < 0$ for the principal branch. It is remarkable that when the branch lines are coincident with the arcs of hyperbolas where $\text{Im}[\chi] = 0$, we obtain a principal branch having $\text{Im}[\chi] \leq 0$ everywhere. In this circumstance the principal branch is called proper branch and the relevant cut plane is called proper sheet. In the following we always assume these branch lines (Fig. 2).

To showcase the branch lines, in Fig. 2 τ_o has been assumed equal to $1 - j$. It is important to observe that on the lower and upper lips of the branch line b_-, the proper branch of $\chi = \sqrt{\tau_o^2 - \alpha^2}$ is real and positive and negative, respectively. Similarly, on the upper and lower lips of the branch line b_+, the proper branch of $\chi = \sqrt{\tau_o^2 - \alpha^2}$ is real and positive and negative, respectively. By recalling that $\arg[z]$ is always between $-\pi$ and $+\pi$, the figure also illustrates the values of $\arg\left[\frac{\alpha + j\chi}{\tau_o}\right]$ on the end points of the lips of the branch lines. These values will be considered in the next section.

2.9.2 The $\alpha = -\tau_o \cos w$ mapping

In cases where the mapping $\chi = \sqrt{\tau_o^2 - \alpha^2}$ is present, it turns out to be useful to introduce the angular complex variable w defined through the mapping:

$$\alpha = \tau_o \cos(w + a) \tag{60}$$

There are many reasons for introducing this mapping. For instance, some specialized analytical methods introduce representations that are different from the Laplace transform. In particular, the Malyuzhinets method introduces the Sommerfeld representations, which may be considered as spectral representations in the w – plane. Furthermore, the introduction of the mapping (60) may help to handle and theoretically manipulate functions that may be hard to work with if defined only in the α – plane.

The parameter a is arbitrary; however, for the sake of convenience, as it will be shown later on, in the following it will be assumed always equal to π. A detailed study of the mapping

$$\alpha = -\tau_o \cos w \tag{61}$$

is very interesting. In particular, it is important to ascertain the properties of the function $\chi = \sqrt{\tau_o^2 - \alpha^2}$ in the w – plane. For instance, since the plus function $F_+(\alpha)$ has only the branch line corresponding to the branch point $\alpha = \tau_o$, if we cut the branch line corresponding to the other branch point $\alpha = -\tau_o$, even though the variable χ changes of sign with the discontinuity, $F_+(\alpha)$ must remain continuous.

Given α, the inverse mapping of (61) introduces infinite values of w given by

$$w(\alpha) = -j\log\left(\frac{\alpha + j\chi}{\tau_o}\right) - \pi = -j\log\left|\frac{\alpha + j\chi}{\tau_o}\right| + \arg\left[\frac{\alpha + j\chi}{\tau_o}\right] + (2n-1)\pi,$$
$$n = 0, \pm 1, \pm 2, \ldots \tag{62}$$

where the argument of a complex number z is assumed to lie in the range:

$$-\pi < \arg[z] \le +\pi$$

No additional branch points with respect to those involved in $\chi = \sqrt{\tau_o^2 - \alpha^2}$ arise from the logarithm. In fact, $\frac{\alpha + j\chi}{\tau_o}$ is never vanishing since α is a finite point:

$$(\alpha + j\chi = 0 \rightarrow \alpha^2 = \tau_o^2 + \alpha^2)$$

Therefore, in the following we assume as branch lines of the mapping the standard branch lines constituted by the arcs of hyperbolas where $\mathrm{Im}[\chi] = 0$ (Fig. 2).

Also, we assume as principal value $w(\alpha)$ the following branch:

$$w(\alpha) = If\left[\arg\frac{\alpha + j\chi}{\tau_o} > -\frac{\pi}{2},\ -j\log\left(\frac{\alpha + j\chi}{\tau_o}\right) - \pi,\ -j\log\left(\frac{\alpha + j\chi}{\tau_o}\right) + \pi\right] \tag{63}$$

where the proper branch of $\chi = \sqrt{\tau_o^2 - \alpha^2}$ and the principal value of $\log[z]$ have been assumed to be:

$$\mathrm{Im}[\chi] \le 0, \quad \log[z] = \log|z| + j\arg[z], \quad (-\pi < \arg[z] \le +\pi)$$

and the function $If[a, b, c]$ gives b if a is true, and c if a is false.

The presence of the function $If[a, b, c]$ has been introduced in order to avoid that the discontinuity of $\arg\frac{\alpha + j\chi}{\tau_o}$ across the branch point $\alpha = -\tau_o$ (Fig. 2) produces a discontinuity in $w(\alpha)$.

When α is real, we have that our principal branch is related to the principal part of the arccos by

$$w(\alpha) = \arccos\left(\frac{\alpha}{\tau_o}\right) - \pi = -\arccos\left(-\frac{\alpha}{\tau_o}\right) \quad \text{Im}[\alpha] = 0 \tag{64}$$

Furthermore, since

$$\chi = \pm\tau_o \sin w \tag{65}$$

on the image r_w of the real axis $\text{Im}[\alpha] = 0$, we have

$$\begin{aligned} w(\alpha) &= -j\log\left(\frac{-\tau_o \cos w + j(\pm\tau_o \sin w)}{\tau_o}\right) - \pi = -j\log(-\cos w + j(\pm\sin w)) - \pi \\ &= -j\log(\cos w \mp j\sin w) = -j\log e^{\mp jw} \end{aligned}$$

This means that only the choice $\chi = -\tau_o \sin w$ yields the identity $w(\alpha) = w$.

Finally, it is important to observe that for the principal value

$$w(-\alpha) = -w(\alpha) - \pi \tag{66}$$

eq. (63) yields the representation of Fig. 3 for the proper sheet in the $w-$ plane.

In this figure, the images r_w and $b_{\pm w}$, of the real axis and of the standard branch lines of the $\alpha -$ plane (Fig. 2), respectively, are reported.

With the mapping defined by (61), both the proper and improper sheets of the $\alpha -$ plane lie on the $w-$plane. For instance, we can observe that the segment of the w-real axis defined by $-\pi \leq w \leq 0$ is the image of the segment $(-\tau_o, \tau_o)$ belonging to the proper sheet, whereas the segment of the real axis defined by $-2\pi \leq w \leq -\pi$ is the image of the segment $(-\tau_o, \tau_o)$ belonging to the improper sheet.

Since the $w-$ plane is image of the proper, of the improper and of other infinite sheets arising from the inversion of the mapping (61), we must be very careful when rewriting in

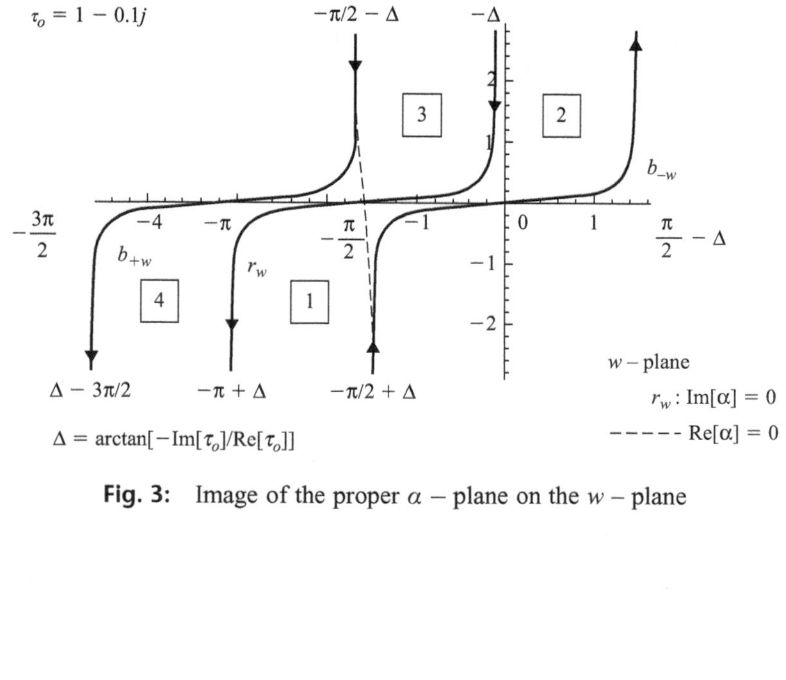

Fig. 3: Image of the proper $\alpha -$ plane on the $w-$ plane

the α – plane equations that have been obtained in the w – plane. For instance, let us suppose that in the w – plane we have an equation that we know to be valid for w representative of a point of α located in the proper sheet; the presence in this equation of the quantity $\chi = -\tau_o \sin w$ introduces the propagation constant $\chi = \sqrt{\tau_o^2 - \alpha^2}$ in the equation rewritten in the α – plane. If we wish to extend the correct equation in the α – plane that is valid also in correspondence of w representative of a point of α located in the improper sheet, we must remember that χ always must be assumed with negative imaginary part. Consequently, we must define $-\tau_o \sin w = \chi$ if w is representative of a point of α located in the proper sheet and $-\tau_o \sin w = -\chi$ if w is representative of a point of α located in the improper sheet. In this way, in the equation rewritten in the α – plane, χ always possesses a negative imaginary part.

Conversely, we cannot assume as correct for all the points of the w – plane the properties that are valid only for the images of the proper sheet. For instance, the function χ in the w – plane is expressed by

$$\sqrt{\tau_o^2 - \alpha^2} = \sqrt{\tau_o^2(1 - \cos^2 w)} = -\tau_o \sin w \tag{67}$$

This expression seems an even function of w (second member) and an odd function of w (third member). The reason for this paradox is that the second member is valid only for values of w that are images of the proper sheet. However, by changing w into $-w$ we are forced to go into the improper sheet and according to the analytic continuation of $\sqrt{\tau_o^2 - \alpha^2} = -\tau_o \sin w$, $\sqrt{\tau_o^2(1 - \cos^2 w)}$ changes sign.

In the following it turns out to be very useful to deal with functions of w defined in the entire w – plane. In general we can always do this through analytic continuation, but we must remember that as it happens in the previous paradox the analytic continuation may involve improper values of χ.

The complete mapping (61) is depicted in Fig. 4. The wave number $\tau_o = \tau_o' + j\tau_o''$ has been assumed complex, with imaginary negative part ($\tau_o'' < 0$). Quadrants 1, 2, 3, 4 of the proper sheet in the α – plane are denoted by the same number encapsulated in a box. For instance, the point $\alpha_o = -\tau_o \cos\varphi_o$ $(0 < \varphi_o < \pi)$ has as proper image $w_o = -\varphi_o$.

Application: Asymptotic evaluation of the integral $I_c = \int_{-\infty}^{\infty} f(\alpha) e^{-j\alpha x} e^{-j\sqrt{k^2-\alpha^2}|y|} d\alpha$

The integral $I_c = \int_{-\infty}^{\infty} f(\alpha) e^{-j\alpha x} e^{-j\sqrt{k^2-\alpha^2}|y|} d\alpha$ is very important since it arises in diffraction problems involving planar stratified structures. Here x and y are the Cartesian coordinates of the observation point. In cylindrical coordinates we have

$$\begin{aligned} x &= \rho \cos\varphi \\ y &= \rho \sin\varphi \end{aligned} \tag{68}$$

Letting $\tau_o = k$ and using the mapping indicated in (61):

$$\begin{aligned} \alpha &= -k \cos w \\ \chi &= -k \sin w \end{aligned} \tag{69}$$

yields the integral

$$I_c = k \int_{r_w} \hat{f}_s(w) e^{jk\rho \cos(w - |\varphi|)} dw \tag{70}$$

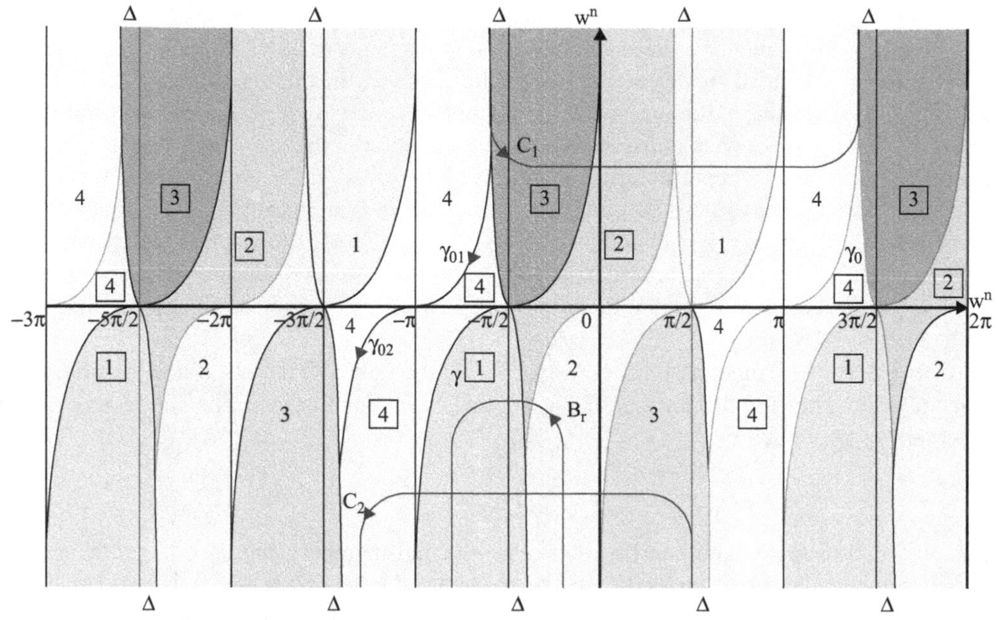

Fig. 4: The complete mapping $\alpha = -\tau_o \cos w$

where

$$\hat{f}_s(w) = f(-k \cos w) \sin w \tag{71}$$

and r_w is the image of the real axis in the w – plane (Fig. 3). Let us consider the points w_s where the function

$$q(w) = j \cos(w - |\varphi|) \tag{72}$$

has a vanishing derivative:

$$q'(w_s) = -j \sin(w_s - |\varphi|) = 0 \tag{73}$$

These points are called *saddle points* since they are the saddle points of the real functions

$$\begin{aligned} u(w) &= u(w_1, w_2) = \text{Re}[q(w)] = \sin(w_1 - |\varphi|) \sinh w_2 \\ v(w) &= v(w_1, w_2) = \text{Im}[q(w)] = \cos(w_1 - |\varphi|) \cosh w_2 \end{aligned} \tag{74}$$

where w_1 and w_2 are the real and imaginary parts of w (Felsen & Marcuvitz, 1973):

$$w = w_1 + j w_2 \tag{75}$$

There is only one relevant saddle point located in the images of the proper η – plane $(-\pi \le w_s \le 0)$ (Fig. 5). It is defined by

$$w_s = -\pi + |\varphi| \Rightarrow w_{1s} = -\pi + |\varphi|, \quad w_{2s} = 0 \tag{76}$$

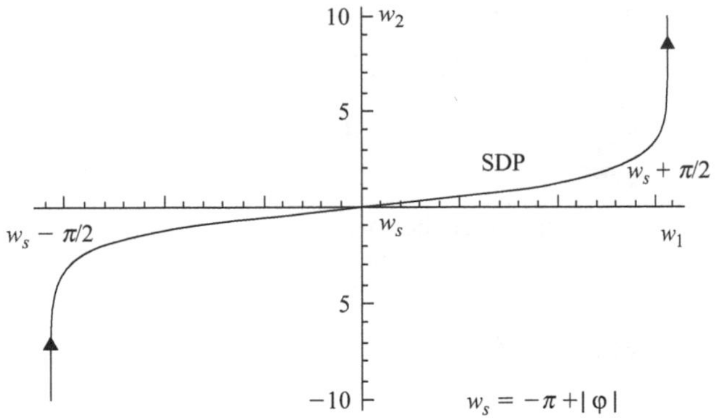

Fig. 5: SDP in the w – plane

Let's consider in the w – plane the *steepest descent path* (SDP) crossing the saddle point (Felsen & Marcuvitz, 1973). This line is defined in such a way that

$$v(w) = v(w_s), \quad u(w) \leq u(w_s) \tag{77}$$

Taking into account that $q(w_s) = j\cos(w_s - |\varphi|) = -j \Rightarrow u(w_s) = 0,\ v(w_s) = -1$ leads to the following definition of the SDP:

$$v(w_1, w_2) = -1, \quad u(w_1, w_2) \leq 0 \tag{78}$$

or

$$\cos(w_1 + \pi - |\varphi|)\cosh(w_2) = 1, \quad \sin(w_1 + \pi - |\varphi|)\sin w_2 \geq 0 \tag{79}$$

Equation (79) yields

$$w_1 = -\pi + |\varphi| + gd(w_2) \tag{80}$$

where $gd(t)$ is the Gudermann function defined by

$$gd(t) = arc\cos\left[\frac{1}{\cosh t}\right]\mathrm{sgn}(t) \tag{81}$$

We oriented the SDP in the direction of increasing values of w_2. Looking at Fig. 5 (see also Fig. 4) we notice that if $|\varphi| > \frac{\pi}{2}$, then points of the SDP are imaged onto the improper sheet of the η – plane. We have

$$I_{SDP} = k\int_{SDP} \hat{f}_s(w)e^{jk\rho\cos(w-|\varphi|)}dw = ke^{-jk\rho}\int_{SDP} \hat{f}_s(w)e^{k\rho u(w)}dw \tag{82}$$

where $u(w)$ is negative or zero on the SDP.

In the far field, the parameter $k\rho \gg 1$; therefore, by taking into account that $u(w) \leq 0$ and that the exponential factor $e^{k\rho u(w)}$ decreases rapidly as $k\rho$ increases, we observe that the major contribution to the integral I_{SDP} is due to points very near the saddle point w_s. If there are no poles of $\hat{f}(w)$ near the saddle point, we obtain the approximate evaluation:

$$I_{SDP} = k\int_{SDP} \hat{f}_s(w)e^{jk\rho\cos(w-|\varphi|)}dw \approx ke^{-jk\rho}\hat{f}_s(w_s)\int_{SDP} e^{k\rho u(w)}dw \tag{83}$$

To evaluate the integral, we introduce the mapping:

$$q(w) - q(w_s) = -s^2 \tag{84}$$

Of course when w moves on the SDP, s is real and will be assumed positive when $w_2 > 0$ and negative when $w_2 < 0$. This yields

$$\int_{SDP} e^{k\rho u(w)}dw = \int_{-\infty}^{\infty} e^{-k\rho s^2}\, w'(s)ds \approx w'(0)\int_{-\infty}^{\infty} e^{-k\rho s^2}\, ds = w'(0)\sqrt{\frac{\pi}{k\rho}} \tag{85}$$

From (84)

$$q'(w)\frac{dw}{ds} = -2s \quad \text{or} \quad \frac{dw}{ds} = \frac{-2s}{q'(w)} \tag{86}$$

Applying Hospital's rule yields

$$w'(0) = \frac{-2}{q''(w_s)w'(0)} \tag{87}$$

or, taking into account that $q''(w_s) = j$,

$$w'(0) = \sqrt{\frac{-2}{q''(w_s)}} = \sqrt{2}\sqrt{j} = \pm\sqrt{2}\; e^{j\frac{\pi}{4}} \tag{88}$$

Looking at Fig. 5, we observe that the sign of $\frac{dw}{ds}$ is positive, so that

$$w'(0) = \sqrt{2}\; e^{j\frac{\pi}{4}} \tag{89}$$

Finally, this yields the following approximation of I_{SDP} defined by (83):

$$I_{SDP} \approx k\sqrt{\frac{2\pi}{k\rho}}e^{-j(k\rho-\pi/4)}\hat{f}_s(-\pi + |\varphi|) = -k\sqrt{\frac{2\pi}{k\rho}}e^{-j(k\rho-\pi/4)}f(k\cos\varphi)\sin|\varphi| \tag{90}$$

Equation (90) cannot be used when poles of $\hat{f}_s(w) = f(-k\cos w)\sin w$ are located near the saddle point. In this case we must use a more general technique, as indicated in Felsen and Marcuvitz (1973) and Senior and Volakis (1995). If the characteristic of $f(\alpha)$ at the pole $\alpha_o = -k\cos\varphi_o$ near $\alpha_s = -k\cos(-\pi + |\varphi|) = k\cos\varphi$ is

$$\frac{R}{\alpha - \alpha_o}$$

we obtain:

$$I_{SDP} \approx \sqrt{\frac{2\pi}{k\rho}} e^{-j(k\rho-\pi/4)}\left[-k f(k\cos\varphi)\sin|\varphi| + \frac{1-F\left(2k\rho\cos^2\frac{\varphi_o+|\varphi|}{2}\right)}{2\cos\frac{\varphi_o+|\varphi|}{2}}R\right] \quad (91)$$

where $F(z)$ is the Kouyoumjian-Pathak transition function defined by

$$F(z) = 2j\sqrt{z}\, e^{jz}\int_{\sqrt{z}}^{\infty} e^{-js^2} ds - 2j\sqrt{z}\sqrt{\pi}\exp\left[j\left(z-\frac{\pi}{4}\right)\right]u\left[Arg\left(\sqrt{z}-\frac{\pi}{4}\right)\right] \quad (92)$$

where the principal branch of $\sqrt{z}$ is: $-\frac{3\pi}{4} < \arg[\sqrt{z}] < \frac{5\pi}{4}$.

The previous equations provide the evaluation of the integral on the SDP. We can relate this evaluation to the original one on the path r_w by warping the contour path r_w into the SDP and taking into account the eventual singularity contributions located in the region between the SDP and r_w.

Sometimes it is preferable to study the warping directly in the α – plane. In this plane the SDP does not change if we change φ into $-\varphi$. Figure 6 illustrates the location of the SDP in the α – plane.

The direction of the SDP in the α – plane is in accordance with the direction of the SDP in the w – plane.

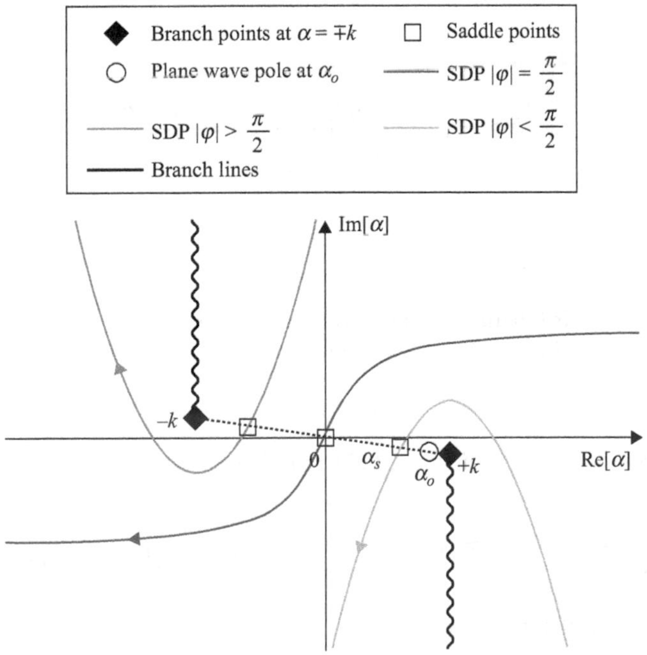

Fig. 6: SDP in the α – plane ($\alpha_s = k\cos\varphi$, $\alpha_o = -k\cos\varphi_o$)

The contribution of the SDP relevant to an isolated first-order saddle point α_s is given by Felsen and Marcuvitz (1973]):

$$\int_{SDP} f(\alpha)e^{\Omega q(\alpha)}d\alpha = \left.\frac{dz}{ds}\right|_{s=0}\sqrt{\frac{\pi}{\Omega}}f(\alpha_s)e^{\Omega q(\alpha_s)} = \pm\sqrt{-\frac{2}{q''(\alpha_s)}}\sqrt{\frac{\pi}{\Omega}}f(\alpha_s)e^{\Omega q(\alpha_s)} \tag{93}$$

where, by looking at the figure, the $\pm$ sign is determined by observing that

$$\left.\frac{d\alpha}{ds}\right|_{s=0} = \pm\sqrt{-\frac{2}{q''(\alpha_s)}}$$

When the saddle point α_s is near a first-order pole α_p, the previous equation is not uniform and we need to modify by using the mapping:

$$s^2 = q(\alpha_s) - q(\alpha)$$

which yields

$$\int_{SDP} f(z)e^{\Omega q(z)}dz = e^{\Omega q(z_s)}\int_{-\infty}^{\infty} G(s)e^{-\Omega s^2}ds$$

where $G(s) = f(\alpha)\frac{d\alpha}{ds}$.

In the s – plane, the pole α_p has the image b:

$$b^2 = q(\alpha_s) - q(\alpha_p) \approx -\frac{q''(\alpha_s)}{2}(\alpha_p - \alpha_s)^2$$

or

$$b \approx \pm\sqrt{-\frac{q''(\alpha_s)}{2}}(\alpha_p - \alpha_s) = \frac{\alpha_p - \alpha_s}{\left.\frac{d\alpha}{ds}\right|_{s=0}}$$

We can write

$$G(s) = \frac{r}{s-b} + T(s)$$

where $T(s)$ is a smooth function on the SDP, and r is the residue of $G(s)$ in $s = b$. Taking into account the definition of $G(s)$, this residue is the same as that of $f(\alpha)$ in $\alpha = \alpha_p$:

$$r = \text{Res}[f(\alpha)]\big|_{\alpha=\alpha_p}$$

Since $T(s)$ is smooth, we can apply eq. (93) and obtain

$$\int_{-\infty}^{\infty} G(s)e^{-\Omega s^2}ds = \pm r\frac{\pi}{j}w\left(\pm b\sqrt{\Omega}\right) + \left(f(\alpha_s)\left.\frac{d\alpha}{ds}\right|_{s=0} + \frac{r}{b}\right)\sqrt{\frac{\pi}{\Omega}} \tag{94}$$

where $w(z)$ is a special function defined in Abramovitz and Stegun (1972, p. 297) and the sign is $\pm$ if $\text{Im}[b] > 0$ or $\text{Im}[b] < 0$, respectively. In practice if $\sqrt{\Omega}|b| > 2$ we can use the non-uniform eq. (93), whereas if $\sqrt{\Omega}|b| < 2$ we must use the uniform eq. (94).

CHAPTER 3

Functions decomposition and factorization

3.1 Decomposition

The general formula for decomposing a complex function vanishing as $\alpha \to \infty$ as a sum of a minus and a plus function $F(\alpha) = F_-(\alpha) + F_+(\alpha)$ is given by the Cauchy equation (Noble, 1958, p. 13):

$$F_+(\alpha) = \frac{1}{2\pi j} \int_{\gamma_1} \frac{F(\alpha')}{\alpha' - \alpha} d\alpha' = \frac{1}{2} F(\alpha) + \frac{1}{2\pi j} P.V. \int_{-\infty}^{\infty} \frac{F(\alpha')}{\alpha' - \alpha} d\alpha', \quad \mathrm{Im}[\alpha] = 0 \qquad (1)$$

where the rightmost integral must be considered a Cauchy principal value (P.V.) integral.

Alternatively, the hook integral symbol may be used in the first integral in (1):

$$F_+(\alpha) = \frac{1}{2\pi j} \int_{-\infty}^{\infty} \frac{F(\alpha')}{\alpha' - \alpha} d\alpha'$$

where the hook on the integral sign indicates that the chosen integration path (e.g., γ_1 in Fig. 1) is the deformed real axis that passes below the pole $\alpha' = \alpha$. Occasionally, in the following we will call the line γ_1 the smile real axis.

The minus decomposed term may be obtained in a similar manner:

$$F_-(\alpha) = -\frac{1}{2\pi j} \int_{\gamma_2} \frac{F(\alpha')}{\alpha' - \alpha} d\alpha' = \frac{1}{2} F(\alpha) - \frac{1}{2\pi j} P.V. \int_{-\infty}^{\infty} \frac{F(\alpha')}{\alpha' - \alpha} d\alpha', \quad \mathrm{Im}[\alpha] = 0 \qquad (2)$$

where the rightmost integral must be considered a Cauchy principal value integral. Occasionally, in the following we will call the line γ_2 in Fig. 1 the frown real axis.

As in the previous case, we may use the symbol

$$F_-(\alpha) = -\frac{1}{2\pi j} \int_{-\infty}^{\infty} \frac{F(\alpha')}{\alpha' - \alpha} d\alpha'$$

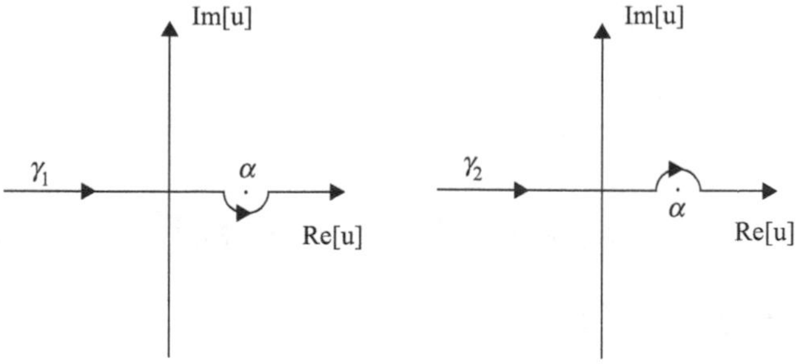

Fig. 1: Integration paths

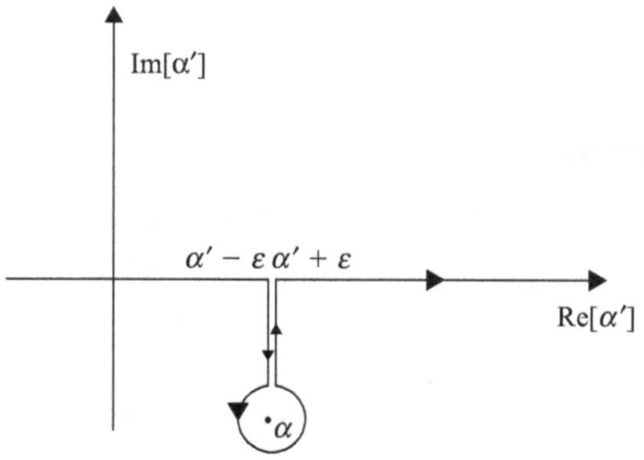

Fig. 2: Modification of the integration path for evaluating $F_+(\alpha)$ when $\mathrm{Im}[\alpha] \leq 0$

where the minus sign accounts for the clockwise sense of the integration path. The second member of the eqs. (1) and (2) holds in the respective regularity half-plane. It is possible to evaluate the integrals $F_+(\alpha)$ and $F_-(\alpha)$ everywhere by analytic continuation. If the integrals are evaluated in closed form, they directly provide the analytical expressions of $F_+(\alpha)$ and $F_-(\alpha)$ everywhere.

On the other hand, if we use a numerical integration for the Cauchy integral (1), the jump at the discontinuity from $\mathrm{Im}[\alpha] \geq 0$ to $\mathrm{Im}[\alpha] \leq 0$ can be taken into account by subtracting the residue at the pole α, that is, $-F(\alpha)$ (Fig. 2). This yields the following representation of $F_+(\alpha)$ when $\mathrm{Im}[\alpha] \leq 0$:

$$F_+(\alpha) = F(\alpha) + \frac{1}{2\pi j}\int_{\gamma_2} \frac{F(\alpha')}{\alpha' - \alpha} d\alpha', \quad \mathrm{Im}[\alpha] \leq 0 \tag{3}$$

The same result can be achieved by applying the second member of eq. (2), valid for $\mathrm{Im}[\alpha] \leq 0$, and

$$F_+(\alpha) = F(\alpha) - F_-(\alpha)$$

A remarkable simplification arises when we are dealing with an even function $F(\alpha)$, that is, $F(\alpha) = F(-\alpha)$. In fact, by replacing α' with $-\alpha'$ in the integral in (1) we obtain

$$F_-(\alpha) = F_+(-\alpha) \tag{4}$$

Several proofs of the fundamental decomposition formulas can be found in Noble (1958). In the following we report a simple proof that is based on the Cauchy formula.

Let us consider the upper region limited by contour $\Gamma_+ = \Gamma_{\infty+} \bigcup \gamma_1$, where $\Gamma_{\infty+}$ is the half-circumference located at ∞. Since the plus-decomposed $F_+(\alpha)$ is regular in this domain and vanishing on $\Gamma_{\infty+}$, the Cauchy formula provides

$$
\begin{aligned}
F_+(\alpha) &= \frac{1}{2\pi j}\int_{\Gamma_+} \frac{F_+(\alpha')}{\alpha' - \alpha} d\alpha' = \frac{1}{2\pi j}\int_{\gamma_1} \frac{F_+(\alpha')}{\alpha' - \alpha} d\alpha' \\
&= \frac{1}{2\pi j}\int_{\gamma_1} \frac{F(\alpha')}{\alpha' - \alpha} d\alpha' - \frac{1}{2\pi j}\int_{\gamma_1} \frac{F_-(\alpha')}{\alpha' - \alpha} d\alpha' \\
&= \frac{1}{2\pi j}\int_{\gamma_1} \frac{F(\alpha')}{\alpha' - \alpha} d\alpha' - \frac{1}{2\pi j}\int_{\Gamma_-} \frac{F_-(\alpha')}{\alpha' - \alpha} d\alpha' = \frac{1}{2\pi j}\int_{\gamma_1} \frac{F(\alpha')}{\alpha' - \alpha} d\alpha' \quad \mathrm{Im}[\alpha] \geq 0
\end{aligned} \tag{5}
$$

where, since $\Gamma_{\infty-}$ is the half-circumference located at infinity in the lower half-plane,

$$\Gamma_- = \Gamma_{\infty-} \bigcup \gamma_1$$

The integral $-\frac{1}{2\pi j}\int_{\Gamma_-} \frac{F_-(\alpha')}{\alpha'-\alpha} d\alpha'$ is null because, being $\mathrm{Im}[\alpha] \geq 0$, the integrand is regular in the interior of the region limited by Γ_-.

3.1.1 Example 1

Let us decompose the function $F(\alpha) = \tau = \sqrt{k^2 - \alpha^2}$, where k represents a propagation constant. Since the function is not vanishing for $|\alpha|$ going to infinity, eq. (1) does not directly apply. However, by rewriting $F(\alpha) = (k^2 - \alpha^2)\frac{1}{\sqrt{k^2-\alpha^2}}$, we can decompose through (1) the function

$$\frac{1}{\sqrt{k^2 - \alpha^2}} = S_-(\alpha) + S_+(\alpha)$$

where

$$S_+(\alpha) = \frac{1}{2\pi j}\int_{\gamma_1} \frac{1}{\sqrt{k^2 - u^2}} \frac{1}{u - \alpha} du \tag{6}$$

To find a closed-form expression for the plus-decomposed function $S_+(\alpha)$, we first evaluate the rightmost integral assuming real values of α, and afterward we extend the resulting expression to the whole complex α – plane by analytic continuation. Before evaluating the integral in (6), it is important to study in detail the two-valued function $\sqrt{k^2 - \alpha^2}$, which has two branch points at $\alpha = \pm k$ (Fig. 3; see also section 2.9.1).

On the arcs of the equilateral hyperbola, the two opposite branches have either vanishing imaginary or vanishing real part, according to the indications of Fig. 3. To have a single-valued function $\tau = \sqrt{k^2 - \alpha^2}$,[1] we introduce artificial barriers called branch lines.

[1] We assume that τ is the branch such that $\tau = k$ when $\alpha = 0$.

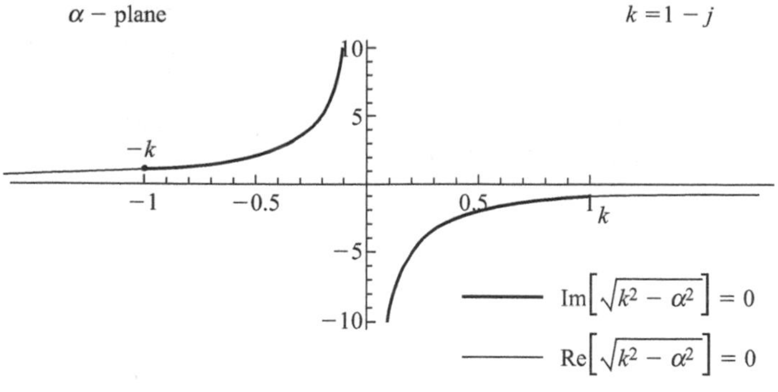

Fig. 3: Standard branch lines of the $\sqrt{k^2 - \alpha^2}$ function

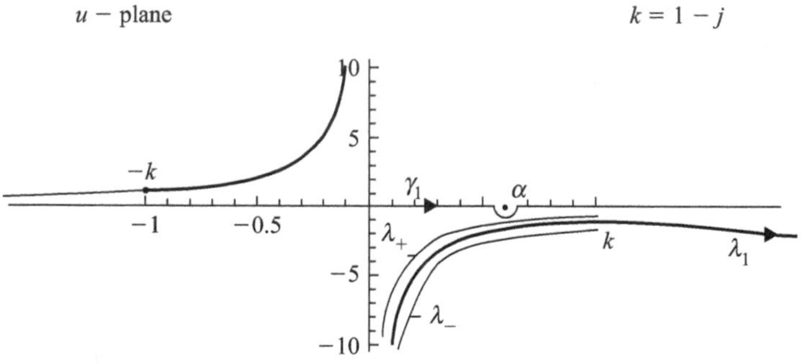

Fig. 4: Deformation of the integration path

In crossing the branch lines, the single-valued function τ changes, with a discontinuity assuming two opposite values on the two lips of the branch line. Branch lines are arbitrary, with the sole limitation of not crossing the real axis. The standard branch lines are the arcs of the equilateral hyperbola where $\text{Im}[\tau] = 0$ (Fig. 3). With this choice, it may be ascertained that $\text{Im}[\tau] \leq 0$ in the entire complex α – plane (section 2.9.1).

Another way to describe the multivalued function $\tau = \sqrt{k^2 - \alpha^2}$ is to imagine that the α – plane consists of two sheets superimposed on each other: by cutting the sheets along the standard branch lines and by imagining that the lower edge of the bottom sheet is joined to the upper edge of the top sheet, it follows that the proper sheet is defined by the sheet where the function $\sqrt{k^2 - \alpha^2}$ assumes the value τ, while the improper sheet is the other sheet where $\sqrt{k^2 - \alpha^2}$ assumes the value $-\tau$.

To evaluate the integral

$$S_+(\alpha) = \frac{1}{2\pi j} \int\limits_{\gamma_1} \frac{1}{\sqrt{k^2 - u^2}} \frac{1}{u - \alpha} du \tag{7}$$

we warp the integration path γ_1 on the line $\lambda = \lambda_+ \bigcup \lambda_-$ made of the two lips of the branch line Γ relevant to the positive branch point $+k$ (Fig. 4). On the lip λ_+ the proper branch is real positive, whereas on the lip λ_- the proper branch is real negative. Taking into account that $\int_{\lambda_+} = -\int_{\lambda_-}$ this yields

$$S_+(\alpha) = \frac{1}{2\pi j}\int\limits_{\gamma_1} \frac{1}{\sqrt{k^2-u^2}}\frac{1}{u-\alpha}du$$

$$= \frac{1}{2\pi j}\int\limits_{\lambda_+} \frac{1}{\sqrt{k^2-u^2}}\frac{1}{u-\alpha}du + \frac{1}{2\pi j}\int\limits_{\lambda_-} \frac{1}{-\sqrt{k^2-u^2}}\frac{1}{u-\alpha}du$$

$$= -\frac{1}{\pi j}\int\limits_{\lambda_-} \frac{1}{\sqrt{k^2-u^2}}\frac{1}{u-\alpha}du \tag{8}$$

By deforming the half contour λ_- into the half straight line λ_1 indicated in Fig. 4,[2] it follows that

$$S_+(\alpha) = -\frac{1}{\pi j}\int\limits_{\lambda_-} \frac{1}{\sqrt{k^2-u^2}}\frac{1}{u-\alpha}du = -\frac{1}{\pi j}\int\limits_{\lambda_1} \frac{1}{\sqrt{k^2-u^2}}\frac{1}{u-\alpha}du \tag{9}$$

By putting $u = k\,x$ on λ_1 we evaluate the integral with MATHEMATICA:

$$S_+(\alpha) = -\frac{1}{\pi j}\int\limits_{\lambda_1} \frac{1}{\sqrt{k^2-u^2}}\frac{1}{u-\alpha}du = -\frac{1}{\pi j}\int\limits_{1}^{\infty} \frac{1}{\sqrt{1-x^2}}\frac{1}{kx-\alpha}dx$$

$$= \frac{\pi - Arc\cos\left(\frac{\alpha}{k}\right)}{\pi\sqrt{k^2-\alpha^2}} = \frac{Arc\cos\left(-\frac{\alpha}{k}\right)}{\pi\sqrt{k^2-\alpha^2}} \tag{10}$$

The closed form thus obtained shows that $S_+(\alpha)$ is a conventional plus function, since its Taylor series expansion about the point $\alpha = -k$ leads to

$$S_+(\alpha) = \frac{Arc\cos\left(-\frac{\alpha}{k}\right)}{\pi\sqrt{k^2-\alpha^2}} = \frac{1}{k\pi} + \frac{\alpha+k}{3k^2\pi} + \frac{2(\alpha+k)^2}{15k^3\pi} + O\left[(\alpha+k)^3\right] \tag{11}$$

and the dominant term near the branch point $\alpha = k$ is given by

$$S_+(\alpha) = \frac{Arc\cos\left(-\frac{\alpha}{k}\right)}{\pi\sqrt{k^2-\alpha^2}} \approx \frac{1}{\sqrt{2k}\sqrt{k-\alpha}}, \qquad \alpha \approx k \tag{12}$$

Due to the important applications of this plus function, we report here some of its alternative expressions:

$$S_+(\alpha) = \frac{Arc\cos\left(-\frac{\alpha}{k}\right)}{\pi\sqrt{k^2-\alpha^2}} = \frac{1}{j\pi\,\tau}\log\frac{j\tau-(k+\alpha)}{j\tau+(k+\alpha)} = \frac{1}{j\pi\,\tau}\log\frac{j\tau-\alpha}{k} \tag{13}$$

---

[2] λ_1 belongs to the straight line that includes the points 0 and k.

$$S_-(\alpha) = S_+(-\alpha) = \frac{Arc\cos\left(\frac{\alpha}{k}\right)}{\pi\sqrt{k^2-\alpha^2}} = \frac{1}{j\pi\,\tau}\log\frac{j\tau-(k-\alpha)}{j\tau+(k-\alpha)} = \frac{1}{j\pi\,\tau}\log\frac{j\tau+\alpha}{k} \tag{14}$$

In all of these expressions, $\tau = \sqrt{k^2-\alpha^2}$ is the value of the proper branch ($\mathrm{Im}[\tau] \le 0$) and the functions $Arc\cos$ and log must be considered as principal values.

Sometimes it is interesting to express the decomposed $S_{+,-}(\alpha)$ in the complex plane w defined by the mapping:

$$\alpha = -k\cos w \quad \tau = -k\sin w \tag{15}$$

This mapping is particularly important for studying angular regions, as it has already been seen in section 2.9.2.

In the w – plane the decomposed $S_{+,-}(\alpha)$ assume the forms

$$S_+(\alpha) = \frac{Arc\cos\left(-\frac{\alpha}{k}\right)}{\pi\sqrt{k^2-\alpha^2}} = S_+(-k\cos w) = \frac{w}{\pi\, k\sin w} \tag{16}$$

$$S_-(\alpha) = \frac{Arc\cos\left(\frac{\alpha}{k}\right)}{\pi\sqrt{k^2-\alpha^2}} = \frac{\pi - Arc\cos\left(-\frac{\alpha}{k}\right)}{\pi\sqrt{k^2-\alpha^2}} = S_-(-k\cos w) = \frac{-\pi - w}{\pi\, k\sin w} \tag{17}$$

Equations (17) and (16) may be obtained in a straightforward way using an ingenious suggestion made by Ament (Noble, 1958, p. 47).

From (13) and (14), the decomposition of $\tau = \sqrt{k^2-\alpha^2}$ follows:

$$\tau = \sqrt{k^2-\alpha^2} = \tau_- + \tau_+ \tag{18}$$

with

$$\tau_+(\alpha) = \frac{\sqrt{k^2-\alpha^2}\,Arc\cos\left(-\frac{\alpha}{k}\right)}{\pi} = \frac{\tau}{j\pi}\log\frac{j\tau-(k+\alpha)}{j\tau+(k+\alpha)} = \frac{\tau}{j\pi}\log\frac{j\tau-\alpha}{k} \tag{19}$$

$$\tau_-(\alpha) = \frac{\sqrt{k^2-\alpha^2}\,Arc\cos\left(\frac{\alpha}{k}\right)}{\pi} = \frac{\tau}{j\pi}\log\frac{j\tau-(k-\alpha)}{j\tau+(k-\alpha)} = \frac{\tau}{j\pi}\log\frac{j\tau+\alpha}{k} \tag{20}$$

It is important to study the asymptotic behavior of the decomposed function as $\alpha \to \infty$. By taking into account that

$$j\tau = \pm\alpha, \quad \text{as } \mathrm{Re}[\alpha] \to \pm\infty$$

it follows that, independently of the sign of $\mathrm{Re}[\alpha]$,

$$\tau_\pm(\alpha) \approx \pm\frac{\alpha}{\pi}\log\left[\left(-\frac{2\alpha}{k}\right)^{\pm 1}\right] = \frac{\alpha}{\pi}\log\left[-\frac{2\alpha}{k}\right], \quad \text{as } \alpha \to \infty \tag{21}$$

3.1.2 Decomposition of an even function

The decomposed functions $F_+(\alpha)$ and $F_-(\alpha)$ may be rewritten as

$$F_+(\alpha) = \frac{1}{2}F(\alpha) + \frac{1}{2\pi j}P.V. \int_{-\infty}^{\infty} \frac{F(u)}{u-\alpha}du$$
$$F_-(\alpha) = +\frac{1}{2}F(\alpha) - \frac{1}{2\pi j}P.V. \int_{-\infty}^{\infty} \frac{F(u)}{u-\alpha}du \tag{22}$$

Since $F(\alpha) = F(-\alpha)$, changing u into $-u$ yields:

$$F_-(\alpha) = +\frac{1}{2}F(\alpha) + \frac{1}{2\pi j}P.V. \int_{-\infty}^{\infty} \frac{F(-u)}{u-(-\alpha)}du = \frac{1}{2}F(-\alpha) + \frac{1}{2\pi j}P.V. \int_{-\infty}^{\infty} \frac{F(u)}{u-(-\alpha)}du$$
$$= F_+(-\alpha) \tag{23}$$

3.1.3 Numerical decomposition

The numerical quadrature of the decomposition formulas requires some care, since it can introduce spurious singularities in the decomposed functions. For instance, if h and A represent the discretization step and the truncating parameter on the quadrature of the integral (1), respectively, we get the approximation

$$F_+(\alpha) = \frac{1}{2}F(\alpha) + \frac{1}{2\pi j}P.V. \int_{-\infty}^{\infty} \frac{F(u)}{u-\alpha}du \approx \frac{1}{2}F(\alpha) + \frac{h}{2\pi j}\sum_{i=-A/h}^{A/h} \frac{F(h\,i)}{h\,i-\alpha} \quad h\,i \neq \alpha \tag{24}$$

We observe that in the exact formula the singularities of $F(\alpha)$ located in upper half-plane are exactly compensated by the analytic continuation of the exact values of the integral. However, this does not happen when the integral is approximate by the sum. Moreover, with the discretization process $F_+(\alpha)$ presents spurious singularities located at $\alpha = h\,i$, $i = -A/h, .. + A/h$. The first circumstance is in some sense acceptable, since by increasing the accuracy of the numerical quadrature one tends to compensate the singularities of $F(\alpha)$ located in upper half-plane. Conversely, the second fact yields a regrettable result. To overcome this problem we rewrite the decomposition formula in the form

$$F_+(\alpha) = \frac{1}{2}F(\alpha) + \frac{1}{2\pi j}P.V. \int_{-\infty}^{\infty} \frac{F(u)}{u-\alpha}du = \frac{1}{2}F(\alpha)$$
$$+ \frac{1}{2\pi j}\lim_{M\to\infty} P.V. \int_{-M}^{M} \frac{F(u)-F(\alpha)}{u-\alpha}du + \frac{1}{2\pi j}F(\alpha)\lim_{M\to\infty} P.V. \int_{-M}^{M} \frac{1}{u-\alpha}du \tag{25}$$

Whence we obtain

$$F_+(\alpha) = \frac{1}{2\pi j} \lim_{M\to\infty} \int_{-M}^{M} \frac{F(u) - F(\alpha)}{u - \alpha} du \tag{26}$$

Consequently, the quadrature of the integral yields

$$F_+(\alpha) \approx \frac{h}{2\pi j} \sum_{i=-A/h}^{A/h} \frac{F(h\,i) - F(\alpha)}{h\,i - \alpha} \tag{27}$$

In this manner, the offending poles $\alpha = h\,i,\ i = -A/h, .. + A/h$ disappear.

By taking into account that Cauchy's formula provides the exact representation

$$F_+(\alpha) = \frac{1}{2} F_+(\alpha) + \frac{1}{2\pi j} P.V. \int_{-\infty}^{\infty} \frac{F_+(u)}{u - \alpha} du \tag{28}$$

we can increase the accuracy of the evaluation of the plus function $F_+(\alpha) \approx F_{o+}(\alpha)$ using

$$F_+(\alpha) \approx \frac{h}{2\pi j} \sum_{i=-A/h}^{A/h} \frac{F_{o+}(h\,i) - F_{o+}(\alpha)}{h\,i - \alpha} \tag{29}$$

We can obtain an alternative quadrature formula by assuming $F(\alpha) = \tilde{F}(\alpha)\tilde{F}_-(\alpha)$, where $\tilde{F}_-(\alpha)$ is an arbitrary minus factor of $F(\alpha)$ vanishing as $\alpha \to \infty$. Now we can write

$$2\pi j\, F_+(\alpha) = \int_{\gamma_1} \frac{F(u)}{u - \alpha} du = \int_{\gamma_1} \frac{\tilde{F}(u)\tilde{F}_-(u)}{u - \alpha} du = \int_{\gamma_1} \frac{(\tilde{F}(u) - \tilde{F}(\alpha))\tilde{F}_-(u)}{u - \alpha} du + \tilde{F}(\alpha) \int_{\gamma_1} \frac{\tilde{F}_-(u)}{u - \alpha} du \tag{30}$$

Since $\tilde{F}_-(\alpha)$ is a minus function that vanishes as $\alpha \to \infty$, the last integral is zero. Furthermore,

$$\int_{\gamma_1} \frac{(\tilde{F}(u) - \tilde{F}(\alpha))\tilde{F}_-(u)}{u - \alpha} du = \int_{-\infty}^{\infty} \frac{(\tilde{F}(u) - \tilde{F}(\alpha))\tilde{F}_-(u)}{u - \alpha} du \tag{31}$$

and we get

$$F_+(\alpha) = \frac{1}{2\pi j} \int_{-\infty}^{\infty} \frac{(\tilde{F}(u) - \tilde{F}(\alpha))\tilde{F}_-(u)}{u - \alpha} du \tag{32}$$

which yields the quadrature equation:

$$F_+(\alpha) \approx \frac{h}{2\pi j} \sum_{i=-A/h}^{A/h} \frac{(\tilde{F}(h\,i) - \tilde{F}(\alpha))\tilde{F}_-(h\,i)}{h\,i - \alpha} \tag{33}$$

It is important to observe that if $\tilde{F}_-(\alpha)$ is chosen such that $\tilde{F}(\alpha) = \tilde{F}_+(\alpha)$ is a plus function, the previous approximation does not introduce spurious poles.

3.1.4 Example 1 revisited

Let us consider the numerical decomposition of the function considered in example 1:

$$\frac{1}{\sqrt{k^2 - \alpha^2}} = S_-(\alpha) + S_+(\alpha)$$

We have

$$S_{a+}(\alpha) \approx \frac{h}{2\pi j} \sum_{i=-A/h}^{A/h} \frac{(\tilde{F}(h\,i) - \tilde{F}(\alpha))\tilde{F}_-(h\,i)}{h\,i - \alpha} \tag{34}$$

where we have assumed

$$\tilde{F}(\alpha) = \tilde{F}_+(\alpha) = \frac{1}{\sqrt{k - \alpha}}, \quad \tilde{F}_-(\alpha) = \frac{1}{\sqrt{k + \alpha}} \tag{35}$$

Figure 5, Fig. 6, and Fig. 7 report the absolute errors $e(\alpha) = \left|\frac{S_+(\alpha) - S_{a+}(\alpha)}{S_+(\alpha)}\right|$ on the real axis and on the half-lines $\alpha = jx$ $(x \geq 0)$ and $\alpha = k + jx$ $(x \geq 0)$ located on the regular half-plane, of the plus functions.

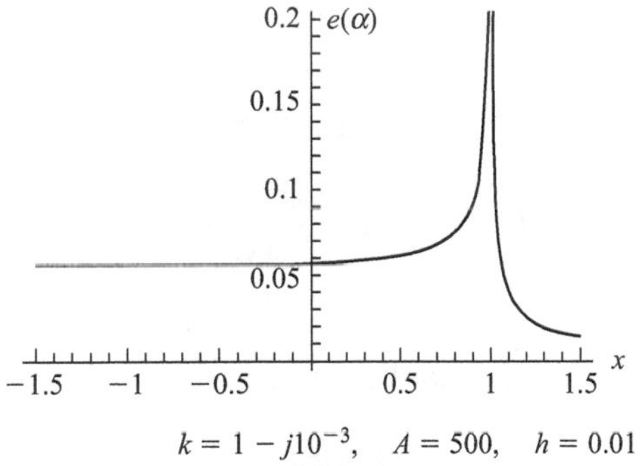

Fig. 5: Relative error on the real axis

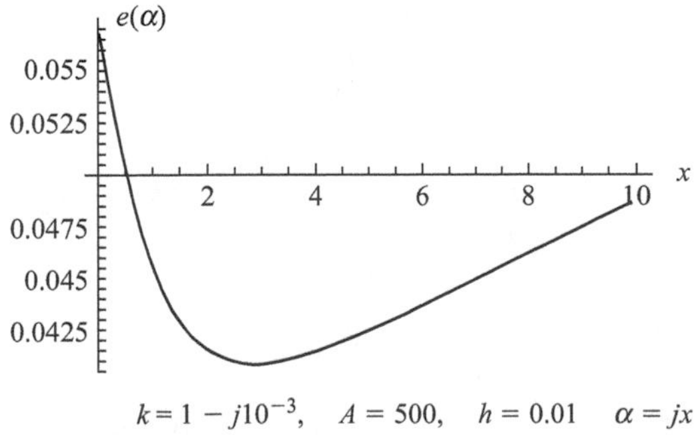

Fig. 6: Relative error on the imaginary axis $\alpha = jx$ $(x \geq 0)$

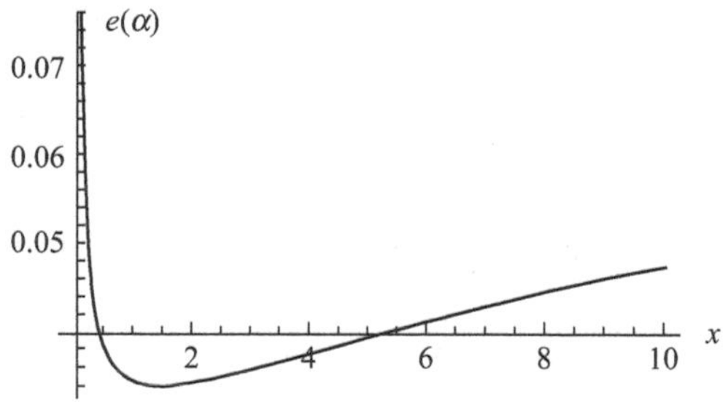

$k = 1 - j10^{-3}, \quad A = 500, \quad h = 0.01 \quad \alpha = 1 + jx$

Fig. 7: Relative error on the line $\alpha = k + jx$ $(x \geq 0)$

3.1.5 The case of meromorphic functions

In dealing with meromorphic functions, instead of using the general eq. (1) we can resort to the Mittag-Leffler decomposition, which is based on the following theorem (Spiegel, 1964, p. 175).

Theorem 1

Let us suppose that the meromorphic function $F(\alpha)$ has simple poles $\alpha_1, \alpha_2, \ldots, \alpha_k, \ldots$ $((\alpha_k \neq 0)$ with residues $\mathrm{Res}[\alpha_1], \mathrm{Res}[\alpha_2], \ldots, \mathrm{Res}[\alpha_k], \ldots$. If, except at the poles, $\lim \frac{F(\alpha)}{\alpha} \to 0$, as $\alpha \to \infty$, then the following expansion holds:

$$F(\alpha) = F(0) + \sum_{k=0}^{\infty} \mathrm{Res}[\alpha_k]\left(\frac{1}{\alpha - \alpha_k} + \frac{1}{\alpha_k}\right) \tag{36}$$

If, in addition, the meromorphic function $F(\alpha)$ is proper, for example, it satisfies $\lim F(\alpha) \to 0$, as $\alpha \to \infty$, then (36) simplifies and becomes

$$F(\alpha) = \sum_{k=1}^{\infty} \mathrm{Res}[\alpha_k]\frac{1}{\alpha - \alpha_k} \tag{37}$$

Equation (37) allows the decomposition of a meromorphic proper function $F(\alpha)$ by separating the poles α_{+k} having negative imaginary part, $\mathrm{Im}[\alpha_{+k}] < 0$, from the poles α_{-k} having positive imaginary part, $\mathrm{Im}[\alpha_{-k}] > 0$. It yields

$$F(\alpha) = \sum_{k=1}^{\infty} \mathrm{Res}[\alpha_k]\frac{1}{\alpha - \alpha_k} = \sum_{k=1}^{\infty} \mathrm{Res}[\alpha_{-k}]\frac{1}{\alpha - \alpha_{-k}} + \sum_{k=1}^{\infty} \mathrm{Res}[\alpha_{+k}]\frac{1}{\alpha - \alpha_{+k}} = S_-(\alpha) + S_+(\alpha) \tag{38}$$

with

$$S_-(\alpha) = \sum_{k=1}^{\infty} \mathrm{Res}[\alpha_{-k}]\frac{1}{\alpha - \alpha_{-k}} \quad S_+(\alpha) = \sum_{k=1}^{\infty} \mathrm{Res}[\alpha_{+k}]\frac{1}{\alpha - \alpha_{+k}} \tag{39}$$

In many applications it is necessary to know the asymptotic behavior of the decomposed functions as $\alpha \to \infty$. For any finite number of poles, we obtain

$$S_{+,-}(\alpha) \approx \alpha^{-1} \tag{40}$$

The case of an infinite number of poles is more critical. To achieve the asymptotic behavior of $S_+(\alpha)$, let us suppose that

$$\mathrm{Res}[\alpha_k] = a\, k^p \left[1 + O\left(\frac{1}{k}\right)\right], \quad (p < 0) \quad \text{and} \quad \alpha_k = -jb\, k\left[1 + O\left(\frac{1}{k}\right)\right] \quad \text{as } k \to \infty \tag{41}$$

It follows that

$$S_+(\alpha) = \sum_{k=1}^{\infty} \left\{ \mathrm{Res}[\alpha_{+k}] \frac{1}{\alpha - \alpha_{+k}} - a\, k^p \frac{1}{\alpha - b\, k} \right\} + \sum_{k=1}^{\infty} a\, k^p \frac{1}{\alpha + jb\, k} \tag{42}$$

In the previous expression, it may be shown that the dominant term as $\alpha \to \infty$ is the last sum. The inverse Laplace transform of this term yields $-j\sum_{k=1}^{\infty} a\, k^p e^{-b\,k\,z}$.

Now we have (Mittra & Lee, 1971, p. 11)[3]:

$$-j\sum_{k=1}^{\infty} a\, k^p e^{-b\,k\,x} \approx -j\,a\Gamma(p+1)\,(bx)^{-(p+1)} \quad \text{as } x \to 0_+ \tag{43}$$

By applying the Laplace transform, we return to the α – plane, and (45) yields

$$S_+(\alpha) \approx -j\,a\Gamma(p+1)b^{-(p+1)} \frac{1}{(-j\alpha)^{-p}} \tag{44}$$

The decomposition formula has been applied many times, and in the literature some slight modifications to both proofs and applications can be found. These are discussed in many texts, in particular in Noble (1958). To conclude this section, we remark again that the general decomposition equations are valid in both scalar and matrix cases.

3.1.6 Decomposition using rational approximants of the function

In the case of rational functions, the Mittag-Leffler expansion involves a finite number of terms (partial fraction expansion). Consequently, we can accomplish the decomposition of arbitrary functions $f(\alpha)$ using the partial fraction expansion of rational approximants of $f(\alpha)$. A very popular technique to obtain rational approximants is based on the Pade representation (see section 6.3.2). However, in many worked examples the Pade approximants do not have a good accuracy so it is preferable to rationalize the function with the interpolation approximant method described in section 6.3.2. For instance, by using this technique for the function

$$\frac{1}{\sqrt{k^2 - \alpha^2}} = S_-(\alpha) + S_+(\alpha)$$

[3] Mittra and Lee limited their proof to the case $-1 < p < 0$.

we obtained the relative error

$$e(\alpha) = \left| \frac{S_+(\alpha) - S_{a+}(\alpha)}{S_+(\alpha)} \right| \tag{45}$$

illustrated in Fig. 8, Fig. 9, and Fig. 10.

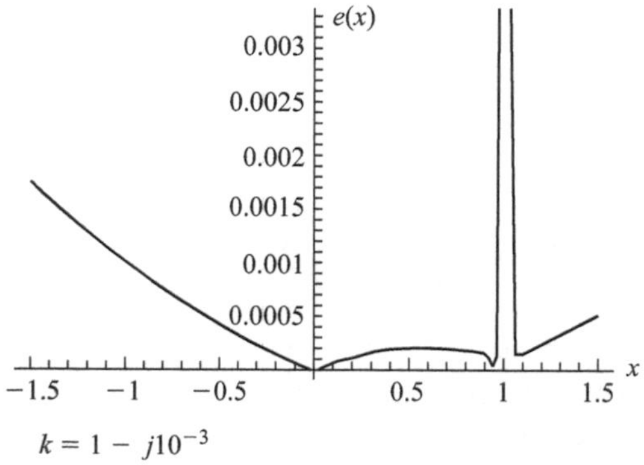

Fig. 8: Error $e(\alpha) = \left|\frac{S_+(\alpha)-S_{a+}(\alpha)}{S_+(\alpha)}\right|$ on the real axis

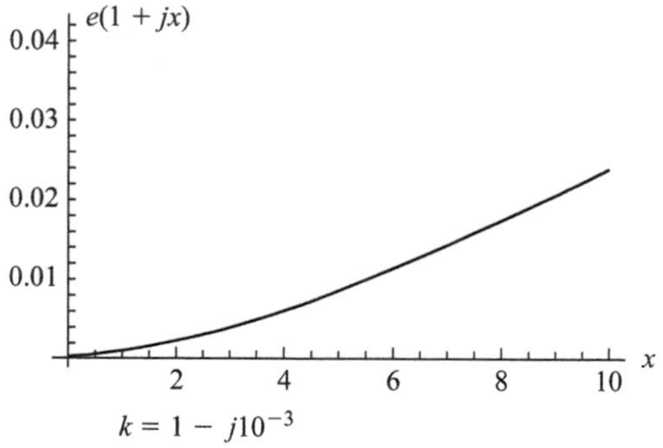

Fig. 9: Error $e(\alpha) = \left|\frac{S_+(\alpha)-S_{a+}(\alpha)}{S_+(\alpha)}\right|$ on the line $\alpha = 1 + jx$

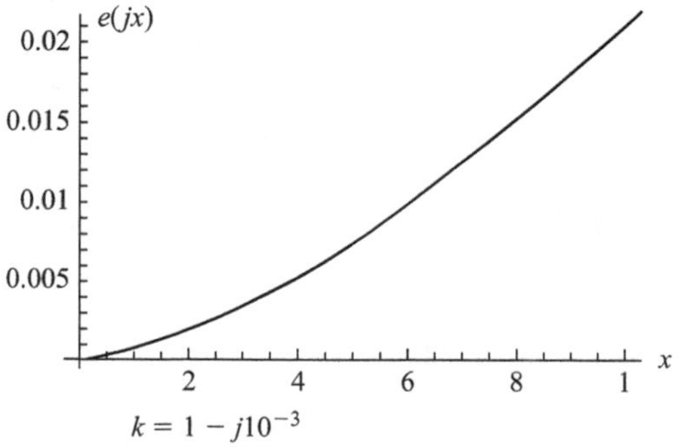

Fig. 10: Error $e(\alpha) = \left|\frac{S_+(\alpha)-S_{a+}(\alpha)}{S_+(\alpha)}\right|$ on the line $\alpha = jx$

3.2 Factorization

3.2.1 General formula for the scalar case

The fundamental idea to factorize a given function $G(\alpha)$ is based on the logarithmic decomposition. By introducing

$$G(\alpha) = e^{\mathrm{Log}[G(\alpha)]} = e^{\psi(\alpha)} \tag{46}$$

the decomposition of $\psi(\alpha) = \mathrm{Log}[G(\alpha)]$ yields

$$\mathrm{Log}[G(\alpha)] = \psi(\alpha) = \psi_-(\alpha) + \psi_+(\alpha) \tag{47}$$

where from (1)

$$\psi_+(\alpha) = \frac{1}{2\pi j}\int_{\gamma_1} \frac{\mathrm{Log}[G(\alpha')]}{\alpha' - \alpha} d\alpha', \quad \psi_-(\alpha) = \psi(\alpha) - \psi_+(\alpha) \tag{48}$$

By taking into account that in the scalar case

$$e^{\psi(\alpha)} = e^{\psi_-(\alpha)+\psi_+(\alpha)} = e^{\psi_-(\alpha)}e^{\psi_+(\alpha)} \tag{49}$$

one obtains the factorization formula:

$$G(\alpha) = G_-(\alpha)G_+(\alpha) \tag{50}$$

where

$$G_+(\alpha) = \exp\left[\frac{1}{2\pi j}\int_{\gamma_1} \frac{\mathrm{Log}[G(\alpha')]}{\alpha' - \alpha} d\alpha\right], \quad G_-(\alpha) = G(\alpha)G_+^{-1}(\alpha) \tag{51}$$

If $G(\alpha)$ is even, we obtain the simplification $G_-(\alpha) = G_+(-\alpha)$.

The last member of (49) is valid even in the matrix case, provided that the matrices $\psi_-(\alpha)$ and $\psi_+(\alpha)$ defined by (48) commute. Unfortunately, $\psi_-(\alpha)$ and $\psi_+(\alpha)$ do commute only in very few special cases.

The integral defining $G_+(\alpha)$ in (51) converges if $G(\alpha) \to 1$ as $\alpha \to \infty$, but this situation does not occur often. In these cases, the usual procedure is to normalize $G(\alpha) = f(\alpha)K(\alpha)$ such that $K(\alpha) \to 1$ as $\alpha \to \infty$, and $f(\alpha)$ may be factorized in other ways.

3.2.2 Example 2

Factorize

$$G(\alpha) = \sqrt{k^2 - \alpha^2}\frac{\alpha^2 + 1}{\alpha^2 + 4} \tag{52}$$

Equation (51) does not apply directly. However, it is possible to overcome the problem by introducing $f(\alpha) = \sqrt{k^2 - \alpha^2}$ and $K(\alpha) = \frac{\alpha^2+1}{\alpha^2+4}$; it follows that

$$G_+(\alpha) = \sqrt{k - \alpha}\, K_+(\alpha) \tag{53}$$

where, through (51) or by inspection, we have $K_+(\alpha) = \frac{\alpha+j}{\alpha+2j}$.

The condition $G(\alpha) \to 1$ as $\alpha \to \infty$ may be weakened if $G(\alpha)$ behaves at least as $|G(\alpha)| \to |\alpha|^p$ as $\alpha \to \infty$ (Noble, 1958, p. 42). For instance, if $G(\alpha)$ is even, then (53) holds again if the integral may be considered as a Cauchy principal-value integral.

$$\psi_+(\alpha) = \lim_{M\to+\infty} \frac{1}{2\pi j} \oint_{-M}^{+M} \frac{\log[G(\alpha)]}{\alpha' - \alpha} d\alpha' \tag{54}$$

Many variants of the factorization formulas can be found in the literature (see, e.g., Mittra & Lee, 1971, p. 113). Here only the most significant one, related to the derivative logarithmic decomposition, is reported. The derivative of eq. (47) yields

$$D[\mathrm{Log}[G(\alpha)]] = \frac{G'(\alpha)}{G(\alpha)} = t_-(\alpha) + t_+(\alpha) \tag{55}$$

where

$$t_\pm(\alpha) = \frac{d}{d\alpha}\psi_\pm(\alpha) \tag{56}$$

This leads to

$$\psi_+(\alpha) = \int_c^\alpha \frac{1}{2\pi j} \int_\gamma \frac{G'(\alpha')}{G(\alpha')(\alpha' - u)} d\alpha'\, du, \quad G_+(\alpha) = e^{\psi_+(\alpha)} \tag{57}$$

where c is a suitable constant. In many cases the introduction of a double integral is balanced by the very simple decomposition of $\frac{G'(\alpha)}{G(\alpha)} = t_-(\alpha) + t_+(\alpha)$.

3.2.3 Example 3

The factorization of $G(\alpha) = \tau = \sqrt{k^2 - \alpha^2}$ introduces the function

$$D[\mathrm{Log}(\tau)] = -\frac{\alpha}{\tau^2} = \frac{1}{2(-k+\alpha)} + \frac{1}{2(k+\alpha)}. \tag{58}$$

It follows that

$$\psi_+(\alpha) = \int \frac{1}{2(-k+\alpha)} d\alpha = \log\sqrt{k-\alpha} \Rightarrow G_+(\alpha) = \sqrt{k-\alpha} \tag{59}$$

3.2.4 Factorization of meromorphic functions

Meromorphic functions can be factorized using the Weirstrass theorem, which factorizes entire functions. More precisely, when the entire function $f(\alpha)$ has an infinite number of simple zeroes $\alpha_1, \alpha_2, \ldots$ the following representation stands:

$$f(\alpha) = f(0) e^{\frac{f'(0)}{f(0)}\alpha} \prod_{n=1}^{\infty} \left(1 - \frac{\alpha}{\alpha_n}\right) e^{\frac{\alpha}{\alpha_n}} \tag{60}$$

The zeroes α_n often behave as

$$\alpha_n = An + B \quad \text{when } n \Rightarrow \infty \tag{61}$$

In these cases, it is convenient to normalize $f(\alpha)$ with respect to the function defined by (Mittra & Lee, 1971, p. 13)

$$\hat{\Gamma}_{A,B}(\alpha) = \prod_{n=1}^{\infty}\left(1 - \frac{\alpha}{An+B}\right)e^{\frac{\alpha}{A\,n}} = \frac{e^{\gamma\frac{\alpha}{A}}\Gamma\left(\frac{B}{A}+1\right)}{\Gamma\left(-\frac{\alpha}{A}+\frac{B}{A}+1\right)} \tag{62}$$

where $\Gamma(\alpha)$ and $\gamma = 0.57721\ldots$ are, respectively, the Euler gamma function and the Euler constant. By normalizing $f(\alpha)$ with $\hat{\Gamma}_{A,B}(\alpha)$ one obtains, after some algebraic manipulations,

$$f(\alpha) = f(0)\, e^{h\alpha}\prod_{n=1}^{\infty}\frac{\left(1-\frac{\alpha}{\alpha_n}\right)}{\left(1-\frac{\alpha}{An+B}\right)}\hat{\Gamma}_{A,B}(\alpha) \tag{63}$$

with

$$h = \frac{f'(0)}{f(0)} + \sum_{n=1}^{\infty}\left(\frac{1}{\alpha_n} - \frac{1}{An}\right) \tag{64}$$

By observing that $\prod_{n=1}^{\infty}\frac{\left(1-\frac{\alpha}{\alpha_n}\right)}{\left(1-\frac{\alpha}{An+B}\right)}$ is bounded as $\alpha \to \infty$, $\alpha \neq An+B$, the asymptotic behavior of $f(\alpha)$ as $\alpha \to \infty$, is identical to that of $\hat{\Gamma}_{A,B}(\alpha)$.

The function $\hat{\Gamma}_{A,B}(\alpha)$ has the following asymptotic behavior as $\alpha \to \infty$, $\alpha \neq An+B$ (Mittra & Lee, 1971, p. 13):

$$\hat{\Gamma}_{A,B}(\alpha) \approx \frac{\Gamma\left(\frac{B}{A}+1\right)}{\sqrt{2\pi}}\left(-\frac{\alpha}{A}\right)^{-\left(\frac{1}{2}+\frac{B}{A}\right)} e^{\frac{\alpha}{A}\log\left(-\frac{\alpha}{A}\right)}e^{-\frac{\alpha}{A}(1-\gamma)}, \quad \alpha \to \infty, \quad \alpha \neq An+B \tag{65}$$

This last equation allows us to study the asymptotic behavior of $f(\alpha)$ as $\alpha \to \infty$.

A meromorphic function $G(\alpha)$ is the ratio of two entire functions. The factorization of $G(\alpha)$ can be obtained by separating, in both numerator and denominator, the zeros and poles located in the two half-planes $\mathrm{Im}[\alpha] > 0$ and $\mathrm{Im}[\alpha] < 0$:

$$G(\alpha) = \hat{G}_-(\alpha)\hat{G}_+(\alpha) \tag{66}$$

where $\hat{G}_-(\alpha)$ and $\hat{G}_+(\alpha)$ are meromorphic functions having zeroes and poles in the half-planes $\mathrm{Im}[\alpha] > 0$ and $\mathrm{Im}[\alpha] < 0$, respectively.

It must be observed that in general $\hat{G}_-(\alpha)$ and $\hat{G}_+(\alpha)$ have nonalgebraic behavior at infinity. To obtain algebraic behavior we may rewrite (66) in the form

$$G(\alpha) = \hat{G}_-(\alpha)e^{w(\alpha)}e^{-w(\alpha)}\hat{G}_+(\alpha) = G_-(\alpha)G_+(\alpha) \tag{67}$$

where $G_-(\alpha) = \hat{G}_-(\alpha)e^{w(\alpha)}$, $G_+(\alpha) = e^{-w(\alpha)}\hat{G}_+(\alpha)$. The entire function $e^{w(\alpha)}$ (and its inverse) is free of zeros, and $w(\alpha)$ is chosen to ensure the algebraic behavior of

$G_-(\alpha)$ and $G_+(\alpha)$. When the zeros have the asymptotic behavior shown in eq. (61), it will be seen later that

$$w(\alpha) = q\alpha \tag{68}$$

where q is a constant.

3.2.5 Example 4

The following example is relevant to the study of the bifurcation of a wave guide a in two wave guides b and c (section 9.2). In this case the required function to factorize is

$$G(\alpha) = \frac{\sin(\tau\, b)\sin(\tau\, c)}{\tau \sin[\tau\,(b+c)]}, \quad \tau = \sqrt{k^2 - \alpha^2}, \quad b > 0, \quad c > 0 \tag{69}$$

Even though $\tau = \sqrt{k^2 - \alpha^2}$ is a two-valued function, $G(\alpha)$ is meromorphic since it is an even function of τ. Before factorizing it, it is important to determine the asymptotic behavior of $G(\alpha)$ as $\alpha \to \infty$. This behavior is algebraic:

$$|G(\alpha)| \propto \frac{e^{|\alpha|b} e^{|\alpha|c}}{|\alpha|\, e^{|\alpha|\,(b+c)}} = \frac{1}{|\alpha|} \quad \text{as } \alpha \to \infty \tag{70}$$

We rewrite

$$G(\alpha) = \frac{bc}{a} \frac{S_b(\alpha) S_c(\alpha)}{S_a(\alpha)} \tag{71}$$

with

$$S_d(\alpha) = \frac{\sin(\tau\, d)}{\tau\, d} \quad d = b, c, \quad a = b + c \tag{72}$$

The zeros of the entire function $S_d(\alpha)$ are given by $\pm\alpha_{dn}$, where

$$\alpha_{dn} = \sqrt{k^2 - \left(\frac{n\pi}{d}\right)^2} \quad n = 1, 2, \ldots, \quad \text{Im}[\alpha_{dn}] < 0, \quad d = b, c, \quad a = b + c \tag{73}$$

The asymptotic behavior of α_{dn} as $n \to \infty$ is given by

$$\alpha_{dn} \approx A_d n + B_d, \quad A_d = -j\frac{\pi}{d}, \quad B_d = 0, \quad d = b, c, \quad a = b + c, \quad \frac{1}{A_a} = \frac{1}{A_b} + \frac{1}{A_c} \tag{74}$$

Taking into account that $S_d(\alpha)$ is even, (60) yields

$$S_d(\alpha) = \frac{\sin(\tau\, d)}{\tau\, d} = \frac{\sin(kd)}{kd} \prod_{n=1}^{\infty} \left(1 - \left(\frac{\alpha}{\alpha_{dn}}\right)^2\right) \tag{75}$$

$$S_d(\alpha) = \frac{\sin(\tau\, d)}{\tau\, d} = \frac{\sin(kd)}{kd} \prod_{n=1}^{\infty} \left(1 + \frac{\alpha}{\alpha_{dn}}\right) e^{-\frac{\alpha}{\alpha_{dn}}} \prod_{n=1}^{\infty} \left(1 - \frac{\alpha}{\alpha_{dn}}\right) e^{\frac{\alpha}{\alpha_{dn}}}$$

$$= \frac{\sin(kd)}{kd} \prod_{n=1}^{\infty} \left(1 + \frac{\alpha}{\alpha_{dn}}\right) e^{-\frac{\alpha}{A_d n}} \prod_{n=1}^{\infty} \left(1 - \frac{\alpha}{\alpha_{dn}}\right) e^{\frac{\alpha}{A_d n}} \tag{76}$$

Notice that when the zeroes are separated into two distinct infinite products, the exponential factors $e^{\mp\frac{\alpha}{\alpha_{dn}}}$ may be substituted by the simpler factors $e^{\mp\frac{\alpha}{A_d n}}$. The presence of an exponential is, however, always necessary to ensure the convergence of the infinite products.

Taking into account (75) and (62), algebraic manipulations yield the following result (non algebraic at infinity) for the plus factorized functions:

$$S_{d+}(\alpha) = \sqrt{\frac{\sin(kd)}{kd}} \prod_{n=1}^{\infty} \frac{\left(1 - \frac{\alpha}{\alpha_{dn}}\right)}{\left(1 - \frac{\alpha}{A_d n}\right)} \hat{\Gamma}_{A_d,0}(\alpha), \quad d = b, c, \quad a = b + c \tag{77}$$

In this formula, the infinite product behaves as a constant as $\alpha \to \infty$. Thus, the asymptotic behavior of $S_{d+}(\alpha)$ coincides with that of $\hat{\Gamma}_{A_d,0}(\alpha)$ in (65):

$$S_{d+}(\alpha) \propto \hat{\Gamma}_{A_d,0}(\alpha) \approx M_d \alpha^{-\frac{1}{2}} e^{\frac{\alpha}{A_d}\log\left(-\frac{\alpha}{A_d}\right)} e^{-\frac{\alpha}{A_d}(1-\gamma)}, \quad \alpha \to \infty \tag{78}$$

Equation (71) yields

$$\hat{G}_+(\alpha) = \sqrt{\frac{bc}{a}} \frac{S_{b+}(\alpha) S_{c+}(\alpha)}{S_{+a}(\alpha)} \tag{79}$$

The asymptotic behavior of $\hat{G}_+(\alpha)$ is given by

$$\hat{G}_+(\alpha) \propto \frac{\alpha^{-\frac{1}{2}} e^{\frac{\alpha}{A_b}\log\left(-\frac{\alpha}{A_b}\right)} e^{-\frac{\alpha}{A_b}(1-\gamma)} \alpha^{-\frac{1}{2}} e^{\frac{\alpha}{A_c}\log\left(-\frac{\alpha}{A_c}\right)} e^{-\frac{\alpha}{A_c}(1-\gamma)}}{\alpha^{-\frac{1}{2}} e^{\frac{\alpha}{A_a}\log\left(-\frac{\alpha}{A_a}\right)} e^{-\frac{\alpha}{A_a}(1-\gamma)}} = \alpha^{-\frac{1}{2}} e^{q\alpha}, \quad \alpha \to \infty \tag{80}$$

where, by taking (76) into account,

$$q = \frac{1}{A_b}\log\frac{1}{A_b} + \frac{1}{A_c}\log\frac{1}{A_c} - \frac{1}{A_a}\log\frac{1}{A_a} = -\frac{j}{\pi}\left(b\log\frac{a}{b} + c\log\frac{a}{c}\right) \tag{81}$$

Equation (80) shows that $\hat{G}_+(\alpha)$ and $\hat{G}_-(\alpha) = \hat{G}_+(-\alpha)$ have exponential behavior as $\alpha \to \infty$. By looking at (67), we define

$$G_-(\alpha) = \hat{G}_-(\alpha)\, e^{q\alpha}, \quad G_+(\alpha) = e^{-q\alpha} \hat{G}_+(\alpha) \tag{82}$$

with q given by (81).

Of course $G_-(\alpha)$ and $G_+(\alpha)$ have algebraic behavior as $\alpha \to \infty$. In fact,

$$G_-(\alpha) \propto \alpha^{-\frac{1}{2}}, \quad G_+(\alpha) \propto \alpha^{-\frac{1}{2}} \tag{83}$$

As a numerical example, let us suppose that

$$\lambda = 1, \quad k = \frac{2\pi}{\lambda}, \quad b = 1.1\frac{\lambda}{2}, \quad c = 1.3\frac{\lambda}{2}, \quad a = b + c = 2.4\frac{\lambda}{2} \tag{84}$$

It is possible to evaluate the infinite product by approximating it with the finite product:

$$\prod_{n=1}^{\infty} \frac{\left(1 - \frac{\alpha}{\alpha_{dn}}\right)}{\left(1 - j\frac{\alpha\, d}{n\pi}\right)} \approx \prod_{n=1}^{N_d} \frac{\left(1 - \frac{\alpha}{\alpha_{dn}}\right)}{\left(1 - j\frac{\alpha\, d}{n\pi}\right)}, \quad d = a, b, c \tag{85}$$

The choice of N_d depends on how many real α_{dn} are present. In fact, real α_{dn} are related to modes that propagate, whereas imaginary α_{dn} involve evanescent modes. In this case we have the following first three α_{dn} ($d = a, b, c$):

$$\begin{aligned} \alpha_{a1} &= 5.71, \quad \alpha_{a2} = 3.47, \quad \alpha_{a3} = -j4.71 \\ \alpha_{b1} &= 2.62, \quad \alpha_{b2} = -j9.54, \quad \alpha_{b3} = -j15.94 \\ \alpha_{c1} &= 4.01, \quad \alpha_{c2} = -j7.35, \quad \alpha_{c3} = -j13.07 \end{aligned} \tag{86}$$

Thus, two modes propagates in waveguide a, one mode in waveguide b, and one mode in waveguide c.

The very small number of propagating modes means that N_d can be assumed small. However, we first assumed high values of N_d by putting

$$N_a = N_c = N_c = 2000 \tag{87}$$

Figure 11a illustrates the absolute value and the argument of $G_+(\alpha)$ for real values of α.

Figure 12 reports the error $e = |G(\alpha) - G_-(\alpha)G_+(\alpha)|$.

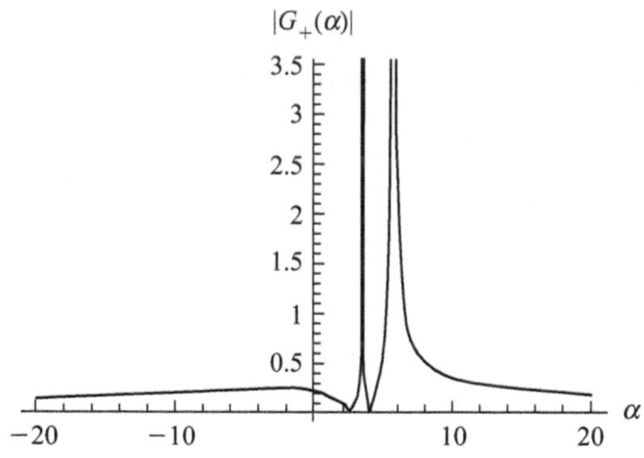

Fig. 11a: Absolute value $|G_+(\alpha)|$ of the factorized function $G_+(\alpha)$

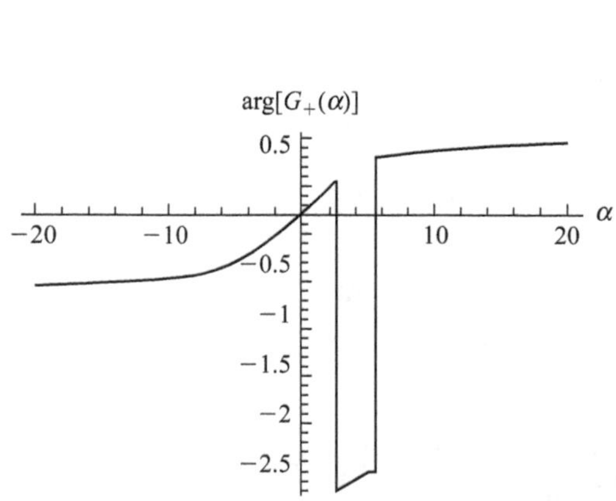

Fig. 11b: Argument $\arg[G_+(\alpha)]$ of the factorized function $G_+(\alpha)$

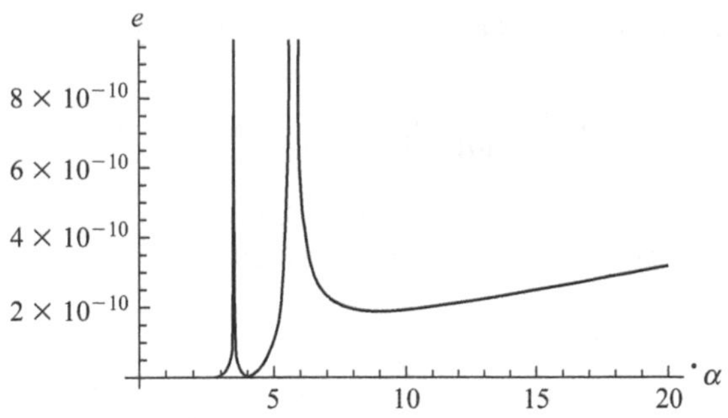

Fig. 12: The error $e = |G(\alpha) - G_-(\alpha)G_+(\alpha)|$

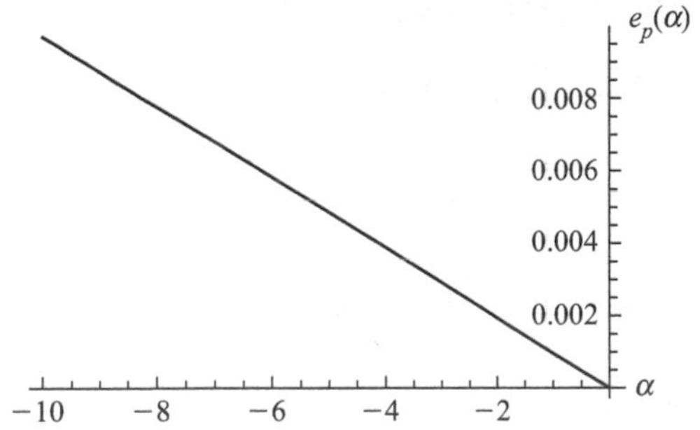

Fig. 13: Truncation error for $N_a = N_c = N_c = 20$

By assuming $N_a = N_c = N_c = 20$ we practically obtain the same plots.

Figure 13 illustrates the relative error when we change $N_a = N_c = N_c = 2000$ to $N_a = N_c = N_c = 20$:

$$e_p(\alpha) = \left| \frac{G_+(\alpha) - G_{1+}(\alpha)}{G_+(\alpha)} \right| \tag{88}$$

where $G_+(\alpha)$ has been evaluated by putting $N_a = N_c = N_c = 2000$, and $G_{1+}(\alpha)$ has been evaluated by putting $N_a = N_c = N_c = 20$.

As a matter of fact, $e = |G(\alpha) - G_-(\alpha)G_+(\alpha)|$ does not depend on q, and a small error e does not assure that the choice of q given by (83) is correct. We ascertained the correctness of q by taking into account that the function $\sqrt{k-\alpha}\, G_+(\alpha)$ and its inverse $\left(\sqrt{k-\alpha}\, G_+(\alpha)\right)^{-1}$ must behave as a bounded nonvanishing constant when $\alpha \to \infty$.

3.2.6 Factorization of kernels involving continuous and discrete spectrum

3.2.6.1 Example 5

Factorization of

$$G(\alpha) = \frac{e^{j\tau\, d}}{\tau \cos[\tau\, d]}, \quad \tau = \sqrt{k^2 - \alpha^2} \tag{89}$$

The asymptotic behavior of $G(\alpha)$ is algebraic and is given by

$$G(\alpha) \propto \frac{e^{|\alpha|d}}{|\alpha|\, e^{|\alpha|d}} = \frac{1}{|\alpha|} \quad \text{as } \alpha \to \infty \tag{90}$$

The factorization of $G(\alpha)$ may be accomplished through the factorization of the factors τ, $e^{j\tau d}$ and $\cos[\tau\, d]$. Since we have from the factorization of τ and the decomposition (13):

$$\tau = \sqrt{k^2 - \alpha^2} = \sqrt{k+\alpha}\sqrt{k-\alpha} \tag{91}$$

$$e^{j\tau\, d} = e^{j\tau_-\, d} e^{j\tau_+\, d} = \exp\left[\frac{\tau\, d}{\pi} \log \frac{j\tau + \alpha}{k}\right] \exp\left[\frac{\tau\, d}{\pi} \log \frac{j\tau - \alpha}{k}\right] \tag{92}$$

it remains only to factorize the entire function $c(\alpha) = \cos[\tau\, d]$. According to (60), first we must obtain all its zeroes $\pm\alpha_n$. These are given by

$$\cos\left(\sqrt{k^2 - \alpha_n{}^2 d}\right) = 0 \Rightarrow \alpha_n = \alpha_{d(n-1/2)}, \quad n = 1, 2, \ldots \tag{93}$$

with

$$\alpha_{d(n)} = \sqrt{k^2 - \left(\frac{n\pi}{d}\right)^2} \to \alpha_n = \sqrt{k^2 - \left(\frac{(n-1/2)\pi}{d}\right)^2}$$

which as $n \to \infty$ assume the values

$$\alpha_n = \alpha_{d(n-1/2)} \approx A_d(n - 1/2) = An + B, \quad A = A_d = -j\frac{\pi}{d}, \quad B = -\frac{1}{2}A, \quad \frac{B}{A} = -\frac{1}{2} \tag{94}$$

From (62) and taking into account that $c(\alpha)$ is even, one obtains

$$\begin{aligned} c(\alpha) = \cos(\tau\, d) &= \cos(kd)\prod_{n=1}^{\infty}\left(1 - \left(\frac{\alpha}{\alpha_n}\right)^2\right) \\ &= \cos(kd)\prod_{n=1}^{\infty}\left(1 + \frac{\alpha}{\alpha_n}\right)e^{-\frac{\alpha}{A\,n}}\prod_{n=1}^{\infty}\left(1 - \frac{\alpha}{\alpha_n}\right)e^{\frac{\alpha}{A\,n}} \end{aligned} \tag{95}$$

Note that the presence of the factors $e^{\mp\frac{\alpha}{A\,n}}$ is necessary to ensure the convergence of the infinite products.

By normalizing with respect to $\hat{\Gamma}_{A,B}(\alpha)$, one obtains the nonalgebraic plus factorized functions

$$\hat{c}_+(\alpha) = \sqrt{\cos(kd)}\prod_{n=1}^{\infty}\frac{\left(1 - \dfrac{\alpha}{\alpha_n}\right)}{\left(1 - \dfrac{\alpha}{An + B}\right)}\hat{\Gamma}_{A,B}(\alpha) \tag{96}$$

$$\hat{G}_+(\alpha) = \frac{\exp\left[\dfrac{\tau\, d}{\pi}\log\dfrac{j\tau - \alpha}{k}\right]}{\sqrt{k-\alpha}\ \hat{c}_+(\alpha)} \tag{97}$$

By taking into account that in our case as $\alpha \to \infty$ (see (65) and (21)),

$$\hat{c}_+(\alpha) \approx \hat{\Gamma}_{A,B}(\alpha) \approx \frac{\Gamma\left(\frac{B}{A}+1\right)}{\sqrt{2\pi}} \left(-\frac{\alpha}{A}\right)^{-\left(\frac{1}{2}+\frac{B}{A}\right)} e^{\frac{\alpha}{A}\log\left(-\frac{\alpha}{A}\right)} e^{-\frac{\alpha}{A}(1-\gamma)} \propto e^{\frac{\alpha}{A}\log\left(-\frac{\alpha}{A}\right)} e^{-\frac{\alpha}{A}(1-\gamma)}$$
$$= e^{\frac{\alpha}{A}\log\left(-\frac{2\alpha}{k}\right)} e^{-\frac{\alpha}{A}\left[\log\left(\frac{2A}{k}\right)+(1-\gamma)\right]} = e^{j\frac{\alpha d}{\pi}\log\left(-\frac{2\alpha}{k}\right)} e^{-j\frac{\alpha d}{\pi}\left[\log\left(-j\frac{2\pi}{kd}\right)+(1-\gamma)\right]} \tag{98}$$

and

$$e^{j\tau_+ d} \propto \exp\left[j\frac{\alpha\, d}{\pi}\log\frac{-2\alpha}{k}\right]$$

it follows that

$$\hat{G}_+(\alpha) \propto \frac{1}{\alpha^{1/2}} e^{q\,\alpha} \tag{99}$$

with the constant q given by

$$q = j\frac{d}{\pi}\left[\log\left(-\frac{j2\pi}{kd}\right) + 1 - \gamma\right] \tag{100}$$

where the principal part of the logarithm is assumed.

The final formulas are

$$G_+(\alpha) = \frac{\exp\left[\frac{\tau\, d}{\pi}\log\frac{j\tau-\alpha}{k} - q\,\alpha\right]}{\sqrt{k-\alpha}\,\sqrt{\cos(kd)}\hat{\Gamma}_{A,B}(\alpha)} \prod_{n=1}^{\infty} \frac{\left(1-\frac{\alpha}{An+B}\right)}{\left(1-\frac{\alpha}{\alpha_n}\right)} \tag{101}$$

or

$$G_+(\alpha) = \frac{\Gamma\left(-\frac{\alpha}{A}+\frac{B}{A}+1\right)\exp\left[\frac{\tau\, d}{\pi}\log\frac{j\tau-\alpha}{k} - q\,\alpha\right]}{\sqrt{k-\alpha}\,\sqrt{\cos(kd)}\, e^{\gamma\frac{\alpha}{A}}\Gamma\left(\frac{B}{A}+1\right)} \prod_{n=1}^{\infty} \frac{\left(1-\frac{\alpha}{An+B}\right)}{\left(1-\frac{\alpha}{\alpha_n}\right)}$$
$$\approx \frac{\Gamma\left(-\frac{\alpha}{A}+\frac{B}{A}+1\right)\exp\left[\frac{\tau\, d}{\pi}\log\frac{j\tau-\alpha}{k} - q\,\alpha\right]}{\sqrt{k-\alpha}\,\sqrt{\cos(kd)}\, e^{\gamma\frac{\alpha}{A}}\Gamma\left(\frac{B}{A}+1\right)} \prod_{n=1}^{N_b} \frac{\left(1-\frac{\alpha}{An+B}\right)}{\left(1-\frac{\alpha}{\alpha_n}\right)} \tag{102}$$

$$G_-(\alpha) = G_+(-\alpha) \tag{103}$$

By assuming $\lambda = 1$, $k = \frac{2\pi}{\lambda}(1-j10^{-6})$, $b = 1.1\frac{\lambda}{2}$ and a truncated product with $N_b = 200$, we obtain an error $G_-(\alpha)G_+(\alpha) - G(\alpha)$ that is vanishing in the range $-100 \leq \alpha \leq 100$. Truly, $e = |G(\alpha) - G_-(\alpha)G_+(\alpha)|$ does not depend on q, and a small error $e = |G(\alpha) - G_-(\alpha)G_+(\alpha)|$ does not assure that the choice of q given by (100) is correct. We ascertained the correctness of q by taking into account that the function $\sqrt{k-\alpha}\,G_+(\alpha)$ and its inverse $\left(\sqrt{k-\alpha}\,G_+(\alpha)\right)^{-1}$ must behave as a bounded nonvanishing constant as $\alpha \to \infty$.

3.2.6.2 Example 6

Factorization of

$$G(\alpha) = e^{-j\tau d} \frac{\sin[\tau\, d]}{\tau\, d} \tag{104}$$

Taking into account the considerations of example 5, we get

$$G_+(\alpha) = \frac{\sqrt{\frac{\sin(kd)}{kd}} \exp\left[-\frac{\tau\, d}{\pi} \log\frac{j\tau - \alpha}{k} + q\,\alpha\right] e^{\gamma\frac{\alpha}{A}} \Gamma\left(\frac{B}{A}+1\right)}{\Gamma\left(-\frac{\alpha}{A}+\frac{B}{A}+1\right)} \prod_{n=1}^{\infty} \frac{\left(1-\frac{\alpha}{\alpha_n}\right)}{\left(1-\frac{\alpha}{An+B}\right)} \tag{105}$$

where

$$A = -j\frac{\pi}{d}, \quad B = 0 \quad q = j\frac{d}{\pi}\left[\log\left(-\frac{j2\pi}{kd}\right) + 1 - \gamma\right] \text{ (see eq. (100))}$$

$$\alpha_n = \sqrt{k^2 - \left(\frac{n\pi}{d}\right)^2} \quad n = 1, 2, \ldots, \quad \mathrm{Im}[\alpha_n] < 0$$

3.3 Decomposition equations in the w – plane

The fundamental formula to decompose a function is based on the Cauchy eqs. (1) and (2):

$$F_+(\alpha) = \frac{1}{2\pi j} \int_{\gamma_1} \frac{F(\alpha')}{\alpha' - \alpha} d\alpha', \quad F_-(\alpha) = -\frac{1}{2\pi j} \int_{\gamma_2} \frac{F(\alpha')}{\alpha' - \alpha} d\alpha' \tag{106}$$

When the function $F(\alpha)$ to be decomposed presents only branch points due to the function $\chi = \sqrt{k^2 - \alpha^2}$, it is sometimes convenient to use the previous equations in the w – plane defined in section 2.9.2.

3.3.1 Evaluation of the plus functions

By starting from $F_+(\alpha) = \frac{1}{2\pi j}\int_{\gamma_1} \frac{F(\alpha')}{\alpha'-\alpha} d\alpha'$, it is possible to warp the line γ_1 into the line γ_o (Fig. 14) surrounding the standard branch line relevant to the branch point $\alpha = +k$. This yields

$$\begin{aligned} F_+(\alpha) &= \frac{1}{2\pi j} \int_{\gamma_o} \frac{F(\alpha')}{\alpha' - \alpha} d\alpha' - \sum_n \frac{R_n}{\alpha_n - \alpha} \\ &= \frac{1}{2\pi j} \int_{\gamma_{o1}} \frac{F(\alpha')}{\alpha' - \alpha} d\alpha' + \frac{1}{2\pi j} \int_{\gamma_{o2}} \frac{F(\alpha')}{\alpha' - \alpha} d\alpha' - \sum_n \frac{R_n}{\alpha_n - \alpha} \end{aligned} \tag{107}$$

where α_n are the poles of $F(\alpha)$ (for the sake of simplicity, we suppose simple) located in the lower half-plane $\mathrm{Im}[\alpha] \leq 0$, and R_n are the corresponding residues. In the w – plane,

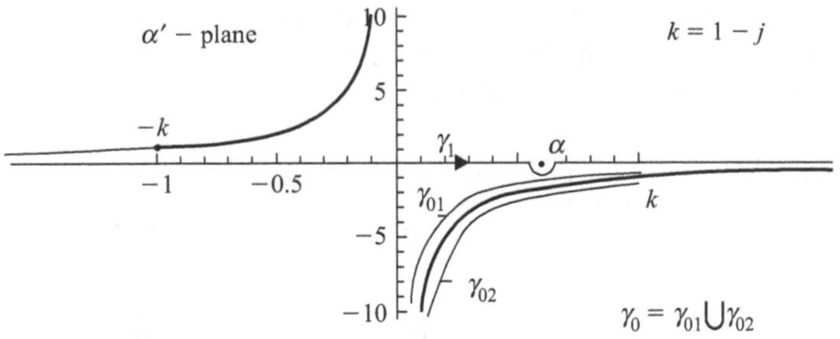

Fig. 14: Deformation of the integration path in the α' – plane

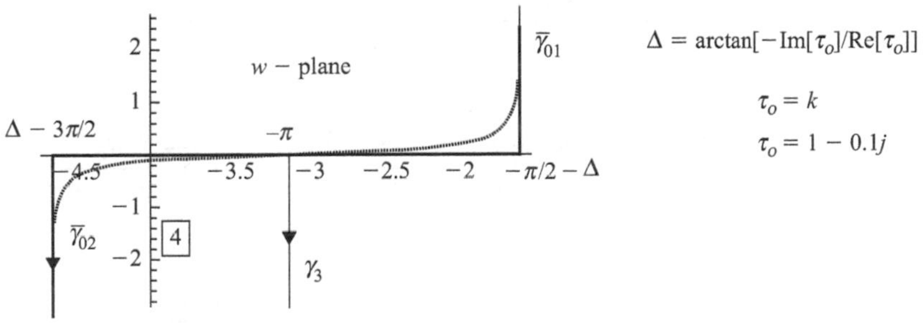

Fig. 15: Deformation of the integration path in the w – plane

$\alpha = -k\cos w$, by assuming that w belongs to the proper image of α (chapter 2, Fig. 3), we have

$$\begin{aligned}
\frac{1}{2\pi j}\int\limits_{\gamma_{o2}} \frac{F(\alpha')}{\alpha'-\alpha}d\alpha' &= -\frac{1}{2\pi j}\int\limits_{\bar{\gamma}_{o2}} \frac{\hat{F}(w')\sin w'}{\cos w' - \cos w}dw' \\
\frac{1}{2\pi j}\int\limits_{\gamma_{o1}} \frac{F(\alpha')}{\alpha'-\alpha}d\alpha' &= -\frac{1}{2\pi j}\int\limits_{\bar{\gamma}_{o1}} \frac{\hat{F}(w')\sin w'}{\cos w' - \cos w}dw' = \frac{1}{2\pi j}\int\limits_{\bar{\gamma}_{o2}} \frac{\hat{F}(-2\pi - w')\sin w'}{\cos w' - \cos w}dw'
\end{aligned} \tag{108}$$

where the notation $F(-k\cos w) = \hat{F}(w)$ has been used.

By deforming $\bar{\gamma}_{o2}$ on γ_3, defined by $w = -\pi + j\,\mathrm{Im}[w]$, $\mathrm{Im}[w] < 0$, (Fig. 15), the poles (located in a proper region of the fourth quadrant) between these two lines do contribute with opposite signs with respect to $-\sum \frac{R_n}{\alpha_n - \alpha}$, so we obtain

$$F_+(\alpha) = -\frac{1}{2\pi j}\int\limits_{\gamma_3} \frac{\left[\hat{F}(w') - \hat{F}(-2\pi - w')\right]\sin w'}{\cos w' - \cos w}dw' - \sum_{n'} \frac{R_n}{\alpha_{n'} - \alpha} \tag{109}$$

where the apex in the summation index implies that the poles to be considered are those in the lower half-plane $\mathrm{Im}[\alpha] < 0$, located outside the subregion of the fourth quadrant between the lower lip of the standard branch line and the half line from k to ∞; this half line is defined by $\alpha = k\cosh u$, with $0 \leq u \leq \infty$.

By looking at the w – plane, it is possible to see that the poles to consider are located in the proper image of the half-plane $\mathrm{Im}[\alpha] \leq 0$ and must be such that $-\pi < \mathrm{Re}[w] < 0$.

By introducing $w' = -\pi + ju$, one obtains

$$\frac{1}{2\pi j}\int_{\gamma_3} \frac{F(\alpha')}{\alpha' - \alpha} d\alpha' = -\frac{1}{2\pi j}\int_0^\infty \frac{[\hat{F}(-\pi + ju) - \hat{F}(-\pi - ju)]\sin(-\pi + ju)}{\cos(-\pi + ju) - \cos w} jdu$$

$$= -\frac{1}{2\pi j}\int_0^\infty \frac{[\hat{F}(-\pi + ju) - \hat{F}(-\pi - ju)]\sinh u}{\cosh u + \cos w} du$$

$$= -\frac{1}{2\pi j}\int_0^\infty \frac{[\hat{F}(-\pi + ju) - \hat{F}(-\pi - ju)]\sinh u}{\cosh u - \dfrac{\alpha}{k}} du \tag{110}$$

that is,

$$F_+(\alpha) = -\frac{1}{2\pi j}\int_0^\infty \frac{[\hat{F}(-\pi + ju) - \hat{F}(-\pi - ju)]\sinh u}{\cosh u + \cos w} du - \sum_{n'} \frac{R_n}{\alpha_{n'} - \alpha}$$

$$= -\frac{1}{2\pi j}\int_0^\infty \frac{[\hat{F}(-\pi + ju) - \hat{F}(-\pi - ju)]\sinh u}{\cosh u - \dfrac{\alpha}{k}} du - \sum_{n'} \frac{R_n}{\alpha_{n'} - \alpha} \tag{111}$$

It must be remembered that, when looking at the w – plane, the poles located in the proper image of the half-plane $\mathrm{Im}[\alpha] \leq 0$ and having $\mathrm{Re}[w] < -\pi$ must not be considered.

Discussion

In the w – plane the poles to be considered are located in the proper image of the half-plane $\mathrm{Im}[\alpha] < 0$, which does not belong to the region between $\bar{\gamma}_{o2}$ and γ_3 (Fig. 15).

In the w – plane the representation (113) stands if $\mathrm{Re}[w] > -\pi$. If the integral has been evaluated in closed form, then according to the analytic continuation this closed form represents the plus function also when $\mathrm{Re}[w] < -\pi$. Conversely, if a numerical integration has been adopted, the jump $\mathrm{Re}[w] > -\pi$, $\mathrm{Re}[w] < -\pi$ has to be studied according to the observations developed in section 3.1. Note that the singularity $u = 0$ that appears in $\hat{F}(-\pi \pm ju)$ is compensated by the function $\sinh u$.

Example

Decomposition of $\frac{1}{\chi} = -\frac{1}{k \sin w} = \hat{F}(w)$.

The last integral of the previous formula provides the following result:

$$\hat{F}_+(w) = -\frac{1}{j\pi\, k}\int_0^\infty \frac{1}{j(\cosh u + \cos w)} du = \frac{1}{\pi\, k}\frac{w}{\sin w} \tag{112}$$

which is in agreement with (16).

The final formula

$$F_+(w) = -\frac{1}{2\pi j}\int_0^{\infty} \frac{\left[\hat{F}(-\pi + ju) - \hat{F}(-\pi - ju)\right]\sinh u}{\cosh u + \cos w} du - \sum_n \frac{R_n}{k\cos w + \alpha_n} \tag{113}$$

shows that the plus functions are even functions in the w – plane, regular at the point $w = 0$ corresponding to the branch point $\alpha = -k$. This is a fundamental property (Daniele, 2001, 2003b, 2004b).

3.3.2 Evaluation of the minus functions

Minus functions may be considered as plus functions evaluated for $\alpha = -\alpha$. By taking into account the property of eq. (66) in chapter 2

$$w(-\alpha) = -w(\alpha) - \pi$$

it follows that the minus function $F_+(-\alpha)$ has the form $\hat{F}_+(-w - \pi)$ where $\hat{F}_+(w)$ is even in w and regular at $w = 0$. Furthermore, the images in the w – plane of the minus functions have the property of being invariant under the substitution of w with $-2\pi - w$, that is, $\hat{F}_+(-w - \pi) = \hat{F}_+(w + \pi) = \hat{F}_+(-(-2\pi - w) - \pi)$.

Starting from $F_-(\alpha) = -\frac{1}{2\pi j}\int_{\gamma_2} \frac{F(\alpha')}{\alpha' - \alpha} d\alpha'$, we warp the integration path γ_2 to the line γ_- (Fig. 16) surrounding the standard branch line relevant to the branch point $\alpha = -k$. This yields

$$F_-(\alpha) = -\frac{1}{2\pi j}\int_{\gamma_2} \frac{F(\alpha')}{\alpha' - \alpha} d\alpha' = \frac{1}{2\pi j}\int_{\gamma_-} \frac{F(\alpha')}{\alpha' - \alpha} d\alpha' - \sum \frac{S_n}{\alpha_n - \alpha} \tag{114}$$

where α_n are the poles of $F(\alpha)$ located in the proper upper half-plane $\mathrm{Im}[\alpha] \geq 0$, and S_n are the corresponding residues.

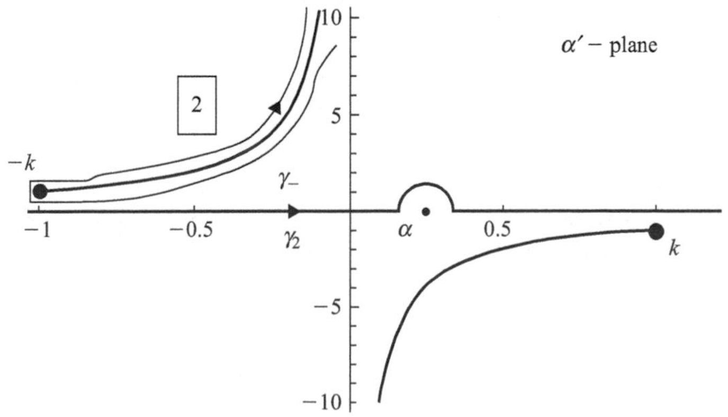

Fig. 16: Deformation of the integration path γ_2 in the α' – plane

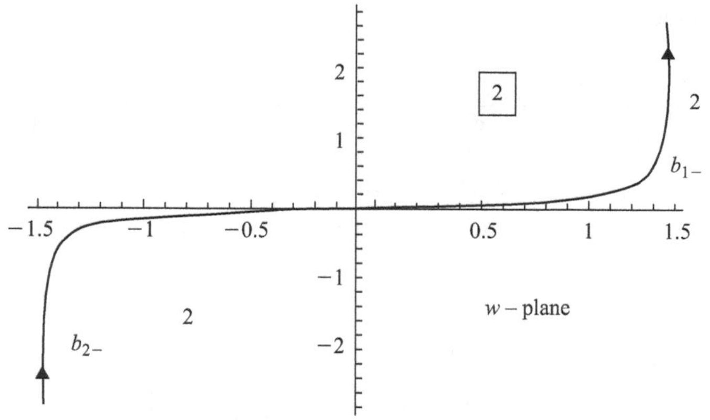

Fig. 17: Deformation of the integration path $b\_$ (image of $\gamma\_$) in the w – plane

Let us now consider the w – plane $\alpha = -\tau_o \cos w$. In this plane, the line $\gamma\_$ has as image the line $b_- = b_{1-} \cup b_{2-}$, and it separates the image of the proper quadrant 2 denoted by the number 2 encapsulated in a square, $\square$, from the image of the improper quadrant 2 denoted by the number 2, as depicted in Fig. 17 (see also chapter 2, Fig. 4). The reported positive direction takes into account that when running on $\gamma\_$ on the α' – plane the proper sheet lies on the left side.

Let us now decompose b_- into b_{1-} (located in $\mathrm{Im}[w] > 0$) and b_{2-} (located in $\mathrm{Im}[w] < 0$). By introducing

$$F(\alpha') = \hat{F}(w'), \quad d\alpha' = \tau_o \sin w' \, dw'$$

this yields

$$\frac{1}{2\pi j}\int_{\gamma_-} \frac{F(\alpha')}{\alpha' - \alpha} d\alpha' = -\frac{1}{2\pi j}\int_{b1-} \frac{\hat{F}(w')}{\cos w' - \cos w} \sin w' \, dw' - \frac{1}{2\pi j}\int_{b2-} \frac{\hat{F}(w')}{\cos w' - \cos w} \sin w' \, dw'$$

The substitution of w' with $-w'$ in the second integral implies that $b_{-2} = -b_{-1}$. Consequently,

$$\frac{1}{2\pi j}\int_{\gamma_-} \frac{F(\alpha')}{\alpha' - \alpha} d\alpha' = -\frac{1}{2\pi j}\int_{b_{1-}} \frac{\hat{F}(w') - \hat{F}(-w')}{\cos w' - \cos w} \sin w' \, dw'$$

By deforming b_{1-} on the imaginary axis $w = j\,\mathrm{Im}[w']$, $\mathrm{Im}[w'] > 0$, the poles, located in a proper region of quadrant 2 between these two lines, do contribute with opposite signs with respect to $-\sum \frac{S_n}{\alpha_n - \alpha}$. By introducing $w' = ju \Rightarrow \sin w' \, dw' = -\sinh u \, du$, one obtains the final results:

$$F_-(\alpha) = \hat{F}_-(w) = -\frac{1}{2\pi j}\int_{\gamma_2} \frac{F(\alpha')}{\alpha' - \alpha} d\alpha' = \frac{1}{2\pi j}\int_0^\infty \frac{\hat{F}(ju) - \hat{F}(-ju)}{\cosh u - \cos w} \sinh u \, du - \sum \frac{S_n}{\alpha_n - \alpha} \tag{115}$$

where the poles to be considered for computation purposes are those located in the half-plane $\mathrm{Im}[\alpha] > 0$ outside the subregion of quadrant 2, between the upper lip of the standard branch line and the half line from $-k$ to $-\infty$ defined by $\alpha = -k\cosh u,\ 0 \le u \le \infty$. Insofar as the w – plane is concerned, the poles to be considered are located in the proper image with $-\pi < \mathrm{Re}[w] < 0$ (Fig. 17).

In the w – plane, eq. (115) is valid if $\mathrm{Re}[w] < 0$. If the integral has been evaluated in closed form, then, according to analytic continuation, this closed form represents the minus function also for $\mathrm{Re}[w] > 0$. Conversely, in the case when a numerical integration has been adopted, the jump from $\mathrm{Re}[w] < 0$ to $\mathrm{Re}[w] > 0$ has to be studied according to the observations developed in section 3.1.

It can be ascertained that $\hat{F}_-(w)$ is a minus function, since it is regular in $w = -\pi(\alpha = \tau_o)$ and $\hat{F}_-(-w-2\pi) = \hat{F}_-(-w)$. In this case also, as in the previously considered one, the singularity at $u = 0$ is *de facto* eliminated by the presence of the function $\sinh u$. However, sometimes, $u = 0$ is a singular point because of the presence of a multiple singularity in the numerator $\hat{F}(ju) - \hat{F}(-ju)$. When this happens, there is a pole at $u = 0$ that cannot be eliminated by the presence of $\sinh u$, and this means that the original integral on b_- does not converge for $w' = 0$. To perform the integral, b_- has to be hooked at $w' = 0$ so that the hook is located in the proper region 2 (Fig. 18).

Example

Factorization of the function $\xi = \sqrt{\tau_o^2 - \left(-\tau_o \cos\left[\frac{\Phi}{\pi}\arccos\left[-\frac{\bar{\eta}}{\tau_o}\right]\right]\right)^2}$.

The factorization of this function is a crucial point for solving diffraction problems involving wedges with aperture angle $2(\pi - \Phi))$ (see chapter 10).

By introducing the $\bar{w}$ – plane, defined by the mapping

$$\bar{\eta} = -\tau_o \cos(\bar{w}) \tag{116}$$

one finds that

$$\xi = -\tau_o \sin\frac{\Phi}{\pi}\bar{w} \tag{117}$$

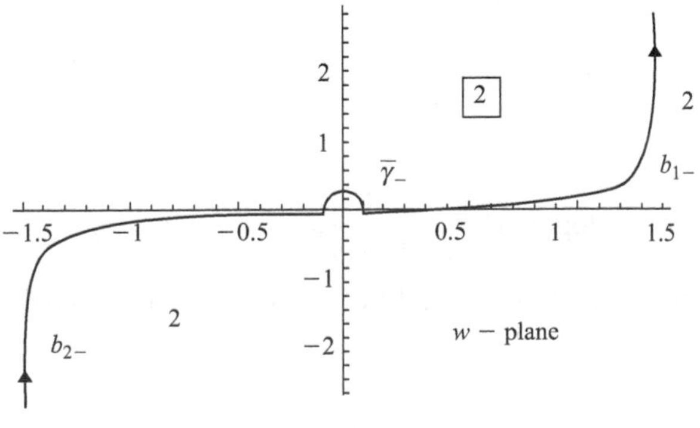

Fig. 18: The line $\bar{\gamma}_-$ obtained by hooking the line b_- at $w' = 0$

The factorization of ξ can be accomplished through the logarithmic decomposition of $g(\overline{\eta})$:

$$g(\overline{\eta}) = \frac{d}{d\overline{\eta}} \log \xi = \frac{d\overline{w}}{d\overline{\eta}} \frac{d}{d\overline{w}} \log \xi = -\frac{d\overline{w}}{d\overline{\eta}} \frac{\Phi}{\pi} \frac{\cos\left(\frac{\Phi}{\pi}\overline{w}\right)}{\sin\left(\frac{\Phi}{\pi}\overline{w}\right)} = -\frac{\Phi}{\pi \tau_o} \frac{\cos\left(\frac{\Phi}{\pi}\overline{w}\right)}{\sin \overline{w} \sin\left(\frac{\Phi}{\pi}\overline{w}\right)} \tag{118}$$

The presence of the double pole at $\overline{w} = 0$ requires the hooking of the line b_- (Fig. 18). The line with the hook will be denoted by $\overline{\gamma}_-$.

Equation (3) provides the minus part of $g(\overline{\eta})$ in the form:

$$g_-(\overline{\eta}) = -\frac{1}{2\pi j} \int_{\gamma_2} \frac{g(\alpha')}{\alpha' - \overline{\eta}} d\alpha' = \frac{1}{2\pi j} \int_{\overline{\gamma}_-} \frac{\hat{g}(\overline{w}') \sin \overline{w}'}{\cos \overline{w}' - \cos \overline{w}} d\overline{w}'$$

$$= \frac{1}{2\pi j} \int_{\overline{\gamma}_-} -\frac{\Phi}{\pi \tau_o} \frac{\cos\left(\frac{\Phi}{\pi}\overline{w}'\right)}{\sin\left(\frac{\Phi}{\pi}\overline{w}'\right)} \frac{1}{\cos \overline{w}' - \cos \overline{w}} d\overline{w}' \tag{119}$$

The contribution of the hooking is given by $-\frac{1}{2}R(0)$, where $R(0)$ is the residue of the integrand at $\overline{w}' = 0$. It follows that

$$g_-(\overline{\eta}) = -\frac{1}{2\pi j} \int_{\overline{\gamma}_1} \frac{\Phi}{\pi \tau_o} \frac{\cos\left(\frac{\Phi}{\pi}\overline{w}'\right)}{\sin\left(\frac{\Phi}{\pi}\overline{w}'\right)} \frac{1}{\cos \overline{w}' - \cos \overline{w}} d\overline{w}'$$

$$= -\frac{1}{2}R(0) + \frac{1}{2\pi j} P.V. \int_{b_-} -\frac{\Phi}{\pi \tau_o} \frac{\cos\left(\frac{\Phi}{\pi}\overline{w}'\right)}{\sin\left(\frac{\Phi}{\pi}\overline{w}'\right)} \frac{1}{\cos \overline{w}' - \cos \overline{w}} d\overline{w}' \tag{120}$$

where P.V. means the Cauchy principal value of the integral.

Since the integrand function is an odd function and the integration path b_- is symmetrical (Fig. 17), the P.V. integral is null, which yields

$$g_-(\overline{\eta}) = -\frac{1}{2}R(0) = \frac{1}{2} \frac{1}{\tau_o} \frac{1}{1 - \cos \overline{w}} = \frac{1}{2} \frac{1}{\tau_o + \overline{\eta}} = \frac{d \log(\xi_-)}{d\overline{\eta}} \tag{121}$$

Consequently,

$$\log(\xi_-) = \int \frac{1}{2} \frac{1}{\tau_o + \overline{\eta}} d\overline{\eta} = \log \sqrt{\tau_o + \overline{\eta}} \tag{122}$$

By taking into account that the factorized functions may be multiplied by a constant factor, we assume

$$\xi_- = \sqrt{\frac{\tau_o + \overline{\eta}}{2}} \tag{123}$$

3.3.4 Use of difference equation for function decomposition

The property of the function $F_+(\alpha)$ in the w – plane allows a decomposition technique based on the use of difference equations. In fact, the decomposition problem

$$F(\alpha) = X_+(\alpha) + Y_+(-\alpha) \tag{124}$$

may be rewritten in the w – plane as

$$\hat{F}(w) = \hat{X}_+(w) + \hat{Y}_+(-w - \pi) = \hat{X}_+(w) + \hat{Y}_+(w + \pi) \tag{125}$$

where $\hat{X}_+(w)$ and $\hat{Y}_+(w)$ are even functions, regular at $w = 0$. By taking into account the properties of these functions, it is possible to eliminate one of the unknowns, for example, by substituting w' with $-w'$ and then subtracting

$$\hat{Y}_+(w + \pi) - \hat{Y}_+(w - \pi) = \hat{F}(w) - \hat{F}(-w) \tag{126}$$

This equation is a difference equation in the unknown $\hat{Y}_+(w)$, and it can be solved with the method described in Osipov and Norris (1999).

3.3.5 The W-H equation as difference equation

In the w – plane, W-H equations may be rewritten in the form

$$\hat{G}(w)\hat{X}_+(w) = \hat{Y}_+(-w - \pi) + \hat{F}_o(w) \tag{127}$$

By using the aforementioned procedure, it is possible to eliminate the unknown $\hat{Y}_+(w)$, thus obtaining the difference equation:

$$\hat{G}(-w - \pi)\hat{X}_+(w + \pi) - \hat{G}(w - \pi)\hat{X}_+(w - \pi) = \hat{F}_o(-w - \pi) - \hat{F}_o(w - \pi) \tag{128}$$

The decomposition-factorization and the solution of difference equation are two different aspects for solving many diffraction problems. However, whereas the decomposition-factorization constitutes a closed mathematical problem, difference equations may involve many solutions; therefore, it is necessary to take into account additional conditions for pinpointing the correct solution.

CHAPTER 4

Exact matrix factorization

4.1 Introduction

The central problem in solving vector Wiener-Hopf equations is the factorization of a $n \times n$ matrix. Even though this problem has been considerably studied in the past, up to now a general method to factorize a $n \times n$ matrix is not known. A discussion of significant advances, achieved in the last few years, appears in Büyükaksoy and Serbest (1993). In this chapter, we outline the most interesting ideas for obtaining explicit matrix Wiener-Hopf factorization.

Before discussing the matrix Wiener-Hopf factorization, it is interesting to mention that this problem has been considered in the more general framework of the vector Riemann-Hilbert problem. An important paper on this approach is the one by Chebotarev (1956), which investigates the conditions under which the vector Riemann problem can be solved by a simple generalization of the formula derived for the solution of the scalar problem. However, these conditions are so restrictive that there seems to be no hope that they could apply to practical cases. Khrapkov (1971) studied the case $n = 2$ in detail, and in particular circumstances he stated explicit factorization formulas. In those years, in Western countries the interest was primarily in matrix factorization relevant to the diffraction by a half-plane with two face impedances. Even though the solution of this problem was obtained by Malyuzhinets in 1951 (Malyuzhinets, 1958b), it is rather surprising that researchers did not succeed in obtaining the related matrix factorization for a long time. In 1975, Rawlins was finally able to solve with an ad hoc technique the particular case constituted by the half-plane with one soft and one hard face. Rawlins's work was substantially improved in 1976, when Hurd introduced a new method (the Wiener-Hopf-Hilbert method), which solves a large class of Wiener-Hopf systems of order two. This method provides also the W-H solution of the diffraction by the two impedances half-plane. A direct factorization formula for the problems where the Hurd method applies was finally obtained in 1978 (Daniele, 1978). Later on, it has been shown that Daniele's method essentially coincides with the results independently obtained by Khrapkov (Luneburg & Hurd, 1984).

In most cases, the Daniele-Khrapkov factorization is nonstandard, due to the singular behavior of the factorized matrices at infinity. Another paper by Daniele (1984a) introduced an approach that overcomes the difficulty that occurs when the regularity conditions at infinity are not satisfied. Generally, this approach yields a classical Jacobi inversion problem (section 4.8.5).

The idea that allows for the direct factorization of a particular class of order two matrices is based on the concept of logarithmic additive decomposition. This was extended by Daniele (1983, 1984b) and Jones (1984) to matrices of arbitrary order. In particular, it is possible to factorize with the logarithmic additive decomposition all the matrices commuting with polynomial matrices (Daniele, 1984b). However, again the factorization formulas involve an essential singularity behavior at infinity, and a cumbersome procedure is necessary to eliminate it (section 4.8.5).

It should be remarked that it may be difficult to recognize whether a given matrix belongs to one of the particular classes of matrices amenable to explicit factorization. In fact, suitable algebraic manipulations may reduce the factorization of a given matrix to the factorization of a matrix having a very different nature. The entries of the matrices may be deeply modified by pre- or post-multiplication with rational matrices. The ability to reduce, when possible, a given matrix to a matrix that can be factorized explicitly requires experience and mathematical skill.

The possibility of obtaining the W-H factorization of a given matrix in closed form remains a challenging and fascinating problem. However, the recent improvements on approximate numerical factorizations (chapters 5 and 6) have overcome this problem from a practical point of view.

4.2 Some possibilities to reduce the order of the kernel matrices

Before discussing in general the methods for solving the matrix factorization problem in closed form, it is interesting to study the possibility of reducing the order of the matrix to be factorized. Even though this study is far from completion, some examples of reduction of order of W-H system can be presented. For instance, if we have matrices of order n such that $G(\alpha)$ or its inverse contains rows constituted by rational elements, we can reduce the order of the matrix with the following procedure.

Let us suppose that the r row of the W-H equations has the form

$$G_{r1}(\alpha)F_{1+}(\alpha) + G_{r2}(\alpha)F_{2+}(\alpha) + \cdots + G_{rn}(\alpha)F_{n+}(\alpha) = F_{r-}(\alpha) \tag{1}$$

where $G_{rs}(\alpha)$ are rational scalars. By separating the plus and minus functions in this equation, we obtain the result (see eq. (15) in chapter 2)

$$P_{r1}(\alpha)F_{1+}(\alpha) + P_{r2}(\alpha)F_{2+}(\alpha) + \cdots + P_{rn}(\alpha)F_{n+}(\alpha) = N(\alpha) \tag{2}$$

where $N(\alpha)$ is a polynomial with a finite number of unknown constants, and $P_{rs}(\alpha)$ are suitable polynomials. From eq. (2) we can eliminate a plus function (e.g., $F_{n+}(\alpha)$), and substituting into the other $n-1$ equations we obtain a W-H system of order $n-1$. We observe that (2) provides a weak representation of $F_{n+}(\alpha)$ (in terms of the other plus functions) that is rational. This means that if there is another row of $G(\alpha)$ with rational elements, we can repeat the previous procedure and eliminate another unknown function (say $F_{(n-1)+}(\alpha)$). We can conclude that the factorization of a matrix having only a row with non-rational coefficients can be reduced to the weak factorization of a scalar. For example, in this manner we can obtain the factorization of a matrix of order two that contains three rational entries.

The reduction of the order holds for matrices having rational eigenvectors. In fact, the Jordan decomposition of $G(\alpha)$ yields

$$G(\alpha) = S(\alpha)J(\alpha)S^{-1}(\alpha) \tag{3}$$

where $S(\alpha)$ is the eigenvectors matrix and $J(\alpha)$ is a Jordan matrix formed by Jordan blocks on the main diagonal constituted by the eigenvalues of $G(\alpha)$ (Gohberg, Lancaster & Rodman, 1982). The rationality of $S(\alpha)$ reduces the problem to the factorization of $J(\alpha)$ (section 4.7). In general, this can be accomplished easily by using the factorization technique for triangular matrices (section 4.3).

If the matrix $S(\alpha)$ were nonrational, there remains some hope of accomplishing a reduction if some eigenvalues present in $J(\alpha)$ are rational. In fact, in the presence of rational eigenvalues we can rewrite the W-H system $G(\alpha) \cdot F_+(\alpha) = F_-(\alpha)$ in the form

$$J(\alpha)S^{-1}(\alpha)F_+(\alpha) = S^{-1}(\alpha)F_-(\alpha) \tag{4}$$

It is possible that one or more scalar W-H equations involved in (4) present the form (2).

Another matrix form that allows us to reduce the order of the factorization of matrices is

$$G(\alpha) = A(\alpha) \otimes 1_m + B(\alpha) \otimes \Gamma(\alpha) \tag{5}$$

where the symbol $\otimes$ means the Kronecker product, $A(\alpha)$ and $B(\alpha)$ are matrices of order n, and 1_m and $\Gamma(\alpha) = diag[\Gamma_i(\alpha)]$ are the identity matrix and an arbitrary diagonal matrix of order m, respectively.

Since given the Kronecker product $C \otimes D$, there exist permutation matrices P and Q such that

$$C \otimes D = P(D \otimes C)Q \tag{6}$$

We obtain

$$G(\alpha) = P \cdot W(\alpha) \cdot Q \tag{7}$$

where

$$W(\alpha) = 1_m \otimes A(\alpha) + \Gamma(\alpha) \otimes B(\alpha) \tag{8}$$

Consequently, since the permutation matrices are constants (their entries are 0 or 1), we reduce the problem to the factorization of $W(\alpha) = 1_m \otimes A(\alpha) + \Gamma(\alpha) \otimes B(\alpha)$, taking into account that $\Gamma(\alpha) = diag[\Gamma_i(\alpha)]$ is diagonal. In extended form we have

$$W(\alpha) = \begin{pmatrix} A(\alpha) + \Gamma_1 B(\alpha) & 0 & 0 \\ 0 & \ddots & 0 \\ 0 & 0 & A(\alpha) + \Gamma_m B(\alpha) \end{pmatrix} \tag{9}$$

In this way, the factorization of $G(\alpha)$ of order $n \cdot m$ is reduced to the factorization of m matrices of order n:

$$W_i(\alpha) = A(\alpha) + \Gamma_i(\alpha)B(\alpha) \tag{10}$$

Sometimes $\Gamma(\alpha)$ is not diagonal but can be cast in the form

$$\Gamma(\alpha) = t(\alpha) \cdot \gamma(\alpha) \cdot t^{-1}(\alpha) \tag{11}$$

where $t(\alpha)$ is rational, and $\gamma(\alpha)$ is diagonal. Taking into account the property

$$[A_1(\alpha) \cdot A_2(\alpha)] \otimes [B_1(\alpha) \cdot B_2(\alpha)] = [A_1(\alpha) \otimes B_1(\alpha)] \cdot [A_2(\alpha) \otimes B_2(\alpha)] \tag{12}$$

and premultiplying by $1_n \otimes t^{-1}(\alpha)$ and postmultiplying by $1_n \otimes t(\alpha)$, we obtain

$$1_n \otimes t^{-1}(\alpha) \cdot G(\alpha) \cdot 1_n \otimes t(\alpha) = A(\alpha) \otimes 1_m + B(\alpha) \otimes \gamma(\alpha) \tag{13}$$

The problem is reduced to the factorization of a matrix where $\Gamma(\alpha)$ is substituted by the diagonal matrix $\gamma(\alpha)$.

In simpler cases we have to factorize:

$$G(\alpha) = B(\alpha) \otimes \Gamma(\alpha) \tag{14}$$

Taking into account (12) and the factorization of $B(\alpha) = B_-(\alpha) \cdot B_+(\alpha)$ and of $\Gamma(\alpha) = \Gamma_-(\alpha) \cdot \Gamma_+(\alpha)$, one finds

$$G(\alpha) = G_-(\alpha) \cdot G_+(\alpha) \tag{15}$$

with

$$G_-(\alpha) = B_-(\alpha) \otimes \Gamma_-(\alpha), \quad G_+(\alpha) = B_+(\alpha) \otimes \Gamma_+(\alpha) \tag{16}$$

4.3 Factorization of triangular matrices

A class of matrix kernels that we are able to explicitly factorize is the triangular matrices one. A matrix $G(\alpha)$ is a triangular matrix if its entries $g_{ij}(\alpha)$ satisfy the following conditions:

$$g_{ij}(\alpha) = 0 \quad \text{for} \quad i < j \text{ lower triangular matrices} \tag{17}$$

or

$$g_{ij}(\alpha) = 0 \quad \text{for} \quad i > j \text{ upper triangular matrices.} \tag{18}$$

The triangular matrices can be factorized with the following method that, without any restriction and for the sake of simplicity, will be detailed for upper triangular matrices only:

$$G(\alpha) = \begin{vmatrix} g_{11} & g_{12} & \cdots & \cdots & g_{1n} \\ 0 & g_{22} & g_{23} & \cdot\cdot & g_{2n} \\ 0 & 0 & g_{33} & \cdots & g_{3n} \\ \cdot\cdot & \cdot\cdot & \cdots & \cdot\cdot & \cdots \\ 0 & 0 & 0 & 0 & g_{nn} \end{vmatrix} \tag{19}$$

The simplest factorization of this class is the one of the order-two matrix $\begin{vmatrix} 1 & g(\alpha) \\ 0 & 1 \end{vmatrix}$. The additive decomposition of $g(\alpha) = g_-(\alpha) + g_+(\alpha)$ yields the explicit factorization:

$$\begin{vmatrix} 1 & g(\alpha) \\ 0 & 1 \end{vmatrix} = \begin{vmatrix} 1 & g_-(\alpha) \\ 0 & 1 \end{vmatrix} \begin{vmatrix} 1 & g_+(\alpha) \\ 0 & 1 \end{vmatrix} \tag{20}$$

This is remarkable since the factorization problem is reduced to an additive decomposition that can always be accomplished.

To factorize matrices of order two in the general case, it is very useful to employ

$$G(\alpha)=\begin{vmatrix} g_{11} & g_{12} \\ 0 & g_{22} \end{vmatrix}=\begin{vmatrix} g_{11-} & 0 \\ 0 & g_{22-} \end{vmatrix}\begin{vmatrix} 1 & (g_{11-})^{-1}g_{12}(g_{22+})^{-1} \\ 0 & 1 \end{vmatrix}\begin{vmatrix} g_{11+} & 0 \\ 0 & g_{22+} \end{vmatrix} \tag{21}$$

where the minus and plus functions follow from the factorization of the scalars g_{11} and g_{22}:

$$g_{11}=g_{11-}g_{11+},\quad g_{22}=g_{22-}g_{22+} \tag{22}$$

Applying (20) to the middle matrix of the rightmost member of (21) yields the factorization

$$G(\alpha)=G_-(\alpha)G_+(\alpha)=\begin{vmatrix} g_{11} & g_{12} \\ 0 & g_{22} \end{vmatrix}=\begin{vmatrix} g_{11-} & 0 \\ 0 & g_{22-} \end{vmatrix}\begin{vmatrix} 1 & g_{e-} \\ 0 & 1 \end{vmatrix}\begin{vmatrix} 1 & g_{e+} \\ 0 & 1 \end{vmatrix}\begin{vmatrix} g_{11+} & 0 \\ 0 & g_{22+} \end{vmatrix} \tag{23}$$

where

$$G_-(\alpha)=\begin{vmatrix} g_{11-} & 0 \\ 0 & g_{22-} \end{vmatrix}\begin{vmatrix} 1 & g_{e-} \\ 0 & 1 \end{vmatrix},\quad G_+(\alpha)=\begin{vmatrix} 1 & g_{e+} \\ 0 & 1 \end{vmatrix}\begin{vmatrix} g_{11+} & 0 \\ 0 & g_{22+} \end{vmatrix} \tag{24}$$

and g_{e-} and g_{e+} follow by the additive decomposition of

$$g_e=(g_{11-})^{-1}g_{12}(g_{22+})^{-1}=g_{e-}+g_{e+} \tag{25}$$

For arbitrary matrices of order $n=k+m$, a given triangular matrix $G(\alpha)$ can be written in the form

$$G(\alpha)=\begin{vmatrix} (g_{11})_{k,k} & (g_{12})_{k,m} \\ (0)_{m,k} & (g_{22})_{m,m} \end{vmatrix}$$

where the square submatrices $(g_{11})_{k,k}$ and $(g_{22})_{m,m}$, respectively, of order k and m are upper triangular matrices. Equation (26) reduces the factorization of $G(\alpha)$ to the factorization of matrices $(g_{11})_{k,k}$ and $(g_{22})_{m,m}$ having inferior order:

$$G(\alpha)=\begin{vmatrix} (g_{11})_{k,k} & (g_{12})_{k,m} \\ (0)_{m,k} & (g_{22})_{m,m} \end{vmatrix}$$

$$=\begin{vmatrix} (g_{11-})_{k,k} & (0)_{k,m} \\ (0)_{m,k} & (g_{22-})_{m,m} \end{vmatrix}\begin{vmatrix} (1)_{k,k} & (g_{11-})_{k,k}^{-1}(g_{12})_{k,m}(g_{22+})_{m,m}^{-1} \\ (0)_{m,k} & (1)_{m,m} \end{vmatrix}\begin{vmatrix} (g_{11+})_{k,k} & (0)_{k,m} \\ (0)_{m,k} & (g_{22+})_{m,m} \end{vmatrix} \tag{26}$$

where $(g_{11})_{k,k}=(g_{11-})_{k,k}\cdot(g_{11-})_{k,k}$ and $(g_{22})_{m,m}=(g_{22-})_{m,m}\cdot(g_{22+})_{m,m}$. In fact, the factorization of the middle matrix of the rightmost member follows from the decomposition of the matrix:

$$(g_e)_{k,m}=(g_{11-})_{k,k}^{-1}(g_{12})_{k,m}(g_{22+})_{m,m}^{-1}=(g_{e-})_{k,m}+(g_{e+})_{k,m} \tag{27}$$

This yields the factorization $G(\alpha) = G_-(\alpha)G_+(\alpha)$:

$$
\begin{aligned}
G_-(\alpha) &= \begin{vmatrix} (g_{11-})_{k,k} & (0)_{k,m} \\ (0)_{m,k} & (g_{22-})_{m,m} \end{vmatrix} \begin{vmatrix} (1)_{k,k} & (g_{e-})_{k,m} \\ (0)_{m,k} & (1)_{m,m} \end{vmatrix} \\
G_+(\alpha) &= \begin{vmatrix} (1)_{k,k} & (g_{e+})_{k,m} \\ (0)_{m,k} & (1)_{m,m} \end{vmatrix} \begin{vmatrix} (g_{11+})_{k,k} & (0)_{k,m} \\ (0)_{m,k} & (g_{22+})_{m,m} \end{vmatrix}
\end{aligned}
\tag{28}
$$

4.4 Factorization of rational matrices

4.4.1 Introduction

The factorization of rational matrices can always be accomplished explicitly (Bart, Goheberg & Kaashoek, 1979). The rational matrices are very important in many electrical engineering problems. For instance, their factorization provides a tool for the solution of the optimal filtering problem (Wiener, 1949) and the impedance synthesis of n-port networks (Newcomb, 1966). Unfortunately, these matrices do not occur in diffraction theory. However, Padé representation allows to approximate an arbitrary function with a rational function, so approximate factorizations of the matrix kernels involved in electromagnetic problems can be accomplished by introducing suitable Padé approximants (Abrahams, 2000) or, more generally, rational approximants of suitable entries of the matrix (see chapter 6).

Every rational matrix $R(\alpha)$ may be rewritten in the form

$$R(\alpha) = \frac{P(\alpha)}{d(\alpha)} \tag{29}$$

where $d(\alpha)$ is a scalar polynomial, and $P(\alpha)$ is a matrix polynomial of the type

$$P(\alpha) = \sum_{i=0}^{l} A_i\, \alpha^i \tag{30}$$

and A_i are $n \times n$ matrices whose entries are constant complex numbers.

Since the factorization of the scalar polynomial $d(\alpha) = d_-(\alpha)d_+(\alpha)$ is straightforward, eq. (29) reduces the factorization of a rational matrix $R(\alpha)$ to the factorization of a matrix polynomial $P(\alpha)$. The fundamental result that allows the factorization of $P(\alpha)$ is the Smith representation (Gohberg, Lancaster & Rodman, 1982) of the polynomial matrices:

$$P(\alpha) = E(\alpha)D(\alpha)F(\alpha) \tag{31}$$

In this representation, $E(\alpha)$ and $F(\alpha)$ are matrix polynomials of order n with constant non-zero determinant, and the matrix $D(\alpha)$ is diagonal:

$$D(\alpha) = Diag[d_r(\alpha)], \quad r = 1, 2, \ldots, n \tag{32}$$

where all the scalar polynomials $d_r(\alpha)$ are divisible by $d_{r-1}(\alpha)$.

The factorization of $D(\alpha)$ is given by

$$D(\alpha) = D_-(\alpha)D_+(\alpha) \tag{33}$$

where, since $d_r(\alpha) = d_{r-}(\alpha)d_{+r}(\alpha)$,

$$D_-(\alpha) = Diag[d_{r-}(\alpha)], \quad r = 1,2,\ldots,n, \quad D_+(\alpha) = Diag[d_{r+}(\alpha)], \quad r = 1,2,\ldots,n \quad (34)$$

The factorization of $P(\alpha)$ follows in the form

$$P(\alpha) = P_-(\alpha)P_+(\alpha) \quad (35)$$

where

$$P_-(\alpha) = E(\alpha)D_-(\alpha) \quad P_+(\alpha) = D_+(\alpha)F(\alpha) \quad (36)$$

By taking into account the property of the determinant of $E(\alpha)$ and $F(\alpha)$, $E^{-1}(\alpha)$ and $F^{-1}(\alpha)$ also are polynomial matrices. Thus, the factorization of the rational matrix $R(\alpha) = R_-(\alpha)R_+(\alpha)$ is given by

$$R_-(\alpha) = \frac{P_-(\alpha)}{d_-(\alpha)}, \quad R_-^{-1}(\alpha) = d_-(\alpha)D_-^{-1}(\alpha)E^{-1}(\alpha) \quad (37)$$

$$R_+(\alpha) = \frac{P_+(\alpha)}{d_+(\alpha)}, \quad R_+^{-1}(\alpha) = d_+(\alpha)F^{-1}(\alpha)D_+^{-1}(\alpha) \quad (38)$$

Even though the Smith representation (31) has a very deep conceptual importance, it is not straightforward. In practical cases, it is better to accomplish the factorization of rational matrices using different procedures. In the following, for the sake of simplicity it will be assumed that the involved poles are simple; however, the presence of multiple poles requires only a slight modification of the presented procedure.

4.4.2 Matching of the singularities

There are different techniques to factorize rational matrices. For instance, those based on the realization theory are very general (Bart, Goheberg & Kaashoek, 1979). In this book, we will not consider these powerful techniques but will limit our considerations to elementary methods.

In general, the factorization of rational matrices can be accomplished by using the ideas of weak factorization introduced in chapter 2, section 2.7. However, if only simple poles are involved, the following technique provides a simpler method.

Let us consider a rational matrix $G(\alpha)$ and its inverse $G^{-1}(\alpha)$ in the form

$$G(\alpha) = \frac{A(\alpha)}{d(\alpha)} \quad (39)$$

$$G^{-1}(\alpha) = \frac{B(\alpha)}{\delta(\alpha)} \quad (40)$$

where $A(\alpha)$ and $B(\alpha)$ are polynomial matrices of order n, and $d(\alpha)$ and $\delta(\alpha)$ are scalars. In the following we will assume

$$\frac{G(\alpha)}{\alpha} \to 0 \quad \text{and} \quad \frac{G^{-1}(\alpha)}{\alpha} \to 0 \quad \text{as } \alpha \to \infty \quad (41)$$

As indicated in chapter 2, section 2.6, the homogeneous solutions of the W-H equation

$$\frac{A(\alpha)}{d(\alpha)} X_{j+}(\alpha) = X_{j-}(\alpha), \quad j = 1, 2, \ldots, n \tag{42}$$

where $\frac{X_{j+,-}(\alpha)}{\alpha} \to 0$, as $\alpha \to \infty$, lead to evaluate factorized matrices in the form:

$$G_-(\alpha) = |X_{1-}(\alpha), X_{2-}(\alpha), \ldots, X_{n-}(\alpha)| \tag{43}$$

$$G_+(\alpha) = |X_{1+}(\alpha), X_{2+}(\alpha), \ldots, X_{n+}(\alpha)|^{-1} \tag{44}$$

Let us introduce the functions:

$$F_{j+}(\alpha) = \frac{X_{j+}(\alpha)}{\alpha - \alpha_p} \tag{45}$$

where α_p has negative imaginary part ($\text{Im}[\alpha_p] < 0$). In the following we will assume that α_p do not belong to the null space of $G(\alpha_p)$ (this happens if $\det[G(\alpha_p)] = 0$ and yields $X_{j-}(\alpha_p) = 0$). Also, we will assume that α_p do not belong to the null space of $G^{-1}(\alpha_p)$ (it happens if $\det[G^{-1}(\alpha_p)] = 0$) and yields $X_{j+}(\alpha_p) = 0$. We did not study the problem of the better choice of α_p. By changing the value of α_p, it has be shown (section 2.6) that the factorized matrices $G_+(\alpha)$ and $G_-(\alpha)$ differ only by a constant matrix.

The homogeneous equation

$$\frac{P(\alpha)}{d(\alpha)} F_{j+}(\alpha) = F_{j-}(\alpha), \quad j = 1, 2, \ldots, n \tag{46}$$

can be rewritten in the form

$$\frac{P(\alpha)}{d(\alpha)} F_{j+}(\alpha) = \frac{X_{j-}(\alpha) - X_{j-}(\alpha_p)}{\alpha - \alpha_p} + \frac{X_{j-}(\alpha_p)}{\alpha - \alpha_p} = F_{j-}(\alpha), \quad j = 1, 2, \ldots, n$$

or

$$\frac{P(\alpha)}{d(\alpha)} F_{j+}(\alpha) = F^s_{j-}(\alpha) + \frac{E_j}{\alpha - \alpha_p} = F_{j-}(\alpha), \quad j = 1, 2, \ldots, n \tag{47}$$

where $E_j = X_{j-}(\alpha_p)$, and the plus $F_{j+}(\alpha)$ and the minus $F^s_{j-}(\alpha) = \frac{X_{j-}(\alpha) - X_{j-}(\alpha_p)}{\alpha - \alpha_p}$ are standard Laplace transforms. In the following we set

$$E_1 = [1, 0, 0, 0, \ldots]^t, \quad E_2 = [0, 1, 0, 0, \ldots]^t, \ldots E_n = [0, 0, 0, 0, \ldots, 1]^t$$

This provides n independent solutions of

$$\frac{P(\alpha)}{d(\alpha)} F_{j+}(\alpha) = F^s_{j-}(\alpha) + \frac{E_j}{\alpha - \alpha_p}, \quad j = 1, 2, \ldots, n \tag{48}$$

For the sake of simplicity, from here onward we will suppress the subscript j and will indicate with

$$\alpha_i \quad i = 1, 2, \ldots, r \text{ zeroes of } d[\alpha] \text{ having } \mathrm{Im}[\alpha_i] > 0 \text{ (minus zeroes)}$$

$$\gamma_i \quad i = 1, 2, \ldots, s \text{ zeroes of } \delta[\alpha] \text{ having } \mathrm{Im}[\alpha_i] < 0 \text{ (plus zeroes)}$$

By taking into account (48), and that the zeroes α_i $i = 1, 2, \ldots, r$ of $d[\alpha]$ that have $\mathrm{Im}[\alpha_i] > 0$ induce poles in $F_-(\alpha)$, one obtains the representation:

$$F_-(\alpha) = \frac{E}{\alpha - \alpha_p} + \sum_i^r \frac{R(\alpha_i)}{\alpha - \alpha_i} \tag{49}$$

Similarly from

$$F_+(\alpha) = \frac{B(\alpha)}{\delta(\alpha)} F_-(\alpha) \tag{50}$$

one obtains the representation

$$F_+(\alpha) = G^{-1}(\alpha_p) \frac{E}{\alpha - \alpha_p} + \sum_i^s \frac{T(\gamma_i)}{\alpha - \gamma_i} \tag{51}$$

The representations (49) and (51) introduce the $r + s$ unknown vectors $R(\alpha_i)$ and $T(\gamma_i)$. Taking into account that they have dimension n, the unknown scalars are $n(r + s)$. We provide $n(r + s)$ equations by evaluating the residues in α_i $(i = 1, 2, \ldots, r)$ in

$$\frac{P(\alpha)}{d(\alpha)} F_+(\alpha) = F_-(\alpha) \tag{52}$$

thus obtaining the $r \cdot n$ scalar equations

$$\mathrm{Residue}[F_-(\alpha)]_{\alpha=\alpha_i} = R(\alpha_i) = \frac{P(\alpha_i)}{d'(\alpha_i)} F_+(\alpha_i) = \quad i = 1, 2, \ldots, r \tag{53}$$

Similarly, from

$$F_+(\alpha) = \frac{B(\alpha)}{\Delta(\alpha)} F_-(\alpha) \tag{54}$$

the $s \cdot n$ scalar equations follow:

$$\mathrm{Residue}[F_+(\alpha)]_{\alpha=\gamma_i} = T(\gamma_i) = \frac{B(\gamma_i)}{\delta'(\gamma_i)} F_-(\gamma_i) = \quad i = 1, 2, \ldots, s \tag{55}$$

By substituting the representations (49) and (51), eqs. (53) and (55) provide the $n(r + s)$ scalar equations that yield the evaluation of the residues $R(\alpha_i)$ $i = 1, 2, \ldots, r$ and $T(\gamma_i)$ $i = 1, 2, \ldots, s$.

In the previous equations, the poles we considered are simple. A slight modification of this procedure allows us to deal with the case where multiple poles are involved.

4.4.2.1 An example of matrix of order three

Let us consider the factorization of the rational matrix:

$$G(\alpha) = \begin{pmatrix} \frac{j}{\alpha^2+1} & 2 & \frac{\alpha^2+4}{\alpha^2+1} \\ 1 & 2 & \frac{\alpha^2+9}{\alpha^2+1} \\ \frac{\alpha^2}{\alpha^2+1} & \frac{1}{\alpha^2+1} & 2 \end{pmatrix} \tag{56}$$

that has as inverse

$$G^{-1}(\alpha) =$$

$$\begin{pmatrix} -\frac{(\alpha^2+1)(4\alpha^4+7\alpha^2-5)}{4\alpha^6+(1-4j)\alpha^4-(3+7j)\alpha^2+5j} & \frac{\alpha^2(\alpha^2+1)(4\alpha^4+7)}{4\alpha^6+(1-4j)\alpha^4-(3+7j)\alpha^2+5j} & -\frac{10(\alpha^2+1)^2}{4\alpha^6+(1-4j)\alpha^4-(3+7j)\alpha^2+5j} \\ \frac{(\alpha^2+1)(\alpha^4-5\alpha^2+2)}{4\alpha^6+(1-4j)\alpha^4-(3+7j)\alpha^2+5j} & \frac{(\alpha^2+1)(\alpha^4+(4-2j)\alpha^2-2j)}{4\alpha^6+(1-4j)\alpha^4-(3+7j)\alpha^2+5j} & -\frac{(\alpha^2+1)(\alpha^4+(5-j)\alpha^2+4-9j)}{4\alpha^6+(1-4j)\alpha^4-(3+7j)\alpha^2+5j} \\ \frac{2\alpha^6+3\alpha^4-1}{4\alpha^6+(1-4j)\alpha^4-(3+7j)\alpha^2+5j} & -\frac{(\alpha^2+1)(2\alpha^4+2\alpha^2-j)}{4\alpha^6+(1-4j)\alpha^4-(3+7j)\alpha^2+5j} & \frac{2(\alpha^2+1)^2(\alpha^2+1-j)}{4\alpha^6+(1-4j)\alpha^4-(3+7j)\alpha^2+5j} \end{pmatrix} \tag{57}$$

The exact factorization of this matrix was accomplished with the method indicated in the previous section by assuming $\alpha_p = -2j$. This procedure involves cumbersome algebraic manipulations that required the use of the computing software MATHEMATICA. The exact plus factorized matrix $G_+(\alpha)$ is reported in the following expression:

$$G_+(\alpha) = \frac{1}{d_+} \begin{vmatrix} n_{11+} & n_{12+} & n_{13+} \\ n_{21+} & n_{22+} & n_{23+} \\ n_{31+} & n_{32+} & n_{33+} \end{vmatrix}$$

where the elements of the matrix are given by

$$\begin{aligned} d_+ = {} & ((-0.184364 + 1.20287\, i) + \alpha)\, ((-0.184363 + 1.20287\, i) + \alpha) \\ & (0.999955\, i + \alpha)(0.999985\, i + \alpha)(1.00006\, i + \alpha)((0.746591 + 0.040993\, i) + \alpha) \\ & ((0.746591 + 0.0409929\, i) + \alpha)((1.02883 + 0.671793\, i) + \alpha)((1.02883 + 0.671794\, i) + \alpha) \end{aligned}$$

$$\begin{aligned} n_{11+} = {} & ((-0.302761 + 0.0427102\, i)((1.20287 + 0.184364\, i) - i\,\alpha)((-0.184364 + 1.20287\, i) + \alpha) \\ & ((-0.152283 + 3.0795\, i) + \alpha)(1.\, i + \alpha)(1.\, i + \alpha)((0.746591 + 0.0409929\, i) + \alpha) \\ & ((0.746591 + 0.0409929\, i) + \alpha)((1.02883 + 0.671793\, i) + \alpha)((1.02883 + 0.671793\, i) + \alpha) \end{aligned}$$

$$\begin{aligned} n_{12-} = {} & (1.96889 - 0.212392\, i)((-0.184364 + 1.20287\, i) + \alpha)((-0.184364 + 1.20287\, i) + \alpha) \\ & (1.\, i + \alpha)(1.\, i + \alpha)((0.108318 + 0.995883\, i) + \alpha)((0.746591 + 0.0409929\, i) + \alpha) \\ & ((0.746591 + 0.0409929\, i) + \alpha)((1.02883 + 0.671793\, i) + \alpha)((1.02883 + 0.671793\, i) + \alpha) \end{aligned}$$

$$\begin{aligned} n_{13+} = {} & (1.73121 + 0.570419\, i)((-0.184364 + 1.20287\, i) + \alpha) \\ & ((-0.184364 + 1.20287\, i) + \alpha)(1.\, i + \alpha)(1.\, i + \alpha)(2.\, i + \alpha)((0.746591 + 0.0409929\, i) + \alpha) \\ & ((0.746591 + 0.0409929\, i) + \alpha)((1.02883 + 0.671793\, i) + \alpha)((1.02883 + 0.671793\, i) + \alpha) \end{aligned}$$

$$n_{21+} = (1.0738 + 0.26881\,i)((-1.219383 + 1.12365\,i) + \alpha)((-0.184364 + 1.20287\,i) + \alpha)$$
$$((-0.184364 + 1.20287\,i) + \alpha)(1.\,i + \alpha)(1.\,i + \alpha)((0.746591 + 0.0409929\,i) + \alpha)$$
$$((0.746591 + 0.0409929\,i) + \alpha)((1.02883 + 0.671793\,i) + \alpha)((1.02883 + 0.671793\,i) + \alpha)$$

$$n_{22+} = (2.03862 - 0.351043\,i)((-0.184364 + 1.20287\,i) + \alpha)((-0.184364 + 1.20287\,i) + \alpha)$$
$$(1.\,i + \alpha)(1.\,i + \alpha)((0.16407 + 1.0472\,i) + \alpha)((0.746591 + 0.0409929\,i) + \alpha)$$
$$((0.746591 + 0.0409929\,i) + \alpha)((1.02883 + 0.671793\,i) + \alpha)((1.02883 + 0.671793\,i) + \alpha)$$

$$n_{23+} = (2.27122 + 1.25689\,i)((-0.184364 + 1.20287\,i) + \alpha)((-0.184364 + 1.20287\,i) + \alpha)$$
$$(1.\,i + \alpha)(1.\,i + \alpha)((0.310885 + 2.56178\,i) + \alpha)((0.746591 + 0.0409929\,i) + \alpha)$$
$$((0.746591 + 0.0409929\,i) + \alpha)((1.02883 + 0.671793\,i) + \alpha)((1.02883 + 0.671793\,i) + \alpha)$$

$$n_{31+} = (0.806631 - 0.0701351\,i)((-0.184364 + 1.20287\,i) + \alpha)((-0.184364 + 1.20287\,i) + \alpha)$$
$$(1.\,i + \alpha)(1.\,i + \alpha)((0.142644 + 0.359436\,i) + \alpha)((0.746591 + 0.0409929\,i) + \alpha)$$
$$((0.746591 + 0.0409929\,i) + \alpha)((1.02883 + 0.671793\,i) + \alpha)((1.02883 + 0.671793\,i) + \alpha)$$

$$n_{32+} = (0.22622 + 0.403998\,i)((-0.184364 + 1.20287\,i) + \alpha)((-0.184364 + 1.20287\,i) + \alpha)$$
$$(1.\,i + \alpha)(1.\,i + \alpha)((0.255013 + 3.42795\,i) + \alpha)((0.746591 + 0.0409929\,i) + \alpha)$$
$$((0.746591 + 0.0409929\,i) + \alpha)((1.02883 + 0.671793\,i) + \alpha)((1.02883 + 0.671793\,i) + \alpha)$$

$$n_{33+} = (2.27122 + 1.25689\,i)((-0.184364 + 1.20287\,i) + \alpha)((-0.184364 + 1.20287\,i) + \alpha)$$
$$(1.\,i + \alpha)(1.\,i + \alpha)((0.310885 + 2.56178\,i) + \alpha)((0.746591 + 0.0409929\,i) + \alpha)$$
$$((0.746591 + 0.0409929\,i) + \alpha)((1.02883 + 0.671793\,i) + \alpha)((1.02883 + 0.671793\,i) + \alpha)$$

where the symbol i represents the imaginary unit j.

It has been verified with MATHEMATICA that the matrix $G_-(\alpha) = G(\alpha) \cdot G_+^{-1}(\alpha)$ is a minus factorized matrix.

4.4.3 The factorization in the framework of the Fredholm equations

The formulation of W-H equations as integral equations points out also an important class of matrix kernels $G(\alpha)$ yielding closed form solutions. In addition to the rational matrices, these are the quasi-rational matrices. They are defined as the meromorphic matrices that, with their inverses, involve a finite number of poles.

Consider the Fredholm eq. (1.54), and let us assume the presence of only simple poles in $G(\alpha)$; let us denote with β_j those poles located in the $\mathrm{Im}[\alpha] > 0$ half-plane. The residue theorem yields

$$\frac{1}{2\pi j}\int_{-\infty}^{\infty} \frac{[G^{-1}(\alpha)G(u) - 1]F_+(u)}{u - \alpha}\,du = \sum_j \frac{G^{-1}(\alpha)R_j F_+(\beta_j)}{\beta_j - \alpha}$$

where R_j is the residue of $G(u)$ in β_j. Consequently, eq. (1.54) can be rewritten in the form

$$F_+(\alpha) = -G^{-1}(\alpha)\left[\sum_j \frac{R_j F_+(\beta_j)}{\beta_j - \alpha} - F_{o+}(\alpha)\right] \tag{58}$$

The poles α_i of $G^{-1}(\alpha)$ that are located in the half-plane $\text{Im}[\alpha] > 0$ are offending. It is required that the vectors $\sum_j \frac{R_j F_+(\beta_j)}{\beta_j - \alpha_i} - F_{o+}(\alpha_i)$ be in the null space of the matrices $W(\alpha_i)$ defined by

$$W(\alpha) = \prod_i (\alpha - \alpha_i) G^{-1}(\alpha)$$

This allows us to obtain the unknowns $F_+(\beta_j)$ that provide, through eq. (58), a full representation of $F_+(\alpha)$.

4.5 Techniques for solving the factorization problem

4.5.1 The logarithmic decomposition

As it happens for the scalar case, the most powerful method for factorizing matrices $G(\alpha)$ is the one relying on the concept of logarithmic decomposition (Heins, 1950a) consisting in the additive decomposition of $\log[G(\alpha)]$ (chapter 3, section 3.2.1):

$$\log[G(\alpha)] = \psi_-(\alpha) + \psi_+(\alpha) \tag{59}$$

The factorization of $G(\alpha) = G_-(\alpha)G_+(\alpha)$ is given by

$$G_-(\alpha) = \exp[\psi_-(\alpha)], \quad G_+(\alpha) = \exp[\psi_+(\alpha)] \tag{60}$$

It is fundamental to observe that the application of eqs. (60) requires that the matrices $\psi_-(\alpha)$ and $\psi_+(\alpha)$ commute. This always occurs for scalar kernels and it is the reason that allows for the general solution of the scalar W-H equation in closed form.

4.5.1.1 Use of the logarithmic decomposition for rational matrices

In this example, we show that rational matrices can be factorized using the logarithmic decomposition. This procedure is very cumbersome but also important since it can be extended to arbitrary nonrational matrices that commute with rational matrices (section 4.9).

To show that in the case of rational matrices the logarithmic decomposition yields two commutative matrices $\psi_-(\alpha)$ and $\psi_+(\alpha)$, let us consider Cayley's theorem:

$$\log\left[\frac{P(\alpha)}{d(\alpha)}\right] = \psi_o(\alpha)1 + \psi_1(\alpha)P(\alpha) + \cdots + \psi_{n-1}(\alpha)P^{n-1}(\alpha) \tag{61}$$

where n is the order of the rational matrix $R(\alpha) = \frac{P(\alpha)}{d(\alpha)}$. If $R(\alpha)$ has distinct eigenvalues $\lambda_i(\alpha)$, the functions $\psi_i(\alpha)$ may be obtained by the Sylvester formula (Pease, 1965, p. 156):

$$\log[R(\alpha)] = \sum_{i=1}^{n} \log[\lambda_i] \prod_{j \neq i}^{n} \frac{R(\alpha) - \lambda_j 1}{\lambda_i - \lambda_j} \tag{62}$$

The decomposition of

$$\psi_i(\alpha) = \psi_{i-}(\alpha) + \psi_{i+}(\alpha) \quad i = 0, 1, \ldots, n-1 \tag{63}$$

yields the following factorization:

$$G(\alpha) = R(\alpha) = \hat{G}_-(\alpha) \cdot \hat{G}_+(\alpha) \tag{64}$$

where

$$\hat{G}_-(\alpha)] = \exp\left[\psi_{o-}(\alpha)1 + \psi_{1-}(\alpha)P(\alpha) + \cdots + \psi_{(n-1)-}(\alpha)P^{n-1}(\alpha)\right] \tag{65}$$

$$\hat{G}_+(\alpha)] = \exp\left[\psi_{o+}(\alpha)1 + \psi_{1+}(\alpha)P(\alpha) + \cdots + \psi_{(n-1)+}(\alpha)P^{n-1}(\alpha)\right] \tag{66}$$

Since $P(\alpha)$ is a polynomial in the variable α, these factorized matrices are regular together with their inverses in the half-planes $\mathrm{Im}[\alpha] \leq 0$ and $\mathrm{Im}[\alpha] \geq 0$, respectively; furthermore, they commute since they are both functions of the same matrix $P(\alpha)$.

However, the previous factorization is in general nonstandard. We cannot use these factorized matrices when they have nonalgebraic behavior as $\alpha \to \infty$. To overcome this difficulty we observe that factorized matrices differ only for the presence of entire factors, so that the factorized matrices $G_-(\alpha)$ and $G_+(\alpha)$ having algebraic behavior have the following form:

$$G_-(\alpha) = \hat{G}_-(\alpha)U(\alpha), \quad G_+(\alpha) = U^{-1}(\alpha)\hat{G}_+(\alpha) \tag{67}$$

The explicit evaluation of the entire matrix $U(\alpha)$ is a formidable task. In the presence of rational matrices or even of matrices having only discrete spectra, we introduce the following technique for accomplishing it (Daniele, 1986). Apparently, no better method seems available. By assuming well-posed problems, we are dealing with factorized matrices $G_{-,+}(\alpha)$ and their inverses whose elements behave as $|\alpha|^p$ $(p < 1)$ for $\alpha \to \infty$. Consequently, it is possible to use Mittag-Leffler expansions for these matrices. By taking into account the presence of simple poles, this yields

$$G_-(\alpha) = \hat{G}_-(\alpha)U(\alpha) = \sum_s \hat{T}_s\left(\frac{1}{\alpha+\beta_s} - \frac{1}{\beta_s}\right)U(-\beta_s) + \hat{G}_-(0)U(0) \tag{68}$$

$$G_+^{-1}(\alpha) = \hat{G}_+^{-1}(\alpha)U(\alpha) = \sum_n \hat{R}_n\left(\frac{1}{\alpha-\alpha_n} + \frac{1}{\alpha_n}\right)U(\alpha_n) + \hat{G}_+^{-1}(0)U(0) \tag{69}$$

In these representations, $-\beta_s$ $(s = 1, 2, \dots)$ and α_n $(n = 1, 2, \dots)$ are the simple poles, with residues $\hat{T}_s$ and $\hat{R}_n$ of $\hat{G}_-(\alpha)$ and $\hat{G}_+^{-1}(\alpha)$, respectively. $U(0)$ is an arbitrary constant matrix that can be set equal to the identity matrix: $U(0) = 1$.

From eqs. (68) and (69) it follows that

$$\begin{aligned} U(\alpha) &= \hat{G}_-^{-1}(\alpha)\left[\sum_s \hat{T}_s\left(\frac{1}{\alpha+\beta_s} - \frac{1}{\beta_s}\right)U(-\beta_s) + \hat{G}_-(0)U(0)\right] \\ U(\alpha) &= \hat{G}_+(\alpha)\left[\sum_n \hat{R}_n\left(\frac{1}{\alpha-\alpha_n} + \frac{1}{\alpha_n}\right)U(\alpha_n) + \hat{G}_+^{-1}(0)U(0)\right] \end{aligned} \tag{70}$$

Since $\hat{G}_+(\alpha)$ and $\hat{R}_n$ are known, the previous equations allow for the evaluation of the entire matrix $U(\alpha)$ through its samples $U(\alpha_n)$ or $U(-\beta_s)$. This evaluation of $U(\alpha)$ is not necessary because we observe that through the third member of eqs. (68) and (69), the samples

$U(\alpha_n)$, $U(-\beta_s)$ and the values $\hat{G}_-(0)$ and $\hat{G}_+^{-1}(0)$ directly provide the standard factorized matrices $G_-(\alpha)$ and $G_+^{-1}(\alpha)$. To obtain a system of equations for the samples $U(\alpha_n)$ and $U(-\beta_s)$, we set $\alpha = \alpha_n$ and $\alpha = -\beta_s$, respectively, in the two eqs. (70). We get the following system:

$$\begin{aligned} U(\alpha_n) &= \hat{G}_-^{-1}(\alpha_n)\left[\sum_s \hat{T}_s\left(\frac{1}{\alpha_n+\beta_s} - \frac{1}{\beta_s}\right)U(-\beta_s) + \hat{G}_-(0)U(0)\right] \\ U(-\beta_s) &= \hat{G}_+(-\beta_s)\left[\sum_n -\hat{R}_n\left(\frac{1}{\beta_s+\alpha_n} - \frac{1}{\alpha_n}\right)U(\alpha_n) + \hat{G}_+^{-1}(0)U(0)\right] \end{aligned} \tag{71}$$

We can reduce the order of this system by elimination. For instance, by eliminating $U(-\beta_s)$ we obtain the following system:

$$U(\alpha_m) = \hat{G}_-^{-1}(\alpha_m)\left[\sum_n Q_n(\alpha_m)\hat{R}_n U(\alpha_n) + V(\alpha_m)\right] \tag{72}$$

where the known functions $Q_n(\alpha)$ and $V(\alpha)$ are given by

$$\begin{aligned} Q_n(\alpha) &= -\sum_s \frac{\alpha}{\alpha_n(\alpha+\beta_s)(\alpha_n+\beta_s)}\hat{T}_s\hat{G}_+(-\beta_s) \\ V(\alpha) &= \left\{\hat{G}_-(0) + \sum_s\left(\frac{1}{\alpha+\beta_s} - \frac{1}{\beta_s}\right)\hat{T}_s\hat{G}_+(-\beta_s)\hat{G}_+^{-1}(0)\right\}U(0) \end{aligned} \tag{73}$$

The solution of the system (72) yields $U(\alpha)$:

$$U(\alpha) = \hat{G}_-^{-1}(\alpha)\left[\sum_n Q_n(\alpha)\hat{R}_n U(\alpha_n) + V(\alpha)\right] \tag{74}$$

When the number of poles involved is finite, the system (71) or (72) can be solved in closed form.

The difficulty of the technique suggested in this section for the factorization of rational matrices is constituted by the evaluation through Cauchy integrals of the decomposition functions indicated in eq. (63). We will show later that these integrals are abelian integrals (Bliss, 2004). However, in the following concrete example these functions do not appear in the final expressions of the standard factorized matrices $G_-(\alpha)$ and $G_+^{-1}(\alpha)$.

Example 1

We apply the aforementioned procedure for the rational matrix considered in the example in section 2.7:

$$R(\alpha) = \begin{vmatrix} 1 & jq\dfrac{n(\alpha)}{p(\alpha)} \\ jq & 1 \end{vmatrix}, \quad [R(\alpha)]^{-1} = \frac{1}{r(\alpha)}\begin{vmatrix} 1 & -jq\dfrac{n(\alpha)}{p(\alpha)} \\ -jq & 1 \end{vmatrix} = \frac{\begin{vmatrix} p(\alpha) & -jq\,n(\alpha) \\ -jq\,p(\alpha) & p(\alpha) \end{vmatrix}}{(1+q^2)(\alpha^2+C^2)},$$

where $C = \sqrt{\frac{B^2+q^2A^2}{1+q^2}}$, $n(\alpha) = \alpha^2 + A^2$, $p(\alpha) = \alpha^2 + B^2$, $r(\alpha) = \det[R(\alpha)] = (1+q^2)\frac{\alpha^2+C^2}{\alpha^2+B^2}$.

We observe that $\hat{R}_-(\alpha)$ presents only the pole $-\beta_1 = jB$ with residue $\hat{T}_1$, and $\hat{R}_+^{-1}(\alpha)$ only the pole $\alpha_1 = -jC$ with residue $\hat{R}_1$. Consequently, system (71) becomes[1]

$$U(-jC) = \hat{R}_-^{-1}(-jC)\left[\hat{T}_1 U(jB)\left(-\frac{1}{j(B+C)} + \frac{1}{jB}\right) + \hat{R}_-(0)\right]$$

$$U(jB) = \hat{R}_+(jB)\left[\hat{R}_1 U(-jC)\left(\frac{1}{j(B+C)} - \frac{1}{jC}\right) + \hat{R}_+^{-1}(0)\right]$$

The elimination of $U(-jC)$ in these equations yields the following equations in the sample $U(jB)$:

$$\left[\mathbf{1} - J_1 J_2 \hat{R}_+(jB)\cdot\hat{R}_1\cdot\hat{R}_-^{-1}(-jC)\cdot\hat{T}_1\right]\cdot U(jB)$$
$$= \hat{R}_+(jB)\cdot\left[\hat{R}_+^{-1}(0) + J_2\hat{R}_1\cdot\hat{R}_-^{-1}(-jC)\cdot\hat{R}_-(0)\right]$$

where

$$J_1 = -\frac{1}{j(B+C)} + \frac{1}{jB}, \quad J_2 = \frac{1}{j(B+C)} - \frac{1}{jC}$$

Taking into account that

$$\hat{R}_+^{-1}(\alpha)\hat{R}_-^{-1}(\alpha) = R^{-1}(\alpha)$$

we observe the following simplification:

$$\hat{R}_1\cdot\hat{R}_-^{-1}(-jC) = \mathrm{Res}\left[R^{-1}(\alpha)\right]\Big|_{\alpha=-jC} = R_C = \frac{1}{r'(-jC)}\begin{vmatrix} 1 & \frac{j}{q} \\ -jq & 1 \end{vmatrix}$$

with

$$\frac{1}{r'(-jC)} = j\frac{(B^2-C^2)^2}{2q^2C(B^2-A^2)}$$

Consequently, we must solve

$$\left[\mathbf{1} - J_1 J_2 \hat{R}_+(jB)\cdot R_C\cdot\hat{T}_1\right]\cdot U(jB) = \hat{R}_+(jB)\cdot\left[\hat{R}_+^{-1}(0) + J_2 R_C\cdot\hat{R}_-(0)\right] \tag{75}$$

This equation does not require the knowledge of the residue $\hat{R}_1$.

To factorize $R(\alpha)$ using the logarithmic decomposition, we observe that

$$\log[R(\alpha)] = \varphi_0(\alpha)\mathbf{1} + \varphi_1(\alpha)R(\alpha)$$

where being λ_1 and λ_2 the two eigenvalues of $R(\alpha)$, $\varphi_0(\alpha)$ and $\varphi_1(\alpha)$ must satisfy

$$\log[\lambda_1] = \varphi_0(\alpha) + \varphi_1(\alpha)\lambda_1, \quad \log[\lambda_2] = \varphi_0(\alpha) + \varphi_1(\alpha)\lambda_2$$

[1] Take into account that $U(0) = 1$.

Solving these, we get

$$\varphi_1(\alpha) = \frac{1}{\lambda_1 - \lambda_2}\log\frac{\lambda_1}{\lambda_2}, \quad \varphi_0(\alpha) = \frac{1}{\lambda_1 - \lambda_2}(\lambda_1 \log \lambda_2 - \lambda_2 \log \lambda_1)$$

and taking into account that

$$\lambda_{1,2} = 1 \pm jq\sqrt{\frac{n(\alpha)}{p(\alpha)}}$$

we obtain

$$\log[R(\alpha)] = \log\sqrt{r(\alpha)}\mathbf{1} + \frac{1}{2\sqrt{n(\alpha)p(\alpha)}}\log\frac{1 + jq\sqrt{\frac{n(\alpha)}{p(\alpha)}}}{1 - jq\sqrt{\frac{n(\alpha)}{p(\alpha)}}}\begin{vmatrix} 0 & n(\alpha) \\ p(\alpha) & 0 \end{vmatrix}$$

This yields the two factorized matrices:

$$\hat{R}_-(\alpha) = \sqrt{r_-(\alpha)}\,\exp\left[\frac{1}{2}t_-(\alpha)\begin{vmatrix} 0 & n(\alpha) \\ p(\alpha) & 0 \end{vmatrix}\right], \quad \hat{R}_+(\alpha) = \hat{R}_-(-\alpha)$$

where, being γ_2 the frown real axis (section 3.1),

$$r_-(\alpha) = \sqrt{1+q^2}\frac{\alpha - jC}{\alpha - jB}, \quad t_-(\alpha) = \int\limits_{\gamma_2}\frac{1}{\sqrt{n(t)p(t)}}\log\frac{1 + jq\sqrt{\frac{n(t)}{p(t)}}}{1 - jq\sqrt{\frac{n(t)}{p(t)}}}\frac{1}{t - \alpha}dt$$

Since for the time being we are interested only in the residue $\hat{T}_1$ at the pole $\alpha = jB$, we warp γ_2 to the closed line that encloses the branch lines relevant to the branch point $\alpha = jB$. Then, the evaluation of the jump of the integrand between the two lips of the branch line yields an integral on the branch line that can be evaluated explicitly:

$$t_-(\alpha) = t_{r-}(\alpha) + 2j\pi\frac{1}{\sqrt{2jB(A^2 - B^2)}}\frac{1}{\sqrt{\alpha - jB}}$$

where $t_{r-}(\alpha)$ is regular at $\alpha = jB$.

Substitution in the exponential present in the expression of $\hat{R}_-(\alpha)$, an algebraic manipulation shows that the residue $\hat{T}_1$ does not depend on $t_{r-}(jB)$, and we get

$$\hat{T}_1 = a\begin{vmatrix} 0 & 1 \\ 0 & 0 \end{vmatrix}, \quad a = j\sqrt{\left(B - \sqrt{\frac{B^2 + q^2A^2}{1+q^2}}\right)\sqrt{1+q^2}\frac{A^2 - B^2}{2B}}$$

To solve eq. (75), the value of $\hat{R}_-(0) = \hat{R}_+(0)$ is required. It is easily obtained since

$$\hat{R}_-(0)\cdot\hat{R}_+(0) = \hat{R}_-^2(0) = R(0) = \sqrt{r(0)}\,\exp\left[\frac{1}{2}t(0)\begin{vmatrix} 0 & n(0) \\ p(0) & 0 \end{vmatrix}\right]$$

whence after algebraic manipulations

$$\hat{R}_-(0) = \frac{1}{2T\sqrt[4]{B^2+q^2A^2}} \begin{vmatrix} \sqrt{B+jqA}+\sqrt{B-jqA} & \frac{A}{B}\left(\sqrt{B+jqA}-\sqrt{B-jqA}\right) \\ \frac{B}{A}\left(\sqrt{B+jqA}-\sqrt{B-jqA}\right) & \sqrt{B+jqA}+\sqrt{B-jqA} \end{vmatrix}$$

where $T = \sqrt{\frac{B}{C\sqrt{1+q^2}}}$.

Finally, the quantity $\hat{R}_+(jB)$ is also required. Unfortunately, the direct evaluation of $\hat{R}_+(jB)$ through the factorized matrix $\hat{R}_+(\alpha)$ introduces nonelementary functions. So we try to obtain this value by taking into account that

$$\hat{R}_-(\alpha)\hat{R}_+(\alpha) = \hat{R}(\alpha)$$

Then

$$\hat{T}_1\hat{R}_+(jB) = \text{Res}[R(\alpha)]|_{\alpha=jB} = R_B = \begin{vmatrix} 0 & q\frac{A^2-B^2}{2B} \\ 0 & 0 \end{vmatrix}$$

This equation does not allows us to obtain $\hat{R}_+(jB)$ because $\hat{T}_1$ is not invertible. To overcome this difficulty, we multiply both members of (75) by $\hat{T}_1$ and obtain

$$\left[\hat{T}_1 - J_1J_2R_B \cdot R_C \cdot \hat{T}_1\right] \cdot U(jB) = R_B \cdot \left[\hat{R}_+^{-1}(0) + J_2R_C \cdot \hat{R}_-(0)\right]$$

Substituting the values of R_B, R_C, $\hat{T}_1$, $\hat{R}_-(0)$ and $\hat{R}_+^{-1}(0)$, the previous equation assumes the form

$$\begin{vmatrix} 0 & 1 \\ 0 & 0 \end{vmatrix} \begin{vmatrix} U_{11}(jB) & U_{12}(jB) \\ U_{21}(jB) & U_{22}(jB) \end{vmatrix} = \begin{vmatrix} e_1 & e_2 \\ 0 & 0 \end{vmatrix}$$

where e_1 and e_2 are known quantities not explicitly reported.

The previous equations provides the evaluation of

$$U_{21}(jB) = e_1, \quad U_{22}(jB) = e_2$$

The other two components $U_{11}(jB)$ and $U_{12}(jB)$ remain indeterminate, but fortunately we can show that they are not required for the evaluation of the standard factorized matrices $R_-(\alpha)$ and $R_+(\alpha)$. In fact, the standard factorized matrices have the form

$$R_-(\alpha) = \hat{R}_-(0) + \hat{T}_1U(jB)\left(\frac{1}{\alpha - jB} + \frac{1}{jB}\right)$$

$$R_+^{-1}(\alpha) = \hat{R}_+^{-1}(0) + \hat{R}_1U(-jC)\left(\frac{1}{\alpha + jC} - \frac{1}{jC}\right)$$

where

$$\hat{T}_1U(jB) = a\begin{vmatrix} e_1 & e_2 \\ 0 & 0 \end{vmatrix}$$

and $\hat{R}_1 U(-jC)$ follows from the equation that relates $U(-jC)$ to $U(jB)$:

$$U(-jC) = \hat{R}_-^{-1}(-jC)\left[\hat{T}_1 U(jB)\left(-\frac{1}{j(B+C)} + \frac{1}{jB}\right) + \hat{R}_-(0)\right]$$

Finally

$$\hat{R}_1 \cdot U(-jC) = R_C \cdot \left[a\begin{vmatrix} e_1 & e_2 \\ 0 & 0 \end{vmatrix}\left(-\frac{1}{j(B+C)} + \frac{1}{jB}\right) + \hat{R}_-(0)\right]$$

For the sake of brevity we do not report the explicit forms of $R_-(\alpha)$ and $R_+(\alpha)$. However, we have verified that the found expressions are standard factorized matrices and satisfy

$$R_-(\alpha) \cdot R_+(\alpha) = R(\alpha)$$

4.6 The factorization problem and the functional analysis

We observe that it is always possible to reduce the factorization problem to an integral equation (section 1.6):

$$F_+(\alpha) + \frac{1}{2\pi j}\int_{-\infty}^{\infty}\left(\frac{G^{-1}(\alpha)G(u)}{u-\alpha} - \frac{1}{u-\alpha}\right)F_+(u)du = \int_{\gamma_1}\frac{G^{-1}(\alpha)F(u)}{u-\alpha}du \tag{76}$$

whence we can resort to the main methods for solving integral equations. They are the iterative method, the Fredholm determinant method and the moment method, which include the Ritz-Garlekin method and the least square method (Ivanov, 1976).

4.6.1 The iterative method

The iterative method is based on the Neumann equations that invert the operator $1 - A$:

$$(1-A)^{-1} = 1 + A + A^2 + \cdots, \quad \|A\| < 1 \tag{77}$$

Equation (77) applies to (76) by assuming

$$A \to A(\alpha, u) = -\frac{1}{2\pi j}\left(\frac{G^{-1}(\alpha)G(u)}{u-\alpha} - \frac{1}{u-\alpha}\right)$$

A factorization iterative algorithm that avoids the formulation in integral equations is indicated in Wiener and Masani (1958) and Jones (1991). Letting

$$G(\alpha) = G_+(\alpha)G_-(\alpha) = \mathbf{1} - A(\alpha)$$
$$G_+^{-1}(\alpha) = \mathbf{1} + R(\alpha)$$

an explicit formula for $R(\alpha)$ is given by Wiener and Masani (1958):

$$R = \{A\}_+ + \{\{A\}_+ A\}_+ + \{\{\{A\}_+ A\}_+ A\}_+ + \cdots \tag{78}$$

where $\{f\}_+$ is the symbol of additive decomposition (section 2.2).

The practical calculation of more than one or two terms in (78) may be prohibitive. Besides, it is regrettable that dealing with rational matrices this series cannot be evaluated in closed form. Thus, further studies are necessary to make (78) effective in practice.

4.6.2 The Fredholm determinant method

The inversion of operators defined in finite spaces (matrices) can be accomplished by using the determinant theory. To extend this method to infinite spaces, firstly we must define in an abstract way the determinant of an arbitrary operator L. Starting by the traditional definition we obtain

$$\det[L] = \exp\{Tr[\log L]\} \tag{79}$$

where $Tr[M]$ is the trace of the operator M.

To invert $L = 1 - \lambda K$ we set

$$L^{-1} = (1-\lambda K)^{-1} = 1 + \lambda K(1-\lambda K)^{-1} = 1 + \lambda \frac{D(\lambda)}{\Delta(\lambda)} \tag{80}$$

where λ is an arbitrary complex variable and

$$D(\lambda) = K(1-\lambda K)^{-1}\det(1-\lambda K)$$
$$\Delta(\lambda) = \det(1-\lambda K)$$

For K constituted by compact operators, the functions $D(\lambda)$ and $\Delta(\lambda)$ are entire on λ. (This is the reason for the normalization with the determinant $\Delta(\lambda)$.) Consequently, we can expand $D(\lambda)$ and $\Delta(\lambda)$ in series that are valid for arbitrary values of λ:

$$\begin{aligned} \Delta(\lambda) &= \Delta_0 - \lambda\Delta_1 + \frac{\lambda^2}{2!}\Delta_2 + \cdots + (-1)^n\frac{\lambda^n}{n!}\Delta_n + \cdots \\ D(\lambda) &= D_0 - \lambda D_1 + \frac{\lambda^2}{2!}D_2 + \cdots + (-1)^n\frac{\lambda^n}{n!}D_n + \cdots \end{aligned} \tag{81}$$

where

$$\Delta_0 = 1, \quad \Delta_1 = Tr[K], \quad \Delta_2 = Tr[K^2] - (Tr[K])^2, \ldots$$
$$D_0 = 1, \quad D_1 = K\,Tr[K] - K^2, \quad D_2 = K(Tr[K])^2 - K\,Tr[K^2] - 2K^2Tr[K] + 2K^3, \ldots$$
$$D_n = K(\Delta_n 1 - n\,D_{n-1})$$

For integral operator M

$$M \cdot f \to \int M(\alpha, u)f(u)du$$

we have

$$\begin{aligned} Tr[M] &= \int M(u,u)du, \\ M^2 \cdot f &\to \int \left[\int M(\alpha,t)M(t,u)dt\right] f(u)du \end{aligned} \tag{82}$$

and so on.

By assuming $\lambda = 1$, $K \rightarrow K(\alpha, u) = \frac{1}{2\pi j}\left(\frac{G^{-1}(\alpha)G(u)}{u-\alpha} - \frac{1}{u-\alpha}\right)$, the previous equations provide the exact solutions of the W-H equations (and the involved factorization problem). However, we are confronted with an infinite series whose coefficients are multiple integrals of the type

$$\underbrace{\int_{-\infty}^{\infty}\int_{-\infty}^{\infty}\cdots\int_{-\infty}^{\infty}}_{r} K(\alpha, t_1)\cdot K(\alpha, t_2)\ldots K(t_r, u)dt_1 dt_2\ldots dt_r, \quad r = 1, 2, \ldots \tag{83}$$

Despite considerable progress in the numerical evaluation of integrals, at present the numerical evaluation of the previous expression is still impractical.

Conversely, the numerical solution of Fredholm equations by using quadrature formulas is possible and constitutes the best method for obtaining efficient approximate solutions of the W-H equations (see chapter 5).

4.6.3 Factorization of meromorphic matrix kernels with an infinite number of poles

The factorization of matrix kernels that are meromorphic is very important in electromagnetic problems involving only infinite discrete spectrum. The concept of the matching of the singularities described in section 4.4.2 and the method reported in section 4.5.1.1 (Daniele, 1986) apply again, but they imply the solution of a system with an infinite number of equations. The solution of this system cannot generally be accomplished in closed form, thus approximate techniques are required.

For instance, the solution of the infinite system of equations can be accomplished by a truncation procedure. Unfortunately, the convergence of the numerical results is guaranteed only when the operator A involved in the solution of the system is a Fredholm operator. To obtain it, functional analysis suggests the decomposition of the operator A in the form

$$A = A_o + C$$

where A_o has an inverse that can be obtained in closed form, and C is a compact operator. This leads to the concept of the dominant equation

$$A_o \cdot x + C \cdot x = y \tag{84}$$

or to the regularized equation

$$(1 + T)x = A_o^{-1}y \tag{85}$$

where $1 + T$ is a Fredholm operator, since $T = A_o^{-1}C$ is a compact operator. The compactness of T legitimates the use of a truncation procedure or other approximate techniques.

In factorizing meromorphic matrices, often the regularization requires the inversion of an infinite matrix A_o whose entries are

$$\frac{1}{\alpha_{-j} - \beta_{-i}}$$

The inversion of this matrix can be obtained in closed form (Mittra & Lee, 1971; Daniele & Zich, 1980; Daniele & Gilli, 1997).

The weak factorization may be generalized to deal with the very important cases in which there are infinite offending zeroes (Büyükaksoy & Serbest, 1993; Idemen, 1977, 1979; Abrahams & Wickham, 1990); for example, the factorization of an arbitrary meromorphic matrix can be considered as a weak factorization problem. In these cases the problem of truncating effectively an infinite number of equations must be considered very carefully.

4.7 A class of matrices amenable to explicit factorization: matrices having rational eigenvectors

All the matrices may be represented by the Jordan form:

$$T(\alpha) = S(\alpha)J(\alpha)S(\alpha)^{-1} \tag{86}$$

where $S(\alpha)$ is the eigenvectors matrix, and $J(\alpha)$ is a Jordan matrix having the form (Pease, 1965):

$$J(\alpha) = quasidiag\{J_{m_1}(\lambda_1), J_{m2}(\lambda_2), \ldots\} \tag{87}$$

In (87) $J_{m_1}(\lambda_i)$ is the $m_i \times m_i$ Jordan block relevant to the eigenvalue $\lambda_i = \lambda_i(\alpha)$ (with multiplicity m_i) of $T(\alpha)$:

$$J_{m_i}(\lambda_i) = \begin{vmatrix} \lambda_i & 1 & 0 & \cdots & 0 \\ 0 & \lambda_i & 1 & \cdots & 0 \\ 0 & 0 & \lambda_i & \cdots & \cdots \\ \cdots & \cdots & \cdots & \cdots & 1 \\ 0 & 0 & 0 & \cdots & \lambda_i \end{vmatrix} \tag{88}$$

The factorization of $T(\alpha)$ can be accomplished explicitly (in a weak form) if the matrix $S(\alpha)$ is rational. In fact, in this case, eq. (86) reduces the factorization of $T(\alpha)$ to that of the Jordan matrix $J(\alpha)$, that, in general, is triangular and can be accomplished with the method reported in section 4.3.

Besides the general method indicated in section 4.3 for triangular matrices, the factorization of the matrix $J(\alpha)$ can be accomplished by the logarithmic decomposition of $\mathrm{Log}[T(\alpha)]$ by taking into account that (Pease, 1965) a generic function of a matrix $f(T)$ is given by

$$f[T(\alpha)] = S(\alpha)\, quasidiag\{f[J_{m_1}(\lambda_1)], f[J_{m2}(\lambda_2)], \ldots\}S^{-1}(\alpha) \tag{89}$$

with

$$f[J_{m_i}(\lambda_i)] = \begin{vmatrix} f(\lambda_i) & \frac{1}{1!}f'(\lambda_i) & \frac{1}{2!}f''(\lambda_i) & \cdots & \frac{1}{(m_i-1)!}f^{(m_i-1)}(\lambda_i) \\ 0 & f(\lambda_i) & \frac{1}{1!}f'(\lambda_i) & \cdots & \frac{1}{(m_i-2)!}f^{(m_i-2)}(\lambda_i) \\ 0 & 0 & f(\lambda_i) & \cdots & \cdots \\ \cdots & \cdots & \cdots & \cdots & 1 \\ 0 & 0 & 0 & \cdots & f(\lambda_i) \end{vmatrix}$$

Since it easy to ascertain that the decomposed matrices of $f[J_{m_i}(\lambda_i)]$

$$f[J_{m_i}(\lambda_i)] = \{f[J_{m_i}(\lambda_i)]\}_- + f[J_{m_i}(\lambda_i)]\}_+ \tag{90}$$

do commute, the following explicit factorization of $T(\alpha)$ follows:

$$\begin{aligned}
\left[T_-(\alpha)\right] &= S(\alpha)\, quasidiag\left\{\exp\left[\{\mathrm{Log}[J_{m_1}(\lambda_1)]\}_-\right], \exp\left[\{Log[J_{m_2}(\lambda_2)]\}_-\right], \ldots\right\} S^{-1}(\alpha) \\
\left[T_+(\alpha)\right] &= S(\alpha)\, quasidiag\left\{\exp\left[\{\mathrm{Log}[J_{m_1}(\lambda_1)]\}_+\right], \exp\left[\{Log[J_{m_2}(\lambda_2)]\}_+\right], \ldots\right\} S^{-1}(\alpha)
\end{aligned} \tag{91}$$

Example 2: Factorization of a particular matrix

The factorization of the following matrix $M(\alpha)$ was considered by Achenbach and Gautesen (1977):

$$M(\alpha) = \frac{1}{\alpha^2 + \eta^2} \begin{vmatrix} \alpha^2 d(\alpha) + \eta^2 \delta(\alpha) & \alpha\eta[\delta(\alpha) - d(\alpha)] \\ \alpha\eta[\delta(\alpha) - d(\alpha)] & \alpha^2 \delta(\alpha) + \eta^2 d(\alpha) \end{vmatrix} \tag{92}$$

where $d(\alpha)$ and $\delta(\alpha)$ are functions of α, and the parameter η is a constant. The eigenvectors are rational:

$$S(\alpha) = \begin{vmatrix} -\dfrac{\alpha}{\eta} & \dfrac{\eta}{\alpha} \\ 1 & 1 \end{vmatrix}, \quad J = \begin{vmatrix} d(\alpha) & 0 \\ 0 & \delta(\alpha) \end{vmatrix}, \quad S^{-1}(\alpha) = \frac{1}{\alpha^2 + \eta^2} \begin{vmatrix} -\alpha\eta & \eta^2 \\ \alpha\eta & \alpha^2 \end{vmatrix} \tag{93}$$

Consequently, it is possible to achieve a weak factorization of the kernel $M(\alpha)$ by factorizing the diagonal matrix J.

Example 3: Matrices of order two having a particular form

The matrices $G(\alpha) = \begin{vmatrix} 1 & a(\alpha) \\ b(\alpha) & 1 \end{vmatrix}$ can be (weakly) factorized in an explicit form if

$$\sqrt{\frac{a(\alpha)}{b(\alpha)}} = \frac{n(\alpha)}{p(\alpha)} \tag{94}$$

where $n(\alpha)$ and $p(\alpha)$ are polynomials. In fact, in this case $S(\alpha)$ is rational since we have

$$S(\alpha) = \begin{vmatrix} -\sqrt{\dfrac{a(\alpha)}{b(\alpha)}} & \sqrt{\dfrac{a(\alpha)}{b(\alpha)}} \\ 1 & 1 \end{vmatrix} = \begin{vmatrix} -\dfrac{n(\alpha)}{p(\alpha)} & \dfrac{n(\alpha)}{p(\alpha)} \\ 1 & 1 \end{vmatrix} \tag{95}$$

4.8 Factorization of a 2×2 matrix

4.8.1 The Hurd method

Let us suppose that the matrix to factorize presents singularities constituted only by branch points $\alpha = \pm k$ (Fig. 1) and has the form (Hurd, 1976; Büyükaksoy & Serbest, 1993)

$$G(\alpha) = \begin{vmatrix} G_1(\alpha) & G_1(\alpha) G_2(\alpha) \\ G_3(\alpha) & -G_2(\alpha) G_3(\alpha) \end{vmatrix} \tag{96}$$

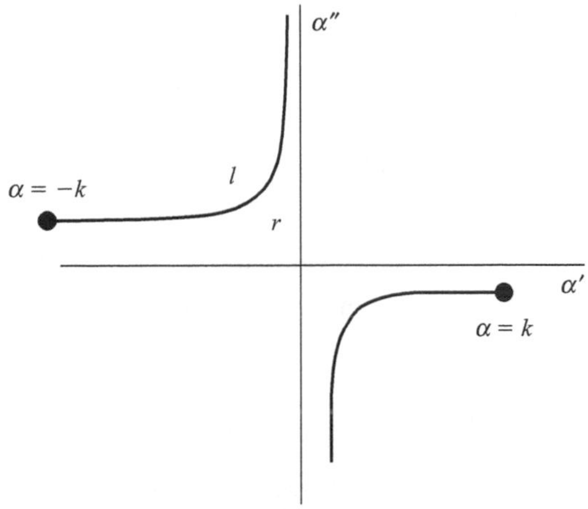

Fig. 1: Branch cuts and branch points of $G(\alpha)$

Additionally, we assume that the two values of $G_2(\alpha)$ on the two lips l and r of the branch line $\alpha = -k$ are opposite. By using the notations

$$G(\alpha)|_{\alpha\in l} = G_l(\alpha), \quad G(\alpha)|_{\alpha\in r} = G_r(\alpha)$$

it follows that

$$G_{2l}(\alpha) = -G_{2r}(\alpha) \tag{97}$$

From the W-H equation $G(\alpha)\cdot F_+(\alpha) = F_-(\alpha)$ we get

$$G_r(\alpha)\cdot F_+(\alpha) = F_{-r}(\alpha), \quad G_l(\alpha)\cdot F_+(\alpha) = F_{-l}(\alpha) \tag{98}$$

where, since $F_+(\alpha)$ is regular in $\mathrm{Im}[\alpha] \geq 0$, we write

$$F_+(\alpha) = F_+(\alpha)|_{\alpha\in l} = F_+(\alpha)|_{\alpha\in r}$$

We can eliminate the unknown $F_+(\alpha)$ in (98) and obtain the following Hilbert problem:

$$G_l(\alpha)G_r^{-1}(\alpha)\cdot F_{-r}(\alpha) = F_{-l}(\alpha) \tag{99}$$

With the specified condition on $G(\alpha)$ we get

$$G_l(\alpha)G_r^{-1}(\alpha) = \begin{pmatrix} 0 & \dfrac{G_{1l}(\alpha)}{G_{3r}(\alpha)} \\ \dfrac{G_{3l}(\alpha)}{G_{1r}(\alpha)} & 0 \end{pmatrix} \tag{100}$$

and the factorization of $G_l(\alpha)G_r^{-1}(\alpha)$ (in the Hilbert problem) reduces to the factorization of scalars.

The Hurd method is very powerful for studying diffraction problems by half-planes. It can be extended to cases where a finite discrete spectrum is present.

4.8.2 The off-diagonal form

A matrix of order two can be rewritten in the following off-diagonal form:

$$G(\alpha) = \begin{vmatrix} 1 & a(\alpha) \\ b(\alpha) & 1 \end{vmatrix} \tag{101}$$

There are many ways to reduce a general matrix of order two to the off-diagonal form; for example, it is possible to rewrite the general matrix $M(\alpha) = \begin{vmatrix} m_{11}(\alpha) & m_{12}(\alpha) \\ m_{21}(\alpha) & m_{22}(\alpha) \end{vmatrix}$ by considering that

$$\begin{aligned} M(\alpha) &= \begin{vmatrix} m_{11}(\alpha) & m_{12}(\alpha) \\ m_{21}(\alpha) & m_{22}(\alpha) \end{vmatrix} \\ &= \begin{vmatrix} m_{11-}(\alpha) & 0 \\ 0 & m_{22-}(\alpha) \end{vmatrix} \begin{vmatrix} 1 & \dfrac{m_{12}(\alpha)}{m_{11-}(\alpha)m_{22+}(\alpha)} \\ \dfrac{m_{21}(\alpha)}{m_{22-}(\alpha)m_{11+}(\alpha)} & 1 \end{vmatrix} \begin{vmatrix} m_{11+}(\alpha) & 0 \\ 0 & m_{22+}(\alpha) \end{vmatrix} \end{aligned} \tag{102}$$

where

$$m_{11}(\alpha) = m_{11-}(\alpha)m_{11+}(\alpha) \quad m_{22}(\alpha) = m_{22-}(\alpha)m_{22+}(\alpha)$$

The factorization of $M(\alpha)$ is thus reduced to that of the central matrix, that has an off-diagonal form, where

$$a(\alpha) = \frac{m_{12}(\alpha)}{m_{11-}(\alpha)m_{22+}(\alpha)} \qquad b(\alpha) = \frac{m_{21}(\alpha)}{m_{22-}(\alpha)m_{11+}(\alpha)} \tag{103}$$

There are many other ways to reduce matrices of order two to off-diagonal form, since it is possible to modify the structure of the original given matrix $M(\alpha)$ to the matrix $W_1(\alpha)M(\alpha)W_2(\alpha)$, where $W_1(\alpha)$ and $W_2(\alpha)$ are (with their inverses) arbitrary minus and plus functions or arbitrary rational functions. When $W_1(\alpha)$ and $W_2(\alpha)$ are rational, the factorization of $M(\alpha) = M_-(\alpha)M_+(\alpha)$ yields to a weak factorization of the given matrix $G(\alpha)$.

The matrices of order two with off-diagonal form have some interesting properties. For example, if the left factorization is known

$$G(\alpha) = \begin{vmatrix} 1 & a(\alpha) \\ b(\alpha) & 1 \end{vmatrix} = \begin{vmatrix} \omega_- & \sigma_- \\ \tau_- & \rho_- \end{vmatrix} \begin{vmatrix} \vartheta_+ & \eta_+ \\ \varsigma_+ & \xi_+ \end{vmatrix} \tag{104}$$

then the right factorization follows (Jones, 1984):

$$G(\alpha) = \begin{vmatrix} 1 & a(\alpha) \\ b(\alpha) & 1 \end{vmatrix} = \begin{vmatrix} \omega_- & \sigma_- \\ \tau_- & \rho_- \end{vmatrix} \begin{vmatrix} \vartheta_+ & \eta_+ \\ \varsigma_+ & \xi_+ \end{vmatrix} = \begin{vmatrix} \xi_+ & \eta_+ \\ \varsigma_+ & \vartheta_+ \end{vmatrix} \begin{vmatrix} \rho_- & \sigma_- \\ \tau_- & \omega_- \end{vmatrix} \tag{105}$$

Before considering explicit factorization techniques for matrices of order two, we remark that the following equation can be useful in many cases:

$$\begin{vmatrix} a & b\frac{f}{g} \\ c\frac{g}{f} & d \end{vmatrix} = \begin{vmatrix} f & 0 \\ 0 & g \end{vmatrix} \cdot \begin{vmatrix} a & b \\ c & d \end{vmatrix} \cdot \begin{vmatrix} 1/f & 0 \\ 0 & 1/g \end{vmatrix} \tag{106}$$

or

$$\begin{vmatrix} a & b \\ c & d \end{vmatrix} = \begin{vmatrix} 1/f & 0 \\ 0 & 1/g \end{vmatrix} \cdot \begin{vmatrix} a & b\frac{f}{g} \\ c\frac{g}{f} & d \end{vmatrix} \cdot \begin{vmatrix} f & 0 \\ 0 & g \end{vmatrix}$$

An off-diagonal matrix of order two $\begin{vmatrix} 1 & a(\alpha) \\ b(\alpha) & 1 \end{vmatrix}$ is called a Daniele matrix when the function $\frac{a(\alpha)}{b(\alpha)}$ is a rational matrix of α, that is,

$$\frac{a(\alpha)}{b(\alpha)} = \frac{n(\alpha)}{p(\alpha)} \tag{107}$$

where $n(\alpha)$ and $p(\alpha)$ are polynomials.

4.8.3 Reduction of matrices commuting with polynomial matrices to the Daniele matrices

Many matrices of order two have the Daniele form. For instance, Hurd (1987) showed that all the matrices of order two commuting with polynomial matrices reduce to matrices having the Daniele form. To obtain this result, let us consider the matrix $W(\alpha)$ commuting with an arbitrary rational matrix

$$R(\alpha) = \frac{P(\alpha)}{d(\alpha)} \tag{108}$$

where $d(\alpha)$ and $P(\alpha) = \begin{vmatrix} l_1(\alpha) & m(\alpha) \\ n(\alpha) & l_2(\alpha) \end{vmatrix}$ are polynomials.

For matrices of order two, Cayley's theorem provides the following representation of the matrix $W(\alpha)$:

$$W(\alpha) = 1 + f(\alpha)P(\alpha) \tag{109}$$

where $f(\alpha)$ is an arbitrary scalar, and $l_1(\alpha), l_2(\alpha), m(\alpha), n(\alpha)$ are arbitrary polynomials in α.

It may be observed that $W(\alpha)$ can be rewritten in the alternative form:

$$W(\alpha) = s(\alpha) \begin{vmatrix} 1 & 0 \\ l(\alpha) & m(\alpha) \end{vmatrix}^{-1} \begin{vmatrix} 1 & a(\alpha) \\ b(\alpha) & 1 \end{vmatrix} \begin{vmatrix} 1 & 0 \\ l(\alpha) & m(\alpha) \end{vmatrix} \tag{110}$$

where

$$s(\alpha) = 1 + \frac{1}{2} f(\alpha)[l_1(\alpha) + l_2(\alpha)], \quad l(\alpha) = \frac{1}{2}[l_1(\alpha) - l_2(\alpha)]$$
$$a(\alpha) = \frac{f(\alpha)}{s(\alpha)}, \quad b(\alpha) = \frac{f(\alpha)}{s(\alpha)}[l^2(\alpha) + m(\alpha)n(\alpha)]$$

This form shows that the factorization of the matrix $W(\alpha)$ reduces to that of the central matrix $\begin{vmatrix} 1 & a(\alpha) \\ b(\alpha) & 1 \end{vmatrix}$. Since $\frac{a(\alpha)}{b(\alpha)} = \frac{1}{[l^2(\alpha)+m(\alpha)n(\alpha)]}$ is rational in α, this central matrix has the Daniele form.

Alternatively, simple algebraic manipulations yield the alternative form

$$W(\alpha) = s(\alpha)1 + f(\alpha)\begin{vmatrix} l(\alpha) & m(\alpha) \\ n(\alpha) & -l(\alpha) \end{vmatrix} \tag{111}$$

Matrices having this form are called Khrapkov matrices since their explicit factorization has been studied in the paper by Khrapkov (1971).

Example 4

Let us examine again the factorization of the matrix considered in Achenbach and Gautesen (1977):

$$M(\alpha) = \frac{1}{\alpha^2 + \eta^2}\begin{vmatrix} \alpha^2 d(\alpha) + \eta^2\delta(\alpha) & \alpha\eta[\delta(\alpha) - d(\alpha)] \\ \alpha\eta[\delta(\alpha) - d(\alpha)] & \alpha^2\delta(\alpha) + \eta^2 d(\alpha) \end{vmatrix} \tag{112}$$

where $d(\alpha)$ and $\delta(\alpha)$ are functions of α and the parameter η is a constant.

We observe that $M(\alpha)$ commutes with the polynomial matrix:

$$P(\alpha) = \begin{vmatrix} -\alpha^2 + \eta^2 & \alpha\eta \\ \alpha\eta & 0 \end{vmatrix} \tag{113}$$

The representation of $M(\alpha)$ follows:

$$M(\alpha) = c(\alpha)[1 + f(\alpha)P(\alpha)] \tag{114}$$

By comparison with (112) this yields

$$c(\alpha) = \frac{\alpha^2\delta(\alpha) + \eta^2 d(\alpha)}{\alpha^2 + \eta^2}, \quad f(\alpha) = \frac{\delta(\alpha) - d(\alpha)}{\alpha^2\delta(\alpha) + \eta^2 d(\alpha)} \tag{115}$$

The factorization of $M(\alpha)$ is reduced to a Daniele matrix where

$$a = \frac{2[\delta(\alpha) - d(\alpha)]}{(\alpha^2 + \eta^2)[\delta(\alpha) + d(\alpha)]}, \quad b = \frac{[\delta(\alpha) - d(\alpha)](\alpha^2 + \eta^2)}{2[\delta(\alpha) + d(\alpha)]} \tag{116}$$

$$\frac{a}{b} = \frac{4}{(\alpha^2 + \eta^2)^2} \tag{117}$$

In this case $\sqrt{\frac{a}{b}}$ is rational, so we can use the technique previously described.

Daniele matrices can be rewritten in the form

$$\begin{vmatrix} 1 & a(\alpha) \\ b(\alpha) & 1 \end{vmatrix} = \begin{vmatrix} 1 & h(\alpha)n(\alpha) \\ h(\alpha)p(\alpha) & 1 \end{vmatrix} \tag{118}$$

where $h(\alpha)$ is a known function, and $n(\alpha)$, $p(\alpha)$ are polynomials.

Using (106), we can put the Daniele matrix in a form such that all the zeroes of the polynomial $n(\alpha)p(\alpha)$ are present only in an off-diagonal entry. For instance,

$$W(\alpha) = \begin{vmatrix} 1 & f(\alpha)n(\alpha)p(\alpha) \\ f(\alpha) & 1 \end{vmatrix} \tag{119}$$

In fact, let α_1 be a zero of $p(\alpha)$. Equation (106) yields

$$\begin{vmatrix} 1 & h(\alpha)n(\alpha) \\ h(\alpha)p(\alpha) & 1 \end{vmatrix} = \begin{vmatrix} \frac{1}{\alpha-\alpha_1} & 0 \\ 0 & 1 \end{vmatrix} \cdot \begin{vmatrix} 1 & h(\alpha)n(\alpha)(\alpha-\alpha_1) \\ \frac{h(\alpha)p(\alpha)}{\alpha-\alpha_1} & 1 \end{vmatrix} \cdot \begin{vmatrix} \alpha-\alpha_1 & 0 \\ 0 & 1 \end{vmatrix} \tag{120}$$

By taking into account that in the second member, the first and third matrices are rational, the factorization of the $\begin{vmatrix} 1 & h(\alpha)n(\alpha) \\ h(\alpha)p(\alpha) & 1 \end{vmatrix}$ reduces to that of

$$\begin{vmatrix} 1 & h(\alpha)n(\alpha)(\alpha-\alpha_1) \\ \frac{h(\alpha)p(\alpha)}{\alpha-\alpha_1} & 1 \end{vmatrix} \tag{121}$$

In this manner, the zero α_1 of $p(\alpha)$ is now present only in the off-diagonal entry $h(\alpha)n(\alpha)(\alpha-\alpha_1)$. By repeating the same procedure for the other zeroes of $p(\alpha)$, we obtain eq. (119) where $f(\alpha)$ differs from $h(\alpha)$ by a constant.

4.8.4 Explicit factorization of Daniele matrices

The Daniele matrices can be factorized explicitly (Daniele, 1978; Rawlins, 1980) by using

$$G(\alpha) = \begin{vmatrix} 1 & a(\alpha) \\ b(\alpha) & 1 \end{vmatrix} = \hat{G}_-(\alpha)\hat{G}_+(\alpha) \tag{122}$$

where

$$\hat{G}_-(\alpha) = \sqrt{g_-(\alpha)}\ \exp\left[\frac{1}{2}t_-(\alpha)\begin{vmatrix} 0 & n(\alpha) \\ p(\alpha) & 0 \end{vmatrix}\right] =$$

$$= \sqrt{g_-(\alpha)}\begin{vmatrix} \cosh\left[\frac{1}{2}t_-(\alpha)\sqrt{n(\alpha)p(\alpha)}\right] & \sqrt{\frac{n(\alpha)}{p(\alpha)}}\sinh\left[\frac{1}{2}t_-(\alpha)\sqrt{n(\alpha)p(\alpha)}\right] \\ \sqrt{\frac{p(\alpha)}{n(\alpha)}}\sinh\left[\frac{1}{2}t_-(\alpha)\sqrt{n(\alpha)p(\alpha)}\right] & \cosh\left[\frac{1}{2}t_-(\alpha)\sqrt{n(\alpha)p(\alpha)}\right] \end{vmatrix} \tag{123}$$

$$\hat{G}_+(\alpha) = \sqrt{g_+(\alpha)}\exp\left[\frac{1}{2}t_+(\alpha)\begin{vmatrix} 0 & n(\alpha) \\ p(\alpha) & 0 \end{vmatrix}\right] =$$

$$= \sqrt{g_+(\alpha)}\begin{vmatrix} \cosh\left[\frac{1}{2}t_+(\alpha)\sqrt{n(\alpha)p(\alpha)}\right] & \sqrt{\frac{n(\alpha)}{p(\alpha)}}\sinh\left[\frac{1}{2}t_+(\alpha)\sqrt{n(\alpha)p(\alpha)}\right] \\ \sqrt{\frac{p(\alpha)}{n(\alpha)}}\sinh\left[\frac{1}{2}t_+(\alpha)\sqrt{n(\alpha)p(\alpha)}\right] & \cosh\left[\frac{1}{2}t_+(\alpha)\sqrt{n(\alpha)p(\alpha)}\right] \end{vmatrix} \tag{124}$$

In these expressions, $g_-(\alpha)$ and $g_+(\alpha)$ are the factorized scalars of the $G(\alpha)$ determinant

$$g(\alpha) = \text{Det}[G(\alpha)] = g_-(\alpha)g_+(\alpha) \tag{125}$$

The functions $t_-(\alpha)$ and $t_+(\alpha)$ follow from the decomposition of the function $t(\alpha)$ defined by

$$t(\alpha) = \frac{1}{\sqrt{n(\alpha)p(\alpha)}}\log\frac{1+\sqrt{a(\alpha)b(\alpha)}}{1-\sqrt{a(\alpha)b(\alpha)}} = t_-(\alpha) + t_+(\alpha) \tag{126}$$

where γ_1 and γ_2 are the smile and frown real axes (section 1.6), respectively, so we have

$$t_+(\alpha) = \frac{1}{2\pi j}\int_{\gamma_1}\frac{t(u)}{u-\alpha}du, \quad t_-(\alpha) = -\frac{1}{2\pi j}\int_{\gamma_2}\frac{t(u)}{u-\alpha}du$$

Usually the decomposition of $t(\alpha)$ requires tedious calculations. An alternative expression is

$$t(\alpha) = \frac{1}{\sqrt{n(\alpha)p(\alpha)}}\log\frac{g(\alpha)}{\left[1-\sqrt{a(\alpha)b(\alpha)}\right]^2} = t_-(\alpha) + t_+(\alpha)$$

This expression put in evidence the branch points present in the logarithm that are the singularities of the determinant $g(\alpha)$ and its inverse $g^{-1}(\alpha)$.

Example 5: Factorization of a rational matrix occurring in diffraction problems

The following matrix is important in diffraction problems in stratified media (section 7.5 in chapter 7):

$$\mathbf{Y_c} = \frac{Y_o}{k\chi}\begin{vmatrix} \tau_o^2 & -\eta\,\alpha_o \\ -\eta\,\alpha_o & \tau^2 \end{vmatrix}$$

where

$$\chi = \sqrt{k^2-\eta^2-\alpha_o^2}, \quad \alpha_o = \sqrt{k^2-\tau_o^2}, \quad \tau = \sqrt{k^2-\eta^2}$$

It can be reduced to Daniele form using

$$\frac{Y_o}{k\chi}\begin{pmatrix} \tau_o & 0 \\ 0 & k+\eta \end{pmatrix}\begin{vmatrix} 1 & -\frac{\eta\,\alpha_o}{\tau_o(k-\eta)} \\ -\frac{\eta\,\alpha_o}{\tau_o(k+\eta)} & 1 \end{vmatrix}\begin{pmatrix} \tau_o & 0 \\ 0 & k-\eta \end{pmatrix}$$

The central matrix $M(\alpha) = \begin{vmatrix} 1 & -\frac{\eta\,\alpha_o}{\tau_o(k-\eta)} \\ -\frac{\eta\,\alpha_o}{\tau_o(k+\eta)} & 1 \end{vmatrix}$ has the form (122) and can be factorized by using (123) and (124):

$$a = -\frac{\eta\,\alpha_o}{\tau_o(k-\eta)}, \quad b = -\frac{\eta\,\alpha_o}{\tau_o(k+\eta)}, \quad n = k+\eta, \quad p = k-\eta$$

$$g = 1 - ab = \frac{k^2\chi^2}{\tau_o^2\tau^2} = g_- g_+, \quad t = \frac{1}{\tau}\log\frac{1+\frac{\alpha_o\eta}{\tau_o\tau}}{1-\frac{\alpha_o\eta}{\tau_o\tau}} = t_- + t_+$$

$$g_\pm = \frac{k}{\tau_o}\frac{\tau_o \mp \eta}{k \mp \eta}$$

To decompose t, we observe branch points $\eta = \pm\tau_o$ (poles of g) and $\eta = \pm k$ (zeroes of g). For the sake of simplicity, we consider only the factorized matrix $M_+(\eta)$. First we warp the smile real axis γ_1 to the closed lines that enclose the branch lines relevant to the branch points $\eta = k$ and $\eta = \tau_o$. Then we evaluate the jump of t on the two lips of branch lines. This evaluation shows that the jump is the same for both branch lines and assumes the value $\frac{2\pi j}{\tau}$. The two branch lines can be connected with an arc located at infinity. In this way, we obtain a line joining the two points τ_o with k that can be warped in the segment $\tau_o - k$. Hence, from the decomposition equation we obtain

$$t_+(\eta) = \frac{1}{2\pi j}\int_{\gamma_1}\frac{1}{\sqrt{k^2-u^2}}\log\frac{1+\frac{\alpha_o u}{\tau_o\sqrt{k^2-u^2}}}{1-\frac{\alpha_o u}{\tau_o\sqrt{k^2-u^2}}}\frac{1}{u-\eta}du = \int_{\tau_o}^{k}\frac{1}{\sqrt{k^2-u^2}}\frac{1}{u-\eta}du$$

The integral in the third member can be evaluated using MATHEMATICA. We get

$$t_+(\eta) = \frac{\pi + 2\arcsin\frac{k}{\eta}}{2\sqrt{\eta^2-k^2}} + \frac{-\log\left[-\frac{2\left(k+\frac{k^2}{\sqrt{k^2-\eta^2}}\right)}{\eta}\right] + \log\left[-\frac{2\left(k^2-\eta\tau_o+\alpha_o\sqrt{k^2-\eta^2}\right)}{\sqrt{k^2-\eta^2}(\eta-\tau_o)}\right]}{\sqrt{k^2-\eta^2}}$$

This yields the following factorized matrix:

$$M_+(\eta) = \sqrt{\frac{k}{\tau_o}\frac{\tau_o-\eta}{k-\eta}}\begin{vmatrix} \cosh\left[\frac{\sqrt{k^2-\eta^2}}{2}t_+(\eta)\right] & \frac{k+\eta}{\sqrt{k^2-\eta^2}}\sinh\left[\frac{\sqrt{k^2-\eta^2}}{2}t_+(\eta)\right] \\ \frac{k-\eta}{\sqrt{k^2-\eta^2}}\sinh\left[\frac{\sqrt{k^2-\eta^2}}{2}t_+(\eta)\right] & \cosh\left[\frac{\sqrt{k^2-\eta^2}}{2}t_+(\eta)\right] \end{vmatrix}$$

To show that this matrix is a rational matrix is a difficult task.

We verified this result by MATHEMATICA. In fact, we numerically observed that:

(a) The matrix $M_+(\eta)$ and its inverse $M_+^{-1}(\eta)$ are plus functions because the series expansion of $M_+(\eta)$ and $M_+^{-1}(\eta)$ about the points $\eta = -k$ and $\eta = -\tau_o$ are regular.

(b) The matrices $M_+(\eta)$ and $M_+^{-1}(\eta)$ are rational functions in $\eta = k$ and $\eta = \tau_o$ because the series expansion of $M_+(\eta)$ and $M_+^{-1}(\eta)$ about these points do not contain irrational powers.

(c) The matrices $M_+(\eta)$ and $M_+^{-1}(\eta)$ are bounded at infinity because $M_+(\eta)$ and $M_+^{-1}(\eta)$ are constant at infinity.

The only difficulty encountered in the performing the numerical evaluation was the impossibility for MATHEMATICA to get a numerical series expansion of the part relevant to the addend of $t_+(\eta)$:

$$t_{2+}(\eta) = \frac{+\log\left[-\frac{2(k^2-\eta\tau_o+\alpha_o\sqrt{k^2-\eta^2})}{\sqrt{k^2-\eta^2}(\eta-\tau_o)}\right]}{\sqrt{k^2-\eta^2}}$$

about the point $\eta = \tau_o$. To overcome this difficulty we operated with algebraic manipulations to simplify the contribution of this term.

4.8.5 The elimination of the offensive behavior for matrices having the Daniele form

To obtain the behavior of the functions $t_\pm(\alpha)$ for $\alpha \to \infty$, it is important to introduce the following identity:

$$\frac{1}{u-\alpha} = -\frac{1}{\alpha} - \frac{u}{\alpha^2} - \cdots - \frac{u^{h-1}}{\alpha^h} + \frac{u^h}{\alpha^{h+1}(u-\alpha)} \tag{127}$$

in the decomposition formulas. By assuming as N the degree of $n(\alpha)p(\alpha)$ and h the smaller integer greater or equal to $N/2-1$, we get the following asymptotic expansion as $\alpha \to \infty$:

$$-t_+(\alpha) = \frac{c_o}{\alpha} + \frac{c_1}{\alpha^2} + \cdots + \frac{c_{h-1}}{\alpha^h} + O\left(\frac{1}{\alpha^{h+1}}\right), \quad t_-(\alpha) = \frac{c_o}{\alpha} + \frac{c_1}{\alpha^2} + \cdots + \frac{c_{h-1}}{\alpha^h} + O\left(\frac{1}{\alpha^{h+1}}\right)$$

where

$$c_i = \int_{-\infty}^{\infty} \frac{t^i}{\sqrt{n(t)p(t)}} \ln\frac{1+\sqrt{a(t)b(t)}}{1-\sqrt{a(t)b(t)}}\, dt, \quad i = 0, \ldots, h-1$$

By taking into account the commutativity of the involved matrices we get

$$\begin{vmatrix} \cosh\left[\frac{1}{2}t_\pm(\alpha)\sqrt{n(\alpha)p(\alpha)}\right] & \sqrt{\frac{n(\alpha)}{p(\alpha)}}\sinh\left[\frac{1}{2}t_\pm(\alpha)\sqrt{n(\alpha)p(\alpha)}\right] \\ \sqrt{\frac{p(\alpha)}{n(\alpha)}}\sinh\left[\frac{1}{2}t_\pm(\alpha)\sqrt{n(\alpha)p(\alpha)}\right] & \cosh\left[\frac{1}{2}t_\pm(\alpha)\sqrt{n(\alpha)p(\alpha)}\right] \end{vmatrix}$$

$$= \begin{pmatrix} \cosh\left[\frac{\sqrt{n(\alpha)p(\alpha)}}{2}\left(\frac{c_o}{\alpha}+\frac{c_1}{\alpha^2}+\cdots+\frac{c_{h-1}}{\alpha^h}\right)\right] & \mp\sqrt{\frac{n(\alpha)}{p(\alpha)}}\sinh\left[\frac{\sqrt{n(\alpha)p(\alpha)}}{2}\left(\frac{c_o}{\alpha}+\frac{c_1}{\alpha^2}+\cdots+\frac{c_{h-1}}{\alpha^h}\right)\right] \\ \mp\sqrt{\frac{p(\alpha)}{n(\alpha)}}\sinh\left[\frac{\sqrt{n(\alpha)p(\alpha)}}{2}\left(\frac{c_o}{\alpha}+\frac{c_1}{\alpha^2}+\cdots+\frac{c_{h-1}}{\alpha^h}\right)\right] & \cosh\left[\frac{\sqrt{n(\alpha)p(\alpha)}}{2}\left(\frac{c_o}{\alpha}+\frac{c_1}{\alpha^2}+\cdots+\frac{c_{h-1}}{\alpha^h}\right)\right] \end{pmatrix}$$

$$\bullet \begin{pmatrix} \cosh\left[\sqrt{n(\alpha)p(\alpha)}\left(O\left(\frac{1}{\alpha^{h+1}}\right)\right)\right] & \sqrt{\frac{n(\alpha)}{p(\alpha)}}\sinh\left[\sqrt{n(\alpha)p(\alpha)}\left(O\left(\frac{1}{\alpha^{h+1}}\right)\right)\right] \\ \sqrt{\frac{p(\alpha)}{n(\alpha)}}\sinh\left[\sqrt{n(\alpha)p(\alpha)}\left(O\left(\frac{1}{\alpha^{h+1}}\right)\right)\right] & \cosh\left[\sqrt{n(\alpha)p(\alpha)}\left(O\left(\frac{1}{\alpha^{h+1}}\right)\right)\right] \end{pmatrix}$$

While the second matrix factor in the second member presents an appropriate behavior as $\alpha \to \infty$, the presence of nonvanishing c_i $(i = 0, \ldots, h-1)$ produces an exponential offending behavior in the first factor. To eliminate it, we introduce suitable rational matrices R having a Daniele form and commuting with $G(\alpha)$ (Daniele, 1984). Applying the factorization (122) we have

$$R = \hat{R}_- \hat{R}_+$$

where, just as occurs in $\hat{G}_\pm$, $\hat{R}_\pm$ present factors having offending behavior as $\alpha \to \infty$. If the offending factors of $\hat{R}_\pm$ compensate those of $\hat{G}_\pm$, then $\hat{G}_\pm \hat{R}_\pm$ are standard factorized matrices. Consequently,

$$G = \hat{G}_- \hat{R}_- \hat{R}_+ R^{-1} \hat{G}_+ = \hat{G}_- \hat{R}_- R^{-1} \hat{R}_+ \hat{G}_+ \qquad (128)$$

reduces the factorization of G to the factorization of the rational matrix R^{-1}.

R^{-1} is a rational matrix that can be factorized according to the general method described in the previous sections. To obtain R, let us consider the matrix

$$r(x,\alpha) = \begin{vmatrix} 1 & \dfrac{n(\alpha)}{x} \\ \dfrac{p(\alpha)}{x} & 1 \end{vmatrix}$$

The factorization through eqs. (123) and (124) yields factorized matrices $\hat{R}_\pm$ that present the offending behavior:

$$\begin{pmatrix} \cosh\left[\dfrac{\sqrt{n(\alpha)p(\alpha)}}{2}\left(\dfrac{r_o(x)}{\alpha}+\dfrac{r_1(x)}{\alpha^2}+\cdots+\dfrac{r_{h-1}(x)}{\alpha^h}\right)\right] & \pm\sqrt{\dfrac{n(\alpha)}{p(\alpha)}}\sinh\left[\dfrac{\sqrt{n(\alpha)p(\alpha)}}{2}\left(\dfrac{r_o(x)}{\alpha}+\dfrac{r_1(x)}{\alpha^2}+\cdots+\dfrac{r_{h-1}(x)}{\alpha^h}\right)\right] \\ \pm\sqrt{\dfrac{p(\alpha)}{n(\alpha)}}\sinh\left[\dfrac{\sqrt{n(\alpha)p(\alpha)}}{2}\left(\dfrac{r_o(x)}{\alpha}+\dfrac{r_1(x)}{\alpha^2}+\cdots+\dfrac{r_{h-1}(x)}{\alpha^h}\right)\right] & \cosh\left[\dfrac{\sqrt{n(\alpha)p(\alpha)}}{2}\left(\dfrac{r_o(x)}{\alpha}+\dfrac{r_1(x)}{\alpha^2}+\cdots+\dfrac{r_{h-1}(x)}{\alpha^h}\right)\right] \end{pmatrix} \qquad (129)$$

where

$$r_i(x) = \int_{-\infty}^{\infty} \frac{t^i}{\sqrt{n(t)p(t)}} \ln\frac{x+\sqrt{n(t)p(t)}}{x-\sqrt{n(t)p(t)}}\,dt \quad i = 0, \ldots, h-1 \qquad (130)$$

To obtain an explicit evaluation of the previous integral, let us consider the derivative

$$\frac{d\,r_i(x)}{dx} = \int_{-\infty}^{\infty} \frac{2t^i}{P(t)-x^2}\,dt \qquad (131)$$

where $P(t) = n(t)p(t)$. Since the integrand function is rational in t, the residue theorem yields terms having the form

$$\frac{2u^{i-1}(x)}{P'(u(x))} \qquad (132)$$

where $u(x)$ is a suitable root of the equation

$$P(t) - x^2 = 0 \tag{133}$$

We can ascertain that the algebraic function $f(u,x) = P(u) - x^2$ is of genus h and a Riemann surface (Springer, Link Encyclopaedia Mathematics) can be associated with it. In this Riemann surface, the quantities $\frac{2u^{i-1}(x)}{P'(u(x))}$ $(i = 1, 2, \ldots, h)$ are abelian differentials of the first kind (Baker, 2006, p. 128). This means that the integration of (131) yields

$$r_i(x) = \int_{\infty}^{x} dx \int_{-\infty}^{\infty} \frac{2t^{i-1}}{P(t) - x^2} dt = A_{i1}\omega_1(x) + A_{i2}\omega_2(x) + \cdots + A_{ih}\omega_h(x) \quad (i = 1, 2, \ldots, h) \tag{134}$$

where $\omega_i(x)$, $(i = 1, 2, \ldots, h)$ constitute a basis of known abelian integrals.

To eliminate the offending behavior, let us introduce the rational matrix:

$$R = \prod_{i=1}^{h} \begin{vmatrix} 1 & \dfrac{n(\alpha)}{x_i} \\ \dfrac{p(\alpha)}{x_i} & 1 \end{vmatrix} \tag{135}$$

To have $\hat{G}_{\pm}\hat{R}_{\pm}$ without offending behavior as $\alpha \to \infty$, we must enforce the following system of nonlinear equations:

$$\begin{gathered} r_o(x_1) + r_o(x_2) + \cdots + r_o(x_h) = -c_o \\ r_1(x_1) + r_1(x_2) + \cdots + r_1(x_h) = -c_1 \\ \cdots\cdots\cdots\cdots\cdots\cdots\cdots\cdots\cdots\cdots \\ r_{h-1}(x_1) + r_{h-1}(x_2) + \cdots + r_{h-1}(x_h) = -c_{h-1} \end{gathered}$$

Taking into account eq. (134), the aforementioned system constitutes a Jacobi inverse problem that is well known in the literature. The existence of its solution is proved by topological methods (see, e.g., Springer, 1957). Several algorithms have been proposed for allowing the construction of the solution of Jacobi's inversion problem (see, e.g., Zverovich, 1971). The conceptual importance of these algorithms is that they provide the solution $\mathbf{x} = (x_i)$ of the nonlinear system by the zeroes of a polynomial of degree equal to the genus h. It means that if $h \leq 4$ we obtain an exact solution constituted by radicals of the coefficients of this polynomial.

4.8.6 A relatively simple case

By taking into account eq. (118), we can assume without loss of generality that $n(\alpha) = 1$. We have a relatively simple case when the polynomial $p(\alpha)$ is biquadratic:

$$p(\alpha) = -(\alpha^4 + A\alpha^2 + B) \tag{136}$$

It involves a genus $h = 1$, and there is the only coefficient

$$c_0 = \int_{-\infty}^{\infty} \frac{1}{\sqrt{p(t)}} \ln \frac{1 + \sqrt{a(t)b(t)}}{1 - \sqrt{a(t)b(t)}} dt \tag{137}$$

that produces offending behavior. To eliminate it, the rational matrix

$$R = \begin{vmatrix} 1 & \frac{1}{x} \\ \frac{p(\alpha)}{x} & 1 \end{vmatrix}$$

yields a similar coefficient

$$r_0(x) = \int_{-\infty}^{\infty} \frac{1}{\sqrt{p(t)}} \ln \frac{x + \sqrt{p(t)}}{x - \sqrt{p(t)}} dt$$

that for a suitable value of x can compensate c_0:

$$r_0(x) + c_o = 0 \tag{138}$$

To obtain an explicit expression for $r_0(x)$, let us consider

$$\begin{aligned} \frac{d\,r_0(x)}{dx} &= \int_{-\infty}^{\infty} \frac{2}{p(t) - x^2} dt = -\int_{-\infty}^{\infty} \frac{2}{t^4 + At^2 + B + x^2} dt \\ &= 2\pi j \left(\frac{2}{4{y_1}^3 + 2A\,y_1} + \frac{2}{4{y_2}^3 + 2A\,y_2} \right) \end{aligned} \tag{139}$$

where the residue theorem has been applied, and y_1 and y_2 are the two solutions of the equation $t^4 + At^2 + B + x^2 = 0$ having negative imaginary part:

$$\begin{aligned} y_1 &= \sqrt{\frac{-A - \sqrt{A^2 - 4B - 4x^2}}{2}} \\ y_2 &= \sqrt{\frac{-A + \sqrt{A^2 - 4B - 4x^2}}{2}} \end{aligned} \tag{140}$$

Substituting in the last member that evaluates $\frac{d\,r_0(x)}{dx}$ yields

$$\frac{d\,r_0(x)}{dx} = -\sqrt{2}\pi j \left(\frac{\frac{1}{\sqrt{-A+\sqrt{A^2-4(B+x^2)}}} - \frac{1}{\sqrt{-A-\sqrt{A^2-4(B+x^2)}}}}{\sqrt{A^2 - 4(B + x^2)}} \right) \tag{141}$$

It is evident that the branch points $x = \pm\frac{1}{2}\sqrt{A^2 - 4B}$ arising from $\sqrt{A^2 - 4(B + x^2)}$ are not present in the x – plane. The only branch points present in $\frac{d\,r_0(x)}{dx}$ are $\pm jB$. Multiplying by

$$\left(\frac{1}{\sqrt{-A + \sqrt{A^2 - 4(B + x^2)}}} + \frac{1}{\sqrt{-A - \sqrt{A^2 - 4(B + x^2)}}} \right)$$

numerator and denominator of the previous equation yields, after an algebraic manipulation[2]:

$$\frac{d\,r_0(x)}{dx} = \frac{j\pi}{\sqrt{2}}\left(\frac{1}{(B+x^2)\sqrt{-\frac{A}{2(B+x^2)}+\frac{1}{\sqrt{B+x^2}}}}\right) \tag{142}$$

Using the substitution

$$z = \sqrt{B+x^2} \tag{143}$$

and taking into account that $r_0(\infty) = 0$, using MATHEMATICA the integral yields

$$r_o(x) = \frac{2\pi}{\sqrt{A+2\sqrt{B}}}\,\mathrm{EllipticF}\left[\arcsin\left(\frac{\sqrt{A+2\sqrt{B}}}{\sqrt{A-2\sqrt{B+x^2}}}\right), \frac{A-2\sqrt{B}}{A+2\sqrt{B}}\right] \tag{144}$$

For a known $r_0(x) = r_0$, the inversion of the previous equation provides the values of x that allow the rational matrix R to eliminate the offending behavior as $\alpha \to \infty$:

$$x = \sqrt{\left(\frac{-A-2\sqrt{B}+A\,\mathrm{sn}\left[\frac{(A+2\sqrt{B})r_o}{2\pi}\middle|\mathrm{m}\right]^2}{2\,\mathrm{sn}\left[\frac{(A+2\sqrt{B})r_o}{2\pi}\middle|\mathrm{m}\right]^2}\right)^2 - b} \tag{145}$$

with

$$m = \frac{A-2\sqrt{B}}{A+2\sqrt{B}} \tag{146}$$

4.8.7 The $\sqrt{a(\alpha)/b(\alpha)}$ rational function of α case

If $\sqrt{\frac{a(\alpha)}{b(\alpha)}}$ is a rational function of α,[3] the following slight modification of the Daniele method allows to obtain directly factorized matrices having always algebraic behavior as $\alpha \to \infty$, thus avoiding the cumbersome above procedure, even if the matrices thus obtained are factorized matrices only in the weak sense.

By introducing

$$\sqrt{\frac{a(\alpha)}{b(\alpha)}} = \frac{n(\alpha)}{p(\alpha)} \tag{147}$$

---

[2] The results is correct if we exclude values of x coincident with the apparent branch points.

[3] This condition can be met by a suitable rational approximation of $\sqrt{a(\alpha)/b(\alpha)}$.

the following representation of $G(\alpha)$ holds:

$$G(\alpha) = \sqrt{g(\alpha)} \exp\left[\frac{1}{2}t(\alpha)\begin{vmatrix} 0 & n^2(\alpha) \\ p^2(\alpha) & 0 \end{vmatrix}\right] = \sqrt{g(\alpha)} \exp\left[\frac{1}{2}n(\alpha)p(\alpha)t(\alpha)\begin{vmatrix} 0 & \frac{n(\alpha)}{p(\alpha)} \\ \frac{p(\alpha)}{n(\alpha)} & 0 \end{vmatrix}\right]$$

$$= \sqrt{g(\alpha)}\begin{vmatrix} \cosh\left[\frac{1}{2}t(\alpha)n(\alpha)p(\alpha)\right] & \frac{n(\alpha)}{p(\alpha)}\sinh\left[\frac{1}{2}t(\alpha)n(\alpha)p(\alpha)\right] \\ \frac{p(\alpha)}{n(\alpha)}\sinh\left[\frac{1}{2}t(\alpha)n(\alpha)p(\alpha)\right] & \cosh\left[\frac{1}{2}t_-(\alpha)n(\alpha)p(\alpha)\right] \end{vmatrix} \tag{148}$$

where $g(\alpha)$ is the determinant of the matrix kernel $G(\alpha)$ and

$$t(\alpha) = \frac{1}{n(\alpha)p(\alpha)} \ln\frac{1+\sqrt{a(\alpha)b(\alpha)}}{1-\sqrt{a(\alpha)b(\alpha)}} \tag{149}$$

The decomposition of $\ln\frac{1+\sqrt{a(\alpha)b(\alpha)}}{1-\sqrt{a(\alpha)b(\alpha)}} = \left\{\ln\frac{1+\sqrt{a(\alpha)b(\alpha)}}{1-\sqrt{a(\alpha)b(\alpha)}}\right\}_- + \left\{\ln\frac{1+\sqrt{a(\alpha)b(\alpha)}}{1-\sqrt{a(\alpha)b(\alpha)}}\right\}_+$ yields the weak factorization of $G(\alpha)$.

4.8.8.1 An alternative method

The Daniele matrix when eq. (146) holds can be rewritten as

$$W(\alpha) = \begin{vmatrix} 1 & f(\alpha)(\alpha - z_1)^2(\alpha - z_2)^2 \ldots (\alpha - z_n)^2 \\ f(\alpha) & 1 \end{vmatrix} \tag{150}$$

where z_i are the zeroes of the polynomials $n(\alpha)$ and $p(\alpha)$.

The following equations reduce the (weak) factorization of $W(\alpha)$ to a matrix very easy to factorize:

$$\begin{vmatrix} 1 & f(\alpha)(\alpha - z_1)^2(\alpha - z_2)^2 \ldots (\alpha - z_n)^2 \\ f(\alpha) & 1 \end{vmatrix}$$

$$= \begin{vmatrix} (\alpha - z_1)(\alpha - z_2)\ldots(\alpha - z_n) & 0 \\ 0 & 1 \end{vmatrix} \cdot \begin{vmatrix} 1 & w(\alpha) \\ w(\alpha) & 1 \end{vmatrix} \cdot \begin{vmatrix} \frac{1}{(\alpha - z_1)(\alpha - z_2)\ldots(\alpha - z_n)} & 0 \\ 0 & 1 \end{vmatrix} \tag{151}$$

where

$$w(\alpha) = f(\alpha)(\alpha - z_1)(\alpha - z_2)\ldots(\alpha - z_n)$$

4.9 The factorization of matrices commuting with rational matrices

4.9.1 Introduction

By taking into account that it is possible to deeply modify the structure of a matrix by suitable algebraic manipulations with rational matrices, in several diffraction problems a given matrix $G(\alpha)$ may be defined by

$$G(\alpha) = R_0(\alpha) + f(\alpha) R_1(\alpha) \tag{152}$$

where $R_0(\alpha)$ and $R_1(\alpha)$ are rational in α, and $f(\alpha)$ is an arbitrary function of α. In this section we show that these matrices are amenable to explicit factorization.

The factorization of $R_0(\alpha) = R_{0-}(\alpha)R_{0+}(\alpha)$ yields

$$G(\alpha) = R_{0-}(\alpha)\lfloor 1 + f(\alpha)R_{0-}(\alpha)^{-1}R_1(\alpha)R_{0+}(\alpha)^{-1}\rfloor R_{0+}(\alpha) \tag{153}$$

By taking into account that $R_{0-}(\alpha)^{-1}R_1(\alpha)R_{0+}(\alpha)^{-1} = \frac{P(\alpha)}{d(\alpha)}$ where $P(\alpha)$ and $d(\alpha)$ are matrix and scalar polynomial functions, respectively, it is possible to reduce the factorization of the matrix $G(\alpha)$ to that of the matrix $W(\alpha) = 1 + \frac{f(\alpha)}{d(\alpha)} P(\alpha)$. This last matrix commutes with the polynomial matrix $P(\alpha)$ and can be explicitly factorized by using the following procedure.

According to Cayley's theorem we have

$$\log[W(\alpha)] = \psi_o(\alpha)1 + \psi_1(\alpha)P(\alpha) + \cdots + \psi_{n-1}(\alpha)P^{n-1}(\alpha) \tag{154}$$

where n is the order of the matrix $W(\alpha)$. If $P(\alpha)$ has distinct eigenvalues $\lambda_i(\alpha)$, the functions $\psi_i(\alpha)$ are obtained through the Sylvester formula (Pease, 1965, p. 156):

$$g[P(\alpha)] = \sum_{i=1}^{n} g[\lambda_i] \prod_{j \neq i}^{n} \frac{P(\alpha) - \lambda_j 1}{\lambda_i - \lambda_j} \tag{155}$$

which provides

$$\log[W(\alpha)] = \log\left[1 + \frac{f(\alpha)}{d(\alpha)} P(\alpha)\right] = g[P(\alpha)] = \sum_{i=1}^{n} \log\left[1 + \frac{f(\alpha)}{d(\alpha)}\lambda_i\right] \prod_{j \neq i}^{n} \frac{P(\alpha) - \lambda_j 1}{\lambda_i - \lambda_j} \tag{156}$$

The decomposition of

$$\psi_i(\alpha) = \psi_{i-}(\alpha) + \psi_{i+}(\alpha) \tag{157}$$

yields the factorized matrices

$$W(\alpha) = \tilde{W}_-(\alpha) \cdot \tilde{W}_+(\alpha) \tag{158}$$

where

$$\begin{aligned} [\tilde{W}_-(\alpha)] &= \exp[\psi_{o-}(\alpha)1 + \psi_{1-}(\alpha)P(\alpha) + \cdots + \psi_{(n-1)-}(\alpha)P^{n-1}(\alpha)] \\ [\tilde{W}_+(\alpha)] &= \exp[\psi_{o+}(\alpha)1 + \psi_{1+}(\alpha)P(\alpha) + \cdots + \psi_{(n-1)+}(\alpha)P^{n-1}(\alpha)] \end{aligned} \tag{159}$$

This factorization process is effective because the factorized matrices $\tilde{W}_-(\alpha)$ and $\tilde{W}_+(\alpha)$ are both functions of the same matrix $P(\alpha)$ and therefore they certainly commute. Furthermore, these factorized matrices are regular together with their inverses in the half-planes $\text{Im}[\alpha] \leq 0$ and $\text{Im}[\alpha] \geq 0$ respectively, since $P(\alpha)$ is a polynomial in α and hence regular on the whole complex α – plane.

When the polynomial matrix $P(\alpha)$ is constant, $P(\alpha) = P_o$, the factorized matrices have algebraic behavior as $\alpha \to \infty$. Conversely, if the polynomial matrix $P(\alpha)$ is not constant, the previous factorized formulas ensure only that the factorized matrices and their inverses are regular in the upper and lower α – half-planes, while their asymptotic behavior as $\alpha \to \infty$ could be nonalgebraic. The nonalgebraic behavior introduces in the Wiener-Hopf procedure a transcendental entire vector function. The general technique to eliminate the offensive asymptotic behavior in the factorized matrices is reported in the next sections. It is very cumbersome, and complicated special functions are present. In these cases, rather than try to obtain an exact factorization, it is perhaps better to use the approximate factorization techniques illustrated in chapters 6 and 7.

Example 6

Consider again the factorization of the matrix (Achenbach & Gautesen, 1977):

$$M(\alpha) = \frac{1}{\alpha^2 + \eta^2} \begin{vmatrix} \alpha^2 d(\alpha) + \eta^2 \delta(\alpha) & \alpha\eta[\delta(\alpha) - d(\alpha)] \\ \alpha\eta[\delta(\alpha) - d(\alpha)] & \alpha^2 \delta(\alpha) + \eta^2 d(\alpha) \end{vmatrix}$$

This matrix is linear in the functions $d(\alpha)$ and $\delta(\alpha)$. Hence, it may be rewritten in the form

$$G(\alpha) = d(\alpha) R_1(\alpha) + \delta(\alpha) R_2(\alpha) \tag{160}$$

where the rational matrices $R_1(\alpha)$ and $R_2(\alpha)$ are given by

$$R_1(\alpha) = \frac{1}{\alpha^2 + \eta^2} \begin{vmatrix} \alpha^2 & -\alpha\eta \\ -\alpha\eta & \eta^2 \end{vmatrix}, \quad R_2(\alpha) = \frac{1}{\alpha^2 + \eta^2} \begin{vmatrix} \eta^2 & \alpha\eta \\ \alpha\eta & \alpha^2 \end{vmatrix} \tag{161}$$

Since the two rational matrices $R_1(\alpha)$ and $R_2(\alpha)$ are singular, eq. (153) does not apply. However, since $R_1(\alpha)$ and $R_2(\alpha)$ commute, from Cayley's theorem a rational matrix $R(\alpha)$ exists such that we can write

$$M(\alpha) = c(\alpha)[1 + f(\alpha) R(\alpha)] \tag{162}$$

which allows us to use the previous procedure.

Matrices of order two having the form (160), where $R_1(\alpha)$ and $R_2(\alpha)$ are singular have been considered in (Meister & Speck, 1989) using a different approach.

4.9.2 Matrix of order two commuting with polynomial matrices

All the matrices of order two commuting with polynomial matrices reduce to matrices having the Daniele form. Consequently, the theory of section 4.8.4 applies, and in general the factorization of these matrices always reduces to a Jacobi problem. However, for the sake of completeness in this section we consider the more direct factorization technique described in 4.9.1.

Let us consider a matrix $G(\alpha)$ that commutes with a polynomial matrix $\mathbf{P}(\alpha)$:

$$\mathbf{P}(\alpha) = \begin{vmatrix} P_{11}(\alpha) & P_{12}(\alpha) \\ P_{21}(\alpha) & P_{22}(\alpha) \end{vmatrix}$$

Since the matrix is of order two, we get

$$\log[G(\alpha)] = \psi_o(\alpha)1 + \psi_1(\alpha)\mathbf{P}(\alpha)$$
$$G_\pm(\alpha) = \exp[\psi_{o\pm}(\alpha)1 + \psi_{1\pm}(\alpha)\mathbf{P}(\alpha)]$$

If the polynomial $\mathbf{P}(\alpha)$ is of first degree, the factorized matrices

$$G_\pm(\alpha) = \exp[\psi_{o\pm}(\alpha)1 + \psi_{1\pm}(\alpha)\mathbf{P}(\alpha)]$$

do not present offending behavior as $\alpha \to \infty$. If $P_{11}(\alpha)$ and $P_{22}(\alpha)$ are polynomials of second degree and $P_{12}(\alpha)$ and $P_{21}(\alpha)$ polynomials of first degree or constants, it is possible to eliminate the offending behavior at $\alpha \to \infty$. In fact, a suitable normalization of the plus unknowns allows to have equal coefficients, say, c in the term containing α^2 in $P_{11}(\alpha)$ and $P_{22}(\alpha)$. Thus, we can reduce the problem to the factorization of a matrix that commutes with the polynomial matrix $\hat{\mathbf{P}}(\alpha) = c\,\alpha^2 1 + \hat{\mathbf{P}}_1(\alpha)$, where $\hat{\mathbf{P}}_1(\alpha)$ is of first degree. Now the logarithm factorized matrices have the offending behavior

$$\exp\left[\pm\frac{1}{\alpha}\hat{P}(\alpha)\int_{-\infty}^{\infty}\hat{\psi}_1(\alpha)d\alpha\right]$$

that can be eliminated by multiplying the logarithm factorized matrices by the exponential scalar $\exp\left[\mp c\,\alpha \int_{-\infty}^{\infty}\varphi_1(\alpha)d\alpha\right]$.

In the more general case, we must eliminate the offending behavior as $\alpha \to \infty$ by using the technique described in section 4.9.6.

Consider now the factorization of $W(\alpha) = 1 + h(\alpha)P(\alpha)$, where $P(\alpha)$ is an arbitrary polynomial matrix of order two and degree two. In this case the nonlinear system reduces to only one equation, and we introduce the rational matrix

$$R(\alpha) = 1 + xP(\alpha) \tag{163}$$

Taking into account the expressions reported later in section 4.9.3, to cancel the offending exponential behaviors we force the unknown parameter x to satisfy the nonlinear equation

$$f(x) = \int_{-\infty}^{\infty}\left(\frac{\log[1+x\lambda_1]}{\lambda_1-\lambda_2} + \frac{\log[1+x\lambda_2]}{\lambda_2-\lambda_1}\right)d\alpha$$
$$= \int_{-\infty}^{\infty}\left(\frac{\log[1+h(\alpha)\lambda_1]}{\lambda_1-\lambda_2} + \frac{\log[1+h(\alpha)\lambda_2]}{\lambda_2-\lambda_1}\right)d\alpha = n \tag{164}$$

where λ_1 and λ_2 are the eigenvalues of $P(\alpha)$. This yields

$$\frac{df(x)}{dx} = \int_{-\infty}^{\infty} \left(\frac{1}{1 + x(\lambda_1 + \lambda_2) + x^2 \lambda_2 \lambda_1} \right) d\alpha \tag{165}$$

The quantities $\lambda_1 + \lambda_2$ and $\lambda_1\lambda_2$ are polynomials, since they are the trace and the determinant of the matrix $P(\alpha)$. Consequently, the integrand function is rational in α, and therefore the residue theorem applies. In some important case the polynomials $\lambda_1 + \lambda_2$ and $\lambda_1\lambda_2$ are of degree 2 and the residue theorem yields

$$\frac{df(x)}{dx} = 2\pi j \frac{1}{\sqrt{x(ax^3 + bx^2 + cx + d)}} \tag{166}$$

where a, b, c, d are constants.

By taking into account that $f(0) = 0$, $f(x)$ can be expressed in terms of elliptic integrals and by inverting them, it is possible to obtain an explicit expression for x in terms of Jacobian elliptic functions. In fact, by introducing

$$ax^3 + bx^2 + cx + d = a(x - x_1)(x - x_2)(x - x_3) \tag{167}$$

we have

$$f(x) = \frac{1}{\sqrt{ax_2}} \frac{2}{(x_1 - x_3)x_1} EllipticF\left[ArcSin\left[\sqrt{\frac{x(x_3 - x_1)}{(x - x_1)}}, \frac{(x_1 - x_2)x_3}{(x_1 - x_3)x_2}\right]\right] \tag{168}$$

Hence

$$f(x) = n \tag{169}$$

has the analytical solution

$$x = \frac{x_1 x_3 \left[JacobiSN\left[\frac{1}{2} n(x_1\sqrt{x_2}(x_1 - x_3), \frac{(x_1-x_2)x_3}{(x_1-x_3)x_2}\right]\right]^2}{x_1 - x_3 + x_3\left[JacobiSN\left[\frac{1}{2} n(x_1\sqrt{x_2}(x_1 - x_3), \frac{(x_1-x_2)x_3}{(x_1-x_3)x_2}\right]\right]^2} \tag{170}$$

The factorization formulas follow from

$$W = \tilde{W}_-\tilde{R}_-R^{-1}\tilde{R}_+\tilde{W}_+ = \tilde{W}_-\tilde{R}_-\left[R^{-1}\right]_-\left[R^{-1}\right]_+\tilde{R}_+\tilde{W}_+ \tag{171}$$

where the factorization of $R^{-1}(\eta) = (1 + xP_2(\eta))^{-1}$ can be obtained by the rational matrices factorization method.

4.9.3 Explicit expression of $\psi_i(\alpha)$ in the general case

According to Sylvester's representation, in the case of simple eigenvalues $\lambda_i = \lambda_i(\alpha)$ of $P(\alpha)$, the matrix $\log[W(\alpha)] = \log[h(\alpha)1 - P(\alpha)]$ may be written as (Pease, 1965)

$$\log[W(\alpha)] = \sum_{i=1}^{n} \log[W(\lambda_i)] \prod_{j \neq i}^{n} \frac{P(\alpha) - \lambda_j 1}{\lambda_i - \lambda_j} \tag{172}$$

Dealing with a matrix of order n, this yields

$$\log[W(\alpha)] = \psi_o(\alpha)1 + \psi_1(\alpha)P(\alpha) + \cdots + \psi_{n-1}(\alpha)P^{n-1}(\alpha) \tag{173}$$

For example, for a matrix $P(\alpha)$ of order two we have

$$\log[h(\alpha)\mathbf{1} - P(\alpha)] = \psi_o(\alpha)1 + \psi_1(\alpha)P(\alpha) \tag{174}$$

where

$$\psi_o(\alpha) = -\frac{\log[h(\alpha) - \lambda_1]\lambda_2}{\lambda_1 - \lambda_2} - \frac{\log[h(\alpha) - \lambda_2]\lambda_1}{\lambda_2 - \lambda_1}$$
$$\psi_1(\alpha) = \frac{\log[h(\alpha) - \lambda_1]}{\lambda_1 - \lambda_2} + \frac{\log[h(\alpha) - \lambda_2]}{\lambda_2 - \lambda_1}$$

For matrices of order three,

$$\log[h(\alpha) - P(\alpha)] = \psi_o(\alpha)1 + \psi_1(\alpha)P(\alpha) + \psi_2(\alpha)P^2(\alpha) \tag{175}$$

where

$$\psi_o(\alpha) = \frac{\log[h(\alpha) - \lambda_1]\lambda_2\lambda_3}{(\lambda_1 - \lambda_2)(\lambda_1 - \lambda_3)} + \frac{\log[h(\alpha) - \lambda_2]\lambda_1\lambda_3}{(\lambda_2 - \lambda_1)(\lambda_2 - \lambda_3)} + \frac{\log[h(\alpha) - \lambda_3]\lambda_1\lambda_2}{(\lambda_3 - \lambda_1)(\lambda_3 - \lambda_2)}$$
$$\psi_1(\alpha) = -\frac{\log[h(\alpha) - \lambda_1](\lambda_2 + \lambda_3)}{(\lambda_1 - \lambda_2)(\lambda_1 - \lambda_3)} - \frac{\log[1 - \lambda_2](\lambda_1 + \lambda_3)}{(\lambda_2 - \lambda_1)(\lambda_2 - \lambda_3)} - \frac{\log[h(\alpha) - \lambda_3](\lambda_1 + \lambda_2)}{(\lambda_3 - \lambda_1)(\lambda_3 - \lambda_2)}$$
$$\psi_2(\alpha) = \frac{\log[h(\alpha) - \lambda_1]}{(\lambda_1 - \lambda_2)(\lambda_1 - \lambda_3)} + \frac{\log[h(\alpha) - \lambda_2]}{(\lambda_2 - \lambda_1)(\lambda_2 - \lambda_3)} + \frac{\log[h(\alpha) - \lambda_3]}{(\lambda_3 - \lambda_1)(\lambda_3 - \lambda_2)}$$

For matrices of order four,

$$\log[h(\alpha)1 - P(\alpha)] = \psi_o(\alpha)1 + \psi_1(\alpha)P(\alpha) + \psi_2(\alpha)P^2(\alpha) + \psi_3(\alpha)P^3(\alpha) \tag{176}$$

where

$$\psi_o(\alpha) = -\frac{\log[h(\alpha) - \lambda_1]\lambda_2\lambda_3\lambda_4}{(\lambda_1 - \lambda_2)(\lambda_1 - \lambda_3)(\lambda_1 - \lambda_4)} - \frac{\log[h(\alpha) - \lambda_2]\lambda_1\lambda_3\lambda_4}{(\lambda_2 - \lambda_1)(\lambda_2 - \lambda_3)(\lambda_2 - \lambda_4)}$$
$$- \frac{\log[h(\alpha) - \lambda_3]\lambda_1\lambda_2\lambda_4}{(\lambda_3 - \lambda_1)(\lambda_3 - \lambda_2)(\lambda_3 - \lambda_4)} - \frac{\log[h(\alpha) - \lambda_4]\lambda_1\lambda_2\lambda_3}{(\lambda_4 - \lambda_1)(\lambda_4 - \lambda_2)(\lambda_4 - \lambda_3)}$$

$$\psi_1(\alpha) = \frac{\log[h(\alpha) - \lambda_1](\lambda_2\lambda_3 + \lambda_2\lambda_4 + \lambda_3\lambda_4)}{(\lambda_1 - \lambda_2)(\lambda_1 - \lambda_3)(\lambda_1 - \lambda_4)} + \frac{\log[h(\alpha) - \lambda_2](\lambda_1\lambda_3 + \lambda_1\lambda_4 + \lambda_3\lambda_4)}{(\lambda_2 - \lambda_1)(\lambda_2 - \lambda_3)(\lambda_2 - \lambda_4)}$$
$$+ \frac{\log[h(\alpha) - \lambda_3](\lambda_1\lambda_2 + \lambda_1\lambda_4 + \lambda_2\lambda_4)}{(\lambda_3 - \lambda_1)(\lambda_3 - \lambda_2)(\lambda_3 - \lambda_4)} + \frac{\log[h(\alpha) - \lambda_4](\lambda_1\lambda_2 + \lambda_1\lambda_3 + \lambda_2\lambda_3)}{(\lambda_4 - \lambda_1)(\lambda_4 - \lambda_2)(\lambda_4 - \lambda_3)}$$

$$\psi_2(\alpha) = -\frac{\log[h(\alpha)-\lambda_1](\lambda_2+\lambda_3+\lambda_4)}{(\lambda_1-\lambda_2)(\lambda_1-\lambda_3)(\lambda_1-\lambda_4)} - \frac{\log[h(\alpha)-\lambda_2](\lambda_1+\lambda_3+\lambda_4)}{(\lambda_2-\lambda_1)(\lambda_2-\lambda_3)(\lambda_2-\lambda_4)}$$
$$-\frac{\log[h(\alpha)-\lambda_3](\lambda_1+\lambda_2+\lambda_4)}{(\lambda_3-\lambda_1)(\lambda_3-\lambda_2)(\lambda_3-\lambda_4)} - \frac{\log[h(\alpha)-\lambda_4](\lambda_1+\lambda_2+\lambda_3)}{(\lambda_4-\lambda_1)(\lambda_4-\lambda_2)(\lambda_4-\lambda_3)}$$

$$\psi_3(\alpha) = \frac{\log[h(\alpha)-\lambda_1]}{(\lambda_1-\lambda_2)(\lambda_1-\lambda_3)(\lambda_1-\lambda_4)} + \frac{\log[h(\alpha)-\lambda_2]}{(\lambda_2-\lambda_1)(\lambda_2-\lambda_3)(\lambda_2-\lambda_4)}$$
$$+\frac{\log[h(\alpha)-\lambda_3]}{(\lambda_3-\lambda_1)(\lambda_3-\lambda_2)(\lambda_3-\lambda_4)} + \frac{\log[h(\alpha)-\lambda_4]}{(\lambda_4-\lambda_1)(\lambda_4-\lambda_2)(\lambda_4-\lambda_3)}$$

The complexity of these expressions is because even though $P(\alpha)$ is rational the eigenvalues $\lambda_i(\alpha)$ are not rational functions of α. If the order of the matrix $P(\alpha)$ does not exceed 4, MATHEMATICA provides analytical expressions for $\lambda_i(\alpha)$.

Alternatively, we can use the following technique to obtain the functions $\psi_i(\alpha)$. This technique is conceptually simpler, since it avoids the introduction of the eigenvalues of $P(\alpha)$. Since

$$W(\alpha) = 1 + \frac{f(\alpha)}{d(\alpha)}P(\alpha) = P(\alpha)\left[1 - \left[-\frac{f(\alpha)}{d(\alpha)}\right]P^{-1}(\alpha)\right] \tag{177}$$

it is not restrictive to refer to the factorization of the matrix:

$$G(\alpha) = 1 - l(\alpha)P^{-1}(\alpha) \tag{178}$$

where $l(\alpha)$ is a scalar known function. We have

$$\log\left[-l(\alpha)P^{-1}(\alpha) + \mathbf{1}\right] = \int_0^{l(\alpha)} \frac{1}{x1 - P(\alpha)}dx \tag{179}$$

Given a matrix $\mathbf{P}$ of order n, we will evaluate the matrix $(x\mathbf{1}-\mathbf{P})^{-1}$ in terms of rational expressions of x. For illustrative purposes we will consider only the case $n=4$. By Cayley's theorem we have

$$(x\mathbf{1}-\mathbf{P})^{-1} = r_o(x)\mathbf{1} + r_1(x)\mathbf{P} + r_2(x)\mathbf{P}^2 + r_3(x)\mathbf{P}^3 \tag{180}$$

The rationality of the four functions $r_i(x)$ $(i = 0, 1, 2, 3)$ is evident, and in the following we will evaluate them explicitly.

Equation (180) can be rewritten as

$$(x\mathbf{1}-\mathbf{P})\left[r_o(x)\mathbf{1} + r_1(x)\mathbf{P} + r_2(x)\mathbf{P}^2 + r_3(x)\mathbf{P}^3\right] = \mathbf{1} \tag{181}$$

Taking into account the characteristic equation

$$\mathbf{P}^4 + a_1\mathbf{P}^3 + a_2\mathbf{P}^2 + a_3\mathbf{P} + a_4\mathbf{1} = 0 \tag{182}$$

eq. (181) yields

$$e_o(x)\mathbf{1} + e_1(x)\mathbf{P} + e_2(x)\mathbf{P}^2 + e_3(x)\mathbf{P}^3 - \mathbf{1} = 0 \tag{183}$$

where

$$\begin{aligned} e_0 &= a_4 r_3 + r_0 x, \quad & e_1 &= a_3 r_3 - r_0 + r_1 x \\ e_2 &= -r_1 + a_2 r_3 + r_2 x, \quad & e_3 &= -r_2 + a_1 r_3 + r_3 x \end{aligned} \tag{184}$$

Equation (183) is equivalent to the following four equations in the unknowns $e_o - 1,\ e_1,\ e_2,\ e_3$:

$$e_o(x) - 1 + e_1(x)\lambda_i + e_2(x)\lambda_i^{\,2} + e_3(x)\lambda_i^{\,3} = 0 \quad (i = 1, \ldots, 4)$$

where λ_i $(i = 1, 2, 3, 4)$ are the eigenvalues of $\mathbf{P}$. Since the determinant of the coefficients is not vanishing,[4] we obtain the values of e_i $(i = 0, 1, 2, 3)$:

$$e_o - 1 = 0, \quad e_1 = 0, \quad e_2 = 0, \quad e_3 = 0$$

Consequently, eq. (184) provide the evaluation of $r_o(x), r_1(x), r_2(x), r_3(x)$. We obtain the following rational expressions:

$$\begin{aligned} r_0(x) &= \frac{x^3 + a_1x^2 + a_2x + a_3}{x^4 + a_1x^3 + a_2x^2 + a_3x + a_4}, \quad & r_1(x) &= \frac{x^2 + a_1x + a_2}{x^4 + a_1x^3 + a_2x^2 + a_3x + a_4} \\ r_2(x) &= \frac{x + a_1}{x^4 + a_1x^3 + a_2x^2 + a_3x + a_4}, \quad & r_3(x) &= \frac{1}{x^4 + a_1x^3 + a_2x^2 + a_3x + a_4} \end{aligned} \tag{185}$$

The Bocher equations provide the evaluation of the coefficients a_i of the characteristic equation in terms of the entries of the matrix $\mathbf{P}$ (182). In fact, we have

$$\begin{aligned} &a_1 = -s_1, \quad a_2 = -\frac{a_1 s_1 + s_2}{2}, \quad a_3 = -\frac{a_2 s_1 + a_1 s_2 + s_3}{3} \\ &a_4 = -\frac{a_3 s_1 + a_2 s_2 + a_1 s_3 + s_4}{4} = (-1)^4 \det[P] \end{aligned} \tag{186}$$

where

$$\begin{aligned} s_1 &= Tr[P], \quad & s_2 &= Tr[P^2] \\ s_3 &= Tr[P^3], \quad & s_4 &= Tr[P^4] \end{aligned}$$

---

[4] This is a Vandermonde determinant.

This shows that for polynomials matrices **P** the coefficients a_i are polynomials of α and the functions

$$r_o(x) = r_o(x, \alpha), \quad r_1(x) = r_1(x, \alpha)$$
$$r_2(x) = r_2(x, \alpha), \quad r_3(x) = r_3(x, \alpha)$$

are rational functions of the algebraic functions

$$f(x, \alpha) = x^4 + a_1(\alpha)x^3 + a_2(\alpha)x^2 + a_3(\alpha)x + a_4(\alpha) \tag{187}$$

Of course the values of x that constitute poles of the function $r_o(x), r_1(x), r_2(x), r_3(x)$ are the zeroes with respect to x of the algebraic equation $f(x, \alpha) = 0$ and represent the eigenvalues of **P**.

From

$$\log[G(\alpha)] = \int_0^{l(\alpha)} \frac{1}{x\mathbf{1} - P(\alpha)} dx \tag{188}$$

it follows that

$$\log\left[-l(\alpha)P^{-1}(\alpha) + \mathbf{1}\right] = \int_0^{l(\alpha)} r_o(x, \alpha)dx\ \mathbf{1} + \int_0^{l(\alpha)} r_1(x, \alpha)dx\ \mathbf{P}(\alpha)$$
$$+ \int_0^{l(\alpha)} r_2(x, \alpha)dx\ \mathbf{P}^2(\alpha) + \int_0^{l(\alpha)} r_3(x, \alpha)dx\ \mathbf{P}^3(\alpha)$$
$$= g_0(\alpha)\mathbf{1} + g_1(\alpha)\mathbf{P}(\alpha) + g_2(\alpha)\mathbf{P}^2(\alpha) + g_3(\alpha)\mathbf{P}^3(\alpha) \tag{189}$$

where

$$g_i(\alpha) = \int_0^{l(\alpha)} r_i(x, \alpha)dx \quad i = 0, 1, 2, 3$$

4.9.4 Asymptotic behavior of the logarithmic representation of $-l(\alpha)P^{-1}(\alpha) + 1$

We have obtained

$$-l(\alpha)P^{-1}(\alpha) + \mathbf{1} = \exp\left[g_o(\alpha)\mathbf{1} + g_1(\alpha)\mathbf{P}(\alpha) + g_2(\alpha)\mathbf{P}^2(\alpha) + g_3(\alpha)\mathbf{P}^3(\alpha)\right] \tag{190}$$

Taking into account that the first member is algebraic as $\alpha \to \infty$,

$$-f(\alpha)P^{-1}(\alpha) + 1 \approx \alpha^{\nu}M \tag{191}$$

where M is a constant matrix, it follows that

$$\begin{aligned} &\mathrm{Exp}[g_o(\alpha)1] \to \alpha^{\nu}, \quad \text{or} \quad g_o(\alpha) \to \nu \log \alpha \\ &\mathrm{Exp}[g_1(\alpha)P(\alpha)] \to \mathrm{Exp}[g_1(\alpha)\alpha^p] \to \alpha^{\nu}, \quad \text{or} \quad g_1(\alpha) \to \frac{\nu \log \alpha}{\alpha^p} \end{aligned} \tag{192}$$

where p is the degree of $P(\alpha)$, and so on. It provides the proof of the convergence of the integrals that will be introduced later.

4.9.5 Asymptotic behavior of the decomposed $\psi_{i\pm}(\alpha)$

Let us consider, for the sake of simplicity, that the degree of $P(\alpha)$ does not exceed 2.[5] The logarithmic decomposition yields

$$\begin{aligned} &\left\{\log\left[-l(\alpha)P^{-1}(\alpha)+1\right]\right\}_{\pm} \\ &\quad = \{g_0(\alpha)\}_{\pm}\mathbf{1} + \{g_1(\alpha)\}_{\pm}\mathbf{P}(\alpha) + \{g_2(\alpha)\}_{\pm}\mathbf{P}^2(\alpha) + \{g_3(\alpha)\}_{\pm}\mathbf{P}^3(\alpha) \end{aligned} \tag{193}$$

To achieve the behavior of the functions $\{g_i(\alpha)\}_{\pm}$, $i = 0, 1, 2, 3$ for $\alpha \to \infty$, it is important to introduce the following identity:

$$\frac{1}{u-\alpha} = -\frac{1}{\alpha} - \frac{u}{\alpha^2} - \frac{u^{N-2}}{\alpha^{N-1}} + \frac{u^{N-1}}{\alpha^{N-1}(u-\alpha)} \tag{194}$$

in the decomposition formulas (section 3.1):

$$\{g_i(\alpha)\}_{\pm} = \pm\frac{1}{2\pi j}\int_{\gamma_{1,2}} \frac{g_i(u)}{u-\alpha}du \quad i = 0, 1, 2, 3 \tag{195}$$

This yields

$$\begin{aligned} \{g_0(\alpha)\}_{\pm} &= \pm\left(-\frac{\psi_0^1}{\alpha}\right) + O\left[\frac{1}{\alpha^2}\right] \\ \{g_1(\alpha)\}_{\pm} &= \pm\left(-\frac{\psi_1^1}{\alpha}\right) + O\left[\frac{1}{\alpha^2}\right] \\ \{g_2(\alpha)\}_{\pm} &= \pm\left(-\frac{\psi_2^1}{\alpha} - \frac{\psi_2^2}{\alpha^2} - \frac{\psi_2^3}{\alpha^3}\right) + O\left[\frac{1}{\alpha^4}\right] \\ \{g_3(\alpha)\}_{\pm} &= \pm\left(-\frac{\psi_3^1}{\alpha} - \frac{\psi_3^2}{\alpha^2} - \frac{\psi_3^3}{\alpha^3} - \frac{\psi_3^4}{\alpha^4} - \frac{\psi_3^5}{\alpha^5}\right) + O\left[\frac{1}{\alpha^6}\right] \end{aligned} \tag{196}$$

---

[5] This case occurs in the problem of diffraction by a half-plane having arbitrary impedance faces immersed in an isotropic medium.

where

$$\psi_0^1(G) = \psi_0^1 = \frac{1}{2\pi j}\int_{-\infty}^{\infty} g_0(u)du$$

$$\psi_1^1(G) = \psi_1^1 = \frac{1}{2\pi j}\int_{-\infty}^{\infty} g_1(u)du$$

$$\psi_2^1(G) = \psi_2^1 = \frac{1}{2\pi j}\int_{-\infty}^{\infty} g_2(u)du, \quad \psi_2^2(G) = \psi_2^2 = \frac{1}{2\pi j}\int_{-\infty}^{\infty} u\, g_2(u)du$$

$$\psi_2^3(G) = \psi_2^3 = \frac{1}{2\pi j}\int_{-\infty}^{\infty} u^2\, g_2(u)du \tag{197}$$

$$\psi_3^1(G) = \psi_3^1 = \frac{1}{2\pi j}\int_{-\infty}^{\infty} g_3(u)du, \quad \psi_3^2(G) = \psi_3^2 = \frac{1}{2\pi j}\int_{-\infty}^{\infty} u\, g_3(u)du$$

$$\psi_3^3(G) = \psi_3^3 = \frac{1}{2\pi j}\int_{-\infty}^{\infty} u^2\, g_3(u)du$$

$$\psi_3^4(G) = \psi_3^4 = \frac{1}{2\pi j}\int_{-\infty}^{\infty} u^3\, g_3(u)du, \quad \psi_3^5(G) = \psi_3^5 = \frac{1}{2\pi j}\int_{-\infty}^{\infty} u^4\, g_3(u)du$$

We observe that

$$\begin{aligned} P(\alpha) &= \mathbf{A}_o + \alpha\mathbf{A}_1 + \alpha^2\mathbf{A}_2 \\ P^2(\alpha) &= \mathbf{B_o} + \alpha\mathbf{B_1} + \alpha^2\mathbf{B_2} + \alpha^3\mathbf{B_3} + \alpha^4\mathbf{B_4} \\ P^3(\alpha) &= \mathbf{C_o} + \alpha\mathbf{C_1} + \alpha^2\mathbf{C_2} + \alpha^3\mathbf{C_3} + \alpha^4\mathbf{C_4} + \alpha^5\mathbf{C}_5 + \alpha^6\mathbf{C}_6 \end{aligned} \tag{198}$$

with $\mathbf{A}_i$, $\mathbf{B}_i$, $\mathbf{C}_i$ constant matrices of order four. Taking these equations into account, we deduce that in the expression

$$\begin{aligned} &\{\log[-l(\alpha)P^{-1}(\alpha) + 1]\}_\pm \\ &\quad = \{g_0(\alpha)\}_\pm \mathbf{1} + \{g_1(\alpha)\}_\pm \mathbf{P}(\alpha) + \{g_2(\alpha)\}_\pm \mathbf{P}^2(\alpha) + \{g_3(\alpha)\}_\pm \mathbf{P}^3(\alpha) \end{aligned}$$

- the term $\{g_o(\alpha)\}_\pm$ does not involve offending exponential behaviors;
- the term $\{g_1(\alpha)\}_\pm$ involves the offending contribution $\mp\alpha^2 \frac{\psi_1^1}{\alpha}\mathbf{A}_2$;
- the term $\{g_2(\alpha)\}_\pm$ involves the offending contributions

$$\mp\alpha^2\frac{\psi_2^1}{\alpha}\mathbf{B}_2 \mp \alpha^3\left(\frac{\psi_2^1}{\alpha} + \frac{\psi_2^2}{\alpha^2}\right)\mathbf{B}_3 \mp \alpha^4\left(\frac{\psi_2^1}{\alpha} + \frac{\psi_2^2}{\alpha^2} + \frac{\psi_2^3}{\alpha^3}\right)\mathbf{B}_4$$

and finally,

- the term $\{g_3(\alpha)\}_\pm$ involves the offending contributions

$$\mp\alpha^2\frac{\psi_3^1}{\alpha}\mathbf{C}_2 \mp \alpha^3\left(\frac{\psi_3^1}{\alpha}+\frac{\psi_3^2}{\alpha^2}\right)\mathbf{C}_3 \mp \alpha^4\left(\frac{\psi_3^1}{\alpha}+\frac{\psi_3^2}{\alpha^2}+\frac{\psi_3^3}{\alpha^3}\right)\mathbf{C}_4$$
$$+\mp\alpha^5\left(\frac{\psi_3^1}{\alpha}+\frac{\psi_3^2}{\alpha^2}+\frac{\psi_3^3}{\alpha^3}+\frac{\psi_3^4}{\alpha^4}\right)\mathbf{C}_5 \mp \alpha^6\left(\frac{\psi_3^1}{\alpha}+\frac{\psi_3^2}{\alpha^2}+\frac{\psi_3^3}{\alpha^3}+\frac{\psi_3^4}{\alpha^4}+\frac{\psi_3^5}{\alpha^5}\right)\mathbf{C}_6$$

These expressions show that there would be no offending behavior if the dominant terms relevant to $\{g_{1,2,3}(\alpha)\}_\pm$ were vanishing:

$$\begin{aligned}&\psi_1^1(G)=0\\&\psi_2^1(G)=0,\quad \psi_2^2(G)=0\quad \psi_2^3(G)=0\\&\psi_3^1(G)=0,\quad \psi_3^2(G)=0,\quad \psi_3^3(G)=0,\quad \psi_3^4(G)=0,\quad \psi_3^5(G)=0\end{aligned}\tag{199}$$

4.9.6 A procedure to eliminate the exponential behavior

The procedure described in this section generalizes the one proposed by Daniele (Daniele, 1984a). If we perform the logarithmic factorization on an arbitrary matrix $R_x(\alpha)$ that commutes with the polynomial matrix $P(\alpha)$ defined already, it is easy to show that

$$\psi_r^s(R_x\cdot G)=\psi_r^s(R_x)+\psi_r^s(G)$$

If $R_x(\alpha)$ is rational depending on suitable parameters x, it suggests the reduction of the factorization of $G(\alpha)$ to that of $G_e(\alpha)=R_x(\alpha)G(\alpha)$ and the choice of the parameters x such that the following equations hold for the matrix $G_e(\alpha)=R_x(\alpha)G(\alpha)$:

$$\begin{aligned}&\psi_1^1(G_e)=\psi_1^1(R_x)+\psi_1^1(G)=0\\&\psi_2^1(G_e)=\psi_2^1(R_x)+\psi_2^1(G)=0,\quad \psi_2^2(G_e)=\psi_2^2(R_x)+\psi_2^2(G)=0\\&\psi_2^3(G_e)=\psi_2^3(R_x)+\psi_2^3(G)=0\\&\psi_3^1(G_e)=\psi_3^1(R_x)+\psi_3^1(G)=0,\quad \psi_3^2(G_e)=\psi_3^2(R_x)+\psi_3^2(G)=0\\&\psi_3^3(G_e)=\psi_3^3(R_x)+\psi_3^3(G)=0,\quad \psi_3^4(G_e)=\psi_3^4(R_x)+\psi_3^4(G)=0\\&\psi_3^5(G_e)=\psi_3^5(R_x)+\psi_3^5(G)=0\end{aligned}\tag{200}$$

If these equations are verified, the logarithmic factorization of $G_e(\alpha)=G_{e-}(\alpha)\cdot G_{e+}(\alpha)$ does not present offensive asymptotic behavior and the factorization of $G(\alpha)=R_x^{-1}(\alpha)\cdot G_e(\alpha)$ can be obtained with the method indicated in section 2.7. Alternatively, taking into account that $G(\alpha)$ and $R_x(\alpha)$ and all their logarithmic factorized matrices commute, we can simplify the factorization of $G(\alpha)$ by rewriting

$$G(\alpha)=G_{e-}(\alpha)R_x^{-1}(\alpha)G_e(\alpha)$$

We obtain

$$G_-(\alpha)=G_{e-}(\alpha)Q_-(\alpha),\quad G_+(\alpha)=Q_+(\alpha)G_{e+}(\alpha)\tag{201}$$

where $Q_-(\alpha)$ and $Q_+(\alpha)$ are the standard factorized matrices of the rational matrix $R_x^{-1}(\alpha)$[6]:

$$R_x^{-1}(\alpha) = Q_-(\alpha) \cdot Q_+(\alpha) \tag{202}$$

Taking into account the nine conditions that must be satisfied, in the following we assume as a possible rational matrix, $R_x(\alpha)$:

$$\mathbf{R}_x(\alpha) = \prod_{j=1}^{9} \left[-x_j \mathbf{P}^{-1}(\alpha) + \mathbf{1} \right]^{n_j} \tag{203}$$

where the scalars x_j and the integer n_j are to be determined.

By setting

$$\psi_1^1\left[-x\,\mathbf{P}^{-1}(\alpha) + \mathbf{1}\right] = f_1(x) = \frac{1}{2\pi j}\int_0^x \int_{-\infty}^{\infty} r_1(x,u)du\,dx$$

$$\psi_2^1\left[-x\,\mathbf{P}^{-1}(\alpha) + \mathbf{1}\right] = f_2(x) = \frac{1}{2\pi j}\int_0^x \int_{-\infty}^{\infty} r_2(x,u)du\,dx$$

$$\psi_2^2\left[-x\,\mathbf{P}^{-1}(\alpha) + \mathbf{1}\right] = f_3(x) = \frac{1}{2\pi j}\int_0^x \int_{-\infty}^{\infty} u\,r_2(x,u)du\,dx$$

$$\psi_2^3\left[-x\,\mathbf{P}^{-1}(\alpha) + \mathbf{1}\right] = f_4(x) = \frac{1}{2\pi j}\int_0^x \int_{-\infty}^{\infty} u^2\,r_2(x,u)du\,dx$$

$$\psi_3^1\left[-x\,\mathbf{P}^{-1}(\alpha) + \mathbf{1}\right] = f_5(x) = \frac{1}{2\pi j}\int_0^x \int_{-\infty}^{\infty} r_3(x,u)du\,dx$$

$$\psi_3^2\left[-x\,\mathbf{P}^{-1}(\alpha) + \mathbf{1}\right] = f_6(x) = \frac{1}{2\pi j}\int_0^x \int_{-\infty}^{\infty} u\,r_3(x,u)du\,dx$$

$$\psi_3^3\left[-x\,\mathbf{P}^{-1}(\alpha) + \mathbf{1}\right] = f_7(x) = \frac{1}{2\pi j}\int_0^x \int_{-\infty}^{\infty} u^2 r_3(x,u)du\,dx$$

$$\psi_3^4\left[-x\,\mathbf{P}^{-1}(\alpha) + \mathbf{1}\right] = f_8(x) = \frac{1}{2\pi j}\int_0^x \int_{-\infty}^{\infty} u^3 r_3(x,u)du\,dx$$

$$\psi_3^5\left[-x\,\mathbf{P}^{-1}(\alpha) + \mathbf{1}\right] = f_9(x) = \frac{1}{2\pi j}\int_0^x \int_{-\infty}^{\infty} u^4 r_3(x,u)du\,dx$$

[6] Notice that $Q_-(\alpha)$ and $Q_+(\alpha)$ must be standard. They can be obtained by the techniques that factorize rational matrices. In general this method yields factorized matrices that do not commute.

and forcing the nine conditions (200), we obtain the following system of nonlinear equations ($q = 9$):

$$
\begin{aligned}
&n_1 f_1(x_1) + n_2 f_1(x_2) + \cdots + n_q f_1(x_q) = -\psi_1^1(G) = -c_1\\
&n_1 f_2(x_1) + n_2 f_2(x_2) + \cdots + n_q f_2(x_q) = -\psi_2^1(G) = -c_2\\
&\ldots\\
&n_1 f_q(x_1) + n_2 f_q(x_2) + \cdots + n_q f_q(x_q) = -\psi_3^5(G) = -c_q
\end{aligned}
\tag{204}
$$

These equations can be rewritten in the abstract form:

$$
\mathbf{F}(\mathbf{x}) = 0 \tag{205}
$$

where

$$
\mathbf{x} = (x_1, x_2, x_3, x_4, \ldots, x_q), \quad \mathbf{F}(\mathbf{x}) = \begin{vmatrix} \sum_{j=1}^{q} n^j f_1(x_j) + c_1 \\ \sum_{j=1}^{q} n^j f_2(x_j) + c_2 \\ \ldots \\ \ldots \\ \sum_{j=1}^{q} n^j f_q(x_j) + c_q \end{vmatrix}
$$

The solution of the nonlinear eq. (205) can be obtained through the iterative process based on the Newton-Raphson formulas:

$$
\mathbf{x}^{(j+1)} = \mathbf{x}^{(j)} - \left[\mathbf{J}(\mathbf{x}^{(j)})\right]^{-1} \mathbf{F}(\mathbf{x}^{(j)}) \tag{206}
$$

where the Jacobian matrix is

$$
J(\mathbf{x}^{(i)}) = \begin{vmatrix} n_1 f_1'(x_1) & n_2 f_1'(x_2) & n_3 f_1'(x_3) & \ldots & n_q f_1'(x_q) \\ n_1 f_2'(x_1) & n_2 f_2'(x_2) & n_3 f_2'(x_3) & \ldots & n_q f_2'(x_q) \\ n_1 f_3'(x_1) & n_2 f_3'(x_2) & n_3 f_3'(x_3) & \ldots & n_q f_3'(x_q) \\ \ldots & \ldots & \ldots & \ldots & \ldots \\ n_1 f_5'(x_1) & n_2 f_5'(x_2) & n_3 f_5'(x_3) & f_5'(x_4) & n_q f_5'(x_q) \end{vmatrix}_{\mathbf{x}=\mathbf{x}^{(i)}}
$$

Despite the complexity of the problem, it must be noted that the Jacobian matrix can be evaluated in closed form by the residue theorem, since it involves integrals $\int_{-\infty}^{\infty}$ of rational functions of u. In fact, the scalars $r_j(x,u)$ are the rational functions:

$$\begin{aligned}
r_0(x) &= \frac{x^3 + a_1x^2 + a_2x + a_3}{x^4 + a_1x^3 + a_2x^2 + a_3x + a_4}, \quad r_1(x) = \frac{x^2 + a_1x + a_2}{x^4 + a_1x^3 + a_2x^2 + a_3x + a_4} \\
r_2(x) &= \frac{x + a_1}{x^4 + a_1x^3 + a_2x^2 + a_3x + a_4}, \quad r_3(x) = \frac{1}{x^4 + a_1x^3 + a_2x^2 + a_3x + a_4}
\end{aligned} \tag{207}$$

with the coefficient $a_i = a_i(u)$, $i = 1, 2, 3, 4$ constituted by polynomials in u that are the coefficients of the characteristic equation $\mathbf{P}^4 + a_1\mathbf{P}^3 + a_2\mathbf{P}^2 + a_3\mathbf{P} + a_4\mathbf{1} = 0$.

We cannot assume $\mathbf{x}^{(0)} = 0$ as the starting point in the Newton-Raphson formulas (206) since $\mathbf{x}^{(0)} = 0$ yields a Jacobian matrix $J(\mathbf{x}^{(0)}) = J(0)$ that is not invertible. To obtain a good starting point $\mathbf{x}^{(0)}$, we approximate the function $f_i(x)$ with the MacLaurin expansion

$$f_i(x) \approx f_i^{(1)}(0)x + \frac{f_i^{(2)}(0)}{2!}x^2 + \cdots + \frac{f_i^{(q)}(0)}{q!}x^q \tag{208}$$

and obtain the starting point $\mathbf{x}^{(0)}$ by solving the algebraic system

$$\begin{aligned}
&f_1^1(0)(n_1x_1^0 + n_2x_2^0 + \cdots + n_qx_q^0) + \cdots + \frac{f_1^q(0)}{q!}\left[n_1(x_1^0)^q + n_2(x_2^0)^q + \cdots + n_q(x_q^0)^q\right] + c_1 = 0 \\
&f_2^1(0)(n_1x_1^0 + n_2x_2^0 + \cdots + n_qx_q^0) + \cdots + \frac{f_2^q(0)}{q!}\left[n_1(x_1^0)^q + n_2(x_2^0)^q + \cdots + n_q(x_q^0)^q\right] + c_2 = 0 \\
&\cdots \\
&f_q^1(0)(n_1x_1^0 + n_2x_2^0 + \cdots + n_qx_q^0) + \cdots + \frac{f_q^q(0)}{q!}\left[n_1(x_1^0)^q + n_2(x_2^0)^q + \cdots + n_q(x_q^0)^q\right] + c_q = 0
\end{aligned} \tag{209}$$

This yields

$$\begin{aligned}
&n_1(x_1^0)^1 + n_2(x_2^0)^1 + \cdots + n_q(x_q^0)^1 = s_1 \\
&n_1(x_1^0)^2 + n_2(x_2^0)^2 + \cdots + n_q(x_q^0)^2 = s_2 \\
&\cdots \\
&n_1(x_1^0)^q + n_2(x_2^0)^q + \cdots + n_q(x_q^0)^q = s_q
\end{aligned} \tag{210}$$

where the quantities s_i are known. Assuming $n_1 = n_2 = \cdots = n_q = N$, the system can be rewritten as

$$\begin{aligned}
&N\left[(x_1^0)^1 + (x_2^0)^1 + \cdots + (x_q^0)^1\right] = s_1 \\
&N\left[(x_1^0)^2 + (x_2^0)^2 + \cdots + (x_q^0)^2\right] = s_2 \\
&\cdots \\
&N\left[(x_1^0)^q + (x_2^0)^q + \cdots + (x_q^0)^q\right] = s_q
\end{aligned} \tag{211}$$

By taking into account the Viete formulas, it is possible to show that this system is equivalent to finding the zeros of a polynomial of degree q. Consequently, explicit expressions of $\mathbf{x}^{(0)}$ are possible in the case $q \leq 4$.

Alternatively, if the component of $\mathbf{x}^{(0)}$ are distinct and ordered according to their absolute values $|x_1^0| > |x_2^0| > \cdots > |x_q^0|$, we can use the approximation

$$\begin{aligned}
&(x_1^0)^1 + (x_2^0)^1 + \cdots + (x_q^0)^1 = \frac{s_1}{N} \\
&(x_1^0)^2 + (x_2^0)^2 + \cdots + (x_{q-1}^0)^2 \approx \frac{s_2}{N} \\
&\cdots \\
&(x_1^0)^{q-1} + (x_2^0)^{q-1} \approx \frac{s_{q-1}}{N} \\
&(x_1^0)^q \approx \frac{s_q}{N}
\end{aligned} \tag{212}$$

that yields immediate expressions for the starting point $\mathbf{x}^{(0)}$ in the Newton-Raphson formulas:

$$\mathbf{x}^{(j+1)} = \mathbf{x}^{(j)} - [\mathbf{J}(\mathbf{x}^{(j)})]^{-1}\mathbf{F}(\mathbf{x}^{(j)}) \tag{213}$$

Notice also that we can choose the integers $n_1, n_2, \ldots, n_q$ such that $|x_1^0| < 1$. For instance, from the first equation we obtain

$$|x_1^0| < \frac{|s_1|}{N} + \frac{|x_2^0|}{N} + \cdots \frac{|x_q^0|}{N} < \frac{|s_1|}{N} + \frac{q-1}{N}|x_1^0|$$

which yields

$$|x_1^0| < \frac{|s_1|}{N - q + 1} < 1$$

provided that $N > |s_1| + q - 1$.

It is convenient to assume large arbitrary integers n_i so that the iteration requires very few steps. However, increasing n_i also increases the difficulty to factorize the rational matrix $R_x^{-1}(\alpha)$ indicated in eq. (202).

4.9.7 On the reduction of the order of the system

Particular forms of the polynomial matrix $P(\alpha)$ allow the order of the system to be reduced. For instance, if in $P(\alpha) = \mathbf{A}_o + \alpha\mathbf{A}_1 + \alpha^2\mathbf{A}_2$ the matrix $\mathbf{A}_2$ is diagonal, it is possible with a suitable normalization of the matrix to be factorized, to obtain $\mathbf{A}_2 = \mathbf{1}$. Consequently, also $\mathbf{B}_4$ and $\mathbf{C}_6$ are unit matrices and the offending behaviors due to the non-vanishing ψ_1^1, ψ_2^3 and ψ_3^5 can be simply eliminated by multiplying the improper factorized by the scalars

$$\exp\left[\pm\alpha(\psi_1^1 + \psi_2^3 + \psi_3^5)\right]$$

In this case, the reduction of the number of the unknowns is three.

In other case it is possible that even though $\mathbf{A}_2 \neq 0$, the matrices $\mathbf{B}_4$ and/or $\mathbf{C}_6$ be vanishing. In these cases again the order of the system is reduced.

If a kernel matrix $G(\alpha)$ commutes with a polynomial matrix $P(\alpha)$ that can be factorized in the form $P(\alpha) = P_1(\alpha)P_2(\alpha)\ldots P_l(\alpha)$, it is possible to reduce the factorization problem of $G(\alpha)$ to the problem of factorizing a matrix commuting with the polynomial matrix $P_2(\alpha)P_3(\alpha)P_4(\alpha)\ldots P_l(\alpha)P_1(\alpha)$ or $P_3(\alpha)P_4(\alpha)\ldots P_l(\alpha)P_1(\alpha)P_2(\alpha)$, and so on. Factorizations of $P(\alpha)$ are always possible. For instance, by using the Smith representation, we have that $P(\alpha) = E(\alpha)D(\alpha)F(\alpha)$. Of course this transformation of the polynomial matrix produces advantages only if the new involved polynomial matrices are of degrees lower than the degree of $P(\alpha)$.

Moreover, we can pre (or post) multiply $P(\alpha)$ by an arbitrary polynomial matrix $P_2(\alpha)$, and the factorization of $G(\alpha)$ is reduced to the factorization of a matrix that commutes with the polynomial $P_2(\alpha)P(\alpha)$ or $P(\alpha)P_2(\alpha)$. However, the degree of the new polynomial matrix is greater that the degree of $P(\alpha)$ and consequently this idea is not effective for reducing the order of the nonlinear system. Also, to resort to the representation of rational matrices in the form (Bart, Gohberg & Kaashoek, 1979) $R^{-1}(\alpha) = C(\alpha\, 1 - A)^{-1}B + D$ can be ineffective since even though the polynomial matrix $\alpha\, 1 - A$ is linear in α, the order of A is obtained by multiplying the order and the degree of $P(\alpha)$. For instance, if $P(\alpha)$ is of order 4 and degree 2 we obtain a matrix $\alpha\, 1 - A$ of order $4 \times 2 = 8$. This implies a Cayley representation involving matrices $P^7(\alpha)$ that increase considerably the number of the unknowns in the nonlinear system. Finally, also the introduction of triangular representation:

$$P(\alpha) = \begin{vmatrix} 1 & 0 \\ P_{21}(\alpha)P_{11}^{-1}(\alpha) & 1 \end{vmatrix} \cdot \begin{vmatrix} P_{11}(\alpha) & 0 \\ 0 & P_{22}(\alpha) - P_{21}(\alpha)P_{11}^{-1}(\alpha)P^{12}(\alpha) \end{vmatrix} \begin{vmatrix} 1 & P_{11}^{-1}(\alpha)P_{12}(\alpha) \\ 0 & 1 \end{vmatrix}$$

is again ineffective in reducing the order of the nonlinear system.

4.9.8 The nonlinear equations as a Jacobi inversion problem

The integral in u present in the functions $f_i(x)$ can be evaluated by the residue theorem. It shows that the functions $f_i(x)$ are abelian integrals involving the algebraic functions $f(x,y) = x^4 + a_1(y)x^3 + a_2(y)x^2 + a_3(y)x + a_4$. Let us suppose that the genus p related to the algebraic function is equal to q and the abelian integrals are of the first kind.

Since the functions $f_{1,2,\ldots,p}(x)$ are constituted by a sum of abelian integrals defined on the same range of integration, they can be expressed as (Bliss, 2004, p. 98):

$$\begin{aligned} f_1(x) &= c_{11}w_1(x) + c_{12}w_2(x) + \cdots + c_{1p}w_p(x) \\ f_2(x) &= c_{21}w_1(x) + c_{22}w_2(x) + \cdots + c_{2p}w_p(x) \end{aligned} \tag{214}$$

$$\ldots$$

$$f_p(x) = c_{p1}w_1(x) + c_{p2}w_2(x) + \cdots + c_{pp}w_p(x) \tag{215}$$

where $w_i(x)$ $i = 1, 2, 3, \ldots, p$ are p independent integrals of the first kind in the Riemann surface of genus p.

Letting $n_1 = n_2 = \cdots = n_q = 1$, the system (215) can be rewritten as

$$\begin{aligned} &w_1(x_1) + w_1(x_2) + \cdots + w_1(x_p) = -m_1 \\ &w_2(x_1) + w_2(x_2) + \cdots + w_2(x_p) = -m_2 \\ &\ldots \\ &w_p(x_1) + w_p(x_2) + \cdots + w_p(x_p) = -m_p \end{aligned} \tag{216}$$

where

$$\begin{vmatrix} m_1 \\ m_2 \\ \ldots \\ m_p \end{vmatrix} = \begin{vmatrix} c_{11} & c_{12} & \cdots & c_{1p} \\ c_{21} & c_{22} & \cdots & c_{2p} \\ \ldots & \ldots & \ldots & \ldots \\ c_{p1} & c_{p2} & \cdots & c_{pp} \end{vmatrix}^{-1} \begin{vmatrix} c_1 \\ c_2 \\ \ldots \\ c_p \end{vmatrix}$$

The new nonlinear system (216) constitutes a Jacobi inversion problem well known in the literature. The conceptual importance of these algorithms is that they provide the solution $\mathbf{x}$ of the nonlinear system by the zeroes of a polynomial of degree equal to the genus p. It means that if $p \leq 4$ we obtain an exact solution constituted by radicals of the coefficients of this polynomial. However, the evaluation of these coefficients is very cumbersome; in general, it involves abelian functions or Theta Riemann function of very complicated argument.

Let us suppose now that the number q of nonlinear equations differs from the genus p of the algebraic equation $f(x, \alpha) = x^4 + a_1(\alpha)x^3 + a_2(\alpha)x^2 + a_3(\alpha)x + a_4(\alpha)$ related to the polynomial matrix $P(\alpha)$. The case $q < p$ reduces to a Jacobi inversion if we introduce $p - n_q$ additional unknowns x_i, $i = n_{q+1}, \ldots, p$ and the rational matrix defined by

$$\mathbf{R}(\alpha) = \prod_{j=1}^{p} \left[-x_j \mathbf{P}^{-1}(\alpha) + \mathbf{1}\right]^{n_j} \tag{217}$$

The previous procedure yields the following system of n_q equations in $p > n_q$ unknowns:

$$\begin{aligned} &w_1(x_1) + w_1(x_2) + \cdots + w_1(x_p) = -m_1 \\ &w_2(x_1) + w_2(x_2) + \cdots + w_2(x_p) = -m_2 \\ &\ldots \\ &w_{n_q}(x_1) + w_{n_q}(x_2) + \cdots + w_{n_q}(x_p) = -m_{n_q} \end{aligned} \tag{218}$$

This system involves only n_q abelian integrals of the first order. As the independent abelian integrals of the first order are p, we can add $p - n_q$ arbitrary equations having the form

$$\begin{aligned} &w_{n_{q+1}}(x_1) + w_{n_{q+1}}(x_2) + \cdots + w_{n_q}(x_p) = -m_{n_{q+1}} \\ &\ldots \\ &w_p(x_1) + w_p(x_2) + \cdots + w_p(x_p) = -m_p \end{aligned} \tag{219}$$

where the quantities m_i, $i = n_{q+1}, \ldots p$ can be arbitrarily chosen. In the case $n_q > p$, we apparently cannot reduce the solution of the nonlinear system to a Jacobi inversion problem.

4.9.9 Weakly factorization of a matrix commuting with a polynomial matrix

In the following we report another technique for eliminating the offending behavior at $\alpha \to \infty$. Let us consider a matrix $G(\alpha)$ commuting with a polynomial matrix $P(\alpha)$:

$$G(\alpha) = g_o(\alpha)1 + g_1(\alpha)P(\alpha) \tag{220}$$

Let us consider a point α_p in the lower half-plane and suppose that for a suitable integer r the matrix $\frac{P(\alpha)}{(\alpha-\alpha_p)^r}$ is bounded. If n is the order of the matrix $G(\alpha)$, then according to Cayley's theorem we have

$$\log[G(\alpha)] = \psi_o(\alpha)1 + \psi_1(\alpha)\frac{P(\alpha)}{(\alpha-\alpha_p)^r} + \psi_{n-1}(\alpha)\left[\frac{P(\alpha)}{(\alpha-\alpha_p)^r}\right]^{n-1} \tag{221}$$

The additive decomposition of the scalars $\psi_i(\alpha)$ yields the following weak factorization:

$$\begin{aligned}\tilde{G}_-(\alpha) &= \exp\left[\psi_{o-}(\alpha)1 + \psi_{1-}(\alpha)\frac{P(\alpha)}{(\alpha-\alpha_p)^r} + \psi_{(n-1)-}(\alpha)\left[\frac{P(\alpha)}{(\alpha-\alpha_p)^r}\right]^{n-1}\right]\\ \tilde{G}_+(\alpha) &= \exp\left[\psi_{o+}(\alpha)1 + \psi_{1+}(\alpha)\frac{P(\alpha)}{(\alpha-\alpha_p)^r} + \psi_{(n-1)+}(\alpha)\left[\frac{P(\alpha)}{(\alpha-\alpha_p)^r}\right]^{n-1}\right]\end{aligned} \tag{222}$$

Now the problem is to eliminate the offending essential singularity α_p that is present in both the factorized matrices $\tilde{G}_-(\alpha)$ and $\tilde{G}_+(\alpha)$.[7] To this end, we rewrite the W-H equation in the form

$$\tilde{G}_+(\alpha)F_+(\alpha) = \tilde{G}_-^{-1}(\alpha)F_-(\alpha) + \tilde{G}_-^{-1}(\alpha)\frac{R_o}{\alpha-\alpha_o}$$

or

$$\begin{aligned}\tilde{G}_+(\alpha)F_+(\alpha) - \tilde{G}_-^{-1}(\alpha_o)\frac{R_o}{\alpha-\alpha_o} &= \tilde{G}_-^{-1}(\alpha)F_-(\alpha) + \left[\tilde{G}_-^{-1}(\alpha) - \tilde{G}_-^{-1}(\alpha_o)\right]\frac{R_o}{\alpha-\alpha_o}\\ &= \frac{R_1}{\alpha-\alpha_p} + \frac{R_2}{(\alpha-\alpha_p)^2} + \frac{R_3}{(\alpha-\alpha_p)^3} + \cdots = \sum_{s=1}^{\infty}\frac{R_s}{(\alpha-\alpha_p)^s}\end{aligned} \tag{223}$$

where the four members represent the same function $f(\alpha)$ that is null at infinity and, except $\alpha = \alpha_p$, is regular for every value of α. The unknowns R_i are the coefficients of the Laurent expansion of $f(\alpha)$. Considering the first and the third member, we have

$$\tilde{G}_+(\alpha)F_+(\alpha) = \sum_{s=1}^{\infty}\frac{R_s}{(\alpha-\alpha_p)^s} + \tilde{G}_-^{-1}(\alpha_o)\frac{R_o}{\alpha-\alpha_o} \tag{224}$$

---

[7] It is easy to verify that $\tilde{G}_-(\alpha)$ and $\tilde{G}_+(\alpha)$ and their inverses possess algebraic behavior as $\alpha \to \infty$.

or

$$F_+(\alpha) = \tilde{G}_+^{-1}(\alpha)\sum_{s=1}^{\infty}\frac{R_s}{(\alpha-\alpha_p)^s} + \tilde{G}_+^{-1}(\alpha)\tilde{G}_-^{-1}(\alpha_o)\frac{R_o}{\alpha-\alpha_o} \tag{225}$$

Now we decompose the known vector $\tilde{G}_+^{-1}(\alpha)\tilde{G}_-^{-1}(\alpha_o)\frac{R_o}{\alpha-\alpha_o}$ as

$$\tilde{G}_+^{-1}(\alpha)\tilde{G}_-^{-1}(\alpha_o)\frac{R_o}{\alpha-\alpha_o} = C(\alpha) + N(\alpha) \tag{226}$$

where $C(\alpha)$ is the known characteristic part:

$$C(\alpha) = \frac{C_1}{\alpha-\alpha_p} + \frac{C_2}{(\alpha-\alpha_p)^2} + \frac{C_3}{(\alpha-\alpha_p)^3} + \cdots = \sum_{s=1}^{\infty}\frac{C_s}{(\alpha-\alpha_p)^s} \tag{227}$$

and $N(\alpha)$ is regular in the essential singularity α_p. By indicating with

$$\frac{G_1}{\alpha-\alpha_p} + \frac{G_2}{(\alpha-\alpha_p)^2} + \frac{G_3}{(\alpha-\alpha_p)^3} + \cdots = \sum_{s=0}^{\infty}\frac{G_s}{(\alpha-\alpha_p)^s} \tag{228}$$

the characteristic part of $\tilde{G}_+^{-1}(\alpha)$, the following representation of

$$\tilde{G}_+^{-1}(\alpha)\sum_{s=1}^{\infty}\frac{R_s}{(\alpha-\alpha_p)^s}$$

is obtained:

$$\tilde{G}_+^{-1}(\alpha)\sum_{s=1}^{\infty}\frac{R_s}{(\alpha-\alpha_p)^s} = \sum_{s=1}^{\infty}\left[\sum_{i=1}^{s} G_{s-i}R_i\right]\frac{1}{(\alpha-\alpha_p)^s} + \text{vector regular in } \alpha = \alpha_p \tag{229}$$

This yields the system

$$\sum_{i=1}^{s} G_{s-i}R_i = C_s, \quad s = 1, 2, \ldots, \infty \tag{230}$$

in the unknowns R_i. The system (230) has a triangular form. It allows for an explicit recurrence expression of R_i.

CHAPTER 5

Approximate solution: The Fredholm factorization

In the natural domain the W-H equations are integral equations defined by a convolution kernel. Generally these equations are of first kind. Procedures to reduce them to Fredholm or Volterra equations have been described in various literature. For instance Noble (1958, p. 230) reduces the W-H eqs. (3a) in chapter 1 to Volterra equations. However, working on the W-H equation in the natural domain does not seem fruitful. It is better to resort to the integral equations defined in the spectral domain.

In particular, the reduction of the W-H equations to Fredholm equations as indicated in chapter 1, section 1.6, is fundamental. In fact, to these authors' opinion, the approximate solutions of these equations (Fredholm factorization) constitute the most powerful and reliable tool for solving the W-H equations when exact factorizations are not possible.

5.1 The integral equations in the α – plane

5.1.1 Introduction

In section 1.6, the W-H equation

$$G(\alpha)F_+(\alpha) = X_-(\alpha) + \frac{R}{\alpha - \alpha_o} \tag{1}$$

has been reduced to the Fredholm equation

$$G(\alpha)F_+(\alpha) + \frac{1}{2\pi j} \cdot \int_{-\infty}^{+\infty} \frac{[G(t) - G(\alpha)]F_+(t)}{t - \alpha} dt = \frac{R}{\alpha - \alpha_o} \tag{2}$$

which applies when $\text{Im}[\alpha_o] < 0$. A slight modification of the procedure indicated in section 1.6 yields the following Fredholm integral equations for the cases where $\text{Im}[\alpha_o] > 0$ and $\text{Im}[\alpha_o] = 0$ (section 5.1.2):

$$F_+(\alpha) + \frac{1}{2\pi j} \cdot \int_{-\infty}^{\infty} \frac{[G^{-1}(\alpha)G(u) - \mathbf{1}]F_+(u)}{u - \alpha} du = G^{-1}(\alpha_o)\frac{R}{\alpha - \alpha_o}, \quad \text{Im}[\alpha_o] > 0 \tag{3}$$

$$F_+(\alpha) + \frac{1}{2\pi j} \cdot \int_{-\infty}^{\infty} \frac{[G^{-1}(\alpha)G(u) - \mathbf{1}]F_+(u)}{u - \alpha} du = \frac{1}{2} G^{-1}(\alpha) \frac{F_o}{\alpha - \alpha_o} + \frac{1}{2} G^{-1}(\alpha_o) \frac{R}{\alpha - \alpha_o}, \quad \mathrm{Im}[\alpha_o] = 0 \quad (4)$$

Similar equations can be obtained for the minus function $F_-(\alpha) = X_-(\alpha) + \frac{R}{\alpha - \alpha_o}$. Setting $G_1(\alpha) = G^{-1}(\alpha)$, we obtain

$$F_-(\alpha) - \frac{1}{2\pi j} \cdot \int_{-\infty}^{\infty} \frac{[G_1^{-1}(\alpha)G_1(u) - \mathbf{1}]F_-(u)}{u - \alpha} du = \frac{R}{\alpha - \alpha_o}, \quad \mathrm{Im}[\alpha_o] < 0$$

$$G_1(\alpha)F_-(\alpha) - \frac{1}{2\pi j} \cdot \int_{-\infty}^{+\infty} \frac{[G_1(t) - G_1(\alpha)]F_-(t)}{t - \alpha} dt = \frac{G_1(\alpha_o)R}{\alpha - \alpha_o}, \quad \mathrm{Im}[\alpha_o] > 0$$

5.1.2 Source pole α_o with positive imaginary part

Let us consider in the W-H equation

$$G(\alpha)F_+(\alpha) = F_-(\alpha) + \frac{F_o}{\alpha - \alpha_o}$$

where the source pole α_o has a positive imaginary part: $\mathrm{Im}[\alpha_o] > 0$. Since in (2) the integrand function is regular at $x = \alpha$, when $\mathrm{Im}[\alpha_o] > 0$ the following equation holds, for continuity:

$$G(\alpha)F_+(\alpha) + \frac{1}{2\pi j} \cdot \int_{\gamma} \frac{[G(x) - G(\alpha)]F_+(x)}{x - \alpha} dx = \frac{F_o}{\alpha - \alpha_o}, \quad \mathrm{Im}[\alpha_o] > 0 \quad (5)$$

where the γ line differs from the real axis as shown in Fig. 1.

Furthermore,

$$G(\alpha_o)\mathrm{Res}[F_+(\alpha)]_{\alpha_o} = F_o \quad (6)$$

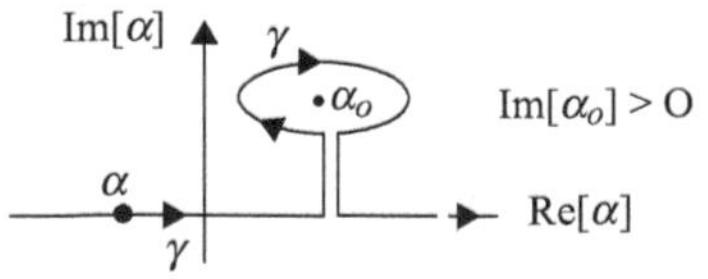

Fig. 1: Deformation of the integration path in presence of poles with positive imaginary parts

It follows that

$$\frac{1}{2\pi j}\cdot\int_{\gamma}\frac{[G(x)-G(\alpha)]F_+(x)}{x-\alpha}dx=\frac{1}{2\pi j}\cdot\int_{-\infty}^{\infty}\frac{[G(x)-G(\alpha)]F_+(x)}{x-\alpha}dx - \text{Res}\left[\frac{[G(x)-G(\alpha)]F_+(x)}{x-\alpha}\right]_{x=\alpha_o} \tag{7}$$

where

$$\text{Res}\left[\frac{[G(x)-G(\alpha)]F_+(x)}{x-\alpha}\right]_{x=\alpha_o}=\frac{[G(\alpha_o)-G(\alpha)]\text{Res}[F_+(x)]_{\alpha_o}}{\alpha_o-\alpha} = \frac{[G(\alpha_o)-G(\alpha)]}{\alpha_o-\alpha}G^{-1}(\alpha_o)F_o \tag{8}$$

By substituting (7) in (5) we obtain

$$G(\alpha)F_+(\alpha)+\frac{1}{2\pi j}\int_{-\infty}^{\infty}\frac{[G(x)-G(\alpha)]F_+(x)}{x-\alpha}dx=G(\alpha)G^{=1}(\alpha_o)\frac{F_o}{\alpha-\alpha_o},\quad \text{Im}[\alpha_o]>0 \tag{9}$$

The same procedure shows that when $\text{Im}[\alpha_o]=0$ the following integral equation stands:

$$G(\alpha)F_+(\alpha)+\frac{1}{2\pi j}\int_{-\infty}^{\infty}\frac{[G(x)-G(\alpha)]F_+(x)}{x-\alpha}dx = \frac{1}{2}\frac{F_o}{\alpha-\alpha_o}+\frac{1}{2}G(\alpha)G^{=1}(\alpha_o)\frac{F_o}{\alpha-\alpha_o},\quad \text{Im}[\alpha_o]=0 \tag{10}$$

5.1.3 Analytical validation of a particular W-H equation

In dealing with the half-plane problem, we have

$$G(\alpha)=\frac{1}{\sqrt{k^2-\alpha^2}} \tag{11}$$

The W-H technique yields the exact solution:

$$F_+(\alpha)=\frac{\sqrt{k+\alpha_o}\sqrt{k-\alpha}}{\alpha-\alpha_o}$$

We now verify that this solution satisfies the Fredholm integral eq. (2).

Introducing the suitable deformation of the integration path reported in Fig. 2, we find

$$\int_{-\infty}^{+\infty}\frac{[G(t)-G(\alpha)]F_+(t)}{t-\alpha}dt=\int_{\lambda_1+\lambda_2}\frac{[G(t)-G(\alpha)]F_+(t)}{t-\alpha}dt=-2\int_{\lambda_1}\frac{1}{\sqrt{k^2-t^2}}\frac{\sqrt{k+\alpha_o}\sqrt{k-t}}{(t-\alpha)(t-\alpha_o)}dt$$

$$=-2\int_{\lambda_3}\frac{1}{\sqrt{k^2-t^2}}\frac{\sqrt{k+\alpha_o}\sqrt{k-t}}{(t-\alpha)(t-\alpha_o)}dt=-2\int_{\lambda_3}\frac{1}{\sqrt{k+t}}\frac{\sqrt{k+\alpha_o}}{(t-\alpha)(t-\alpha_o)}dt$$

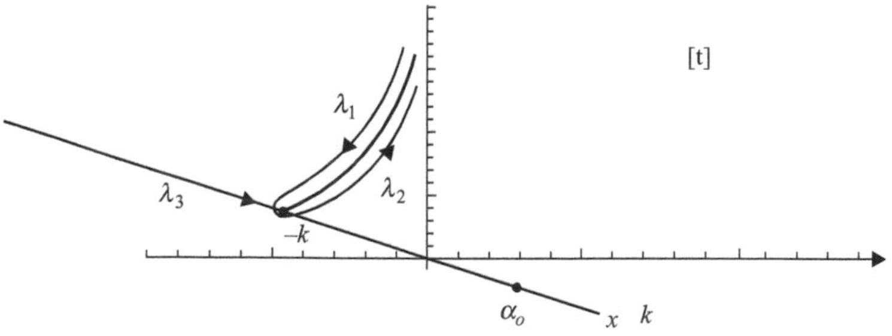

Fig. 2: Deformation of the original integration path

which with the substitution $t = -ku$ yields

$$\int_{-\infty}^{+\infty} \frac{[G(t) - G(\alpha)]F_+(t)}{t-\alpha} dt = -2 \int_{\lambda_3} \frac{1}{\sqrt{k+t}} \frac{\sqrt{k+\alpha_o}}{(t-\alpha)(t-\alpha_o)} dt$$

$$= -\frac{2}{k} \int_1^{\infty} \frac{1}{\sqrt{1-u}} \frac{\sqrt{1+\alpha_o/k}}{(u+\alpha/k)(u+\alpha_o/k)} du = \frac{2j\pi}{\alpha-\alpha_o} - \frac{2j\pi\sqrt{k+\alpha_o}}{\sqrt{k+\alpha}(\alpha-\alpha_o)}$$

This proves the validity of the Fredholm equation. Conversely, a numerical check is not easily obtained. First of all $\frac{G(t)-G(\alpha)}{t-\alpha}$ is indeterminate as $t = \alpha$; hence, it is necessary to evaluate the limit for $t \to \alpha$. For (11) it yields

$$\frac{G(t) - G(\alpha)}{t-\alpha} \approx G'(\alpha) = \frac{\alpha}{(k^2-\alpha^2)^{3/2}} \quad \text{as } t \to \alpha$$

A difficulty arises from the fact that singularities of the kernel (e.g., the branch points) can be located near the real axis. Finally, the presence of an integration defined over an infinite support could require a large number of interpolating points in the quadrature formulas. In general, the previous considerations prevent efficient numerical solutions of the original Fredholm integral eq. (2). However, we will modify this equation by deforming the integration line in the $\alpha -$ plane or by introducing a different complex planes where the kernel presents a more suitable behavior. In this way the Fredholm factorization technique becomes very efficient and provides a reliable tool for solving the W-H equations.

5.1.4 A property of the integral in the Fredholm equation

For the previous example, we observe that $\int_{-\infty}^{+\infty} \frac{[G(t)-G(\alpha)]F_+(t)}{t-\alpha} dt$ does not involve the singularities of $F_+(\alpha)$. For instance, we will show that the integral

$$\frac{1}{2\pi j} \cdot \int_{-\infty}^{\infty} \frac{[G(u) - G(\alpha)]F_+(u)}{u-\alpha} du$$

present in the Fredholm equation, does not involve the pole α_o of the source that is present in $F_+(\alpha)$. In fact, setting

$$F_+(x) = \frac{T}{x-\alpha_o} + \text{regular term at } \alpha = \alpha_o$$

yields

$$\frac{1}{2\pi j}\cdot\int_{-\infty}^{\infty}\frac{[G(u)-G(\alpha)]}{u-\alpha}\frac{T}{u-\alpha_o}du$$

$$= \frac{1}{2\pi j}\cdot\int_{-\infty}^{\infty}\frac{[G(u)-G(\alpha)]}{(u-\alpha)(u-\alpha_o)}Tdu = \frac{1}{2\pi j(\alpha-\alpha_o)}\cdot\int_{-\infty}^{\infty}\left(\frac{G(u)-G(\alpha)}{u-\alpha}-\frac{G(u)-G(\alpha)}{u-\alpha_o}\right)Tdu$$

$$= \frac{1}{2\pi j(\alpha-\alpha_o)}\cdot\int_{-\infty}^{\infty}\left[\frac{G(u)-G(\alpha)}{u-\alpha}-\frac{G(u)-G(\alpha_o)}{u-\alpha_o}-\frac{G(\alpha_o)-G(\alpha)}{u-\alpha_o}\right]Tdu$$

$$= \frac{1}{2\pi j(\alpha-\alpha_o)}P.V.\int_{-\infty}^{\infty}[m(u,\alpha)-m(u,\alpha_o)]\cdot Tdu + \frac{G(\alpha_o)-G(\alpha)}{2\pi j(\alpha-\alpha_o)}\cdot P.V.\int_{-\infty}^{\infty}-\frac{1}{u-\alpha_o}Tdu$$

where

$$m(u,\alpha) = \frac{G(u)-G(\alpha)}{u-\alpha} \tag{12}$$

and both integrals are to be interpreted as principal values (P.V.) in the singular point $x = \pm\infty$. Now we have

$$P.V.\int_{-\infty}^{\infty}\frac{1}{u-\alpha_o}du = P.V.\int_{-\infty}^{\infty}\frac{u+\alpha_o}{u^2-\alpha_o^2}du = P.V.\int_{-\infty}^{\infty}\frac{u}{u^2-\alpha_o^2}du + \int_{-\infty}^{\infty}\frac{\alpha_o}{u^2-\alpha_o^2}du$$

$$= \alpha_o\int_{-\infty}^{\infty}\frac{1}{u^2-\alpha_o^2}du = -\alpha_o 2\pi j\frac{1}{2\alpha_o} = -\pi j$$

which yields

$$\frac{1}{2\pi j}\int_{-\infty}^{\infty}\frac{[G(u)-G(\alpha)]}{u-\alpha}\cdot\frac{T}{u-\alpha_o}du = \frac{1}{2\pi j(\alpha-\alpha_o)}P.V.\int_{-\infty}^{\infty}[m(u,\alpha)-m(u,\alpha_o)]\cdot Tdu$$

$$+\frac{G(\alpha_o)-G(\alpha)}{2(\alpha-\alpha_o)}\cdot T \tag{13}$$

Clearly the last expression shows that the integral is regular at $\alpha = \alpha_o$.

The physical interpretation of the integral present in the Fredholm equation depends on the particular application. For instance, in diffraction problems the pole α_o denotes a geometrical optics contribution. The obtained result shows that the integral in the Fredholm equation always has the physical meaning of a diffracted field.

5.1.5 Numerical solution of the Fredholm equations

5.1.5.1 *Introduction*

To obtain accurate numerical results of the Fredholm factorization method, we first observe that the integral eq. (2) could involve a kernel

$$M(\alpha, t) = G^{-1}(\alpha)\frac{G(\alpha) - G(t)}{\alpha - t}$$

that is not a compact operator. Fortunately, we have a criterion to ascertain the compactness of $M(\alpha, t)$. In fact, if $G(\alpha)$ and $G^{-1}(\alpha)$ exist and are bounded on the real axis as $\alpha \to \pm\infty$, the integral eq. (2) is of second kind (Daniele 2004a). In our applications we are dealing with equations where the critical points constituted by the singularity α_o of the source or the structural singularities of the kernel are near the integration lines. These circumstances could damage the convergence of the numerical solution and procedures that overcome these problems are indicated in the literature (Kantorovich & Krylov, 1967). The fundamental expedient to improve the numerical solution of a Fredholm equation related to a W-H equation is to warp the contour path constituted by the real $\alpha -$ axis into a modified line that is located sufficiently far from the critical points. This expedient was used with success to factorize kernel matrices in several different problems (Daniele, 2004a; Daniele & Lombardi, 2006, 2007).

Let us indicate with $\lambda(y)$, $-\infty < y < \infty$ a line on the $t -$ plane located far enough from the singularities of $G(\alpha)$ and $G^{-1}(\alpha)$ and such that there are no singularities of the Fredholm kernel $M(\alpha, t)$ in the region of the $t -$ plane between the real axis and this line. For instance, in wedge problems the straight line (Fig. 3)

$$\lambda(y) = e^{j\frac{\pi}{4}}y, \quad -\infty < y < \infty \tag{14}$$

satisfies this condition.

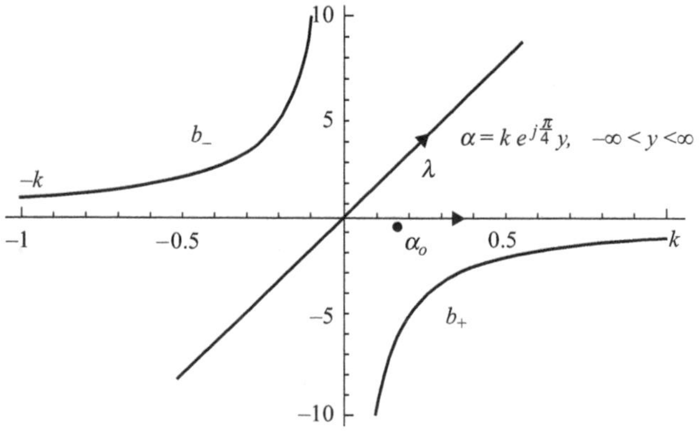

Fig. 3: Warping of the real axis on the line $\lambda(y) = k\, e^{j\frac{\pi}{4}}y$, $-\infty < y < \infty$

Other contours are possible, but usually mapping seems the best one (Daniele, 2004a). In general, the deformation of the real axis to the line $\alpha(y) = \lambda(y)$ reduces the Fredholm eq. (2) to

$$G(\alpha(y))F_+(\alpha(y)) = -\frac{1}{2\pi j}\cdot\int_{-\infty}^{+\infty}\frac{[G(t(x)) - G(\alpha(y))]F_+(t(x))}{t(x) - \alpha(y)}t'(x)dx + \frac{F_o}{\alpha(y) - \alpha_o}, \quad -\infty < t < \infty \tag{15}$$

where $t(x) = \lambda(x)$.

The sampled version of this equation is

$$G(\alpha(y)F_+(\alpha) + \frac{h}{2\pi j}\cdot\sum_{i=-A/h}^{A/h} m(\alpha(y), t(h\,i))F_+(t(h\,i))t'(hi) = \frac{F_o}{\alpha(y) - \alpha_o} \tag{16}$$

where A and h are discretization parameters, and

$$m(\alpha, t) = \text{If}\left[\alpha == t, G'(\alpha), \frac{G(\alpha) - G(t))}{\alpha - t}\right]$$

We define the following matrices of order $2A + 1$:

$$M = \frac{h}{2\pi j}Table[m[\alpha(y), t(h\,i)], \{y, -A, A, h\}, \{i, -A/h, A/h\}] \tag{17}$$

$$D = DiagonalMatrix[Table[G[\alpha(y)], \{y, -A, A, h\}] \tag{18}$$

and the vector

$$s = Table\left[\frac{F_o}{\alpha(y) - \alpha_1}, \{y, -A, A, h\}\right] \tag{19}$$

The vector

$$u = LinearSolve[D + M, s] \tag{20}$$

defines the samples $F_+(t(h\,i))$, $i = 0, \pm 1, \pm 2, \ldots, \pm A/h$. It yields the approximate solution:

$$F_{a+}(\alpha(y)) = G^-(\alpha(y))\left[-\frac{h}{2\pi j}\cdot\sum_{i=-A/h}^{A/h} m(\alpha(y), t(h\,i))F_+(t(h\,i))t'(hi) + \frac{F_o}{\alpha(y) - \alpha_0}\right] \tag{21}$$

or, in the α – plane

$$F_{a+}(\alpha) = G^{-1}(\alpha)\left[-\frac{h}{2\pi j}\cdot\sum_{i=-A/h}^{A/h} m(\alpha, t(h\,i))F_+(h\,i)t'(hi) + \frac{F_o}{\alpha - \alpha_0}\right] \tag{22}$$

$$F_{a-}(\alpha) = -\frac{h}{2\pi j} \cdot \sum_{i=-A/h}^{A/h} m(\alpha, t(h\,i))F_{+}(t(h\,i))t'(hi) + \frac{F_o}{\alpha - \alpha_0} \tag{23}$$

Generally, the numerical solution of Fredholm integral equation of second kind obtained by using the aforementioned simple schemes is very efficient. For instance, theoretical considerations (Kantorovich & Krylov, 1967) ascertain that diminishing h and increasing A assures that the numerical solution converges to the exact solution of the integral equation.

5.1.5.2 *Some applications of the Fredholm factorization*

Introduction

In the previous section we stated that the Fredholm factorization involves compact kernels provided that both $G(\alpha)$ and $G^{-1}(\alpha)$ exist and are bounded as $\alpha \to \pm\infty$. With the exception of the kernels relevant to the longitudinal modified W-H equations, we experienced that $G(\alpha)$ can always be normalized in order to accomplish these conditions. For example, consider the following W-H kernels:

(a) $G(\alpha) = \dfrac{e^{j\tau\, d}}{\tau \cos[\tau\, d]}$

(b) $G(\alpha) = \dfrac{\sin(\tau\, b)\, \sin(\tau\, c)}{\tau \sin[\tau\, a]}, \quad b + c = a$

Multiplying them times the well-known function $\tau = \sqrt{k^2 - \alpha^2}$ yields the W-H kernels

(a) $g(\alpha) = \dfrac{e^{j\tau\, d}}{\cos[\tau\, d]}$

(b) $g(\alpha) = \dfrac{\sin(\tau\, b)\, \sin(\tau\, c)}{\sin[\tau\, a]}, \quad b + c = a$

which satisfy the conditions that assure that the Fredholm kernels $M(\alpha, t)$ are compact.

The most significant expedient to improve the numerical solution of the Fredholm equation related to a W-H equation is to warp the contour path constituted by the real $\alpha -$ axis into a modified line far from the singularities and/or zeros of the kernels. In the following we apply these considerations to kernels (a) and (b).

The singularities of these kernels are as follows:

Case (a): branch points: $\alpha = \pm k$, poles of $G(\alpha)$:

$$\pm\alpha_n = \pm\sqrt{k^2 - \left(\frac{(n - 1/2)\pi}{d}\right)^2} \quad n = 1, 2, \ldots \tag{24}$$

Case (b): branch points: $\alpha = \pm k$, poles of $G(\alpha)$:

$$\pm\alpha_{an} = \pm\sqrt{k^2 - \left(\frac{n\pi}{a}\right)^2} \quad n = 1, 2, \ldots \tag{25}$$

poles of $G^{-1}(\alpha)$ or zeroes of $G(\alpha)$:

$$\pm\alpha_{bn} = \pm\sqrt{k^2 - \left(\frac{n\pi}{b}\right)^2} \quad n = 1, 2, \ldots \qquad \pm\,\alpha_{cn} = \pm\sqrt{k^2 - \left(\frac{n\pi}{c}\right)^2} \quad n = 1, 2, \ldots \tag{26}$$

Let us observe that in both examples the singularities of the kernel are located in the standard branch lines b_+ and b_- (Fig. 3) so that $\pm k$ and some $\pm\alpha_n$ are very near the real axis for vanishing values of the imaginary part of k.

Factorization of kernel (a)

First we normalize the kernel

$$g(\alpha) = \frac{e^{j\sqrt{k^2-\alpha^2}\,d}}{\cos\left[\sqrt{k^2-\alpha^2}\,d\right]} \tag{27}$$

and then obtain numerical solution of the Fredholm equation for two different integration lines. To be realistic, we introduce a propagation constant k with very low imaginary part: $k = \frac{2\pi}{\lambda}(1 - j10^{-8})$. This involves some singularities of the kernel very near the real axis. However, we overcome this problem by warping the original line constituted by the real axis into another line far from the singularities. To estimate the accuracy of this approximate solution, a physically important parameter is the residue of the minus function $F_-(\alpha)$ in the pole $-\alpha_{d(1)} = -\alpha_1$ (section 3.2.6). We have

$$\text{Res}[F_-(\alpha)]\big|_{-\alpha_1} = \text{Res}[g(\alpha)]\big|_{-\alpha_1} F_+(-\alpha_1) \approx \text{Res}[g(\alpha)]\big|_{-\alpha_1} F_{a+}(-\alpha_1) \tag{28}$$

Consequently, the accuracy of the numerical solution can be obtained by comparing $F_{a+}(-\alpha_1)$ with the exact value $F_+(-\alpha_1)$. We obtain

$$m(\alpha, t) = \text{If}\left[\alpha == t, m_o(\alpha), \frac{g(\alpha) - g(t))}{\alpha - t}\right] \tag{29}$$

where

$$m_o(\alpha) = g'(\alpha) = -\frac{jd\alpha}{\sqrt{k^2-\alpha^2}\,\cos\sqrt{k^2-\alpha^2}\,d}$$

- First integration contour: the numerical solution is considered on the original line constituted by the real axis.

  Figure 4 illustrates the obtained results. The accuracy of the results is not satisfying since we obtained

  $$F_{a+}(-\alpha_1) = -0.03738 - j0.00291$$

  $$F_+(-\alpha_1) = -0.04823 + j0.01040$$

- Second integration contour: the numerical solution is considered on the line $\lambda(y) = e^{j\frac{\pi}{4}}y$, $-\infty < y < \infty$ (Fig. 3)

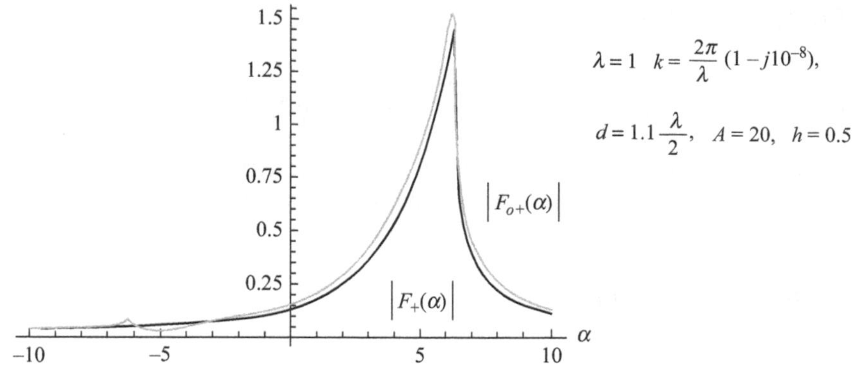

Fig. 4: Comparison between exact and approximate solutions, where the integration line is the real axis

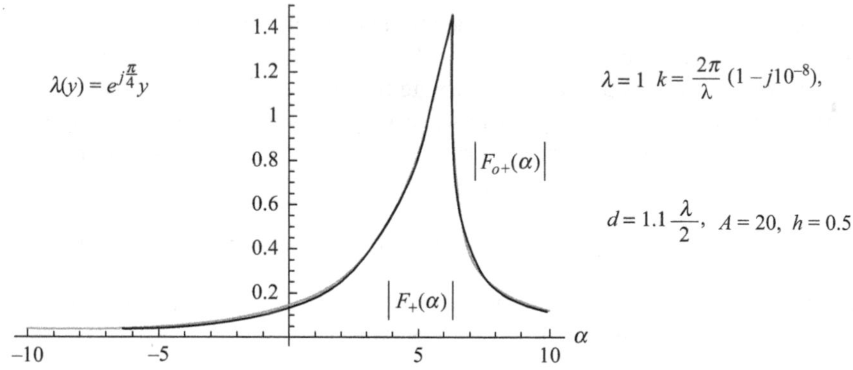

Fig. 5: Comparison between the exact and the approximate solution, where the integration line is $\lambda(y) = e^{j\frac{\pi}{4}}y$

Figure 5 illustrates the obtained results. The accuracy is improved: compare with Fig. 4. Now we have

$$F_{a+}(-\alpha_1) = -0.0427751 + \text{j}\,0.017618$$
$$F_+(-\alpha_1) = -0.04823 + j0.01040$$

Factorization of kernel (b)

The exact factorization of the kernel $G(\alpha) = \frac{\sin(\tau\, b)\,\sin(\tau\, c)}{\tau\,\sin[\tau\, a]}$, $b + c = a$ was considered in section 3.2.5. First, we use the normalized kernel

$$g(\alpha) = \frac{\sin(\tau\, b)\,\sin(\tau\, c)}{\sin[\tau\, a]}, \quad b + c = a \tag{30}$$

and then obtain the numerical solution of the Fredholm equation for two different contour lines. Again, to be realistic we introduce a propagation constant k with very low imaginary part: $k = \frac{2\pi}{\lambda}(1 - j10^{-8})$, which involves some singularities of the kernel very near the real axis. However, we overcome this problem by warping the original line constituted by the real axis into another line far from the singularities.

To estimate the accuracy of the approximate solution, a physically important parameter is the residue of the minus function $F_-(\alpha)$ in the pole $-\alpha_{a1} = -\alpha_1$ (see section 3.2.5). We have

$$\text{Res}[F_-(\alpha)]\big|_{-\alpha_1} = \text{Res}[g(\alpha)]\big|_{-\alpha_1} F_+(-\alpha_1) \approx \text{Res}[g(\alpha)]\big|_{-\alpha_1} F_{a+}(-\alpha_1) \tag{31}$$

The accuracy of the numerical solution can be obtained by comparing $F_{a+}(-\alpha_1)$ with the exact value $F_+(-\alpha_1)$. Another parameter that estimates the accuracy is the residue of the minus function $F_-(\alpha)$ in the poles α_{b1} and α_{c1} (see section 3.2.5). We have

$$\text{Res}[F_+(\alpha)]\big|_{\alpha_{b1}} = \text{Res}\left[G^{-1}(\alpha)\right]\big|_{\alpha_{b1}} F_-(\alpha_{b1}) \approx \text{Res}\left[G^{-1}(\alpha)\right]\big|_{\alpha_{b1}} F_{a-}(\alpha_{b1}) \tag{32}$$

$$\text{Res}[F_+(\alpha)]\big|_{\alpha_{c1}} = \text{Res}\left[G^{-1}(\alpha)\right]\big|_{\alpha_{c1}} F_-(\alpha_{c1}) \approx \text{Res}\left[G^{-1}(\alpha)\right]\big|_{\alpha_{c1}} F_{a-}(\alpha_{c1}) \tag{33}$$

Consequently, the accuracy of the numerical solution can be obtained by comparing $F_{a-}(\alpha_{b,c1})$ with the exact value $F_-(\alpha_{b,c1})$. We obtain

$$m(\alpha, t) = \text{If}\left[\alpha == t, m_o(\alpha), \frac{g(\alpha) - g(t))}{\alpha - t}\right]$$

$$m_o(\alpha) = g'(\alpha)$$

The expression of $m_o(\alpha)$ is not reported here.

- First integration contour: the numerical solution is considered on the original line constituted by the real axis. Even though we now work with a compact kernel, the Fredholm solution is not sufficiently accurate and therefore is not reported here.
- Second integration line: the numerical solution is considered on the line $\lambda(y) = e^{j\frac{\pi}{4}}y$, $-\infty < y < \infty$ (Fig. 3).

Figure 6 illustrates the obtained results. Using a compact kernel and solving on the line $\lambda(y) = e^{j\frac{\pi}{4}}y$, $-\infty < y < \infty$ considerably improves the accuracy of the numerical results.

Observe the peaks for negative α in the approximate solution. They are due to the poles $-\alpha_{b1}$ and $-\alpha_{c1}$ very near the real axis that are spuriously present in the approximate solution $F_{a+}(\alpha)$. Now we have

$$\begin{aligned} F_+(-\alpha_{a1}) &= -0.0766365 - j0.134256 \\ F_{a+}(-\alpha_{a1}) &= -0.0735487 - j0.131697 \\ F_-(\alpha_{b1}) &= -0.300055 - j0.0625272 \\ F_{a-}(\alpha_{b1}) &= -0.294696 - j0.0644443 \\ F_-(\alpha_{c1}) &= -0.58009 - j0.053066 \\ F_{a-}(\alpha_{c1}) &= -0.582995 - j0.0543879 \end{aligned}$$

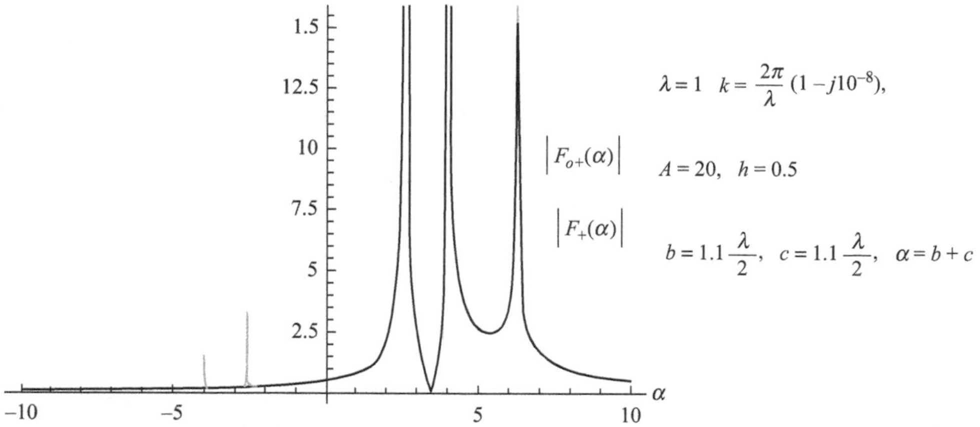

Fig. 6: Comparison between the exact and the approximate solution, where the integration line is $\lambda(y) = e^{j\frac{\pi}{4}}y$

Examples on rational functions

Let us consider the solution of the W-H equation

$$G(\alpha)F_+(\alpha) = F_-(\alpha) + \frac{1}{\alpha - \alpha_0}\begin{vmatrix} 1 \\ 0 \end{vmatrix}$$

where the kernel is the rational matrix factorized in section 2.7:

$$G(\alpha) = \begin{vmatrix} 1 & jq\dfrac{\alpha^2 + A_n{}^2}{\alpha^2 + B^2} \\ jq & 1 \end{vmatrix} = G_-(\alpha) \cdot G_+(\alpha)$$

and $G_-(\alpha)$ and $G_+(\alpha)$ are given in section 2.7.

In the numerical solution it has been assumed that

$$A_n = 1, \quad B = 2, \quad q = 0.5, \quad A = 5, \quad h = 0.1 \quad \alpha_o = 1 - j10^{-1}$$

We obtained an approximate solution $F_{a+}(\alpha)$ by Fredholm factorization using different paths:

$$\lambda(x) = x$$
$$\lambda(x) = xe^{j\pi/4}$$
$$\lambda(x) = x + j \arctan x$$

In all cases the plots relevant to the approximate solution and the exact solution

$$F_+(\alpha) = \frac{G_+^{-1}(\alpha)G_-^{-1}(\alpha_p)}{\alpha - \alpha_p}\begin{vmatrix} 1 \\ 1 \\ \frac{1}{2} \end{vmatrix}$$

are indistinguishable.

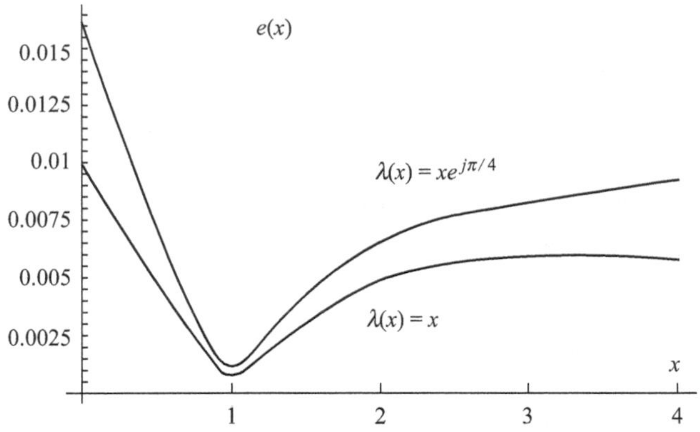

Fig. 7: The error $e(\alpha) = \left|\frac{F_{+1}(\alpha)-F_{a+1}(\alpha)}{F_{+1}(\alpha)}\right|$

Figure 7 illustrates the error $e(\alpha) = \left|\frac{F_{+1}(\alpha)-F_{a+1}(\alpha)}{F_{+1}(\alpha)}\right|$ relevant to the first components of $F_+(\alpha)$.

The error relevant to using the path $\lambda(x) = x + j\arctan x$ is larger and has not been considered here (Daniele, 2004a).

Similar errors are present in the Fredholm factorization of the rational matrix of order three considered in section 4.4.2.1 (see Daniele, 2004a; Daniele & Lombardi, 2007).

5.1.6 Analytic continuation outside the integration line

Equations (22) and (23) provide the representations of $F_+(\alpha)$ and $F_-(\alpha)$ in the complex α – plane through an analytic continuation of the numerical solutions. The validity of these representations start from the points of the integration line of the Fredholm equation and extend until we cross the singularities present in them. Unfortunately, beside the structural singularities present in the kernel there are additional spurious singularities due to the approximation of the integral with sums. Now we discuss this item by observing that in the exact representation of $F_+(\alpha)$:

$$F_+(\alpha) = -\frac{1}{2\pi j}\cdot\int_{-\infty}^{+\infty}\frac{G^{-1}(\alpha)\cdot[G(\alpha)-G(t)]F_+(t)}{\alpha-t}dt + G^{-1}(\alpha)\cdot\frac{R}{\alpha-\alpha_o},\quad \mathrm{Im}[\alpha_o]<0 \quad (34)$$

there is a compensation of the minus structural singularities (poles or branch points located in the upper half-plane $\mathrm{Im}[\alpha] > 0$) in the second member. To show this, let us assume the presence of the zero α_c of $\det[G(\alpha)]$ in $\mathrm{Im}[\alpha] > 0$. Looking at

$$F_+(\alpha) = -\frac{1}{2\pi j}\cdot\int_{-\infty}^{+\infty}\frac{G^{-1}(\alpha)\cdot[G(\alpha)-G(t)]F_+(t)}{\alpha-t}dt + G^{-1}(\alpha)\cdot\frac{R}{\alpha-\alpha_o},\quad \mathrm{Im}[\alpha_o]<0 \quad (35)$$

the residue in α_c of the first term of the second member is given by

$$R_{\alpha_c} = -\frac{1}{2\pi j}\mathrm{Res}[G^{-1}(\alpha)]_{\alpha_c} \cdot \int_{-\infty}^{+\infty} \frac{-G(t)F_+(t)}{\alpha_c - t}dt = \frac{1}{2\pi j}\mathrm{Res}[G^{-1}(\alpha)]_{\alpha_c} \cdot \int_{-\infty}^{+\infty} \frac{F_-(t)}{\alpha_c - t}dt \quad (36)$$

Taking into account that $F_-(t)$ presents in the half-plane $\mathrm{Im}[\alpha] < 0$ only the pole α_o with residue R, the previous equation yields

$$R_{\alpha_c} = -\frac{1}{2\pi j}\mathrm{Res}[G^{-1}(\alpha)]_{\alpha_c} \cdot \int_{-\infty}^{+\infty} \frac{-G(t)F_+(t)}{\alpha_c - t}dt = \frac{1}{2\pi j}\mathrm{Res}[G^{-1}(\alpha)]_{\alpha_c} \cdot \int_{-\infty}^{+\infty} \frac{F_-(t)}{\alpha_c - t}dt$$
$$= -\mathrm{Res}[G^{-1}(\alpha)]_{\alpha_c} \cdot \frac{R}{\alpha_c - \alpha_o} \quad (37)$$

Since the second term presents a residue $\mathrm{Res}[G^{-1}(\alpha)]_{\alpha_c} \cdot \frac{R}{\alpha_c - \alpha_o}$ in the pole α_c that compensates the previous one, it follows that the exact expression of $F_+(\alpha)$ given by (57) does not present the spurious pole α_c located in the upper half-plane $\mathrm{Im}[\alpha] > 0$. Consequently, $F_+(\alpha)$ is indeed a plus function. The same reasoning can be used when a branch line is present in $G^{-1}(\alpha)$ in the upper half-plane $\mathrm{Im}[\alpha] > 0$. In fact, the jump on the two lips in the first term of the second member of (57) yields

$$\Delta\left[\frac{1}{2\pi j}\int_{-\infty}^{+\infty} \frac{G^{-1}(\alpha) \cdot G(t) \cdot F_+(t)}{\alpha - t}dt\right] = \Delta\left[G^{-1}(\alpha)\right]\frac{1}{2\pi j}\int_{-\infty}^{+\infty} \frac{F_-(t)}{\alpha - t}dt$$
$$= -\Delta\left[G^{-1}(\alpha)\right]\frac{R}{\alpha - \alpha_o} \quad (38)$$

which compensates the jump in the second term: $\Delta[G^{-1}(\alpha)]\frac{R}{\alpha - \alpha_o}$. These compensations for offending minus singularities present in the exact representations derive from the relationship

$$\frac{1}{2\pi j} \cdot \int_{-\infty}^{+\infty} \frac{G(t)F_+(t)}{\alpha_c - t}dt = -\frac{R}{\alpha_c - \alpha_o} \quad (39)$$

that holds if α_c is located in the upper half-plane $\mathrm{Im}[\alpha] > 0$.

Conversely, in the approximate representation

$$\frac{h}{2\pi j} \cdot \sum_{i=-A/h}^{i=A/h} \frac{G(h\, i)F_+(h\, i)}{\alpha_c - h\, i} \approx -\frac{R}{\alpha_c - \alpha_o} \quad (40)$$

the presence of the sum instead of the integral does not assure this compensation. The offending presence of minus singularities in plus functions $F_+(\alpha)$ follows. Consequently, the quantity

$$\frac{h}{2\pi j} \cdot \sum_{i=-A/h}^{i=A/h} \frac{G(h\, i)F_+(h\, i)}{\alpha - h\, i} \approx -\frac{R}{\alpha - \alpha_o} \quad (41)$$

for every value of α in $\text{Im}[\alpha] > 0$ represents an accurate indicator of the presence of offending singularities. Similar considerations can be done for the minus function $F_-(\alpha)$. Sometimes in application problems the evaluation of the unknowns is required in a region of the $\alpha-$ plane where the representations (22) and (23) are not valid, for instance, in improper sheets or points near the singularities. In these cases we can resort to analytic continuation. The analytic continuation of numerical results is an old yet very difficult problem of applied mathematics that can be approached in various ways. For instance, by introducing the mapping $\alpha = -k\cos w$, the W-H equation for the impenetrable wedge problem (section 1.3.1) assumes the form of difference equations

$$\hat{G}(w)\hat{X}_+(w) = \hat{Y}_+(-w-\Phi) \tag{42}$$

where the plus functions are even functions of w. Taking into account the aforementioned property of the plus functions, it is possible to eliminate the unknown $\hat{Y}_+(w)$, thus obtaining the relation

$$\hat{X}_+(w+2\Phi) = \hat{G}^{-1}(-w-2\Phi)\hat{G}(w)\hat{X}_+(w) \tag{43}$$

Starting from the values of $\hat{X}_+(w)$ known via eqs. (22) and (23), this last equation allows us to evaluate $\hat{X}_+(w)$ and $\hat{Y}_+(-w-\Phi)$ for every value of w. When it is important to evaluate residues of $F_+(\alpha)$ and $F_-(\alpha)$ in poles located in structural singularities defined by the zeroes of $\det[G(\alpha)]$ and $\det[G^{-1}(\alpha)]$, we can resort to the W-H equation

$$\text{Res}[F_+(\alpha)]_{\alpha=\alpha_i} = \text{Res}\left[G^{-1}(\alpha)\right]_{\alpha=\alpha_i} F_-(\alpha_i) \tag{44}$$

Since we can evaluate $\text{Res}[G^{-1}(\alpha)]_{\alpha=\alpha_i}$ exactly, the accuracy of the evaluation of $\text{Res}[F_+(\alpha)]_{\alpha=\alpha_i}$ depends on the accuracy of the evaluation of $F_-(\alpha_i)$ through the representation given by (44). Since α_i is a point located in the regular region of $F_-(\alpha)$, usually no problems are encountered in obtaining a good evaluation of $F_-(\alpha_i)$ with this representation.

5.2 The integral equations in the $w-$ plane

Let us consider the W-H equation

$$G(\alpha)F_+(\alpha) = F_-(\alpha) + \frac{F_o}{\alpha - \alpha_o} \tag{45}$$

If $F_+(\alpha)$ is a standard plus functions ($\text{Im}[\alpha_o] < 0$), it yields the following Fredholm integral equation:

$$G(\alpha)F_+(\alpha) + \frac{1}{2\pi j}\cdot\int_{-\infty}^{+\infty}\frac{[G(x)-G(\alpha)]F_+(x)}{x-\alpha}dx = \frac{F_o}{\alpha-\alpha_o}, \quad \text{Im}[\alpha_o] < 0 \tag{46}$$

It is convenient to deform the contour path constituted by the real axis into the straight line λ_α shown in Fig. 8. If there are no poles of $G(\alpha)$ and $F_+(\alpha)$ in the shaded region, one finds

$$\frac{1}{2\pi j}\cdot\int_{-\infty}^{+\infty}\frac{[G(x)-G(\alpha)]F_+(x)}{x-\alpha}dx = \frac{1}{2\pi j}\cdot\int_{\lambda_\alpha}\frac{[G(x)-G(\alpha)]F_+(x)}{x-\alpha}dx \tag{47}$$

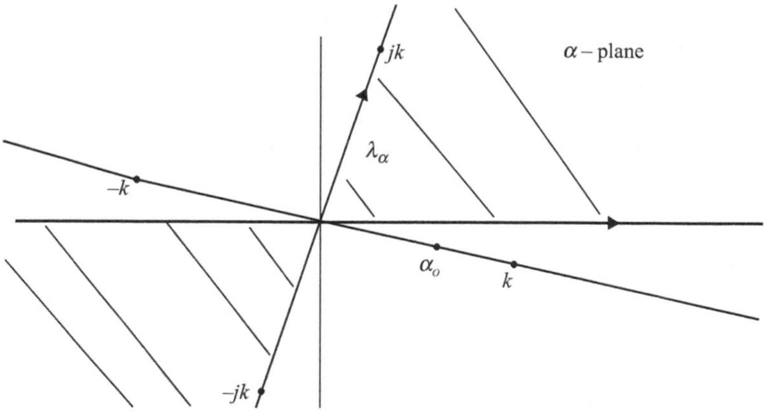

Fig. 8: Deformation of the contour path

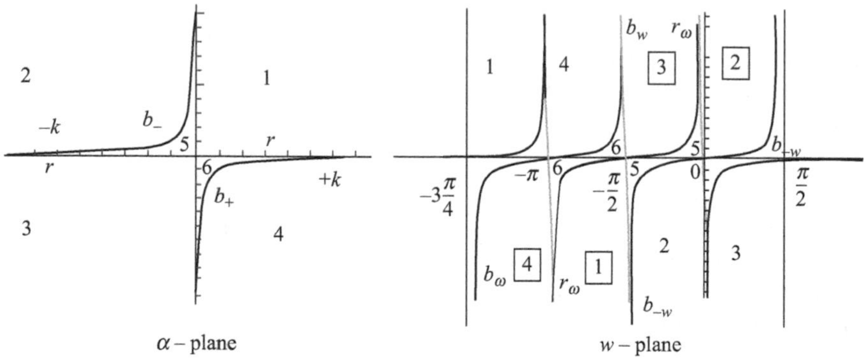

Fig. 9: Integration paths in the w – plane

In wedge problems, the kernel $M(\alpha, t)$ presents a more suitable behavior in the w – plane defined by (see section 2.9.2):

$$\alpha = -\tau_o \cos w \;\; x = -\tau_o \cos w' \tag{48}$$

In the w – plane, λ_w and r_w (Fig. 9) are the images of the line λ_α and of the real axis of the α – plane, respectively.

Furthermore,

$$\alpha_o = -\tau_o \cos w_o = -\tau_o \cos \varphi_o \tag{49}$$

where $w_o = -\varphi_o$, $\pi/2 < \varphi_o < \pi$. In the w – plane the Fredholm equation becomes

$$\hat{G}(w)\hat{F}_+(w) + \frac{1}{2\pi j}\int_{\lambda_w} \frac{\left[\hat{G}(w') - \hat{G}(w)\right]\hat{F}_+(w')}{-\tau_o \cos w' + \tau_o \cos w} \tau_o \sin w' dw' = -\frac{F_o}{\tau_o \cos w - \tau_o \cos w_o} \tag{50}$$

By introducing

$$w' = -\frac{\pi}{2} + ju \tag{51}$$

we have that $dw' = jdu$, $\cos w' = j \sinh u$, $\sin w' = -\cosh u$,

$$\hat{G}(w') = H(u), \quad \hat{F}_+(w) = Y(u)$$

$$\hat{G}(w)\hat{F}_+(w) + \frac{1}{2\pi j}\int_{+\infty}^{-\infty} \frac{[H(u) - \hat{G}(w)]\,Y(u)}{-j\sinh u + \cos w}(-\cosh u)\,jdu = -\frac{F_o}{\tau_o \cos w - \tau_o \cos w_o} \tag{52}$$

or

$$\hat{G}(w)\hat{F}_+(w) + \frac{1}{2\pi j}\int_{-\infty}^{+\infty} \frac{[H(u) - \hat{G}(w)]\,Y(u)}{-j\sinh u + \cos w}\cosh u\,jdu = -\frac{F_o}{\tau_o \cos w - \tau_o \cos w_o} \tag{53}$$

This provides the representation

$$\hat{F}_+(w) = -\frac{\hat{G}^{-1}(w)}{2\pi j}\int_{-\infty}^{+\infty} \frac{[H(u) - \hat{G}(w)]\,Y(u)}{-j\sinh u + \cos w}\cosh u\,jdu - \frac{\hat{G}^{-1}(w)F_o}{\tau_o \cos w - \tau_o \cos w_o} \tag{54}$$

Let us now derive the equations valid for the samples $Y(h\,i)$. By putting $w = -\frac{\pi}{2} + jt$ in the equation

$$\hat{G}(w)\hat{F}_+(w) + \frac{1}{2\pi j}\int_{-\infty}^{+\infty} \frac{[H(u) - \hat{G}(w)]\,Y(u)}{-j\sinh u + \cos w}\cosh u\,jdu = -\frac{F_o}{\tau_o \cos w - \tau_o \cos w_o} \tag{55}$$

we find

$$H(t)Y(t) + \frac{1}{2\pi j}\int_{-\infty}^{+\infty} \frac{[H(u) - H(t)]}{-\sinh u + \sinh t}\cosh u\,Y(u)du = -\frac{F_o}{\tau_o(j\sinh t - \cos w_o)} \tag{56}$$

or

$$H(t)Y(t) + \frac{1}{2\pi j}\int_{-\infty}^{+\infty} m(t,u)Y(u)du = -\frac{F_o}{\tau_o(j\sinh t - \cos w_o)} \tag{57}$$

where

$$\hat{F}_+(w) = Y(u)$$

$$m(t,u) = \mathrm{If}\left[u == t, \lim\frac{[H(u) - H(t)]}{-\sinh u + \sinh t}\cosh u, \frac{[H(u) - H(t)]}{-\sinh u + \sinh t}\right]$$

with the limit evaluated as $u = t$.

The sampled form of the previous Fredholm equation is

$$H(hr)Y_a(hr) + \frac{h}{2\pi j}\sum_{i=-A/h}^{A/h} m(hr,hi)Y_a(hi) = -\frac{F_o}{\tau_o(j\sinh hr - \cos w_o)},$$
$$r = 0, \pm 1, \pm 2, \ldots, \pm\frac{A}{h} \tag{58}$$

The solution of the previous equation in $2\frac{A}{h}+1$ unknowns $Y_a(h\,i)$, $(i = 0, \pm 1, \pm 2, \ldots, \pm\frac{A}{h})$ yields the representation

$$Y_a(t) = -\frac{H(t)^{-1}}{2\pi j}h\sum_{i=-A/h}^{A/h} m(t,hi)Y_a(hi) - \frac{H(t)^{-1}F_o}{\tau_o(j\sinh t) - \tau_o\cos w_o} \tag{59}$$

which in turn, in the w – plane, yields the approximate plus function

$$\hat{F}_{a+}(w) = Y_a(u) \tag{60}$$

or

$$\hat{F}_{a+}(w) = -\frac{\hat{G}(w)^{-1}}{2\pi j}h\sum_{i=-A/h}^{A/h} m\left[-j\left(w+\frac{\pi}{2}\right), hi\right]Y_a(hi) - \frac{\hat{G}(w)^{-1}F_o}{\tau_o(\cos w - \cos w_o)} \tag{61}$$

5.3 Additional considerations on the Fredholm equations

5.3.1 Presence of poles of the kernel in the warped region

For illustrative purposes, let us consider the presence of a pole α_s in the shaded region of Fig. 8. This pole may be present in the plus function $F_+(\alpha)$ (zero of $\det[G^{-1}(\alpha)]$) or in the minus function $F_-(\alpha)$ (zero of $\det[G(\alpha)]$). By following the same procedure used in deriving (61), it is possible to obtain directly the following integral equation on the λ_α line:

(a) if α_s is a pole of $F_-(\alpha)$:

$$G(\alpha)F_+(\alpha) + \frac{1}{2\pi j}\cdot\int_{\lambda_\alpha}\frac{[G(x) - G(\alpha)]F_+(x)}{x-\alpha}dx = \frac{F_o}{\alpha - \alpha_o} + \frac{X_o}{\alpha - \alpha_s}, \quad \mathrm{Im}[\alpha_o] < 0 \tag{62}$$

where X_o is unknown. It should be noted that the last unknown term is present only if $\mathrm{Im}[\alpha_s] > 0$

(b) if α_s is a pole of $F_+(\alpha)$:

$$G(\alpha)F_+(\alpha) + \frac{1}{2\pi j}\cdot\int_{\lambda_\alpha}\frac{[G(x) - G(\alpha)]F_+(x)}{x-\alpha}dx = \frac{F_o}{\alpha - \alpha_o} + \frac{Y_o}{\alpha - \alpha_s}, \quad \mathrm{Im}[\alpha_o] < 0 \tag{63}$$

where Y_o is unknown. It should be noted that the last unknown term is present only if $\mathrm{Im}[\alpha_s] < 0$

In order to evaluate X_o one can use the sampled form of the equation:

$$\frac{1}{2\pi j}\cdot\int\limits_{\lambda_a}\frac{F_+(x)}{x-\alpha_s}dx = 0 \tag{64}$$

and similarly, to evaluate Y_o one can use the sampled form of the equation:

$$\frac{1}{2\pi j}\cdot\int\limits_{\lambda_a}\frac{F_+(x)}{x+\alpha_s}dx = 0 \tag{65}$$

5.3.2 The Fredholm factorization for particular matrices

Matrices having the form $G(\alpha)=\begin{pmatrix} A(\alpha) & B(\alpha)e^{+j\alpha L} \\ C(\alpha)e^{-j\alpha L} & D(\alpha)\end{pmatrix}$ where L is a real quantity are note amenable to the method described in this chapter. They will be considered in chapter 6.

5.3.3 The Fredholm equation relevant to a modified kernel

An important characteristic of the Fredholm factorization discussed in the previous sections is constituted by the fact that it is not necessary to regularize the operator having as symbol $G(\alpha)$ by performing preliminary factorizations.

To improve the accuracy of the factorization, sometimes it is convenient to generalize the Fredholm factorization by introducing a slight modification of the technique discussed above. Let us rewrite $G(\alpha)$:

$$G(\alpha) = G_m(\alpha)G_p(\alpha) \tag{66}$$

where the inverse of $G_m(\alpha)$ is regular in the lower half-plane $\mathrm{Im}[\alpha] \leq 0$. By repeating the procedure that yields the Fredholm equation we obtain

$$G_p(\alpha)F_+(\alpha) + \frac{1}{2\pi j}\cdot\int\limits_{-\infty}^{+\infty}\frac{[G_p(t)-G_p(\alpha)]F_+(t)}{t-\alpha}dt = \frac{G_m^{-1}(\alpha_o)\cdot R}{\alpha-\alpha_o} \quad \mathrm{Im}[\alpha_o] < 0 \tag{67}$$

This is another Fredholm equation that sometimes can be more convenient than the original one. For instance, if $G_m(\alpha)$ and $G_p(\alpha)$ are the standard factorized matrices of

$$G(\alpha) = G_-(\alpha)G_+(\alpha)$$

then the integral vanishes, and one obtains the expected result:

$$F_+(\alpha) = \frac{G_+^{-1}(\alpha)\cdot G_-^{-1}(\alpha_o)\cdot R}{\alpha-\alpha_o} \tag{68}$$

In this case we say that $G_m(\alpha) = G_-(\alpha)$ produces a total regularization of the matrix kernel $G(\alpha)$. In general, we can choose the matrix $G_m(\alpha)$ to have a partial regularization of $G(\alpha)$. This expedient could simplify the numerical solution of the integral equation.

CHAPTER 6

Approximate solutions: Some particular techniques

All the powerful methods (e.g., iterative, moment) developed in functional analysis can be effectively used for solving W-H equations or for obtaining factorized matrices. Furthermore, the observation that it is possible to factorize matrices with rational elements suggests approximating the kernels with rational approximants.

6.1 The Jones method for solving modified W-H equations

6.1.1 Introduction

In the following we present two powerful methods due to Jones that allow us to solve the modified W-H equations, avoiding their reduction to classical W-H systems. The Jones methods are very popular. In fact, they were used and continue to be used in very numerous applications. The only inconvenient of these methods is the requirement of a preliminary factorization of the kernel $G(\alpha)$.

6.1.2 Longitudinal modified W-H equation

Let us rewrite the longitudinal modified W-H equation in its original form

$$G(\alpha)F(\alpha) + e^{j\alpha L}F_+(\alpha) + F_-(\alpha) = F_o(\alpha) \quad (L > 0) \tag{1}$$

where $F(\alpha)$ is an entire function having the following properties:

- $F(\alpha)$ is vanishing as $\alpha \to \infty$ in the half-plane $\mathrm{Im}[\alpha] > 0$ and grows exponentially as $\alpha \to \infty$ in the half-plane $\mathrm{Im}[\alpha] < 0$
- $F(\alpha)e^{-j\alpha L}$ is vanishing as $\alpha \to \infty$ in the half-plane $\mathrm{Im}[\alpha] < 0$ and grows exponentially as $\alpha \to \infty$ in the half-plane $\mathrm{Im}[\alpha] > 0$

The factorization of $G(\alpha) = G_-(\alpha) \cdot G_+(\alpha)$ yields

$$G_+(\alpha)F(\alpha) + e^{j\alpha L}G_-^{-1}(\alpha)F_+(\alpha) + G_-^{-1}F_-(\alpha) = G_-^{-1}\alpha)F_o(\alpha) \tag{2}$$

Separating the plus and minus parts in (2) yields

$$G_+(\alpha)F(\alpha) + \frac{1}{2\pi j}\int_{\gamma_1} \frac{e^{j\beta L}G_-^{-1}(\beta)F_+(\beta)}{\beta-\alpha}d\beta = S_{1+}(\alpha) \tag{3}$$

$$G_-^{-1}(\alpha)F_-(\alpha) - \frac{1}{2\pi j}\int_{\gamma_2} \frac{e^{j\beta L}G_-^{-1}(\beta)F_+(\beta)}{\beta-\alpha}d\beta = S_{1-}(\alpha) \tag{4}$$

where $S_{1-}(\alpha)$ and $S_{1+}(\alpha)$ follows from the decomposition of $G_-^{-1}(\alpha)F_o(\alpha)$:

$$G_-^{-1}(\alpha)F_o(\alpha) = S_{1-}(\alpha) + S_{1+}(\alpha) \tag{5}$$

and γ_1 and γ_2 are the smile and frown real axis paths, respectively.

Equation (1) can be rewritten as

$$G(\alpha)e^{-j\alpha L}F(\alpha) + F_+(\alpha) + e^{-j\alpha L}F_-(\alpha) = e^{-j\alpha L}F_o(\alpha) \tag{6}$$

We observe from the characteristics of $F(\alpha)$ that the function $e^{-j\alpha L}F(\alpha)$ is a minus function since it is vanishing in the half-plane $\text{Im}[\alpha] < 0$.

The (right) factorization of $G(\alpha) = g_+(\alpha) \cdot g_-(\alpha)$ yields

$$g_-(\alpha)e^{-j\alpha L}F(\alpha) + g_+^{-1}(\alpha)F_+(\alpha) + g_+^{-1}(\alpha)e^{-j\alpha L}F_-(\alpha) = g_+^{-1}(\alpha)e^{-j\alpha L}F_o(\alpha) \tag{7}$$

Separating the plus and minus parts in (7), we get:

$$g_+^{-1}(\alpha)F_+(\alpha) + \frac{1}{2\pi j}\int_{\gamma_1} \frac{g_+^{-1}(\beta)e^{-j\beta L}F_-(\beta)}{\beta-\alpha} = S_{2+}(\alpha) \tag{8}$$

$$g_-(\alpha)e^{-j\alpha L}F(\alpha) - \frac{1}{2\pi j}\int_{\gamma_1} \frac{g_+^{-1}(\beta)e^{-j\beta L}F_-(\beta)}{\beta-\alpha} = S_{2-}(\alpha) \tag{9}$$

where $S_{2-}(\alpha) + S_{2+}(\alpha)$ follows from the decomposition of $g_+^{-1}(\alpha)e^{-j\alpha L}F_o(\alpha)$

Now we show that the system of eqs. (4) and (8)

$$\begin{aligned} G_-^{-1}(\alpha)F_-(\alpha) - \frac{1}{2\pi j}\int_{\gamma_2} \frac{e^{j\beta L}G_-^{-1}(\beta)F_+(\beta)}{\beta-\alpha}d\beta = S_{1-}(\alpha) \\ g_+^{-1}(\alpha)F_+(\alpha) + \frac{1}{2\pi j}\int_{\gamma_1} \frac{g_+^{-1}(\beta)e^{-j\beta L}F_-(\beta)}{\beta-\alpha}d\beta = S_{2+}(\alpha) \end{aligned} \tag{10}$$

can be reduced to a Fredholm system of the second kind. In fact, by changing β into $-\beta$ in the first equation of (10) and α into $-\alpha$ in the second equation we get

$$G_-^{-1}(\alpha)F_-(\alpha) + \frac{1}{2\pi j}\int_{\gamma_1} \frac{e^{-j\beta L}G_-^{-1}(-\beta)F_+(-\beta)}{\beta+\alpha}d\beta = S_{1-}(\alpha) \tag{11}$$

$$g_+^{-1}(-\alpha)F_+(-\alpha) + \frac{1}{2\pi j}\int_{\gamma_1} \frac{e^{-j\beta L}g_+^{-1}(\beta)F_-(\beta)}{\beta+\alpha}d\beta = S_{2+}(-\alpha) \tag{12}$$

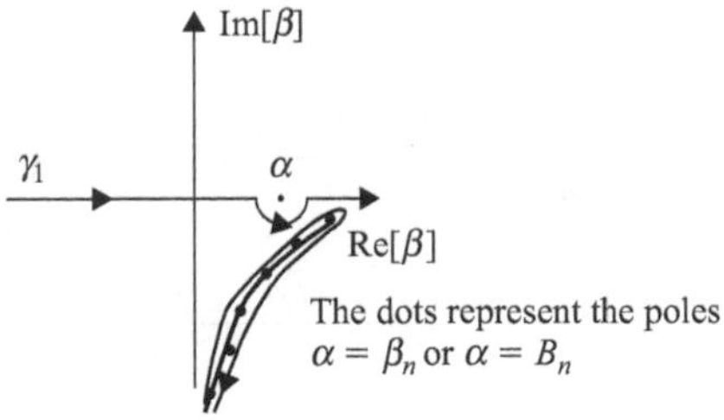

Fig. 1: Discrete spectrum of $G_-^{-1}(-\beta)$ and $g_+^{-1}(\beta)$

The advantage of this system consists in the presence of the factor $e^{-j\beta L}$ that is readily vanishing for values of β located in the lower half-plane $\mathrm{Im}[\beta] < 0$. To appreciate this fact, let us consider the presence of only a discrete spectrum in $G_-^{-1}(-\beta)$ and $g_+^{-1}(\beta)$ (Fig. 1). Taking into account the regularity of $F_+(-\beta)$ and $F_-(\beta)$ in the half-plane $\mathrm{Im}[\beta] < 0$, we obtain

$$\frac{1}{2\pi j}\int_{\gamma_1}\frac{e^{-j\beta L}G_-^{-1}(-\beta)F_+(-\beta)}{\beta+\alpha}d\beta = -\sum_m \frac{e^{-jB_mL}R(-B_m)F_+(-B_m)}{B_m+\alpha} \tag{13}$$

$$\frac{1}{2\pi j}\int_{\gamma_1}\frac{e^{-j\beta L}g_+^{-1}(\beta)F_-(\beta)}{\beta+\alpha} = -\sum_m \frac{e^{-j\beta_mL}r(\beta_m)F_-(\beta_m)}{\beta_m+\alpha} \tag{14}$$

where B_m and $R(-B_m)$ are the poles and the residues of $G_-^{-1}(-\beta)$, respectively, whereas β_m and $r(\beta_m)$ are the poles and the residues of $g_+^{-1}(\beta)$, respectively.

Substituting and letting $\alpha = \beta_n$ in (11) and $\alpha = B_n$ in (12), we get the system in the unknowns $F_-(\beta_m)$ and $F_+(-B_m)$:

$$\begin{aligned} G_-^{-1}(\beta_n)F_-(\beta_n) - \sum_m \frac{e^{-jB_m L}R(-B_m)F_+(-B_m)}{B_m+\beta_n} &= S_{1-}(\beta_n) \\ g_+^{-1}(-B_n)F_+(-B_n) - \sum_m \frac{e^{-j\beta_m L}r(\beta_m)F_-(\beta_m)}{\beta_m+B_n} &= S_{2+}(-B_n) \end{aligned} \tag{15}$$

The operators involving in the sums are compact. Moreover, the presence of the exponential $e^{-jB_m L}$ and $e^{-j\beta_m L}$ allow truncate the sums to be truncated to a very few terms.

In the presence of only a continuous spectrum (Fig. 2) we can warp the smile real axis γ_1 to the closed line γ that encloses the branch line Γ.

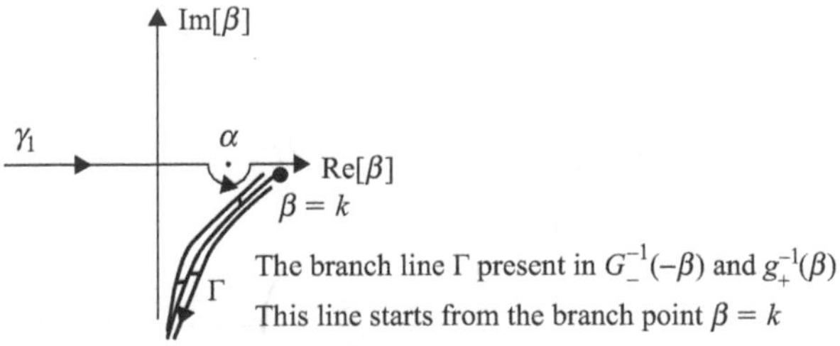

Fig. 2: Continuous spectrum of $G_-^{-1}(-\beta)$ and $g_+^{-1}(\beta)$

We obtain the Fredholm equations

$$
\begin{aligned}
G_-^{-1}(\alpha)F_-(\alpha) + \frac{1}{2\pi j}\int_\Gamma \frac{e^{-jtL}M(t)Y_-(t)}{t+\alpha}dt &= S_{1-}(\alpha) \\
g_+^{-1}(-\alpha)Y_-(\alpha) + \frac{1}{2\pi j}\int_\Gamma \frac{e^{-jtL}N(t)F_-(t)}{t+\alpha} &= S_{2+}(-\alpha) \quad \alpha \in \Gamma
\end{aligned}
\tag{16}
$$

where $M(t)$ and $N(t)$ are the jumps of $G_-^{-1}(-\beta)$ and $g_+^{-1}(\beta)$ on the branch line Γ, and $Y_-(t) = F_+(-t)$.

The Fredholm equations can be solved by quadrature and again the presence of e^{-jtL} allows us to assume a very small integration interval. Moreover, the presence of the rapidly vanishing factor e^{-jtL} suggests to represent the unknowns $F_-(t)$ and $Y_-(t)$ in Taylor series about the branch point $t = k$:

$$
\begin{aligned}
N(t)F_-(t) &= \sum_i A_i(t+k)^i \\
M(t)Y_-(t) &= \sum_i B_i(t+k)^i
\end{aligned}
\tag{17}
$$

Substituting (17) into (16) leads to the Whittaker functions defined by

$$
W_{i-1/2}(z) = \int_0^\infty \frac{u^i e^{-u}}{u+z}du \tag{18}
$$

The representation (17) for evaluating the integrals in the Fredholm eqs. (16) introduces very few constant unknowns A_i and B_i. These unknowns must satisfy a system of linear equations that derives by suitable manipulations on eqs. (16).

6.1.3 Transversal modified W-H equation

By factorizing the kernel $G(\alpha) = G_-(\alpha) \cdot G_+(\alpha)$ in the transversal modified W-H equation

$$
G(\alpha)F_+(\alpha) + H(\alpha)F_+(-\alpha) + F_-(\alpha) = F_{0+}(\alpha)
$$

we find

$$
G_+(\alpha)F_+(\alpha) + G_-^{-1}(\alpha)\cdot H(\alpha)F_+(-\alpha) + G_-^{-1}(\alpha)\cdot F_-(\alpha) = G_-^{-1}(\alpha)\cdot F_{0+}(\alpha) \tag{19}
$$

Separating in the plus and minus functions

$$
G_+(\alpha)F_+(\alpha) + \frac{1}{2\pi j}\int_{\gamma_1} \frac{G_-^{-1}(\beta)\cdot H(\beta)F_+(-\beta)}{\beta-\alpha}d\beta = S_+(\alpha) \tag{20}
$$

$$
G_-^{-1}(\alpha)\cdot F_-(\alpha) - \frac{1}{2\pi j}\int_{\gamma_2} \frac{G_-^{-1}(\beta)\cdot H(\beta)F_+(-\beta)}{\beta-\alpha}d\beta = S_-(\alpha) \tag{21}
$$

where $S_-(\alpha)$ and $S_+(\alpha)$ follow from the decomposition of $G_-^{-1}(\alpha)F_{o+}(\alpha)$:

$$G_-^{-1}(\alpha)F_{o+}(\alpha) = S_{1-}(\alpha) + S_{1+}(\alpha) \tag{22}$$

Changing α into $-\alpha$ in (20) yields the Fredholm equation:

$$G_+(-\alpha)X_-(\alpha) + \frac{1}{2\pi j}\int_{\gamma_1} \frac{G_-^{-1}(\beta)\cdot H(\beta)X_-(\beta)}{\beta+\alpha}d\beta = S_+(-\alpha) \tag{23}$$

where $X_-(\alpha) = F_+(-\alpha)$.

In alternative to the quadrature method, as we did for the longitudinal W-H equations, it is interesting to warp the smile real axis γ_1 to enclose all the singularities of $G_-^{-1}(\beta)\cdot H(\beta)\cdot X_-(\beta)$ located in the half-plane $\mathrm{Im}[\beta] < 0$. We observe that, since $G_-^{-1}(\beta)$ and $X_-(\beta)$ are regular in this half-plane, the singularities involved are only those of $H(\beta)$. For instance, if $H(\beta)$ presents only a discrete spectrum constituted by the simple poles β_m as indicated in Fig. 1, we can rewrite eq. (23) in the form

$$G_+(-\alpha)X_-(\alpha) - \sum_m \frac{G_-^{-1}(\beta_m)\cdot R(\beta_m)X_-(\beta_m)}{\beta_m+\alpha} = S_+(-\alpha) \tag{24}$$

where $R(\beta_m)$ are the residues of $H(\beta)$ in β_m. Putting $\alpha = \beta_n$ in (24) yields the system

$$G_+(-\beta_n)X_-(\beta_n) - \sum_m \frac{G_-^{-1}(\beta_m)\cdot R(\beta_m)X_-(\beta_m)}{\beta_m+\beta_n} = S_+(-\beta_n) \tag{25}$$

which is very convenient and has been used in numerous applications presented in the literature.

The same considerations apply when $H(\beta)$ presents only a continuous spectrum. We obtain the Fredholm equation

$$G_+(-\alpha)X_-(\alpha) + \frac{1}{2\pi j}\int_{\Gamma} \frac{G_-^{-1}(\beta)\cdot M(\beta)X_-(\beta)}{\beta+\alpha}d\beta = S_+(-\alpha), \quad \alpha\in\Gamma \tag{26}$$

that is easily solved by quadrature.

6.2 The Fredholm factorization for particular matrices

Sometimes the kernel $G(\alpha)$ has a form that does not allows the reduction to Fredholm integral equations, as indicated in chapter 5. For instance, let us consider the kernel

$$G(\alpha) = \begin{pmatrix} A(\alpha) & B(\alpha)e^{+j\alpha L} \\ C(\alpha)e^{-j\alpha L} & D(\alpha) \end{pmatrix} \tag{27}$$

associated with the W-H equation.

$$G(\alpha)F_+(\alpha) = X_-(\alpha) + \frac{R}{\alpha-\alpha_p}, \quad \mathrm{Im}[\alpha_p] < 0. \tag{28}$$

First, we observe that it is not restrictive to consider the parameter L as nonnegative. In fact, the following equation holds:

$$\begin{pmatrix} A(\alpha) & B(\alpha)e^{-j\alpha L} \\ C(\alpha)e^{+j\alpha L} & D(\alpha) \end{pmatrix} = \begin{pmatrix} 0 & 1 \\ 1 & 0 \end{pmatrix} \begin{pmatrix} D(\alpha) & C(\alpha)e^{+j\alpha L} \\ B(\alpha)e^{-j\alpha L} & A(\alpha) \end{pmatrix} \begin{pmatrix} 0 & 1 \\ 1 & 0 \end{pmatrix}$$

The reason for the difficulties in factorizing (27) is essentially due to the presence of $e^{+j\alpha L}$. This factor does not allow us to close the line γ_1 (section 1.6) with a semicircle located at ∞ in the lower half-plane α. Consequently, it is impossible to deduce straightforwardly the Fredholm equation relevant to the W-H matrix kernels. To overcome these difficulties we propose two methods:

(a) Without loss of generality we can reduce the factorization of matrix (27) to that of the matrix $\begin{pmatrix} g(\alpha) & e^{+j\alpha L} \\ e^{-j\alpha L} & d(\alpha) \end{pmatrix}$. Next we rewrite

$$\begin{pmatrix} g(\alpha) & e^{+j\alpha L} \\ e^{-j\alpha L} & d(\alpha) \end{pmatrix} = G_o(\alpha) + \begin{pmatrix} 0 & 0 \\ 0 & d(\alpha) \end{pmatrix}$$

$$= G_{o-}(\alpha)\left(1 + G_{o-}^{-1}(\alpha)\begin{pmatrix} 0 & 0 \\ 0 & d(\alpha) \end{pmatrix} G_{o+}^{-1}(\alpha)\right) G_{o+}(\alpha)$$

where

$$G_o(\alpha) = \begin{pmatrix} g(\alpha) & e^{+j\alpha L} \\ e^{-j\alpha L} & 0 \end{pmatrix} = G_{o-}(\alpha) G_{o+}(\alpha)$$

can be factorized with the Jones method (section 6.1.2). The previous equation reduces the problem to the factorization of the matrix $\left(1 + G_{o-}^{-1}(\alpha)\begin{pmatrix} 0 & 0 \\ 0 & d(\alpha) \end{pmatrix} G_{o+}^{-1}(\alpha)\right)$, and this can be accomplished with the Fredholm factorization described in chapter 5.

(b) Alternatively, Abrahams and Wickham (1990) suggested rewriting the matrix $G(\alpha)$ in the form

$$G(\alpha) = \begin{pmatrix} A(\alpha) & B(\alpha)e^{+j\alpha L} \\ C(\alpha)e^{-j\alpha L} & D(\alpha) \end{pmatrix} = \begin{pmatrix} 1 & 0 \\ 0 & e^{-j\alpha L} \end{pmatrix} \cdot M(\alpha) \cdot \begin{pmatrix} 1 & 0 \\ 0 & e^{j\alpha L} \end{pmatrix} \tag{29}$$

where the central matrix

$$M(\alpha) = \begin{pmatrix} A(\alpha) & B(\alpha) \\ C(\alpha) & D(\alpha) \end{pmatrix}$$

can be factorized analytically or numerically:

$$M(\alpha) = M_-(\alpha) M_+(\alpha) \tag{30}$$

Equation (30) yields

$$G(\alpha) = \tilde{G}_-(\alpha) \cdot \tilde{G}_+(\alpha) \tag{31}$$

where

$$\tilde{G}_-(\alpha) = \begin{pmatrix} 1 & 0 \\ 0 & e^{-j\alpha L} \end{pmatrix} \cdot M_-(\alpha), \quad \tilde{G}_+(\alpha) = M_+(\alpha) \cdot \begin{pmatrix} 1 & 0 \\ 0 & e^{j\alpha L} \end{pmatrix} \tag{32}$$

$$\tilde{G}_-^{-1}(\alpha) = M_-^{-1}(\alpha)\begin{pmatrix} 1 & 0 \\ 0 & e^{+j\alpha L} \end{pmatrix}, \quad \tilde{G}_+^{-1}(\alpha) = \begin{pmatrix} 1 & 0 \\ 0 & e^{-j\alpha L} \end{pmatrix} \cdot M_+^{-1}(\alpha) \tag{33}$$

The factorized matrices $\tilde{G}_-(\alpha)$ and $\tilde{G}_+(\alpha)$ and their inverses are regular in the lower and upper half-planes $\mathrm{Im}[\alpha] \leq 0$ and $\mathrm{Im}[\alpha] \geq 0$, respectively. However, they do not constitute standard factorized matrices since $\tilde{G}_-(\alpha)$ grows exponentially in the half-plane $\mathrm{Im}[\alpha] \geq 0$ and $\tilde{G}_+(\alpha)$ grows exponentially in the half-plane $\mathrm{Im}[\alpha] \leq 0$. To overcome this problem, we rewrite (28) in the form

$$\tilde{G}_+(\alpha)F_+(\alpha) - \tilde{G}_-^{-1}(\alpha_p)\frac{R}{\alpha - \alpha_p} = \tilde{G}_-^{-1}(\alpha)X_-(\alpha) + \left[\tilde{G}_-^{-1}(\alpha) - \tilde{G}_-^{-1}(\alpha_p)\right]\frac{R}{\alpha - \alpha_p} = w(\alpha) \tag{34}$$

Since the first member is regular in the half-plane $\mathrm{Im}[\alpha] \geq 0$ and the second member is regular in the half-plane $\mathrm{Im}[\alpha] \leq 0$, it follows that the function $w(\alpha)$ is entire. Moreover, it is vanishing in $\mathrm{Im}[\alpha] \geq 0$ and conversely grows exponentially in $\mathrm{Im}[\alpha] \leq 0$.

Multiplying (34) by $e^{-j\alpha L}$ yields that $e^{-j\alpha L}w(\alpha)$ is vanishing in $\mathrm{Im}[\alpha] \leq 0$ and conversely grows exponentially in $\mathrm{Im}[\alpha] \geq 0$. These properties of the entire function $w(\alpha)$ will be utilized later.

It is possible to derive the Fredholm integral equation for $w(\alpha)$ by using the result

$$F_+(\alpha) = \tilde{G}_+^{-1}(\alpha)w(\alpha) + \frac{\tilde{G}_+^{-1}(\alpha) \cdot \tilde{G}_-^{-1}(\alpha_p) \cdot R}{\alpha - \alpha_p} \tag{35}$$

By integrating on the "frown" real axis γ_2 we rewrite:

$$\frac{1}{2\pi j}\int_{\gamma_2} \frac{F_+(u)}{u - \alpha}du = \frac{1}{2\pi j}\int_{\gamma_2} \frac{\tilde{G}_+^{-1}(u)w(u)}{u - \alpha}du + \frac{1}{2\pi j}\int_{\gamma_2} \frac{\tilde{G}_+^{-1}(u) \cdot \tilde{G}_-^{-1}(\alpha_o)}{u - \alpha} \cdot \frac{R}{u - \alpha_p}du \tag{36}$$

Now the first member is vanishing since we can close the line γ_2 with a half-circle of radius $\to \infty$ located in the half-plane $\mathrm{Im}[u] \geq 0$, where $F_+(u)$ is regular and vanishing as $u \to \infty$. It follows that

$$\frac{1}{2\pi j}\int_{\gamma_2} \frac{\tilde{G}_+^{-1}(u)w(u)}{u - \alpha}du = -\frac{1}{2\pi j}\int_{\gamma_2} \frac{\tilde{G}_+^{-1}(u) \cdot \tilde{G}_-^{-1}(\alpha_p)}{u - \alpha} \cdot \frac{R}{u - \alpha_p}du \tag{37}$$

or

$$-\frac{1}{2}\tilde{G}_+^{-1}(\alpha)w(\alpha) + \frac{1}{2\pi j}P.V.\int_{-\infty}^{\infty} \frac{\tilde{G}_+^{-1}(u)w(u)}{u - \alpha}du = -\frac{1}{2\pi j}\int_{\gamma_2} \frac{\tilde{G}_+^{-1}(u) \cdot \tilde{G}_-^{-1}(\alpha_p)}{u - \alpha} \cdot \frac{R}{u - \alpha_p}du \tag{38}$$

Taking into account that $e^{-j\alpha L}w(\alpha)$ is regular and is vanishing in $\mathrm{Im}[\alpha] \leq 0$, we have

$$\frac{1}{2\pi j}\int\limits_{\gamma_1}\frac{e^{-juL}w(u)}{u-\alpha}du = 0 \tag{39}$$

where γ_1 is the smile real axis (section 1.6). Equation (39) can be rewritten as

$$\frac{\tilde{G}_+^{-1}(\alpha)e^{j\alpha L}}{2\pi j}\int\limits_{\gamma_1}\frac{e^{-juL}w(u)}{u-\alpha}du = \frac{1}{2}\tilde{G}_+^{-1}(\alpha)w(\alpha) + \frac{1}{2\pi j}P.V\int\limits_{-\infty}^{\infty}\frac{\tilde{G}_+^{-1}(\alpha)e^{j(\alpha-u)L}w(u)}{u-\alpha}du = 0 \tag{40}$$

Subtracting (38) from (40) yields

$$\tilde{G}_+^{-1}(\alpha)w(\alpha) + \frac{1}{2\pi j}\int\limits_{-\infty}^{\infty}\frac{[\tilde{G}_+^{-1}(\alpha)e^{j(\alpha-u)L} - \tilde{G}_+^{-1}(u)]w(u)}{u-\alpha}du$$

$$= \frac{1}{2\pi j}\int\limits_{\gamma_2}\frac{\tilde{G}_+^{-1}(u)\cdot\tilde{G}_-^{-1}(\alpha_p)}{u-\alpha}\cdot\frac{R}{u-\alpha_p}du \tag{41}$$

or

$$w(\alpha) + \frac{1}{2\pi j}\int\limits_{-\infty}^{\infty}\frac{[e^{j(\alpha-u)L}\mathbf{1} - \tilde{G}_+(\alpha)\tilde{G}_+^{-1}(u)]w(u)}{u-\alpha}du$$

$$= \frac{1}{2\pi j}\int\limits_{\gamma_2}\frac{\tilde{G}_+(\alpha)\tilde{G}_+^{-1}(u)\cdot\tilde{G}_-^{-1}(\alpha_p)}{u-\alpha}\cdot\frac{R}{u-\alpha_p}du \tag{42}$$

For evaluating the second member we observe that because of the presence of $\tilde{G}_+^{-1}(u)$ that grows exponentially for $\mathrm{Im}[u] \geq 0$, we cannot close the line γ_2 with an infinite half-circle located in the upper half-plane. Conversely, we can close γ_2 with an infinite half-circle located in the lower half-plane. Taking into account that this line encloses the poles α and α_p and all the singularities of $\tilde{G}_+^{-1}(u)$, one has that

$$\frac{1}{2\pi j}\int\limits_{\gamma_2}\frac{\tilde{G}_+(\alpha)\tilde{G}_+^{-1}(u)\cdot\tilde{G}_-^{-1}(\alpha_o)}{u-\alpha}\cdot\frac{R}{u-\alpha_p}du$$

$$= \frac{\tilde{G}_+(\alpha)[\tilde{G}_+^{-1}(\alpha_p) - \tilde{G}_+^{-1}(\alpha)]\cdot\tilde{G}_-^{-1}(\alpha_p)R}{\alpha-\alpha_p} + \frac{1}{2\pi j}\int\limits_{\gamma_2}\frac{\tilde{G}_+(\alpha)\tilde{G}_+^{-1}(u)\cdot\tilde{G}_-^{-1}(\alpha_p)}{u-\alpha}\cdot\frac{R}{u-\alpha_p}du \tag{43}$$

where λ_2 is a closed line that encloses only the singularities of $\tilde{G}_+^{-1}(u)$. The first term of the second member is the sum of the residues in the poles $u = \alpha$ and $u = \alpha_p$; it is interesting to observe that this term is regular in $\alpha = \alpha_p$. Now we show that eq. (42) is a Fredholm equation of second kind when the matrix $M_+^{-1}(\alpha)$ reduces on the real axis to a constant matrix A_o as $\alpha \to \pm\infty$. First, taking into account that

$$\tilde{G}_+^{-1}(\alpha) = \begin{pmatrix} 1 & 0 \\ 0 & e^{-j\alpha L} \end{pmatrix} \cdot M_+^{-1}(\alpha) \tag{44}$$

we rewrite (42) in the form

$$w(\alpha) + \frac{1}{2\pi j}\tilde{G}_+(\alpha) \int_{-\infty}^{\infty} \frac{\left[\begin{pmatrix} e^{j\alpha L} & 0 \\ 0 & 1 \end{pmatrix}(M_+^{-1}(\alpha)) - \begin{pmatrix} e^{juL} & 0 \\ 0 & 1 \end{pmatrix}(M_+^{-1}(u))\right] e^{-juL} w(u)}{u-\alpha} du$$

$$= \frac{1}{2\pi j}\int_{\gamma_2} \frac{\tilde{G}_+(\alpha)\tilde{G}_+^{-1}(u) \cdot \tilde{G}_-^{-1}(\alpha_p)}{u-\alpha} \cdot \frac{R}{u-\alpha_p} du \tag{45}$$

or

$$w(\alpha) + \frac{1}{2\pi j}\tilde{G}_+(\alpha) \int_{-\infty}^{\infty} \frac{\left[\begin{pmatrix} e^{j\alpha L} & 0 \\ 0 & 1 \end{pmatrix}(M_+^{-1}(\alpha) - A_o) - \begin{pmatrix} e^{juL} & 0 \\ 0 & 1 \end{pmatrix}(M_+^{-1}(u) - A_o)\right] e^{-juL} w(u)}{u-\alpha} du$$

$$+ \frac{1}{2\pi j}\tilde{G}_+(\alpha) \int_{-\infty}^{\infty} \frac{\left[\begin{pmatrix} e^{j\alpha L} & 0 \\ 0 & 1 \end{pmatrix} - \begin{pmatrix} e^{juL} & 0 \\ 0 & 1 \end{pmatrix}\right] e^{-juL} A_o w(u)}{u-\alpha} du$$

$$= \frac{1}{2\pi j}\int_{\gamma_2} \frac{\tilde{G}_+(\alpha)\tilde{G}_+^{-1}(u) \cdot \tilde{G}_-^{-1}(\alpha_p)}{u-\alpha} \cdot \frac{R}{u-\alpha_p} du \tag{46}$$

The last integral of the first member can be evaluated in the form

$$\frac{1}{2\pi j}\tilde{G}_+(\alpha) \int_{-\infty}^{\infty} \frac{\left[\begin{pmatrix} e^{j\alpha L} & 0 \\ 0 & 1 \end{pmatrix} - \begin{pmatrix} e^{juL} & 0 \\ 0 & 1 \end{pmatrix}\right] e^{-juL} A_o w(u)}{u-\alpha} du$$

$$= \frac{1}{2\pi j}\tilde{G}_+(\alpha) \int_{\gamma_1} \begin{pmatrix} e^{j\alpha L} & 0 \\ 0 & 1 \end{pmatrix} \frac{e^{-juL} A_o w(u)}{u-\alpha} du + \frac{1}{2\pi j}\tilde{G}_+(\alpha) \int_{\gamma_1} \frac{\begin{pmatrix} e^{juL} & 0 \\ 0 & 1 \end{pmatrix} e^{-juL} A_o w(u)}{u-\alpha} du \tag{47}$$

where γ_1 is the smile real axis. Taking into account the properties of $w(\alpha)$, the smile real axis can be closed with an infinite half-circle located in the lower half-plane; it follows that the first integral of the second member of (47) is vanishing. Conversely, in the second integral of the second member of (47) we close the smile real axis with an infinite half-circle located in the upper half-plane, which yields

$$\frac{1}{2\pi j}\tilde{G}_+(\alpha)\int_{\gamma_1}\frac{\begin{pmatrix} e^{juL} & 0 \\ 0 & 1\end{pmatrix} e^{-juL}A_o w(u)}{u-\alpha} = \tilde{G}_+(\alpha)\begin{pmatrix} 1 & 0 \\ 0 & e^{-j\alpha L}\end{pmatrix} A_o w(\alpha) = M_+(\alpha)A_o w(\alpha) \quad (48)$$

Substituting these results in (46) we obtain

$$\begin{aligned}&[1+M_+(\alpha)A_o]w(\alpha)\\ &+\frac{1}{2\pi j}\tilde{G}_+(\alpha)\int_{-\infty}^{\infty}\frac{\left[\begin{pmatrix} e^{j\alpha L} & 0 \\ 0 & 1\end{pmatrix}(M_+^{-1}(\alpha)-A_o)-\begin{pmatrix} e^{juL} & 0 \\ 0 & 1\end{pmatrix}(M_+^{-1}(u)-A_o)\right]e^{-juL}w(u)}{u-\alpha}du\\ &=\frac{1}{2\pi j}\int_{\gamma_2}\frac{\tilde{G}_+(\alpha)\tilde{G}_+^{-1}(u)\cdot\tilde{G}_-^{-1}(\alpha_p)}{u-\alpha}\cdot\frac{R}{u-\alpha_p}du\end{aligned} \quad (49)$$

The previous equation is of the second kind, since we can show the compactness of the operator

$$\frac{\left[\begin{pmatrix} e^{j\alpha L} & 0 \\ 0 & 1\end{pmatrix}(M_+^{-1}(\alpha)-A_o)-\begin{pmatrix} e^{juL} & 0 \\ 0 & 1\end{pmatrix}(M_+^{-1}(u)-A_o)\right]}{u-\alpha}$$

by adopting the considerations used in Daniele (2004b).

The presence of unknown $w(\alpha)$ that grows exponentially in the lower half-plane does not allow to deform the real axis to use the expedients in section 5.1.5.1. However, we can consider the properties of $w(\alpha)$ to find an alternative way to the quadrature method (Abrahams and Wickham, 1990).

Let us consider the inverse transform $\hat{w}(x)$ of $w(\alpha)$:

$$\hat{w}(x) = \frac{1}{2\pi j}\int_{-\infty}^{\infty} w(\alpha)e^{-j\alpha x}d\alpha \quad (50)$$

For $x < 0$ we can close the real axis with an infinite half-circle located in the upper half-plane, yielding

$$\hat{w}(x) = 0, \quad \text{for } x < 0$$

Similarly, by rewriting (50) in the form

$$\hat{w}(x) = \frac{1}{2\pi j}\int_{-\infty}^{\infty} w(\alpha)e^{-j\alpha L}e^{j\alpha(L-x)}d\alpha \quad (51)$$

and taking into account the asymptotic behavior of $w(\alpha)e^{-j\alpha L}$ in the lower half-plane, in the range $L - x < 0$ or $x > L$, we can close the contour with an infinite half-circle located in the lower half-plane. This yields

$$\hat{w}(x) = 0, \quad \text{for } x > L \tag{52}$$

Hence, the function $\hat{w}(x)$ is not vanishing only in the interval $0 < x < L$, and its Fourier transform $w(\alpha)$ can be represented by the sampling theorem. More directly, we observe that the function

$$\frac{w(\alpha)}{e^{j\alpha L} - 1}$$

is a meromorphic function vanishing at infinity. The Mittag-Leffler expansion of this function yields the series (sampling theorem):

$$\frac{w(\alpha)}{e^{j\alpha L} - 1} = \sum_{n=-\infty}^{\infty} \frac{w(\alpha_n)}{jL(\alpha - \alpha_n)} \tag{53}$$

where $\alpha_n = n\frac{2\pi}{L}$, thus leading to the following representation of $w(\alpha)$ through the sampling $w(\alpha_n)$:

$$w(\alpha) = \sum_{n=-\infty}^{\infty} \frac{(e^{j\alpha L} - 1)w(\alpha_n)}{jL(\alpha - \alpha_n)} \tag{54}$$

Substituting in the integral of the Fredholm eq. (49) and forcing $\alpha = \alpha_m (m = 0, \pm 1, \pm 2, \ldots)$ yields a system of infinite unknowns $w(\alpha_n)$ that according to the compactness of the kernel can be solved by truncation.

A simplification of the integral eq. (49) can be accomplished by observing that by suitably choosing $P_-(\alpha)$, $Q_-(\alpha)$, $R_-(\alpha)$ and $S_-(\alpha)$, we can rewrite (Abrahams & Wickham, 1990):

$$\tilde{G}_-(\alpha) = \begin{pmatrix} p_-(\alpha) & q_-(\alpha) \\ r_-(\alpha)e^{-j\alpha L} & s_-(\alpha)e^{-j\alpha L} \end{pmatrix} = \begin{pmatrix} P_-(\alpha) & 0 \\ 0 & Q_-(\alpha) \end{pmatrix} \cdot \begin{pmatrix} R_-(\alpha) & 1 \\ S_-(\alpha)e^{-j\alpha L} & e^{-j\alpha L} \end{pmatrix} \tag{55}$$

and similarly

$$\tilde{G}_+^{-1}(\alpha) = \begin{pmatrix} P_+(\alpha) & 0 \\ 0 & Q_+(\alpha) \end{pmatrix} \cdot \begin{pmatrix} 1 & R_+(\alpha) \\ e^{-j\alpha L} & S_+(\alpha)e^{-j\alpha L} \end{pmatrix} \tag{56}$$

By introducing the plus unknown $\begin{pmatrix} P_+(\alpha) & 0 \\ 0 & Q_+(\alpha) \end{pmatrix}^{-1} F_+(\alpha)$ and the minus unknown $\begin{pmatrix} P_-(\alpha) & 0 \\ 0 & Q_-(\alpha) \end{pmatrix}^{-1} X_-(\alpha)$, it is shown that it is not restrictive to assume that the factorized matrices $\tilde{G}_-(\alpha)$ and $\tilde{G}_+^{-1}(\alpha)$ present the form

$$\tilde{G}_-(\alpha) = \begin{pmatrix} R_-(\alpha) & 1 \\ S_-(\alpha)e^{-j\alpha L} & e^{-j\alpha L} \end{pmatrix} \tag{57}$$

$$\tilde{G}_+^{-1}(\alpha) = \begin{pmatrix} 1 & R_+(\alpha) \\ e^{-j\alpha L} & S_+(\alpha)e^{-j\alpha L} \end{pmatrix} \tag{58}$$

For instance, this yields

$$M_+^{-1}(\alpha) = \begin{pmatrix} 1 & R_+(\alpha) \\ 1 & S_+(\alpha) \end{pmatrix} \tag{59}$$

and

$$A_o = \begin{pmatrix} 1 & R_+(\infty) \\ 1 & S_+(\infty) \end{pmatrix} \tag{60}$$

The integrand of (49) becomes

$$\begin{aligned} &\begin{pmatrix} e^{j\alpha L} & 0 \\ 0 & 1 \end{pmatrix}(M_+^{-1}(\alpha) - A_o) - \begin{pmatrix} e^{juL} & 0 \\ 0 & 1 \end{pmatrix}(M_+^{-1}(u) - A_o)e^{-juL}w(u) \\ &= \begin{pmatrix} 0 & e^{j\alpha L}[R_+(\alpha) - R_+(\infty)] - e^{juL}[R_+(u) - R_+(\infty)] \\ 0 & S_+(\alpha) - S_+(u) \end{pmatrix} e^{-juL}w(u) \\ &= \begin{vmatrix} e^{j\alpha L}[R_+(\alpha) - R_+(\infty)] - e^{juL}[R_+(u) - R_+(\infty)] \\ S_+(\alpha) - S_+(u) \end{vmatrix} e^{-juL}w_2(u) \end{aligned} \tag{61}$$

where $w_2(\alpha)$ is the second component of the vector $w(\alpha) = \begin{vmatrix} w_1(\alpha) \\ w_2(\alpha) \end{vmatrix}$. This means that the system of the two integral equations

$$\begin{aligned} &[1 + M_+(\alpha)A_o]w(\alpha) \\ &+ \frac{1}{2\pi j}\tilde{G}_+(\alpha)\int_{-\infty}^{\infty} \frac{\left[\begin{pmatrix} e^{j\alpha L} & 0 \\ 0 & 1 \end{pmatrix}(M_+^{-1}(\alpha) - A_o) - \begin{pmatrix} e^{juL} & 0 \\ 0 & 1 \end{pmatrix}(M_+^{-1}(u) - A_o)\right]e^{-juL}w(u)}{u - \alpha}du \\ &= \frac{1}{2\pi j}\int_{\gamma_2} \frac{\tilde{G}_+(\alpha)\tilde{G}_+^{-1}(u) \cdot \tilde{G}_-^{-1}(\alpha_p)}{u - \alpha} \cdot \frac{R}{u - \alpha_p}du \end{aligned} \tag{62}$$

can be reduced to two uncoupled scalar equations.

Method (b) can also be applied to solve longitudinal modified W-H equations. In fact, these equations are equivalent to a vector system with kernel (section 1.5.2):

$$\overline{G}(\alpha) = \begin{pmatrix} G(\alpha) & e^{j\alpha L} \\ e^{-j\alpha L} & 0 \end{pmatrix} \tag{63}$$

6.3 Rational approximation of the kernel

6.3.1 Pade approximants

An approximate factorization can be obtained using Pade representations of the kernel (Abrahams, 2000). MATHEMATICA provides Pade approximants very easily. For instance, the instruction

$$G_p(\alpha) = Pade[G(\alpha), \{\alpha, 0, m, n\}] \tag{64}$$

expresses the scalar kernel $G(\alpha)$ in terms of a rational function $G_p(\alpha)$ where the order of the polynomials that constitute the numerator and the denominator are m and n, respectively. The constant 0 means that this Pade approximant starts from the point $\alpha = 0$, where the approximation error is vanishing. The Pade approximants $G_p(\alpha)$ are rational functions and may be factorized in closed form

$$G_p(\alpha) = G_{p-}(\alpha) G_{p+}(\alpha_o) \tag{65}$$

To ascertain the accuracy of this approximation let us consider the half-plane problem (section 1.1.1) and assume

$$k = 1 - j0.01, \quad \varphi_o = \frac{2}{3}\pi \Rightarrow \alpha_o = k/2 \tag{66}$$

Choosing $m = 20$ and $n = 22$, MATHEMATICA provides very quickly a Pade approximant $G_p(\alpha)$ presenting the error $e(\alpha) = \left|\frac{G(\alpha)-G_p(\alpha)}{G(\alpha)}\right|$ reported in Fig. 3.

The error is very small in the visible spectrum $-1 \le \alpha \le 1$ (Fig. 3a); however, it increases considerably out of this interval (Fig. 3b) because for $\alpha \to \infty$, $G_p(\alpha) \to \alpha^{-2}$, whereas $G(\alpha) \to \alpha^{-1}$. It is interesting to locate the position of the poles of the Pade approximant $G_p(\alpha)$. These are shown in Fig. 4.

Since the structural spectrum is constituted by the two branch points $\pm k$, these poles are spurious. However, Fig. 4 shows that their location may simulate the two branch lines actually present in the problem. Fig. 5 reports the relative error $e_p(\alpha) = \left|\frac{F_+(\alpha)-F_{p+}(\alpha)}{F_+(\alpha)}\right|$

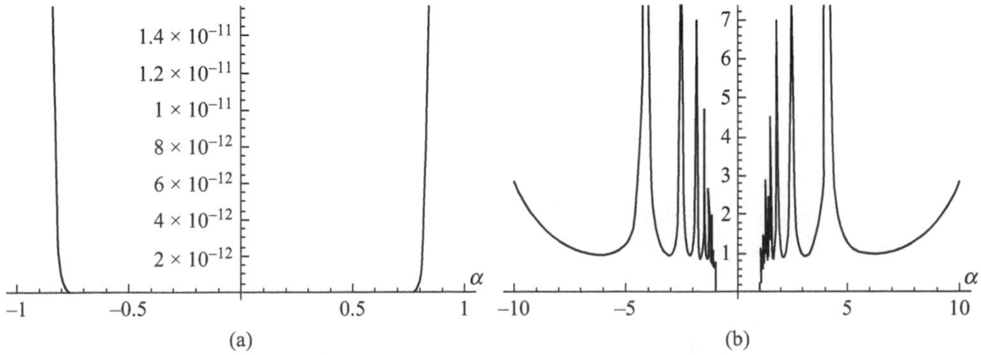

Fig. 3: Relative error of the Pade approximant for $k = 1 - j0.01$, $\phi_o = \frac{2}{3}\pi \Rightarrow \alpha_o = k/2$ (a) $-1 \le \alpha \le 1$, (b) $-10 \le \alpha \le 10$

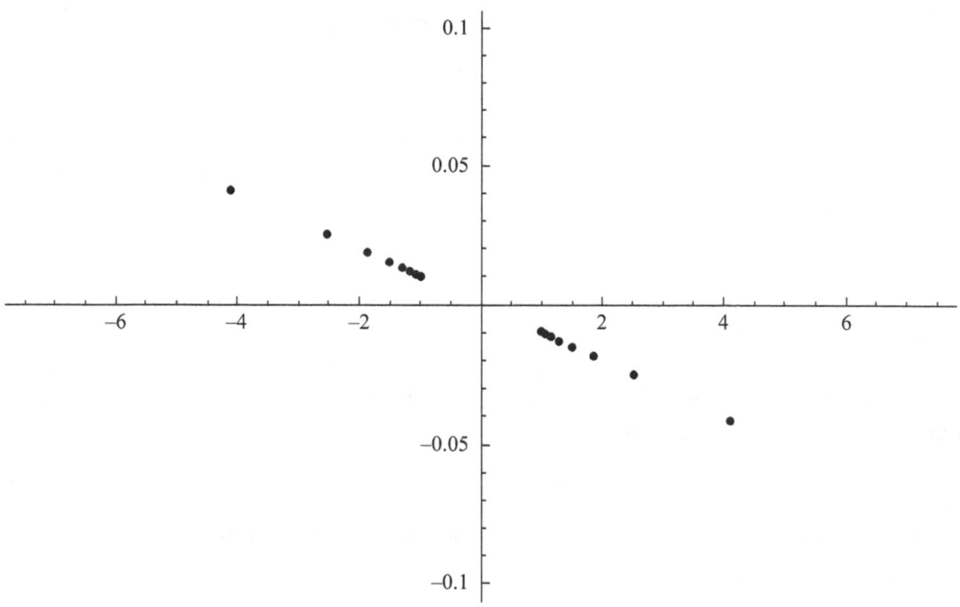

Fig. 4: Poles of the Pade approximant

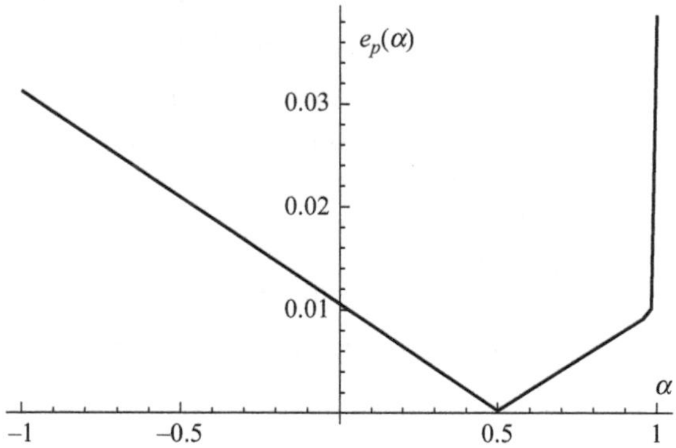

Fig. 5: Relative error $e_p(\alpha) = \left|\frac{F_+(\alpha) - F_{p+}(\alpha)}{F_+(\alpha)}\right|$

The scattering pattern from the half-plane is reported in Fig. 6, which shows that in practice there is no difference between the exact solution and the Pade-approximated solution.

By reducing the Pade parameters m and n, the relative error $e_p(\alpha) = \left|\frac{F_+(\alpha) - F_{p+}(\alpha)}{F_+(\alpha)}\right|$ increases. For instance, by choosing $m = 8$ and $n = 10$ the error is doubled. However, the scattering pattern does not change dramatically.

Sometimes the parameter k, related to the branch point, can be arbitrary chosen. For instance, in the considered example, the scattering pattern does not depend on k. In these cases, since the rational approximation is better when the spectrum is far from the real axis α, it is convenient to assume k with a strong imaginary part, such as $k = 1 - j$. Figure 7 illustrates the strong reduction of the relative error $e_p(\alpha) = \left|\frac{F_+(\alpha) - F_{p+}(\alpha)}{F_+(\alpha)}\right|$.

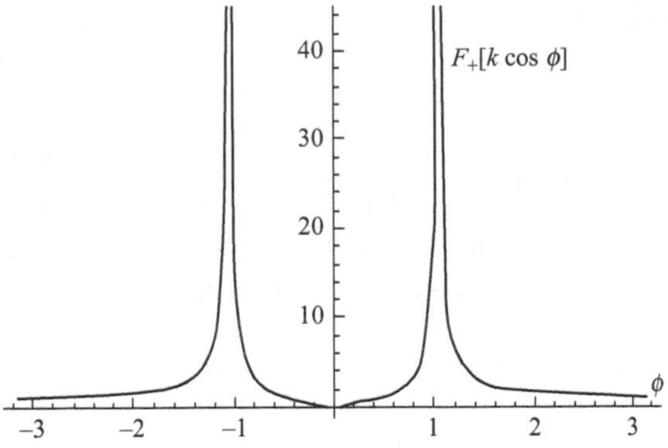

Fig. 6: Scattering pattern

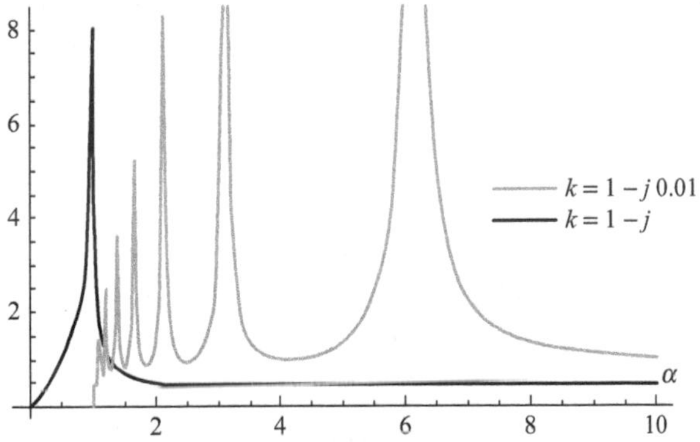

Fig. 7: Different values of k: comparison of the relative errors

6.3.2 An interpolation approximant method

The Pade approximant is accurate only at α points near the starting point α_s; for instance, in the previous examples $\alpha_s = 0$ has been chosen. It is possible to improve the approximation accuracy on a larger α range by using sophisticated techniques, such as the multiple point Pade approximation. However, it is much more effective to introduce an interpolation method. The main idea of this approach is to force the function

$$i[\alpha] = \left[G(0) + \sum_{i=1}^{m} b_i \alpha^i\right] - \left[1 + \sum_{i=m+1}^{m+n} b_i \alpha^{i-m}\right] G(\alpha) \tag{68}$$

to vanish in a set of α_i points suitably chosen:

$$i[\alpha_i] = 0 \quad i = 1, 2, \ldots, m+n \tag{69}$$

The $n+m$ previous equations yield a system of $n+m$ order for the b_i, $i = 1, 2, \ldots, m+n$ unknowns. The solution of the system provides the following rational approximation for the kernel $G(\alpha)$:

$$G(\alpha) \approx G_i(\alpha) = \frac{G(0) + b_1\alpha + b_2\alpha^2 + \cdots + b_m\alpha^m}{1 + b_{m+1}\alpha + b_{m+2}\alpha^2 + \cdots + b_{m+n}\alpha^n} \tag{70}$$

In this way, it is possible to obtain a rational approximant valid on a large range of α. In general, the rational approximations provided by this technique show a better accuracy with respect to those obtained by the Pade approximant. From here on, in the applications we will refer to rational representations obtained with the interpolation method.

6.3.2.1 *Presence of a discrete infinite spectrum*

In this case, the major error in the rational approximation of the kernel arises from the singularities near the real axis α. To strongly improve the approximation it is convenient to rationalize only the part of the kernel containing the spectrum far from the real axis. For instance, let us consider the factorization of the kernel previously considered in example 4 of section 3.2.4:

$$G(\alpha) = \frac{\sin(\tau\, b)\sin(\tau\, c)}{\tau \sin[\tau\,(b+c)]} \tag{71}$$

Introducing the same numerical data of that section, namely,

$$\lambda = 1, \quad k = \frac{2\pi}{\lambda}, \quad b = 1.1\frac{\lambda}{2}, \quad c = 1.3\frac{\lambda}{2}, \quad a = b + c = 2.4\frac{\lambda}{2} \tag{72}$$

for vanishing values of $\mathrm{Im}[k]$, we have two real zeroes $\alpha = \alpha_{b1}$, $\alpha = \alpha_{c1}$ and two real poles $\alpha = \alpha_{a1}$ and $\alpha = \alpha_{a2}$. Consequently, it is convenient to consider the rational approximants of the function

$$M(\alpha) = G(\alpha)\frac{(\alpha^2 - \alpha_{a1}^2)(\alpha^2 - \alpha_{a2}^2)}{(\alpha^2 - \alpha_{b1}^2)(\alpha^2 - \alpha_{c1}^2)} = \frac{\sin(\tau\, b)\sin(\tau\, c)}{\tau \sin[\tau\,(b+c)]}\frac{(\alpha^2 - \alpha_{a1}^2)(\alpha^2 - \alpha_{a2}^2)}{(\alpha^2 - \alpha_{b1}^2)(\alpha^2 - \alpha_{c1}^2)} \tag{73}$$

which presents only an imaginary spectrum : $\pm\alpha_{ai}$ $(i = 3, 4, \ldots)$, $\pm\alpha_{bi}$ $(i = 2, 3, 4, \ldots)$ and $\pm\alpha_{ci}$ $(i = 2, 3, 4, \ldots)$. The Pade approximant of $M(\alpha)$

$$Ma(\alpha\_) = Pade[M(\alpha), \{\alpha, 0, 20, 22\}] \tag{74}$$

yields the approximate factorized function

$$G_{a+}(\alpha) = M_{a+}(\alpha)\frac{(\alpha - \alpha_{b1})(\alpha - \alpha_{c1})}{(\alpha - \alpha_{a1})(\alpha - \alpha_{a2})} \tag{75}$$

The plots reporting the absolute values and the argument of the exact $G_+(\alpha)$ and of the approximate $G_{a+}(\alpha)$ for a large range of values of α are indistinguishable. The following figures report the relative errors defined by

$$e_{abs}(\alpha) = \frac{|G_+(\alpha)| - |G_{a+}(\alpha)|}{|G_+(\alpha)|} \tag{76}$$

$$e_+(\alpha) = \left|\frac{G_+(\alpha) - G_{a+}(\alpha)}{G_+(\alpha)}\right| \tag{77}$$

The relative error on the absolute values of the exact and approximate factorized functions is very negligible (Fig. 8); for instance, for $\alpha = 2$ the relative error is $e_{abs}(2) = 6.08\ 10^{-11}$. The biggest relative error is for $e_+(\alpha) = \left|\frac{G_+(\alpha) - G_{a+}(\alpha)}{G_+(\alpha)}\right|$ (Fig. 9), and it arises from the relative error on the phase. This can be ascertained in Fig. 10, which illustrates the arguments of the exact $G_+(\alpha)$ and approximate $G_{a+}(\alpha)$.

6.3.2.2 Presence of continuous and discrete spectra

In dealing with truncated waveguides, the kernels involve continuous and discrete spectra. In these cases our experience suggests that the best way to obtain rational approximants is based on the interpolation method considered in section 6.2.3. In the following, the approximate rational factorization of the kernel

$$G(\alpha) = \frac{e^{j\tau d}}{\tau \cos[\tau d]}, \quad \tau = \sqrt{k^2 - \alpha^2} \tag{78}$$

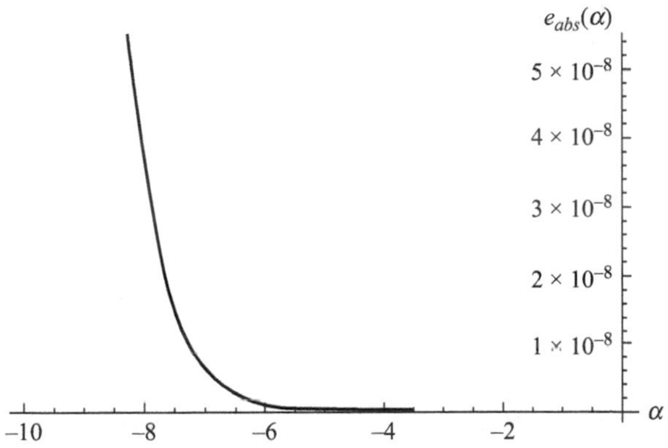

Fig. 8: Relative error on the absolute values of the plus factorized function

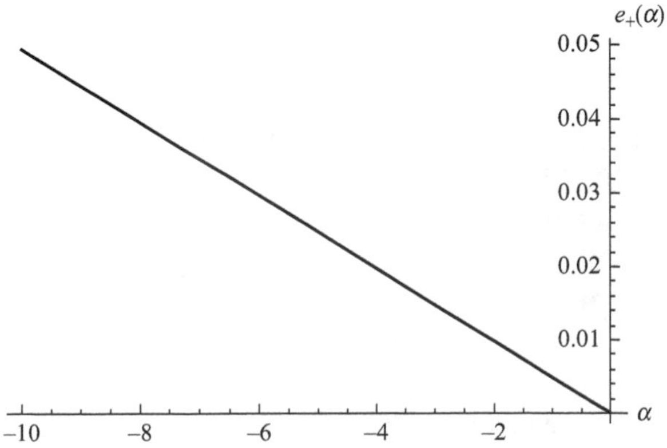

Fig. 9: Relative error on the plus factorized function

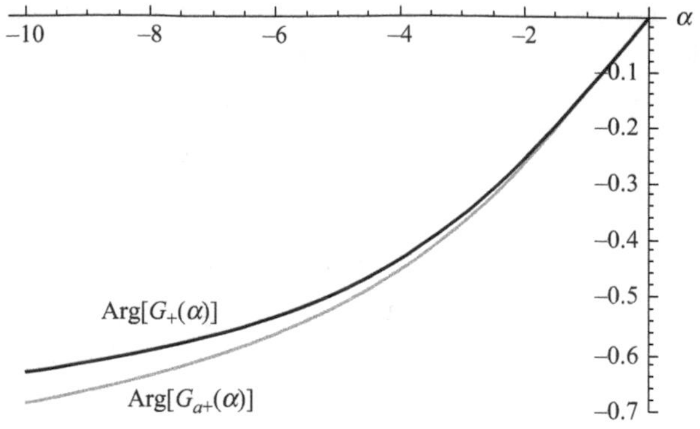

Fig. 10: Comparison between the exact and the approximate phase of the factorized plus function

will be considered. This kernel has been exactly factorized in section 3.2.6 (example 5); the exact plus factorized function is

$$G_+(\alpha) = \frac{\Gamma\left(-\frac{\alpha}{A} + \frac{B}{A} + 1\right)\exp\left[\frac{\tau d}{\pi}\log\frac{j\tau-\alpha}{k} - q\,\alpha\right]}{\sqrt{k-\alpha}\,\sqrt{\cos(kd)}\,e^{\gamma\frac{\alpha}{A}}\Gamma\left(\frac{B}{A}+1\right)}\prod_{n=1}^{\infty}\frac{\left(1-\frac{\alpha}{An+B}\right)}{\left(1-\frac{\alpha}{\alpha_n}\right)}$$

$$\approx \frac{\Gamma\left(-\frac{\alpha}{A} + \frac{B}{A} + 1\right)\exp\left[\frac{\tau d}{\pi}\log\frac{j\tau-\alpha}{k} - q\,\alpha\right]}{\sqrt{k-\alpha}\,\sqrt{\cos(kd)}\,e^{\gamma\frac{\alpha}{A}}\Gamma\left(\frac{B}{A}+1\right)}\prod_{n=1}^{N_b}\frac{\left(1-\frac{\alpha}{A\,n+B}\right)}{\left(1-\frac{\alpha}{\alpha_n}\right)} \qquad (79)$$

To compare the exact and approximate factorizations, let us assume that $\lambda = 1$, $k = \frac{2\pi}{\lambda}(1 - j10^{-2})$, $b = 1.1\frac{\lambda}{2}$.

Since $G(\alpha)$ is an even function, the following even function has to be forced to be vanishing in the suitably chosen points α_i:

$$i[\alpha] = \left[G(0) + \sum_{i=1}^{n} b_i\alpha^{2i}\right] - \left[1 + \sum_{i=n+1}^{2n+1} b_i\alpha^{2(i-n)}\right]G(\alpha) \qquad (80)$$

To obtain the $2n + 1$ unknown b_i, we forced $i[m] = 0$ in $m = 1, 2, \ldots, 2n + 1$.

By assuming $n = 10$, the solution b_i $(i = 1, 2, \ldots, 21)$ of the previous system, obtained with MATHEMATICA, provides the following rational approximation of the kernel $G(\alpha)$:

$$G(\alpha) \approx G_a(\alpha) = \frac{G(0) + b_1\alpha^2 + b_2\alpha^4 + \cdots + b_{10}\alpha^{20}}{1 + b_{11}\alpha^2 + b_{12}\alpha^4 + \cdots + b_{21}\alpha^{22}} \qquad (81)$$

The factorization of the rational kernel $G_a(\alpha) = G_{a-}(\alpha)G_{a+}(\alpha)$ has been accomplished by MATHEMATICA. By assuming $N_b = 200$ in the truncated product of $G_+(\alpha)$, Fig. 11 and Fig. 12 report the relative errors $e(\alpha) = \left|\frac{G_+(\alpha) - G_{a+}(\alpha)}{G_+(\alpha)}\right|$ and $e_{ph}(\alpha) = \frac{\arg[G_+(\alpha)] - \arg[G_{a+}(\alpha)]}{\arg[G_+(\alpha)]}$.

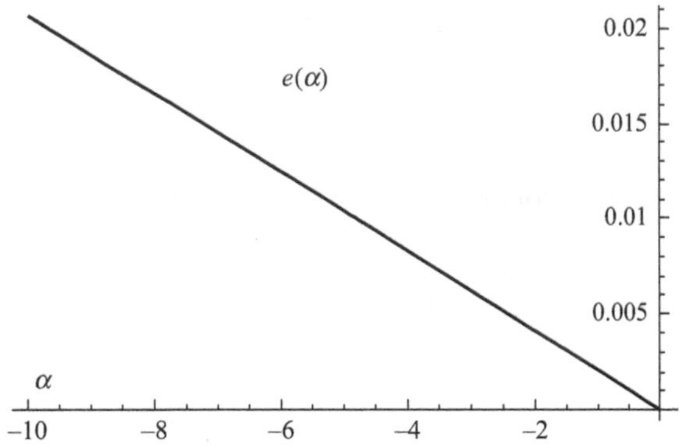

Fig. 11: Relative error on the plus factorized function

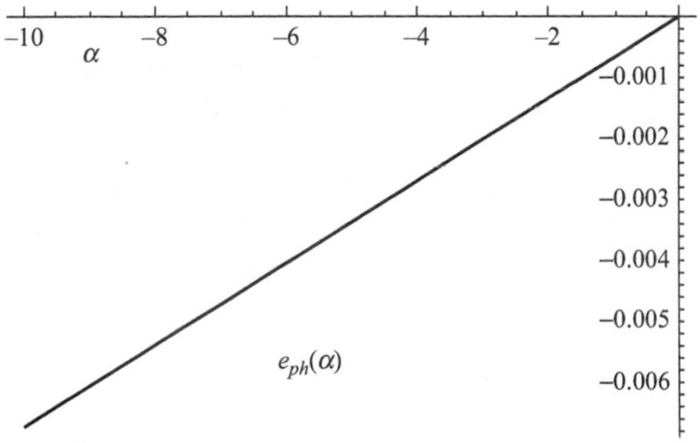

Fig. 12: Relative error on the phase of the plus factorized function

6.4 Moment method

6.4.1 Introduction

In this section we consider general aspects of this method, which is very powerful for solving the arbitrary linear equation

$$A \cdot x = y \tag{82}$$

This equation is defined in arbitrary Hilbert space. A is a known linear operator, x is an unknown vector, and y is a known vector.

Let us introduce a set of vectors ψ_n, which we call expansion functions. We represent x in the form

$$x = \sum_n x_n \psi_n \tag{83}$$

This representation is exact if the set ψ_n $(n = 1, 2, \ldots)$ is a basis in the Hilbert space. However, it remains very useful in every case, provided that the vectors ψ_n are chosen appropriately.

Substituting (83) into (82) yields

$$\sum_n x_n A \cdot \psi_n = y \tag{84}$$

Let us introduce a set of vectors φ_m $(m = 1, 2, \ldots)$ that we call test functions. The scalar product with φ_m yields

$$\sum_n A_{mn} x_n = y_m \tag{85}$$

where

$$A_{mn} = \varphi_m \cdot A \cdot \psi_n \tag{86}$$

$$y_m = \varphi_m \cdot y \tag{87}$$

The solution of (85) furnishes the values of x_n that represent x through (83).

When the set of expansion functions coincides with the set of test functions, the moment method is also called the Garlekin method.

Equation (83) can be defined in the natural domain or in the spectral domain.

In this book we consider only formulations in the spectral domain. We can apply the moment method directly to the Fredholm eqs. (1.54), or more simply we can start from the W-H eq. (1.4):

$$G(\alpha)F_+(\alpha) = F_-^s(\alpha) + F_{o+}(\alpha) \tag{88}$$

To solve this equation with the moment method we recall the Parseval theorem:[1]

$$u \cdot v = \int_{-\infty}^{\infty} u(z)v(z)dz = \frac{1}{2\pi}\int_{-\infty}^{\infty} \tilde{u}(-\alpha)\tilde{v}(\alpha)d\alpha$$

and use the representation

$$F_+(\alpha) = \sum_i C_i \tilde{\psi}_{+1}(\alpha) \tag{89}$$

where the constants C_i are unknown and the set of plus functions $\tilde{\psi}_{+i}(\alpha)$ (expansion functions) is known. A projection with the set of plus test functions φ_{+j} provides the equations for the unknowns:

$$\sum_i G_{ji}\, C_i = N_j \tag{90}$$

[1] The symbol ~ stands for Fourier transform.

where

$$G_{ji} = \varphi_{+j} \cdot G \cdot \psi_{+i} = \frac{1}{2\pi} \int_{-\infty}^{\infty} \tilde{\varphi}_{+j}(-\alpha) \cdot G(\alpha) \cdot \tilde{\psi}_{+i}(\alpha)\, d\alpha \tag{91}$$

$$N_j = \varphi_{+j} \cdot F_{o+} = \frac{1}{2\pi} \int_{-\infty}^{\infty} \tilde{\varphi}_{+j}(-\alpha) \cdot F_{o+}(\alpha)\, d\alpha \tag{92}$$

The minus function $X_-(\alpha)$ does not appear, since in performing the projection with the test functions φ_{+j}, Parseval's theorem provides the equations

$$\int_{-\infty}^{\infty} \tilde{\varphi}_{+j}(-\alpha) \cdot X_-(\alpha) d\alpha = 2\pi\, \varphi_{+j} \cdot x_- = 0 \tag{93}$$

It is important to observe that there are many advantages to formulate the moment method in the spectral domain rather than in the natural domain. For instance:

- The entries of the matrix G_{ji} require only the evaluation of a single (and not double) integral.[2]
- The matrix kernel $G(\alpha)$ is always known in analytical form.
- The solution obtained in the spectral domain is very suitable for obtaining the asymptotic evaluation of far fields.

6.4.2 Stationary properties of the solutions with the moment method

The inverse A^{-1} of the operator A provides the solution of eq. (82):

$$x = A^{-1} \cdot y \tag{94}$$

However, in many applications, a function h requires the functional

$$w = h \cdot x \tag{95}$$

or more generally a function of w:

$$f(w) = f(h \cdot x) \tag{96}$$

To evaluate the functional w with moment method, let us suppose that the set of ψ_n and φ_m are bi-orthogonal:[3]

$$\varphi_m \cdot \psi_n = \delta_{mn}$$

[2] Compare eq. (93) with eq. (86).

[3] We can obtain a set φ_m bi-orthogonal to the set of ψ_n by using the Gram-Schmidt normalization.

The operator A can be represented by the matrix

$$A = \begin{vmatrix} A_{11} & \dots & A_{1N} & A_{1,N+1} & \dots \\ \dots & \dots & \dots & \dots & \dots \\ A_{N1} & \dots & A_{NN} & A_{N,N+1} & \dots \\ A_{N+1,1} & \dots & A_{N+1,N} & A_{N+1,N+1} & \dots \\ \dots & \dots & \dots & \dots & \dots \end{vmatrix}$$

where

$$A_{m,n} = \varphi_m \cdot A \cdot \psi_n \tag{97}$$

With the use of N moments, this expression can be rewritten as

$$A = A_o + \delta A \tag{98}$$

where

$$A_o = \begin{vmatrix} A_{11} & \dots & A_{1N} & 0 & \dots \\ \dots & \dots & \dots & \dots & \dots \\ A_{N1} & \dots & A_{NN} & 0 & \dots \\ 0 & \dots & 0 & 0 & \dots \\ \dots & \dots & \dots & \dots & \dots \end{vmatrix}, \quad \delta A = \begin{vmatrix} 0 & \dots & 0 & A_{1,N+1} & \dots \\ \dots & \dots & \dots & \dots & \dots \\ 0 & \dots & 0 & A_{N,N+1} & \dots \\ A_{N+1,1} & \dots & A_{N+1,N} & A_{N+1,N+1} & \dots \\ \dots & \dots & \dots & \dots & \dots \end{vmatrix}$$

With the moment method, eq. (83) is approximated by

$$A_o \cdot x_o = y_o \tag{99}$$

where y_o has the same first N component $y_m = \varphi_m \cdot y \ (m = 1, 2, \dots, N)$ of y while the other components are all vanishing. In the following we will indicate the error with δx:[4]

$$\delta x = x - x_o \tag{100}$$

To evaluate the error in the functional $w = h \cdot x$, we introduce the vector h_o in the Euclidean space E_N. It contains only the first N nonvanishing components $\varphi_m \cdot h \ (m = 1, 2, \dots, N)$. We have

$$w = h \cdot x = (h_o + \delta h) \cdot (x_o + \delta x) = h_o \cdot x_o + h_o \cdot \delta x + \delta h \cdot x_o + \delta h \cdot \delta x \tag{101}$$

Since δh has nonvanishing components for $m = N + 1, N + 2, \dots$ and x_o has nonvanishing components for $m = 1, 2, \dots, N$, we get

$$\delta h \cdot x_o = 0$$

To evaluate $h_o \cdot \delta x$, let us consider the operator A_o^t, the transpose of the matrix A_o in the space E_N. Indicate with x_o' the solution of

$$A_o^t \cdot x_o' = h_o \tag{102}$$

We get

$$h_o \cdot \delta x = (A_o^t \cdot x_o') \cdot \delta x = x_o' \cdot A_o \cdot \delta x \tag{103}$$

Taking into account (99), from

$$A \cdot x = (A_o + \delta A) \cdot (x_o + \delta x) = y_o + \delta y$$

---

[4] We observe that δx has components $\varphi_m \cdot \delta x$ that are not vanishing also for $m = N + 1, N + 2, \dots$

we get

$$A_o \cdot \delta x = -\delta A \cdot x_o - \delta A \cdot \delta x + \delta y \tag{104}$$

Taking into account that $x'_o \cdot \delta A \cdot x_o = 0$ and $x'_o \cdot \delta y = 0$, substituting (104) into (103) yields

$$h_o \cdot \delta x = x'_o \cdot (-\delta A \cdot x_o - \delta A \cdot \delta x + \delta y) = -x'_o \cdot \delta A \cdot \delta x \tag{105}$$

Substituting (105) into (101) evaluates the functional $w = h \cdot x$ in the form

$$w = h \cdot x = h_o \cdot x_o - x'_o \cdot \delta A \cdot \delta x + \delta h \cdot \delta x \tag{106}$$

This means that the evaluation of the functional $w = h \cdot x$ with $h_o \cdot x_o = h \cdot x_o$, where x_o is the approximate solution obtained with the moment method is stationary, since the error $-x'_o \cdot \delta A \cdot \delta x + \delta h \cdot \delta x$ is of second order.

In many applications it is sufficient to use only one moment $\varphi_1, \ \psi_1$. The moment eq. (85) reduces to

$$\varphi_1 \cdot A \cdot \psi_1 \, x_1 = \varphi_1 \cdot y \tag{107}$$

From (83) we get

$$x = \frac{\varphi_1 \cdot y}{\varphi_1 \cdot A \cdot \psi_1} \psi_1 \tag{108}$$

$$h \cdot x = \frac{\varphi_1 \cdot y}{\varphi_1 \cdot A \cdot \psi_1} h \cdot \psi_1 \tag{109}$$

In the natural domain $x \to x(z), \ a < z < b$, (108) becomes

$$h \cdot x = \frac{\varphi_1 \cdot y}{\varphi_1 \cdot A \cdot \psi_1} h \cdot \psi_1 = \frac{\int_a^b \varphi_1(z) y(z) dz \int_a^b h(z) \psi_1(z) dz}{\int_a^b \int_a^b \varphi_1(z) A(z, z') \psi_1(z') dz' \, dz} \tag{110}$$

Equation (110) can be onerous since the evaluation of a double integral is required.

However, in the problems considered in this book the operator A is of convolution type:

$$A(z, z') = A(z - z') \to \tilde{A}(\alpha) = \int_{-\infty}^{\infty} A(u) e^{ju\alpha} du$$

In the spectral domain (110) yields

$$h \cdot x = \frac{\varphi_1 \cdot y}{\varphi_1 \cdot A \cdot \psi_1} h \cdot \psi_1 = \frac{1}{2\pi} \frac{\int_{-\infty}^{\infty} \tilde{\varphi}_1(-\alpha) \tilde{y}(\alpha) d\alpha \int_{-\infty}^{\infty} \tilde{h}(-\alpha) \tilde{\psi}_1(\alpha) d\alpha}{\int_{-\infty}^{\infty} \tilde{\varphi}_1(-\alpha) \tilde{A}(\alpha) \tilde{\psi}_1(\alpha) d\alpha} \tag{111}$$

and all the integrals to be evaluated are simple.

To conclude this section, we observe that if $w = h \cdot x$ is stationary in $x = x_o$, a generic function $f(w) = f(h \cdot x)$ is also stationary in the same point, provided that $f'(w)|_{w=h \cdot x_o}$ is regular. In fact we have

$$\delta f(w) = f'(w)\delta w$$

whence if δw is a differential of second order, $\delta f(w)$ is also a differential of second order.

6.4.2.1 *Stationary properties of the eigenvalues evaluated with the moment method*

In many applications the operator A depends on a parameter β that must be evaluated to satisfy the homogeneous equation

$$A(\beta) \cdot x = 0 \tag{112}$$

By using the moment method with expansion functions ψ_n and test functions φ_n we get

$$\sum_n A_{mn}(\beta)x_n = y_m$$

Hence,

$$\det[A(\beta)] = 0 \tag{113}$$

where $A(\beta)$ is the matrix that has the entries $A_{mn}(\beta)$.

Let us indicate with β the exact values that satisfy the previous equation. The approximate values of $A_o(\beta)$ produce an approximate value β_o that satisfies

$$\det[A_o(\beta_o)] = 0 \tag{114}$$

We have

$$\det[A(\beta)] = \det[A(\beta_o) + \delta\beta A'(\beta_o)] = \det[A(\beta_o)] + \delta\beta H + O(\delta\beta^2) = 0 \tag{115}$$

where H derives from the evaluation of the determinant of the sum of the two matrices $A(\beta_o)$ and $\delta\beta A'(\beta_o)$. Letting

$$A(\beta) = \begin{vmatrix} A_o(\beta) & A_{o\varepsilon}(\beta) \\ A_{\varepsilon o}(\beta) & A_{\varepsilon\varepsilon}(\beta) \end{vmatrix}$$

and using the Laplace formula for evaluating the determinant of a matrix, it can be shown that if the element of $A_{o\varepsilon}(\beta)$, $A_{\varepsilon o}(\beta)$ and $A_{\varepsilon\varepsilon}(\beta)$ are of order ε we get

$$\det[A(\beta_o)] = \det[A_o(\beta_o)]\,\det[A_{\varepsilon\varepsilon}(\beta_o)] + O(\varepsilon^2) = 0 \tag{116}$$

Taking into account that $\det[A_o(\beta_o)] = 0$ and substituting (116) into (115) yields

$$\delta\beta H = O(\delta\beta^2) + O(\varepsilon^2)$$

which shows that the error $\delta\beta = \beta - \beta_o$ is of second order.

6.4.3 An electromagnetic example: the impedance of a wire antenna in free space

Let us consider the wire antenna shown in Fig. 13, supplied by a voltage V_o. Indicate with $I(z)$ the total current at position z along the wire. The current density is defined by

$$\mathbf{J}(z) = I(z)\,\delta(x)\,\delta(y)\,\hat{z} \tag{117}$$

The electric field $\mathbf{E}$ radiated by the antenna can be expressed in the form

$$\mathbf{E} = -j\omega\, A\,\hat{z} + \frac{\nabla\nabla\cdot}{j\omega\varepsilon_o\mu_o}\hat{z}\, A \tag{118}$$

where the potential A is given by

$$A = \frac{\mu_o}{4\pi}\int_{-L/2}^{L/2} \frac{e^{-jk_oR}}{R} I(z')dz' \tag{119}$$

and $R = \sqrt{\rho^2 + (z - z')^2}$ is the distance between the observation point and the source point. On the wire surface, by indicating with a the radius of the wire, we have

$$R = \sqrt{a^2 + (z - z')^2}$$

By forcing E_z to vanish on the wire surface and the source voltage V_o to be concentrated at $z = 0$, we get

$$\int_{-L/2}^{L/2} g(z, z')I(z')dz' = -V_o\delta(z) \tag{120}$$

where

$$g(z, z') = \frac{Z_o}{j4\pi k_o}\left(\frac{d^2}{dz^2} + k_o^2\right)\frac{e^{-jk_o\sqrt{a^2+(z-z')^2}}}{\sqrt{a^2 + (z - z')^2}}$$

Equation (120) is an integro-differential equation. To obtain an integral equation we can consider Hallen's approach (Hallen, 1962), which reduces (120) to a second-order ordinary

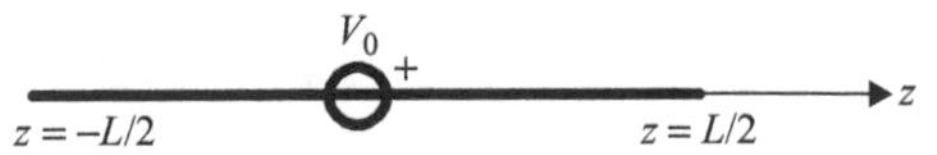

Fig. 13: A wire antenna in free space

differential equation in the unknown $\int_{-L/2}^{L/2} \frac{e^{-jk_o\sqrt{a^2+(z-z')^2}}}{\sqrt{a^2+(z-z')^2}} I(z')dz'$. Solving this equation we obtain Hallen's integral equation:

$$\int_{-L/2}^{L/2} \frac{e^{-jk_o\sqrt{a^2+(z-z')^2}}}{\sqrt{a^2+(z-z')^2}} I(z')dz' = -2\pi j \frac{V_o}{Z_o} \sin|k_o z| + C\cos k_o z \tag{121}$$

where C must be determined by imposing the boundary condition $I(L/2) = 0$.

The solution of Hallen's integral equation has been extensively studied. If we are interested only in the admittance of the wire antenna defined by

$$Y_a = \frac{I(0)}{V_o} \tag{122}$$

we observe that by letting

$$x = I(z), \quad h = \frac{\delta(z)}{V_o} \tag{123}$$

the admittance is simply the functional $h \cdot x$. Therefore, we can proceed directly to the moment solution of (120), thereby avoiding the introduction of Hallen's equation.

For the sake of the brevity, in the following we consider only the half-wave wire: $L = \lambda/2$ where $\lambda = \frac{2\pi}{k_o}$ is the wavelength.

Taking into account that the sinusoidal profile for $I(z)$ is a physically acceptable approximation, we use only one moment:

$$\psi_1(z) = \varphi_1(z) = \cos\frac{2\pi}{\lambda} z \tag{124}$$

From (120) we get

$$Y_a = h \cdot x = \frac{\varphi_1 \cdot y}{\varphi_1 \cdot g \cdot \psi_1} h \cdot \psi_1 = -\frac{\int_{-\lambda/4}^{-\lambda/4} \varphi_1(z) V_o \delta(z) dz \int_{-\lambda/4}^{-\lambda/4} \frac{\delta(z)}{V_o} \psi_1(z) dz}{\int_{-\lambda/4}^{-\lambda/4}\int_{-\lambda/4}^{-\lambda/4} g(z,z')\varphi_1(z)\psi_1(z')dz\,dz'}$$

$$= -\frac{\varphi_1(0)\psi_1(0)}{\int_{-\lambda/4}^{-\lambda/4}\int_{-\lambda/4}^{-\lambda/4} g(z,z')\varphi_1(z)\psi_1(z')dz\,dz'}$$

The stationary expression of the impedance of a half-wave wire antennas is

$$Z_a = -\int_{-\lambda/4}^{-\lambda/4}\int_{-\lambda/4}^{-\lambda/4} g(z,z')\varphi_1(z)\psi_1(z')dz\,dz'$$

Taking into account that $Z_o = 377\Omega$, we can evaluate numerically the double integral and obtain

$$Z_a = 73.2 + j42.5 \quad (\Omega)$$

The numerical integration is cumbersome; hence, taking into account that the kernel $g(z, z') = g(z - z')$ is of convolution type it is preferable to work in the spectral domain. This will be done in section 8.4.

6.5 Comments on the approximate methods for solving W-H equations

In the previous sections, we described approximate methods for solving W-H equations. In practical terms, we experienced that the more powerful ones are those based on rational approximants of the kernel and on the quadrature method for solving Fredholm integral equations. These techniques provide approximate factorized kernels that are accurate in their regularity regions. This limitation can be overcome by analytic continuation. For instance, taking into account that

$$G(\alpha) = G_-(\alpha)G_+(\alpha) \tag{125}$$

approximate values of $G_+^{\pm 1}(\alpha)$ at points where $\text{Im}[\alpha] < 0$ may be evaluated by

$$G_+^{\pm 1}(\alpha) = (G_-^{-1}(\alpha)G(\alpha))^{\pm 1} \tag{126}$$

Consequently, since $G^{\pm 1}(\alpha)$ is exactly known everywhere, the problem of evaluating the plus factorized matrix outside its regularity half-plane is reduced to the evaluation of $G_-^{\pm 1}(\alpha)$ in its regularity half-plane $\text{Im}[\alpha] < 0$.

The power of the previously introduced approximate methods depends on the spectrum of the kernel; for example, if there is not an infinite discrete spectrum together with a continuous spectrum, we suggest working on the Fredholm equation in the w – plane, which in this case provides very accurate approximate solutions. Conversely, if only an infinite discrete spectrum is present, then accurate solutions can be obtained by the rational approximant of kernels and, in particular, we found that the interpolation methods work better than the Pade approximants.

In all cases, the difficulties arise mainly from the presence of singularities near the real axis, and therefore it is always convenient to try to eliminate these singularities using appropriate algebraic manipulations.

PART 2

Applications

CHAPTER 7

The half-plane problem

7.1 Wiener-Hopf solution of discontinuity problems in plane-stratified regions

The study the electromagnetic propagation in stratified planar regions (Fig. 1) is very important since it provides all the fundamental concepts needed for the study of wave propagation. There are very excellent books devoted to this topic (Felsen & Marcuvitz, 1973, chapter 5; Brekhovskikh, 1960). In this book, we examine some aspects of the problem that are not considered in the cited works. For instance, we introduce planar discontinuities in some sections of the stratified regions and develop a unified theory for this problem that is based on the spatial Fourier transforms of the fields and on the Wiener-Hopf formulation.

In this section, for illustrative purposes we consider only stratified isotropic media in which

$$\boldsymbol{\varepsilon}_i = \varepsilon_i\,\mathbf{1}, \quad \boldsymbol{\mu}_i = \mu_i\,\mathbf{1}, \quad \boldsymbol{\xi} = 0 = \boldsymbol{\zeta}$$

and $\mathbf{1}$ is the unity dyadic. The general formulation that involves arbitrary linear stratified media will be presented in section 7.10.1.

The fundamental equations for studying stratified regions of isotropic media are the following transverse field equations (Felsen & Marcuvitz, 1973, p. 186):

$$\begin{aligned} -\frac{\partial}{\partial y}\mathbf{E}_t &= j\omega\mu_i\left(\mathbf{1} + \frac{\nabla_t\nabla_t}{k_i^2}\right)\cdot\mathbf{H}_t\times\hat{y} + \mathbf{M}_t\times\hat{y} + \frac{\nabla_t J_y}{j\omega\varepsilon_i} \\ -\frac{\partial}{\partial y}\mathbf{H}_t &= j\omega\varepsilon_i\left(\mathbf{1} + \frac{\nabla_t\nabla_t}{k_i^2}\right)\cdot\hat{y}\times\mathbf{E}_t + \hat{y}\times\mathbf{J}_t + \frac{\nabla_t M_y}{j\omega\mu_i} \end{aligned} \tag{1}$$

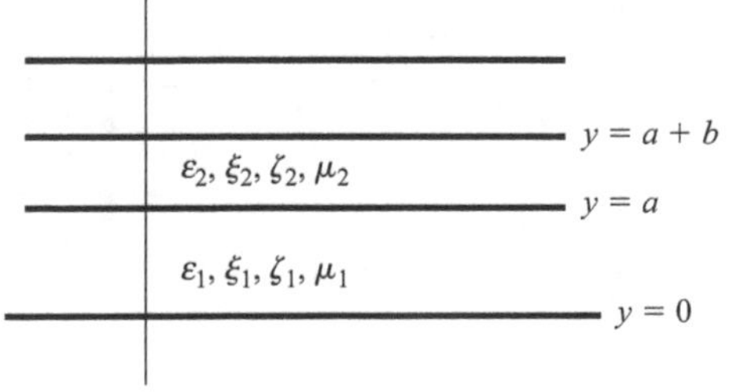

Fig. 1: Arbitrary plane-stratified linear regions

where

$$\mathbf{E}_t = E_z\hat{z} + E_x\hat{x}, \quad \mathbf{H}_t = H_z\hat{z} + H_x\hat{x}, \quad \nabla_t = \nabla - \hat{y}\frac{\partial}{\partial y},$$
$$\mathbf{M}_t = M_z\hat{z} + M_x\hat{x}, \quad \mathbf{J}_t = J_z\hat{z} + J_x\hat{x}, \quad k_i = \omega\sqrt{\mu_i\varepsilon_i}$$

7.2 Spectral transmission line in homogeneous isotropic regions

Let us consider time harmonic electromagnetic fields with a time dependence specified by the factor $e^{j\omega t}$, which is omitted. These fields are considered in an isotropic homogeneous layer with permettivity ε_o and permeability μ_o.

It is convenient to consider the free-source layer indicated in Fig. 2a as a piece of wave guide (Felsen & Marcuvitz, 1973, pp. 183–207) having normal direction y and an unbounded section in the transverse plane (x, z). For the sake of simplicity, we assume that only the components E_z, H_x, and H_y are nonzero and that there are no variations of these fields in the z-direction ($\partial/\partial z = 0$):

$$E_z = E_z(x,y), \quad H_x = H_x(x,y), \quad H_y = H_y(x,y)$$

Since the guide presents an unbounded cross section, we are dealing with a modal continuous spectrum (Felsen & Marcuvitz, 1973) that can be represented through the Fourier transform of the transverse field:

$$\begin{aligned} V(\eta,y) &= \int_{-\infty}^{\infty} E_z(x,y)e^{j\eta x}dx \\ I(\eta,y) &= \int_{-\infty}^{\infty} H_x(x,y)e^{j\eta x}dx \end{aligned} \tag{2}$$

Applying the transverse field eq. (1) yields the following transmission equations in the voltage $V(\eta,y)$ and current $I(\eta,y)$:

$$\begin{aligned} -\frac{d}{dy}V(\eta,y) &= j\tau Z_c I(\eta,y) \\ -\frac{d}{dy}I(\eta,y) &= j\tau Y_c V(\eta,y) \end{aligned} \tag{3}$$

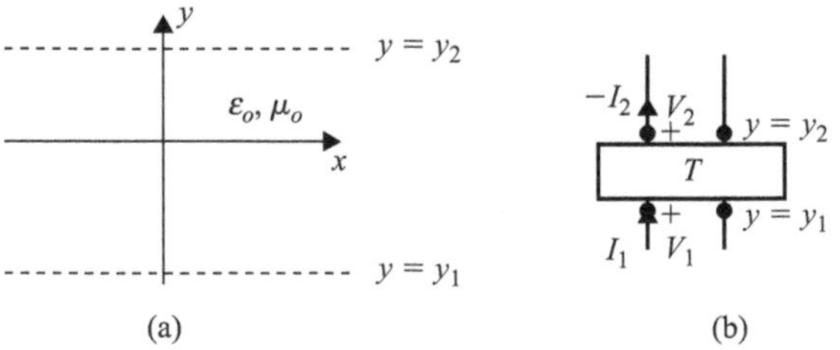

Fig. 2: (a) Layer of isotropic homogeneous medium, (b) two-port relevant to the layer

where being $Z_o = \sqrt{\frac{\mu_o}{\varepsilon_o}}$ and $k = \omega\sqrt{\mu_o \varepsilon_o}$ the characteristic impedance and the propagation constant of the medium, respectively, the propagation constant τ and the characteristic impedance Z_c of the line are defined by

$$\tau = \tau(\eta) = \sqrt{k^2 - \eta^2} \tag{4}$$

$$Z_c = \frac{\omega\mu_o}{\tau} = Z_c(\eta) = \frac{k\,Z_o}{\sqrt{k^2 - \eta^2}} \qquad Y_c = \frac{1}{Z_c} = \frac{\tau}{\omega\mu_o} \tag{5}$$

7.2.1 Circuital considerations

Solving the transmission line eq. (3) in the layer bounded by the section $y = y_1$ and $y = y_2$ (Fig. 2a) yields

$$V_1 = \cos(\tau d)V_2 + jZ_c \sin(\tau d)(-I_2) \tag{6}$$

$$I_1 = jY_c \sin(\tau d)V_2 + \cos(\tau d)(-I_2) \tag{7}$$

where $d = y_2 - y_1$, $V_{1,2} = V_{1,2}(\eta) = V(\eta, y_{1,2})$, $I_{1,2} = I_{1,2}(\eta) = \pm I(\eta, y_{1,2})$.

A circuit representation of the layer is constituted by the two-port shown in Fig. 2b, where **T** is the transmission matrix of the two-port:

$$T = \begin{pmatrix} \cos(\tau d) & jZ_c \sin(\tau d) \\ jY_c \sin(\tau d) & \cos(\tau d) \end{pmatrix} \tag{8}$$

If the transmission matrix **T** is known, circuit theory provides the following impedance matrix and admittance matrix of the two-port:

$$Z = \begin{pmatrix} -j\cot(\tau d)\,Z_c & -jZ_c \dfrac{1}{\sin(\tau d)} \\ -jZ_c \dfrac{1}{\sin(\tau d)} & -j\cot(\tau d)\,Z_c \end{pmatrix} \tag{9}$$

$$Y = \begin{pmatrix} -j\cot(\tau d)\,Y_c & +jY_c \dfrac{1}{\sin(\tau d)} \\ +jY_c \dfrac{1}{\sin(\tau d)} & -j\cot(\tau d)\,Y_c \end{pmatrix} = Z^{-1} \tag{10}$$

The reciprocity property of the two-port provide some simple circuit representations. For instance, Fig. 3 illustrates the Π and T representations, whose parameters are

$$Y_1 = Y_2 = jY_c \tan\frac{\tau d}{2}$$

$$Y_3 = -jY_c \frac{1}{\sin \tau d}$$

$$Z_1 = Z_2 = -jZ_c \tan\frac{\tau d}{2}$$

$$Z_3 = -jZ_c \frac{1}{\sin \tau d}$$

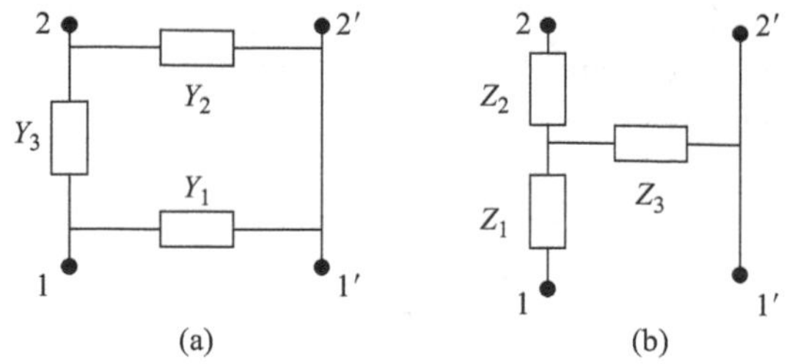

Fig. 3: (a) Π representation, (b) T representation

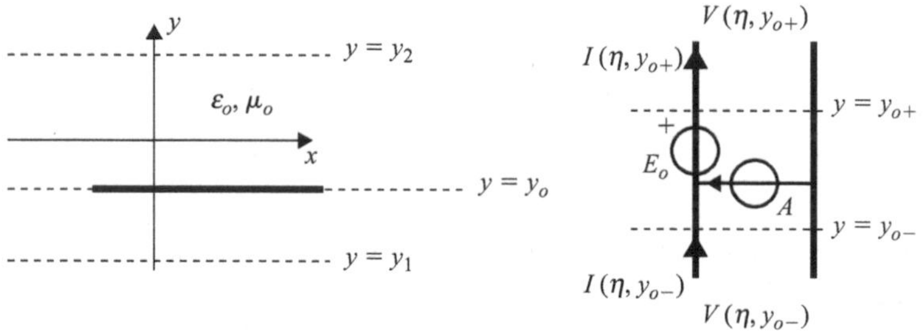

Fig. 4: Circuit representation of planar discontinuity

To conclude this section, we observe that if the length d of the layer is infinitely long the impedance seen by the initial section of the layer is the characteristic impedance of the two-port. Of course this impedance is coincident with the impedance Z_c of the transmission line.

7.2.2 Jump of voltage or current in a section where it is present a discontinuity

Let us assume that a planar discontinuity is present in the section $y = y_o$ (Fig. 4).

We can model this discontinuity with current and voltage generators given by

$$A = A(\eta) = I(\eta, y_{o+}) - I(\eta, y_{o-}) = \int_{-\infty}^{\infty} [H_x(x, y_{o+}) - H_x(x, y_{o-})] e^{j\eta x} dx \tag{11}$$

$$E_o = E_o(\eta) = V(\eta, y_{o+}) - V(\eta, y_{o-}) = \int_{-\infty}^{\infty} [E_z(x, y_{o+}) - E_z(x, y_{o-})] e^{j\eta x} dx \tag{12}$$

7.2.3 Jump of voltage or current in a section where a concentrated source is present

The circuit description of a concentrated source can be obtained using eq. (1).

In essence, as for planar discontinuities, concentrated sources at the point $y = y_o, x = 0$ introduce in the circuit representation voltage or current generators located at the section $y = y_o$ of the transmission line. For instance, for a line-source electrical current $\mathbf{J} = I_o\, \delta(x)\delta(y - y_o)\hat{z}$, eqs. (1) and (2) yield

$$-\frac{d}{dy}V(\eta, y) = j\tau Z_c I(\eta, y)$$
$$-\frac{d}{dy}I(\eta, y) = j\tau Y_c V(\eta, y) + I_o\delta(y - y_o)$$

Similarly, for a line-source magnetic current $M = V_o\, \delta(x)\delta(y - y_o)\hat{z}$, we get

$$-\frac{d}{dy}V(\eta, y) = j\tau Z_c I(\eta, y) - V_o\delta(y - y_o)$$
$$-\frac{d}{dy}I(\eta, y) = j\tau Y_c V(\eta, y)$$

7.3 Wiener-Hopf equations in the Laplace domain

In the presence of remote sources it is convenient to introduce Laplace transforms instead of Fourier transforms. In fact, in these cases the Fourier transforms exist only for real values of the propagation constants; furthermore, when the sources are remote, impulsive functions are present. The Laplace transforms are always analytic functions in the complex variable η.

For a given function $\psi(x)$ defined on the entire x axis, we define the plus and minus Laplace transforms as the analytic continuation of the integrals

$$\Psi_+(\eta) = \int_0^\infty \psi(x)e^{j\eta x}dx, \quad \Psi_-(\eta) = \int_{-\infty}^0 \psi(x)e^{j\eta x}dx$$

The total Laplace transform $\Psi(\eta)$ is given by

$$\Psi(\eta) = \Psi_-(\eta) + \Psi_+(\eta)$$

The circuit considerations developed in the previous section hold also for the total Laplace transform. However, it is important to remember that the inverse Laplace transforms have the form

$$\psi_-(x) = \frac{1}{2\pi}\int_{B-} \Psi_-(\eta)e^{-j\eta x}d\eta$$

$$\psi_+(x) = \frac{1}{2\pi}\int_{B_+} \Psi_+(\eta)e^{-j\eta x}d\eta$$

where B_- is a horizontal line located under the singularities of $\Psi_-(\eta)$, and B_+ is a horizontal line located above the singularities of $\Psi_+(\eta)$.

To put in evidence the difference between Laplace and Fourier transforms, let us consider a field having an x dependence described by[1]

$$f(x) = e^{-j\eta_o x}$$

The Fourier transform requires real values of η_o and yields

$$\int_{-\infty}^{\infty} e^{-j\eta_o x} e^{j\eta x} dx = 2\pi\delta(\eta - \eta_o)$$

Conversely, in the domain of the Laplace transforms, $F_+(\eta)$ is the analytic continuation of

$$F_+(\eta) = \int_{0}^{\infty} e^{-j\eta_o x} e^{j\eta x} dx = j\frac{1}{\eta - \eta_o}$$

whereas $F_-(\eta)$ is the analytic continuation of

$$F_-(\eta) = \int_{-\infty}^{0} e^{-j\eta_o x} e^{j\eta x} dx = -j\frac{1}{\eta - \eta_o}$$

It follows that the complete Laplace transform $F(\eta)$ of a plane wave is null:

$$F(\eta) = F_-(\eta) + F_+(\eta) = -j\frac{1}{\eta - \eta_o} + j\frac{1}{\eta - \eta_o} = 0$$

Notice, however, that if the term $-j\frac{1}{\eta-\eta_o}$ is interpreted as a minus functions, if $\mathrm{Im}[\eta_o] < 0$, the inverse transformation introduces a Bromwich line B_- such that

$$\frac{1}{2\pi}\int_{B_-} -j\frac{1}{\eta - \eta_o} e^{-j\eta x} d\eta = e^{-j\eta_o x} u(-x)$$

Conversely, if the term $j\frac{1}{\eta-\eta_o}$ is interpreted as a plus functions, if $\mathrm{Im}[\eta_o] < 0$, the inverse transformation introduces a Bromwich line B_+ such that

$$\frac{1}{2\pi}\int_{B_+} j\frac{1}{\eta - \eta_o} e^{-j\eta x} d\eta = e^{-j\eta_o x} u(x)$$

Hence, despite the zero value of $F(\eta) = F_-(\eta) + F_+(\eta)$, the inverse transformation of $F_-(\eta)$ and $F_+(\eta)$ reproduces exactly the field due to the remote source. Since the total Laplace transforms for the primary field in presence of remote sources are vanishing, we can deduce the Wiener-Hopf equations in the spectral domain by ignoring the presence of the remote sources. This always yields Wiener-Hopf equations in the homogeneous form discussed in section 2.4.2. Let us remember that the Laplace transform $\Psi(\eta)$ does exist also for complex values of η_o. This fact allows us to introduce small losses in the media to avoid singularities on the real axis $\mathrm{Im}[\eta] = 0$.

---

[1] The dependence $e^{-j\eta_o x}$ is typical of remote source (incident plane wave, incident mode).

The aforementioned considerations raise the following question: what is the integration path B in the inversion equation $\psi(x) = \frac{1}{2\pi}\int_B \Psi(\eta)e^{-j\eta x}d\eta$?

This problem does not exist if we work with Fourier transform since in this case the integration line always is the real axis. However, in the Laplace domain, for the presence of a pole η_o due to a remote source, nonconventional plus or minus functions are involved. Consequently, in the inversion equation we must assume B as a Bromwich line. For instance, if the pole η_o is not conventional for the plus part of source term, we must set $B = B_+$, that is, a horizontal line above the pole η_o; conversely, if the pole is not conventional for the minus part of source term, we must set $B = B_-$, that is, a horizontal line below the pole η_o. For the sake of brevity, the justification of this rule is not reported here.

7.4 The PEC half-plane problem

7.4.1 E-polarization case

The planar discontinuity is constituted by a PEC half-plane located at $y = 0$. The voltage

$$V = V(\eta, 0_+) = V(\eta, 0_-) = \int_{-\infty}^{\infty} E_z(x, 0_+)e^{j\eta x}dx = \int_{-\infty}^{\infty} E_z(x, 0_-)e^{j\eta x}dx$$
$$= \int_{0}^{\infty} E_z(x, 0_+)e^{j\eta x}dx = V_+(\eta) \tag{13}$$

is continuous since $E_z(x, y)$ is zero on the half-plane $y = 0, x \leq 0$. It means that the voltage generator indicated in Fig. 4 vanishes. Moreover, the vanishing of $E_z(x, y)$ on the half-plane implies that $V = V_+(\eta)$ is a plus function.

Taking into account that the continuity of $H_x(x, y)$ on the aperture $y = 0, x \geq 0$ implies

$$A = I(\eta, 0_+) - I(\eta, 0_-) = \int_{-\infty}^{\infty} [H_x(x, 0_+) - H_x(x, 0_-)]e^{j\eta x}dx$$
$$= \int_{-\infty}^{0} [H_x(x, 0_+) - H_x(x, 0_-)]e^{j\eta x}dx = A_-(\eta) \tag{14}$$

the function $A = A_-(\eta)$ is a minus function. According to the boundary condition on a PEC surface, $-A = -A_-(\eta)$ represents the Fourier transform of the total physical current induced on both the faces of the PEC half-plane.

Since the layers $y \geq 0$ and $y \leq 0$ are unbounded, we have to consider the impedance Z_c in both directions. This entails the circuit representation of Fig. 5. Analysis of this circuit leads to

$$V = Z_c || Z_c A = \frac{Z_c}{2} A \tag{15}$$

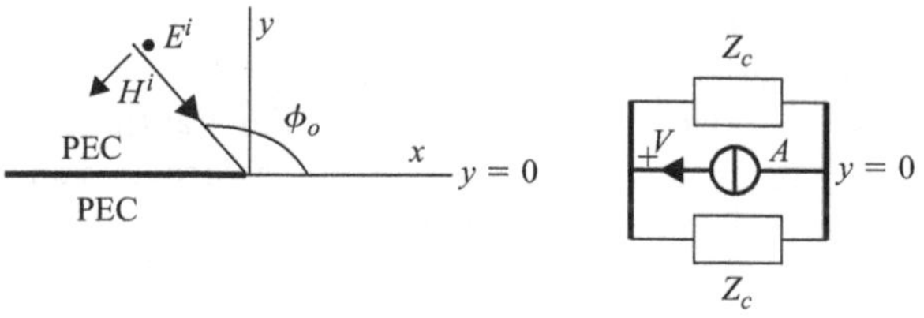

Fig. 5: Half-plane problem and its circuit representation

or taking into account that

$$\tau = \tau(\eta) = \sqrt{k^2 - \eta^2} \quad Z_c = \frac{\omega\mu_o}{\tau} = Z_c(\eta) = \frac{k\,Z_o}{\sqrt{k^2 - \eta^2}} \quad Y_c = \frac{1}{Z_c} = \frac{\tau}{\omega\mu_o}$$

and $V = V_+(\eta)$, $A = A_-(\eta)$

$$\frac{2\sqrt{k^2 - \eta^2}}{\omega\mu} V_+(\eta) = A_-(\eta) \tag{16}$$

The homogeneous W-H eq. (16) can be rewritten in the normal form of section 2.1 by taking into account the source constituted by the incident plane wave:

$$E_z^i(x, y) = E_o e^{jk\rho\cos(\varphi - \varphi_o)} \tag{17}$$

On the aperture $y = 0,\ x \geq 0$ or $\varphi = 0,\ \rho = x$ this provides the voltage

$$V_+^i(\eta) = \int_0^\infty E_o e^{jk\rho\cos\varphi_o} e^{j\eta x} dx = j\frac{E_o}{\eta - \eta_o} \tag{18}$$

where $\eta_o = -k\cos\varphi_o$.

Taking into account that there is no reflected wave contribution on the aperture, $V_+^s(\eta) = V_+(\eta) - V_+^i(\eta)$ does not present the pole $\eta_o = -k\cos\varphi_o$. From eq. (16), it follows that the characteristic part of $A = A_-(\eta)$ at the pole $\eta_o = -k\cos\varphi_o$ is

$$\frac{2\sqrt{k^2 - \eta_o{}^2}}{\omega\mu} j\frac{E_o}{\eta - \eta_o}$$

Notice that by letting $\eta = -k\cos w$ (section 2.9.2) $\eta_o = -k\cos\varphi_o$ corresponds to $w_o = -\varphi_o$ in the w – plane, which yields the determination $\sqrt{k^2 - \eta_o{}^2} = -k\sin w_o = k\sin\varphi_o$. Consequently, we have

$$\frac{2\sqrt{k^2 - \eta_o{}^2}}{\omega\mu} j\frac{E_o}{\eta - \eta_o} = 2j\frac{E_o \sin\varphi_o}{Z_o(\eta - \eta_o)} \tag{19}$$

and the normal form of the W-H eq. (16) is

$$\frac{2\sqrt{k^2 - \eta^2}}{\omega\mu} V_+(\eta) = A_-^d(\eta) + 2j\frac{E_o \sin\varphi_o}{Z_o(\eta - \eta_o)} \tag{20}$$

where $A_-^d(\eta)$ is regular at the pole $\eta_o = -k\cos\varphi_o$.

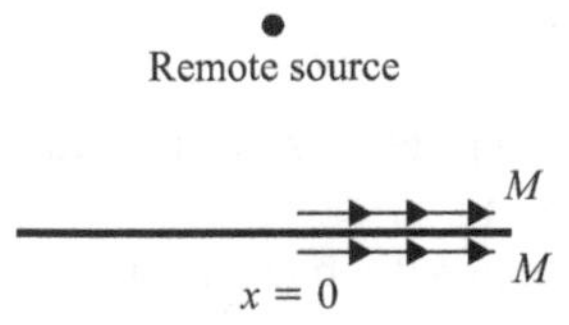

Fig. 6: Equivalence theorem

The source term $A_-^p = 2j\frac{E_o \sin\varphi_o}{Z_o(\eta-\eta_o)}$ in (20) can be obtained directly by observing that it represents the primary contribution for the total current induced on the half-plane. In fact, (Fig. 6) the total current induced on the PEC half-plane is the sum of the contribution of the remote source in presence of a full PEC plane in $y = 0$ (i.e., the primary field) plus the contribution of the equivalent magnetic current M present on the aperture. Since the latter contribution does not present any pole, the characteristic part of $A_-(\eta)$ is the minus Fourier transform of the current induced by the primary field:

$$H_x(x, 0_+) = \underset{\substack{\Uparrow \\ M}}{H_x^d(x, 0_+)} + \underset{\substack{\Uparrow \\ \text{Primary field}}}{H_x^g(x, 0_+)} \tag{21}$$

Taking into account the reflected field on the plane $y = 0$ (Fig. 6):

$$H_x^g(x, 0_+) = -2\frac{E_o}{Z_o}\sin\varphi_o e^{jk\rho\cos(\pi-\varphi_o)} = -2\frac{E_o}{Z_o}\sin\varphi_o e^{jk\,x\cos\varphi_o} \tag{22}$$

we get

$$A_-^p = \int_{-\infty}^{0} H_x^g(x, 0_+)e^{j\eta x}dx = 2j\frac{E_o\sin\varphi_o}{Z_o(\eta-\eta_o)} \tag{23}$$

with $\eta_o = -k\cos\varphi_o$.

The solution of the W-H eq. (20) (see section 2.4.2) is

$$V_+(\eta) = \frac{1}{\sqrt{k-\eta}\sqrt{k+\eta_o}}j\frac{kE_o\sin\varphi_o}{\eta-\eta_o}, \quad A_-(\eta) = \frac{2\sqrt{k+\eta}}{Z_o\sqrt{k+\eta_o}}j\frac{E_o\sin\varphi_o}{\eta-\eta_o} \tag{24}$$

Since the factorized functions can be chosen within a constant factor, we can choose arbitrarily the branch of $\tau_+(\eta) = \sqrt{k-\eta}$ and let $\sqrt{k+\eta} = \frac{\sqrt{k^2-\eta^2}}{\sqrt{k-\eta}}$. In the following, if not indicated otherwise we will adopt as branch the one that satisfies the condition $\tau_+(0) = \sqrt{k}$ for real values of k. In the w – plane (section 2.9.2) we have

$$\tau_+(\eta) = \sqrt{k-\eta} = \sqrt{2k}\cos\frac{w}{2}, \quad \tau_-(\eta) = \sqrt{k+\eta} = -\sqrt{2k}\sin\frac{w}{2} \tag{25}$$

7.4.2 Far-field contribution

The solution of the W-H equations allows us to evaluate the total field by using the equivalence theorem indicated in Fig. 6.

The effects of the equivalent magnetic currents M are given by

$$\begin{aligned} V^M(\eta, y) &= V_+(\eta)e^{-j\tau y}, \quad \mathrm{I}^M(\eta, y) = \frac{V_+(\eta)}{Z_c}e^{-j\tau y} \quad y > 0 \\ V^M(\eta, y) &= V_+(\eta)e^{+j\tau y}, \quad \mathrm{I}^M(\eta, y) = -\frac{V_+(\eta)}{Z_c}e^{j\tau y} \quad y < 0 \end{aligned} \tag{26}$$

where the plus W-H unknown $V_+(\eta)$ is expressed by (24):

$$V_+(\eta) = \frac{1}{\sqrt{k-\eta}\sqrt{k+\eta_o}} j\frac{kE_o \sin\varphi_o}{\eta - \eta_o} \tag{27}$$

Consequently, we obtain

$$\begin{aligned} V(\eta, y) &= V^P(\eta, y) + V^M(\eta, y) \\ \mathrm{I}(\eta, y) &= \mathrm{I}^P(\eta, y) + \mathrm{I}^M(\eta, y) \end{aligned} \tag{28}$$

where according to the equivalence theorem $V^P(\eta, y)$ and $\mathrm{I}^P(\eta, y)$ represent the primary field, that is, the Laplace transform of the field due to the source in the presence of the PEC plane located at $y = 0$. For sources constituted by plane waves for $y > 0$ in the spectral domain, both $V^P(\eta, y)$ and $\mathrm{I}^P(\eta, y)$ are null. So it is more convenient to rephrase (28) in the natural domain. For instance, for the first of (28),

$$E_z(x, y) = E_z^i(x, y) + E_z^r(x, y) + E_z^M(x, y) \tag{29}$$

where $E_z^i(x, y) + E_z^r(x, y)$ is the primary field (the incident plus the reflected wave). Note that according to the equivalence theorem, for $y < 0$ we have $E_z^i(x, y) = 0$, $E_z^r(x, y) = 0$.

The inverse Fourier transform yields the following expression for the longitudinal field:

$$E_z^M(x, y) = \frac{jkE_o \sin\varphi_o}{2\pi} \int_{B_+} \frac{e^{-j\eta x} e^{-j\sqrt{k^2-\eta^2}|y|}}{\sqrt{k-\eta}\sqrt{k+\eta_o}(\eta - \eta_o)} d\eta \tag{30}$$

where the line B_+ is a straight line above the pole η_o. Algebraic manipulations allow us to reduce this integral to special functions involving Fresnel integrals (Mittra & Lee, 1971). However, in different problems as such those considered in the next sections, it is not possible to obtain the inverse Fourier transforms in closed form. Approximate evaluation techniques must be implemented. In particular, for the evaluation of far fields, the saddle point method (Felsen & Marcuvitz, 1973) is particularly important.

Using the mapping indicated in section 2.9.2,

$$\begin{aligned} \eta &= -k\cos w \\ \tau &= -k\sin w \end{aligned} \tag{31}$$

and considering the cylindrical coordinates

$$x = \rho \cos \varphi$$
$$y = \rho \sin \varphi \tag{32}$$

yields the integral

$$I_c = k \int_{r_w} \hat{f}_s(w) e^{jk\rho \cos(w - |\varphi|)} dw \tag{33}$$

where

$$\hat{f}_s(w) = f(-k \cos w) \sin w = \frac{jE_o \sin \varphi_o \sin w}{2\pi k 2 \cos\frac{w}{2}\left[-\sin\left(\frac{-\varphi_o}{2}\right)\right](-\cos w + \cos \varphi_o)}$$
$$= \frac{jE_o \cos\frac{\varphi_o}{2} \sin\frac{w}{2}}{\pi k(-\cos w + \cos \varphi_o)} \tag{34}$$

and r_w is the image of the real axis in the $w-$ plane (chapter 2, Fig. 3). We can warp the contour path r_w in (33) into the SDP. Since the eventual pole contribution provides the geometric optics contribution, it means that the SDP contribution has the fundamental physical meaning of representing the diffracted far field. Taking into account the application in section 2.9.2, we obtain the following contribution of the SDP; the contribution is obtained according to the direction of SDP indicated in the application of section 2.9. Observe that this direction is opposite with respect to that of r_w.

$$E_z^{SDP} = -j\sqrt{\frac{2}{\pi k\rho}} \frac{E_o \cos\frac{\varphi_o}{2} \cos\frac{\varphi}{2}}{\cos \varphi + \cos \varphi_o} e^{-j(k\rho + \pi/4)} \quad k\rho >> 1 \tag{35}$$

Equation (35) cannot be used when the observation point approaches boundaries of the incident and reflected waves. In this case the pole is near the saddle point, and the approximation (2.90) is no longer valid. In this case we must use a more general technique indicated in Felsen and Marcuvitz (1973) and Volakis and Senior (1995); see also section 2.9.

The previous equations provide the evaluation of the integral on the SDP. We can relate this evaluation to the original one on the path r_w by warping the contour path r_w in (33) into the SDP and taking into account the eventual singularity contributions located in the region between the SDP and r_w.

Sometimes it is preferable to study the warping directly in the $\eta-$ plane. In this plane, the SDP does not change if we change φ into $-\varphi$. Figure 7 illustrates the location of the SDP in the $\eta-$ plane.

The direction of the SDP in the $\eta-$ plane is according to the direction of the SDP in the $w-$ plane. With this orientation, for every value of φ, $-\pi < \varphi \leq \pi$ we obtain

$$E_z^{SDP} = -j\sqrt{\frac{2}{\pi k\rho}} \frac{E_o \cos\frac{\varphi_o}{2} \cos\frac{\varphi}{2}}{\cos \varphi + \cos \varphi_o} e^{-j(k\rho + \pi/4)} \quad k\rho >> 1$$

If $\cos \varphi_o$ is positive, the pole η_o lies in the second quadrant. However, the Bromwich path B_+ is always located above η_o. Figure 7 shows that by warping the path B_+ on the SDP,

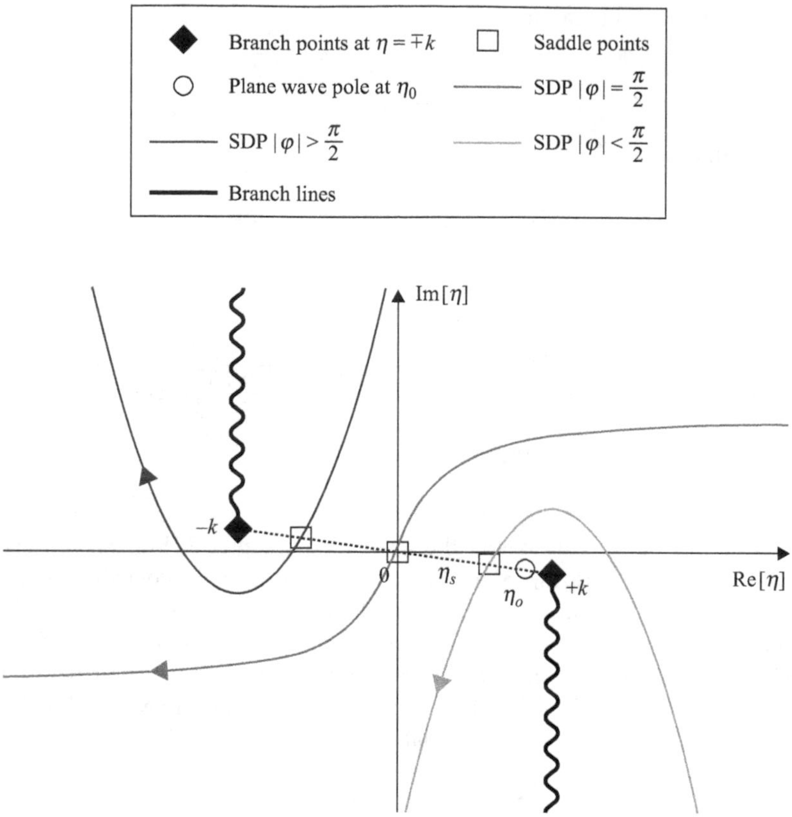

Fig. 7: The steepest descent path (SDP) in the η – plane ($\eta_s = k\cos\varphi$, $\eta_o = -k\cos\varphi_o$)

the pole η_o is always captured clockwise for $\bar{\eta}_o = \frac{\eta_o}{k} < \frac{\eta_s}{k} = \bar{\eta}_s$. The warping yields the following result:

$$E_z^M(x,y) + E_z^{SDP}(x,y) = -2\pi j \frac{jkE_o \sin\varphi_o}{2\pi} \text{Res}\left[\frac{e^{-j\eta x} e^{-j\sqrt{k^2-\eta^2}|y|}}{\sqrt{k-\eta}\sqrt{k+\eta_o}(\eta-\eta_o)}\right]_{\eta_o} u(\bar{\eta}_s - \bar{\eta}_o)$$

$$= E_o e^{-j\eta_o x} e^{-j\tau_o|y|} u(\bar{\eta}_s - \bar{\eta}_o) = \begin{cases} -E_z^r(x,y)u(\bar{\eta}_s - \bar{\eta}_o) & y > 0 \\ E_z^i(x,y)u(\bar{\eta}_s - \bar{\eta}_o) & y < 0 \end{cases} \quad (36)$$

where by taking into account that $w_o = -\varphi_o$

$$\tau_o = -k\sin w_o = k\sin\varphi_o$$

and the boundary conditions

$$E_z^i(x,0) + E_z^r(x,0) = 0$$

we obtain

$$E_z(x,y) = E_z^i(x,y) + E_z^r(x,y) + E_z^M(x,y) = E_z^i(x,y) + E_z^r(x,y)u(\bar{\eta}_o - \bar{\eta}_s) + E_z^{dif}(x,y), \quad y > 0$$
$$E_z(x,y) = +E_z^M(x,y = E_z^i(x,y)u(\bar{\eta}_s - \bar{\eta}_o) + E_z^{dif}(x,y), \quad y < 0$$

(37)

with

$$E_z^{dif} = -E_z^{SDP} = j\sqrt{\frac{2}{\pi k\rho}}\frac{E_o \cos\frac{\varphi_o}{2}\cos\frac{\varphi}{2}}{\cos\varphi + \cos\varphi_o}e^{-j(k\rho+\pi/4)} \quad k\rho >> 1, \quad \varphi \neq \pm(\pi - \varphi_o)$$

The previous equations show that, according to geometrical optics, for $y > 0$ there is no reflected wave when $0 < \varphi < \pi - \varphi_o$. Conversely, in the region $y < 0$ there is no incident wave when $-\pi < \varphi < \varphi_o - \pi$ (deep shadow region).

7.5 Skew incidence

Let us consider an incident plane wave impinging on the PEC half-plane at a skew incidence angle β:

$$E_z^i = E_o e^{j\tau_o\rho\cos(\varphi-\varphi_o)}e^{-j\alpha_o z} \quad H_z^i = H_o e^{j\tau_o\rho\cos(\varphi-\varphi_o)}e^{-j\alpha_o z} \tag{38}$$

where k is the propagation constant and

$$\alpha_o = k\cos\beta, \quad \tau_o = k\sin\beta = \sqrt{k^2 - \alpha_o^2}, \quad \rho\cos(\varphi - \varphi_o) = x\cos\varphi_o + y\sin\varphi_o$$

Using Maxwell's equations yields the following other components of the incident plane wave:

$$\begin{aligned}
E_x^i &= e^{j\tau_o\rho\cos(\varphi-\varphi_o)}e^{-j\alpha_o z}\frac{E_o\alpha_o\cos\varphi_o + Z_o H_o k\sin\varphi_o}{\tau_o} \\
E_y^i &= e^{j\tau_o\rho\cos(\varphi-\varphi_o)}e^{-j\alpha_o z}\frac{E_o\alpha_o\sin\varphi_o - Z_o H_o k\cos\varphi_o}{\tau_o} \\
H_x^i &= e^{j\tau_o\rho\cos(\varphi-\varphi_o)}e^{-j\alpha_o z}\frac{-E_o k Y_o\sin\varphi_o + \alpha_o H_o\cos\varphi_o}{\tau_o} \\
H_y^i &= e^{j\tau_o\rho\cos(\varphi-\varphi_o)}e^{-j\alpha_o z}\frac{E_o k Y_o\cos\varphi_o + H_o\alpha_o\sin\varphi_o}{\tau_o}
\end{aligned} \tag{39}$$

where Z_o and $Y_o = Z_o^{-1}$ are the impedance and admittance of the medium.

To derive the Wiener-Hopf equations, let us consider the transverse fields

$$\mathbf{E}_t = \hat{z}E_z + \hat{x}E_x, \quad \mathbf{H}_t = \hat{z}H_z + \hat{x}H_x \tag{40}$$

and their Fourier transforms

$$\begin{aligned}
\mathbf{V}(\eta, y) &= e^{j\alpha_o z}\int_{-\infty}^{\infty}\hat{y}\times\mathbf{E}_t(x,y,z)e^{j\eta x}dx \\
\mathbf{I}(\eta, y) &= e^{j\alpha_o z}\int_{-\infty}^{\infty}\mathbf{H}_t(x,y,z)e^{j\eta x}dx
\end{aligned} \tag{41}$$

In the following we will also use matrix notations; for instance,

$$V = \begin{vmatrix} V_1 \\ V_2 \end{vmatrix} = \begin{vmatrix} -\tilde{E}_x \\ \tilde{E}_z \end{vmatrix}, \quad I = \begin{vmatrix} I_1 \\ I_2 \end{vmatrix} = \begin{vmatrix} \tilde{H}_z \\ \tilde{H}_x \end{vmatrix}$$

where $\tilde{F}(\eta)$ means the Fourier transform of $F(x)$.

By substitution in the transverse eq. (1) and by taking into account that in the Fourier domain $\nabla_t = -j\,\boldsymbol{\sigma}$ with $\boldsymbol{\sigma} = \eta\,\hat{x} + \alpha_o\hat{z}$, we get

$$\frac{d\,\mathbf{V}}{dz} = -\mathbf{Z}\cdot\mathbf{I}$$
$$\frac{d\,\mathbf{I}}{dz} = -\mathbf{Y}\cdot\mathbf{V}$$

where

$$\mathbf{Z} = j\omega\mu\left(\mathbf{1}_t - \frac{\boldsymbol{\sigma}\times\hat{y}\,\boldsymbol{\sigma}\times\hat{y}}{k^2}\right), \quad \mathbf{Y} = j\omega\varepsilon\left(\mathbf{1}_t - \frac{\boldsymbol{\sigma}\,\boldsymbol{\sigma}}{k^2}\right)$$

The matrices $\mathbf{Y}$ and $\mathbf{Z}$ commute

$$\mathbf{Y}\cdot\mathbf{Z} = \mathbf{Z}\cdot\mathbf{Y} = -\chi^2\mathbf{1}_t$$

where $\chi = \sqrt{k^2 - \eta^2 - \alpha_o^2}$.

These equations represent a transmission line with propagation constant χ and (dyadic) characteristic impedance:[2]

$$\mathbf{Z}_c = \frac{1}{j\chi}\mathbf{Z} = \frac{\omega\mu}{\chi}\hat{\boldsymbol{\sigma}}\hat{\boldsymbol{\sigma}} + \frac{\chi}{\omega\varepsilon}\hat{\boldsymbol{\beta}}\hat{\boldsymbol{\beta}} = \frac{Z_o}{k\sqrt{k^2 - \alpha_o^2 - \eta^2}}\begin{vmatrix} k^2 - \eta^2 & \eta\,\alpha_o \\ \eta\,\alpha_o & \tau_o^2 \end{vmatrix}$$

where[3]

$$\tau_o = \sqrt{k^2 - \alpha_o^2}, \quad \sigma = \sqrt{\alpha_o^2 + \eta^2}, \quad \hat{\boldsymbol{\sigma}} = \frac{\boldsymbol{\sigma}}{\sigma}, \quad \hat{\boldsymbol{\beta}} = \hat{y}\times\hat{\boldsymbol{\sigma}}$$

The function $\mathbf{V}_+(\eta) = \mathbf{V}(\eta, 0)$ is a plus function since $\mathbf{E}_t(x, 0, z) = 0$ on the PEC half-plane $y = 0$. Furthermore, we have

$$\begin{aligned} \mathbf{Y}_c\cdot\mathbf{V}_+(\eta) &= \mathbf{I}(\eta, 0_+) \\ \mathbf{Y}_c\cdot\mathbf{V}_+(\eta) &= -\mathbf{I}(\eta, 0_-) \end{aligned} \tag{42}$$

where, with the adopted order for the components of the field, the matrix admittance $\mathbf{Y}_c$ is given by

$$\mathbf{Y}_c = \mathbf{Z}_c^{-1} = \frac{Y_o}{k\sqrt{k^2 - \eta^2 - \alpha_o^2}}\begin{vmatrix} \tau_o^2 & -\eta\,\alpha_o \\ -\eta\,\alpha_o & k^2 - \eta^2 \end{vmatrix} \tag{43}$$

Summing (42) yields the homogeneous W-H equation

$$2\mathbf{Y}_c\cdot\mathbf{V}_+(\eta) = \mathbf{A}_-(\eta) \tag{44}$$

where $\mathbf{A}_-(\eta)$ is the Fourier transform of the total current induces on the PEC half-plane:

$$\mathbf{A}_-(\eta) = \mathbf{I}(\eta, 0_+) - \mathbf{I}(\eta, 0_-) \tag{45}$$

---

[2] The evaluation of the characteristic impedances for an arbitrary linear medium will be considered in section 7.10.2.

[3] The deduction of the last member requires some algebraic manipulations.

The factor $e^{j\tau_o\rho\cos(\varphi-\varphi_o)}$ evaluated at $\varphi = 0$ produces in the plus function $\mathbf{V}_+(\eta)$ a pole arising from the integral

$$\int_0^\infty e^{j\tau_o\cos\varphi_o x}e^{j\eta x}dx = j\frac{1}{\eta-\eta_o} \tag{46}$$

where $\eta_o = -\tau_o\cos\varphi_o$.

Taking into account that

$$\begin{aligned} E_z^i &= E_o e^{j\tau_o\,\rho\cos(\varphi-\varphi_o)}e^{-j\alpha_o z} \\ E_x^i &= e^{j\tau_o\,\rho\cos(\varphi-\varphi_o)}e^{-j\alpha_o z}\frac{E_o\alpha_o\cos\varphi_o + Z_oH_ok\sin\varphi_o}{\tau_o} \end{aligned} \tag{47}$$

we obtain that the residue $\mathbf{T}_o$ of $\mathbf{V}_+(\eta)$ at this pole is known and is given by

$$\mathbf{T}_o = jE_o\hat{x} - \frac{E_o\alpha_o\cos\varphi_o + Z_oH_ok\sin\varphi_o}{\tau_o}\hat{z} \tag{48}$$

From eq. (44), the residue of $\mathbf{A}_-(\eta)$ at the pole $\eta_o = -\tau_o\cos\varphi_o$ is given by

$$\text{Res}[\mathbf{A}_-(\eta)]_{\eta=\eta_o} = 2Y_c(\eta_o)\cdot T_o = R_o = |R_{o1}, R_{o2}|^t \tag{49}$$

where

$$R_{o1} = -2jH_o, \quad R_{o2} = -\frac{2j\alpha_o\cos\varphi_o}{\tau_o}H_o + \frac{2jk\sin\varphi_o}{\tau_oZ_o}E_o \tag{50}$$

Thus, we obtain the following W-H equation in normal form:

$$2\mathbf{Y}_c\cdot\mathbf{V}_+(\eta) = \mathbf{A}_-^s(\eta) + \frac{R_o}{\eta-\eta_o} \tag{51}$$

with $\mathbf{A}_-^s(\eta)$ a conventional minus function regular at $\eta - \eta_o$.

The factorization of the kernel

$$G(\eta) = 2\mathbf{Y}_c = 2\frac{Y_o}{k\sqrt{k^2-\eta^2-\alpha_o^2}}\begin{vmatrix} \tau_o^2 & -\eta\,\alpha_o \\ -\eta\,\alpha_o & k^2-\eta^2 \end{vmatrix} \tag{52}$$

can be accomplished in several ways. For example, we can rewrite

$$G(\eta) = \frac{Y_o}{k\sqrt{\tau_o+\eta}}\begin{pmatrix} \tau_o & 0 \\ 0 & k+\eta \end{pmatrix}\begin{vmatrix} 1 & -\dfrac{\eta\,\alpha_o}{\tau_o(k-\eta)} \\ -\dfrac{\eta\,\alpha_o}{\tau_o(k+\eta)} & 1 \end{vmatrix}\begin{pmatrix} \tau_o & 0 \\ 0 & k-\eta \end{pmatrix} \tag{53}$$

The factorization problem has been reduced to the factorization of the central matrix $\begin{vmatrix} 1 & -\frac{\eta\,\alpha_o}{\tau_o(k+\eta)} \\ -\frac{\eta\,\alpha_o}{\tau_o(k-\eta)} & 1 \end{vmatrix}$. This matrix is both rational and has a Daniele form, so both techniques indicated in chapter 4 apply and we get a straightforward factorization of $G(\eta)$.[4]

[4] See, for instance, section 4.8.4.

A more direct way to solve the W-H eq. (51) derives from the scalar form of these equations:

$$2\frac{Y_o}{k\sqrt{k^2-\eta^2-\alpha_o^2}}\hat{V}_{1+} = A_{1-}^s + \frac{R_{o1}}{\eta-\eta_o} \tag{54}$$

$$2\frac{Y_o}{k\sqrt{k^2-\eta^2-\alpha_o^2}}\hat{V}_{2+} = A_{2-}^s + \frac{R_{o2}}{\eta-\eta_o} \tag{55}$$

where

$$\hat{V}_{1+} = \tau_o^2 V_{1+} - \eta\,\alpha_o V_{2+} \tag{56}$$

$$\hat{V}_{2+} = -\eta\,\alpha_o V_{1+} + (k^2-\eta^2)V_{2+} \tag{57}$$

or, by inverting:

$$V_{1+} = \frac{(k^2-\eta^2)\hat{V}_{1+} + \eta\,\alpha_o\hat{V}_{2+}}{\tau_o^2-\eta^2} \tag{58}$$

$$V_{2+} = \frac{\eta\,\alpha_o\hat{V}_{1+} + \tau_o^2\hat{V}_{2+}}{\tau_o^2-\eta^2} \tag{59}$$

Equations (54) and (55) are not coupled W-H equations. The considerations of section 2.5 apply since the plus functions $\hat{V}_{2+}$ do not vanish for $\eta \to \infty$; hence, it is not a Laplace transform.

The factorization of the scalar $y_c = 2\frac{Y_o}{k\sqrt{k^2-\eta^2-\alpha_o^2}} = y_{c-}(\eta)y_{c+}(\eta)$ immediately yields the following solution for eq. (54):

$$\hat{V}_{1+}(\eta) = \frac{1}{y_{c+}(\eta)y_{c-}(\eta_o)}\frac{R_{o1}}{\eta-\eta_o}$$

For eq. (55), an entire function must be introduced defined by

$$y_{c+}(\eta)\hat{V}_{2+} - \frac{R_{o2}}{y_{c-}(\eta_o)(\eta-\eta_o)}$$
$$= \frac{1}{y_{c-}(\eta)}A_{2-}^s + \frac{R_{o2}}{y_{c-}(\eta)(\eta-\eta_o)} - \frac{R_{o2}}{y_{c-}(\eta_o)(\eta-\eta_o)} = w = w_o$$

and, by taking into account the asymptotic behavior of $\frac{1}{y_{c-}(\eta)}$ and the Laplace transform A_{2-}^s, is a constant w_o.

We get

$$\hat{V}_{2+}(\eta) = \frac{1}{y_{c+}(\eta)y_{c-}(\eta_o)}\frac{R_{o1}}{\eta-\eta_o} + \frac{w_o}{y_{c+}(\eta)}$$

To obtain w_o we observe that $V_{1+}(\eta)$ and $V_{2+}(\eta)$ defined by eqs. (58) and (59) must be regular at $\eta = -\tau_o$. This yields the same equation

$$\alpha_o^2\hat{V}_{1+}(-\tau_o) - \tau_o\,\alpha_o\hat{V}_{2+}(-\tau_o) = 0$$
$$-\tau_o\,\alpha_o\hat{V}_{1+}(-\tau_o) + \tau_o^2\hat{V}_{2+}(-\tau_o) = 0$$

that provides the value of w_o.

Usually in literature, for physical interpretation reasons, an alternative weak factorization (Luneburg & Serbest, 2000) is preferred and will be considered next.

The matrix

$$\mathbf{Y}_c = \frac{Y_o}{k\sqrt{k^2-\eta^2-\alpha_o^2}}\begin{vmatrix} \tau_o^2 & -\eta\,\alpha_o \\ -\eta\,\alpha_o & k^2-\eta^2 \end{vmatrix}$$

can be rewritten:

$$\mathbf{Y}_c = \frac{Y_o}{k\,\tau}t^{-1}(\eta)\cdot\begin{vmatrix} k^2 & 0 \\ 0 & \tau^2 \end{vmatrix}\cdot t(\eta) = \mathbf{Y}_{wc-}(\eta)\cdot\mathbf{Y}_{wc+}(\eta)$$

where $\tau = \sqrt{\tau_o^2-\eta^2}$ and $t(\eta) = \begin{pmatrix} \eta & -\alpha_o \\ \alpha_o & \eta \end{pmatrix}$.

We obtain the following weak factorized matrices:

$$\mathbf{Y}_{wc-}(\eta) = \sqrt{\frac{Y_o}{k}}\frac{t^{-1}(\eta)}{\sqrt{\tau_o+\eta}}\cdot\begin{vmatrix} k & 0 \\ 0 & \tau_o+\eta \end{vmatrix} \quad \mathbf{Y}_{wc+}(\eta) = \sqrt{\frac{Y_o}{k}}\begin{vmatrix} k & 0 \\ 0 & \tau_o-\eta \end{vmatrix}\frac{t(\eta)}{\sqrt{\tau_o-\eta}} \tag{60}$$

The offending poles arise from

$$t^{-1}(\eta) = \frac{t_a(\eta)}{\eta^2+\alpha_o^2} \tag{61}$$

where $t_a(\eta) = \begin{pmatrix} \eta & \alpha_o \\ -\alpha_o & \eta \end{pmatrix}$, and they are $\eta_{1,2} = \pm j\alpha_o$.

The W-H equation can be rewritten in the form

$$2\mathbf{Y}_{cw+}(\eta)\cdot\mathbf{V}_+(\eta) = \mathbf{Y}_{cw-}^{-1}(\eta)\mathbf{A}_-(\eta)$$

or

$$2\mathbf{Y}_{cw+}(\eta)\cdot\mathbf{V}_+(\eta) - \mathbf{Y}_{cw-}^{-1}(\eta_o)\cdot\frac{R_o}{\eta-\eta_o} = \mathbf{Y}_{cw-}^{-1}(\eta)\mathbf{A}_-(\eta) - \mathbf{Y}_{cw-}^{-1}(\eta_o)\cdot\frac{R_o}{\eta-\eta_o} = w \tag{62}$$

Since both member of the last equation are regular in $\mathrm{Im}[\eta] \leq 0$ and $\mathrm{Im}[\eta] \geq 0$, respectively, it seems that the offensive poles $\eta_{1,2} = \pm j\alpha_o$ do not produce any effect. However, the asymptotic behavior of both members of the equation shows that the entire vector is constant. To evaluate this constant, we must eliminate offending poles present both in $\mathbf{V}_+(\eta)$ and $\mathbf{A}_-(\eta)$. We have

$$\begin{aligned} \mathbf{V}_+(\eta) &= \frac{\sqrt{\tau_o-\eta}}{2\sqrt{\frac{Y_o}{k}}}\frac{t_a(\eta)}{\eta^2+\alpha_o^2}\cdot\begin{vmatrix} k & 0 \\ 0 & \tau_o-\eta \end{vmatrix}^{-1}\cdot\left[w+\mathbf{Y}_{cw-}^{-1}(\eta_o)\cdot\frac{R_o}{\eta-\eta_o}\right] \\ \mathbf{A}_-(\eta) &= \sqrt{\frac{Y_o}{k}}\frac{1}{\sqrt{\tau_o+\eta}}\cdot\frac{t_a(\eta)}{\eta^2+\alpha_o^2}\cdot\begin{vmatrix} k & 0 \\ 0 & \tau_o+\eta \end{vmatrix}\cdot\left[w+\mathbf{Y}_{cw-}^{-1}(\eta_o)\cdot\frac{R_o}{\eta-\eta_o}\right] \end{aligned} \tag{63}$$

To eliminate the offending pole $\eta_1 = j\alpha_o$ in $\mathbf{V}_+(\eta)$, we must have

$$\begin{vmatrix} k & 0 \\ 0 & \tau_o - \eta \end{vmatrix}^{-1} \cdot \left[w + \mathbf{Y}_{cw-}^{-1}(\eta_o) \cdot \frac{R_o}{\eta - \eta_o} \right]_{\eta = j\alpha_o} = c_1 u_1 \tag{64}$$

where $u_1 = \begin{vmatrix} j \\ 1 \end{vmatrix}$ is the null space of $t_a(j\alpha_o)$. Similarly, to eliminate the offending pole $\eta_2 = -j\alpha_o$ in $\mathbf{A}_-(\eta)$, we must have

$$\begin{vmatrix} k & 0 \\ 0 & \tau_o + \eta \end{vmatrix} \cdot \left[w + \mathbf{Y}_{cw-}^{-1}(\eta_o) \cdot \frac{R_o}{\eta - \eta_o} \right]_{\eta = -j\alpha_o} = c_2 u_2 \tag{65}$$

where $u_2 = \begin{vmatrix} -j \\ 1 \end{vmatrix}$ is the null space of $t_a(-j\alpha_o)$.

It follows that

$$\begin{aligned} w + \mathbf{Y}_{cw-}^{-1}(\eta_o) \cdot \frac{R_o}{j\alpha_o - \eta_o} &= c_1 \begin{vmatrix} \frac{1}{k} & 0 \\ 0 & \frac{1}{\tau_o - j\alpha_o} \end{vmatrix} u_1 = c_1 \begin{vmatrix} j\frac{1}{k} \\ \frac{1}{\tau_o - j\alpha_o} \end{vmatrix} \\ w - \mathbf{Y}_{cw-}^{-1}(\eta_o) \cdot \frac{R_o}{j\alpha_o + \eta_o} &= c_2 \begin{vmatrix} -j\frac{1}{k} \\ \frac{1}{\tau_o - j\alpha_o} \end{vmatrix} \end{aligned} \tag{66}$$

The previous equations provide four equations in the unknowns $w = \begin{vmatrix} w_1 \\ w_2 \end{vmatrix}$, $c_{1,2}$. First we obtain $c_{1,2}$ by eliminating w:

$$c_1 \begin{vmatrix} j\frac{1}{k} \\ \frac{1}{\tau_o - j\alpha_o} \end{vmatrix} + c_2 \begin{vmatrix} +j\frac{1}{k} \\ \frac{1}{-\tau_o + j\alpha_o} \end{vmatrix} = -\mathbf{Y}_{cw-}^{-1}(\eta_o) \cdot \frac{2j\alpha_o R_o}{\alpha_o^2 + \eta_o^2}$$

We obtain

$$\begin{vmatrix} c_1 \\ c_2 \end{vmatrix} = -[\mathbf{Y}_{cw-}(\eta_o) \cdot M]^{-1} \cdot \frac{2j\alpha_o R_o}{\alpha_o^2 + \eta_o^2} \tag{67}$$

where

$$M = \begin{vmatrix} j\frac{1}{k} & j\frac{1}{k} \\ \frac{1}{\tau_o - j\alpha_o} & -\frac{1}{\tau_o - j\alpha_o} \end{vmatrix}$$

Once $c_{1,2}$ are known, we get w by one of the eq. (66).

It is easy to show that imposing the absence of the offending poles present in $\mathbf{V}_+(\eta)$ and $\mathbf{A}_-(\eta)$ is equivalent to imposing the absence of the poles $\eta_{1,2} = \pm j\alpha_o$ both in $\mathbf{V}_+(\eta)$ and $\mathbf{A}_-(\eta)$.

7.6 Diffraction by an impedance half plane

7.6.1 Deduction of W-H equations in diffraction problems by impenetrable half-planes

The problem considered in this section has been studied by several authors. We will derive different W-H equations that are present in the literature. These equations can be obtained using the general theory of stratified media.

With reference to Fig. 8, we introduce the following Fourier transform of electromagnetic components located in the (*x*, *z*) plane (we omit the factor $e^{-j\alpha_o z}$):

$$\mathbf{V}(\eta, y) = \int_{-\infty}^{\infty} \begin{vmatrix} -E_x(x,y) \\ E_z(x,y) \end{vmatrix} e^{j\eta x} dx \; \mathbf{I}(\eta, y) = \int_{-\infty}^{\infty} \begin{vmatrix} H_z(x,y) \\ H_x(x,y) \end{vmatrix} e^{j\eta x} dx \tag{68}$$

For half-planes immersed in free space we have

$$\mathbf{V}(\eta, 0_+) = \mathbf{V}_a(\eta) = \mathbf{Z}_c \mathbf{I}(\eta, 0_+) = \mathbf{Z}_c \mathbf{I}_a(\eta) \tag{69}$$

$$\mathbf{V}(\eta, 0_-) = \mathbf{V}_b(\eta) = -\mathbf{Z}_c \mathbf{I}(\eta, 0_-) = -\mathbf{Z}_c \mathbf{I}_b(\eta) \tag{70}$$

or

$$\mathbf{V}_a(\eta) = \mathbf{Z}_c \mathbf{I}_a(\eta), \quad \mathbf{V}_b(\eta) = -\mathbf{Z}_c \mathbf{I}_b(\eta)$$

where (section 7.5):

$$\mathbf{Z}_c = \frac{1}{\omega\varepsilon\xi} \begin{vmatrix} k^2 - \eta^2 & \eta\,\alpha_o \\ \eta\,\alpha_o & \tau_o^2 \end{vmatrix}$$

Using (69) and (70) and taking into account the boundary conditions:

$$\mathbf{V}_{a-}(\eta) = -Z_a \mathbf{I}_{a-}(\eta), \quad \mathbf{V}_{b-}(\eta) = Z_b \mathbf{I}_{b-}(\eta) \tag{71}$$

where

$$V_{a-}(\eta) = \{V(\eta, 0_+)\}_-, \quad V_{b-}(\eta) = \{V(\eta, 0_-)\}_-$$
$$\mathbf{I}_{a-}(\eta) = \{\mathbf{I}(\eta, 0_+)\}_-, \quad \mathbf{I}_{b-}(\eta) = \{\mathbf{I}(\eta, 0_-)\}_-$$

we get the following W-H equations for the impenetrable half-plane problems:

$$\begin{aligned} \mathbf{V}_+(\eta) - \mathbf{Z}_c \mathbf{I}_+(\eta) &= (\mathbf{Z}_c + \mathbf{Z}_a)\mathbf{I}_{a-}(\eta) \\ \mathbf{V}_+(\eta) + \mathbf{Z}_c \mathbf{I}_+(\eta) &= -(\mathbf{Z}_c + \mathbf{Z}_b)\mathbf{I}_{b-}(\eta) \end{aligned} \tag{72}$$

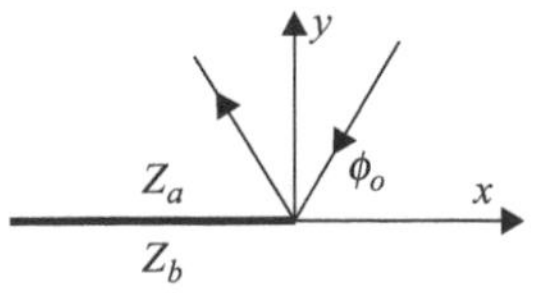

Fig. 8: Impenetrable half-plane

where

$$\mathbf{V}_+(\eta) = \{\mathbf{V}(\eta, 0_\pm)\}_+, \quad \mathbf{I}_+(\eta) = \{\mathbf{I}(\eta, 0_\pm)\}_+$$

Multiplying the second eq. (72) by $\mathbf{Y}_c$ yields the system

$$\begin{vmatrix} 1 & -Z_c \\ Y_c & 1 \end{vmatrix} \cdot \begin{vmatrix} V_+ \\ I_+ \end{vmatrix} = \begin{vmatrix} 1 + Z_c \cdot Y_a & 0 \\ 0 & 1 + Y_c \cdot Z_b \end{vmatrix} \cdot \begin{vmatrix} Z_a \cdot I_{a-} \\ -I_{b-} \end{vmatrix} \tag{73}$$

Taking into account that

$$\begin{vmatrix} 1 & -Z_c \\ Y_c & 1 \end{vmatrix}^{-1} = \frac{1}{2} \begin{vmatrix} 1 & Z_c \\ -Y_c & 1 \end{vmatrix}$$

yields the inverse of the kernel matrix

$$\frac{1}{2} \begin{vmatrix} 1 & Z_c \\ -Y_c & 1 \end{vmatrix} \cdot \begin{vmatrix} 1 + Z_c \cdot Y_a & 0 \\ 0 & 1 + Y_c \cdot Z_b \end{vmatrix} = \frac{1}{2} \begin{vmatrix} 1 + Z_c \cdot Y_a & (Z_c + Z_b) \\ -(Y_c + Y_a) & 1 + Y_c \cdot Z_b \end{vmatrix} = \frac{1}{2} + \frac{1}{2} \begin{vmatrix} Z_c \cdot Y_a & Z_c + Z_b \\ -(Y_c + Y_a) & Y_c \cdot Z_b \end{vmatrix}$$
$$= \frac{1}{2} \begin{vmatrix} 1 & Z_b \\ -Y_a & 1 \end{vmatrix} + \frac{1}{2} \begin{vmatrix} Z_c \cdot Y_a & Z_c \\ -Y_c & Y_c \cdot Z_b \end{vmatrix} = \frac{1}{2} \begin{vmatrix} 1 & Z_b \\ -Y_a & 1 \end{vmatrix} + \frac{1}{2} \begin{vmatrix} Z_c & 0 \\ 0 & Y_c \end{vmatrix} \begin{vmatrix} Y_a & 1 \\ -1 & Z_b \end{vmatrix}$$

This matrix reduces to the factorization of

$$1 + \begin{vmatrix} Z_c & 0 \\ 0 & Y_c \end{vmatrix} \begin{vmatrix} Y_a & 1 \\ -1 & Z_b \end{vmatrix} \begin{vmatrix} 1 & Z_b \\ -Y_a & 1 \end{vmatrix}^{-1}$$

We can evaluate the matrix $\begin{vmatrix} 1 & Z_b \\ -Y_a & 1 \end{vmatrix}^{-1}$ by taking into account that

$$\begin{vmatrix} 1 & Z_b \\ -Y_a & 1 \end{vmatrix} = \begin{vmatrix} 1 & 0 \\ -Y_a & 1 \end{vmatrix} \begin{vmatrix} 1 & 0 \\ 0 & 1 + Y_a Z_b \end{vmatrix} \begin{vmatrix} 1 & Z_b \\ 0 & 1 \end{vmatrix}$$

This yields

$$\begin{vmatrix} 1 & Z_b \\ -Y_a & 1 \end{vmatrix}^{-1} = \begin{vmatrix} 1 & -Z_b \\ 0 & 1 \end{vmatrix} \begin{vmatrix} 1 & 0 \\ 0 & (1 + Y_a Z_b)^{-1} \end{vmatrix} \begin{vmatrix} 1 & 0 \\ Y_a & 1 \end{vmatrix}$$

whence

$$\begin{vmatrix} Y_a & 1 \\ -1 & Z_b \end{vmatrix} \begin{vmatrix} 1 & Z_b \\ -Y_a & 1 \end{vmatrix}^{-1} = \begin{vmatrix} Y_a & 1 \\ -1 & Z_b \end{vmatrix} \begin{vmatrix} 1 & -Z_b \\ 0 & 1 \end{vmatrix} \begin{vmatrix} 1 & 0 \\ 0 & (1 + Y_a Z_b)^{-1} \end{vmatrix} \begin{vmatrix} 1 & 0 \\ Y_a & 1 \end{vmatrix}$$
$$= \begin{vmatrix} [1 + (1 - Y_a Z_b)(1 + Y_a Z_b)^{-1}] Y_a & (1 - Y_a Z_b)(1 + Y_a Z_b)^{-1} \\ -(1 + Z_a Y_b)^{-1} & 2(Y_a + Y_b)^{-1} \end{vmatrix}$$

We reduce the problem to the right factorization of the matrix

$$W = 1 + \begin{vmatrix} Z_c & 0 \\ 0 & Y_c \end{vmatrix} \begin{vmatrix} [1+(1-Y_aZ_b)(1+Y_aZ_b)^{-1}]Y_a & (1-Y_aZ_b)(1+Y_aZ_b)^{-1} \\ -(1+Z_aY_b)^{-1} & 2(Y_a+Y_b)^{-1} \end{vmatrix} = W_+ \cdot W_- \tag{74}$$

We can obtain different W-H equations by introducing the plus functions defined by

$$\begin{aligned} E_{a+}(\eta) &= V_a(\eta) + Z_a I_a(\eta) \\ E_{b+}(\eta) &= V_b(\eta) - Z_b I_b(\eta) \end{aligned} \tag{75}$$

Equations (69), (70), and (75) yield the following result:

$$V_a(\eta) = \mathbf{Z}_c(\eta)[\mathbf{Z}_c(\eta) + \mathbf{Z}_a]^{-1}\mathbf{E}_{a+}(\eta) \tag{76}$$

$$V_b(\eta) = \mathbf{Z}_c(\eta)[\mathbf{Z}_c(\eta) + \mathbf{Z}_b]^{-1}\mathbf{E}_{b+}(\eta) \tag{77}$$

$$I_a(\eta) = [\mathbf{Z}_c(\eta) + \mathbf{Z}_a]^{-1}\mathbf{E}_{a+}(\eta) \tag{78}$$

$$I_b(\eta) = -[\mathbf{Z}_c(\eta) + \mathbf{Z}_b]^{-1}\mathbf{E}_{b+}(\eta) \tag{79}$$

Subtracting (76) and (77) and successively (78) and (79) we get the W-H equations:

$$G(\eta)\begin{vmatrix} \mathbf{E}_{a+}(\eta) \\ \mathbf{E}_{b+}(\eta) \end{vmatrix} = \begin{vmatrix} X_-(\eta) \\ Y_-(\eta) \end{vmatrix} \tag{80}$$

where

$$G(\eta) = \begin{vmatrix} \mathbf{Z}_c(\eta)[\mathbf{Z}_c(\eta)+\mathbf{Z}_a]^{-1} & -\mathbf{Z}_c(\eta)[\mathbf{Z}_c(\eta)+\mathbf{Z}_b]^{-1} \\ [\mathbf{Z}_c(\eta)+\mathbf{Z}_a]^{-1} & [\mathbf{Z}_c(\eta)+\mathbf{Z}_b]^{-1} \end{vmatrix}$$

and the minus functions are defined by

$$X_-(\eta) = V_{a-}(\eta) - V_{b-}(\eta) \quad Y_-(\eta) = I_{a-}(\eta) - I_{b-}(\eta) \tag{81}$$

For the presence of the relationship

$$\begin{aligned} \mathbf{Z}_c(\eta)[\mathbf{Z}_c(\eta)+\mathbf{Z}_a]^{-1} &= (1+\mathbf{Z}_a\mathbf{Z}_c^{-1}(\eta))^{-1} \\ \mathbf{Z}_c(\eta)[\mathbf{Z}_c(\eta)+\mathbf{Z}_b]^{-1} &= (1+\mathbf{Z}_b\mathbf{Z}_c^{-1}(\eta))^{-1} \end{aligned}$$

the following equation holds also in presence of anisotropic $\mathbf{Z}_a$ and $\mathbf{Z}_b$:

$$\begin{aligned} G(\eta) &= \begin{vmatrix} \mathbf{Z}_c(\eta)[\mathbf{Z}_c(\eta)+\mathbf{Z}_a]^{-1} & -\mathbf{Z}_c(\eta)[\mathbf{Z}_c(\eta)+\mathbf{Z}_b]^{-1} \\ [\mathbf{Z}_c(\eta)+\mathbf{Z}_a]^{-1} & [\mathbf{Z}_c(\eta)+\mathbf{Z}_b]^{-1} \end{vmatrix} \\ &= \begin{vmatrix} \frac{1}{2}\left\{1+\mathbf{Z}_a[\mathbf{Z}_c(\eta)]^{-1}\right\} & \frac{1}{2}\{\mathbf{Z}_a+\mathbf{Z}_c(\eta)\} \\ -\frac{1}{2}\left\{1+\mathbf{Z}_b[\mathbf{Z}_c(\eta)]^{-1}\right\} & \frac{1}{2}\{\mathbf{Z}_b+\mathbf{Z}_c(\eta)\} \end{vmatrix}^{-1} \end{aligned} \tag{82}$$

where **1** is the identity matrix of order two. Consequently, the system

$$G(\eta)\cdot\begin{vmatrix} E_{a+}(\eta) \\ E_{b+}(\eta) \end{vmatrix} = \begin{vmatrix} X_-(\eta) \\ Y_-(\eta) \end{vmatrix}$$

yields

$$G_i(\eta)\cdot\begin{vmatrix} X_-(\eta) \\ Y_-(\eta) \end{vmatrix} = \begin{vmatrix} E_{a+}(\eta) \\ E_{b+}(\eta) \end{vmatrix}$$

where

$$G_i(\eta) = G^{-1}(\eta) = \begin{vmatrix} \frac{1}{2}\left\{\mathbf{1} + \mathbf{Z}_a[\mathbf{Z}_c(\eta)]^{-1}\right\} & \frac{1}{2}\{\mathbf{Z}_a + \mathbf{Z}_c(\eta)\} \\ -\frac{1}{2}\left\{\mathbf{1} + \mathbf{Z}_b[\mathbf{Z}_c(\eta)]^{-1}\right\} & \frac{1}{2}\{\mathbf{Z}_b + \mathbf{Z}_c(\eta)\} \end{vmatrix} \tag{83}$$

7.6.2 Presence of isotropic impedances Z_a and Z_b

The diffraction at skew incidence ($\alpha_o \neq 0$) by a half-plane having isotropic surface impedance has been solved by Bucci and Franceschetti (1976), who used the Malyuzhinets-Sommerfeld method, and by Senior (1978) and Luneburg and Serbest (2000), who used the Wiener-Hopf technique. The solution accomplished in Bucci and Franceschetti (1976) appears doubtful. Perplexities arise from the observation that solution of difference equations are not unique, so we must be very careful to impose on the possible solutions of the difference equation all the physical properties of the true solution. On the other hand, the Wiener-Hopf solution is related to a closed mathematical problems that always provides a unique solution. Concerning this problem, the W-H solution described in Luneburg and Serbest (2000) is very interesting. In this work, an important matrix transform is introduced that reduces the order of the W-H system by four a two. This transform is defined by the polynomial matrix $t(\eta) = \begin{vmatrix} \eta & -\alpha_o \\ \alpha_o & \eta \end{vmatrix}$.

For the presence of isotropic impedances the following simplifications arise in the kernel $G_i(\eta)$.

First let us observe that $G_i(\eta)$ depends only on the matrix

$$\mathbf{Z}_c = \frac{1}{\omega\varepsilon\xi}\begin{vmatrix} k^2-\eta^2 & \eta\,\alpha_o \\ \eta\,\alpha_o & \tau_o^2 \end{vmatrix} = \frac{1}{\xi}P_{z2}(\eta) \quad \text{or}$$

$$\mathbf{Y}_c = \mathbf{Z}_o^{-1} = \frac{1}{\omega\mu\xi}\begin{vmatrix} \tau_o^2 & -\eta\,\alpha_o \\ -\eta\,\alpha_o & k^2-\eta^2 \end{vmatrix} = \frac{1}{\xi}P_{y2}(\eta)$$

The key point to reduce the order of matrices to be factorized is the observation that the polynomial matrix $t(\eta) = \begin{vmatrix} \eta & -\alpha_o \\ \alpha_o & \eta \end{vmatrix}$ diagonalizes $\mathbf{Z}_c$ and $\mathbf{Y}_c$. In fact

$$\mathbf{Z}_c = t^{-1}(\eta)\cdot\begin{vmatrix} \frac{\xi}{\omega\varepsilon} & 0 \\ 0 & \frac{k^2}{\omega\varepsilon\xi} \end{vmatrix}\cdot t(\eta), \quad \mathbf{Y}_c = t^{-1}(\eta)\cdot\begin{vmatrix} \frac{\omega\varepsilon}{\xi} & 0 \\ 0 & \frac{\xi}{\omega\mu} \end{vmatrix}\cdot t(\eta)$$

or

$$\begin{vmatrix} k^2-\eta^2 & \eta\,\alpha_o \\ \eta\,\alpha_o & \tau_o^2 \end{vmatrix} = t^{-1}(\eta)\cdot d(\eta)\cdot t(\eta)$$

where

$$d(\eta) = \begin{vmatrix} \xi^2 & 0 \\ 0 & k^2 \end{vmatrix}$$

It follows that putting $T(\eta) = \begin{vmatrix} t(\eta) & 0 \\ 0 & t(\eta) \end{vmatrix}$ yields the matrix $T(\eta)\cdot G_i(\eta)\cdot T^{-1}(\eta)$ that has the form

$$T(\eta)\cdot G_i(\eta)\cdot T^{-1}(\eta) = \begin{vmatrix} x & 0 & x & 0 \\ 0 & x & 0 & x \\ x & 0 & x & 0 \\ 0 & x & 0 & x \end{vmatrix}$$

where x means a nonzero element. By introducing the matrix P defined by

$$P = \begin{vmatrix} 1 & 0 & 0 & 0 \\ 0 & 0 & 1 & 0 \\ 0 & 1 & 0 & 0 \\ 0 & 0 & 0 & 1 \end{vmatrix}$$

we can derive the following quasi diagonal matrix:

$$G_t(\eta) = P\cdot T(\eta)\cdot G_i(\eta)\cdot T^{-1}(\eta)\cdot P = \begin{vmatrix} G_{t1}(\eta) & 0 \\ 0 & G_{t2}(\eta) \end{vmatrix}$$

where the matrices of order two $G_{t1}(\eta)$ and $G_{t2}(\eta)$ are defined by

$$G_{t1}(\eta) = \frac{1}{2}\begin{vmatrix} \left(1+\dfrac{Z_a\omega\varepsilon}{\xi}\right) & \dfrac{Z_a+\frac{\xi}{\omega\varepsilon}}{Z_o} \\ -\left(1+\dfrac{Z_b\omega\varepsilon}{\xi}\right) & \dfrac{Z_b+\frac{\xi}{\omega\varepsilon}}{Z_o} \end{vmatrix}, \quad G_{t1}(\eta) = \frac{1}{2}\begin{vmatrix} \left(1+\dfrac{Z_a\omega\varepsilon\xi}{k^2}\right) & \dfrac{Z_a+\frac{k^2}{\omega\varepsilon\xi}}{Z_o} \\ -\left(1+\dfrac{Z_b\omega\varepsilon\xi}{k^2}\right) & \dfrac{Z_b+\frac{k^2}{\omega\varepsilon\xi}}{Z_o} \end{vmatrix}$$

In the following, the property $P^2 = 1$ will prove useful, where 1 is the unity matrix of order four. We rewrite the previous equation in the form

$$G_t(\eta) = P\cdot T(\eta)\cdot G_i(\eta)\cdot T^{-1}(\eta)\cdot P = \begin{vmatrix} G_{t1}(\eta) & 0 \\ 0 & G_{t2}(\eta) \end{vmatrix}$$

$$G_i(\eta) = T^{-1}(\eta)\cdot P\cdot G_t(\eta)\cdot P\cdot T(\eta).$$

This equation reduces the factorization of the matrix of order four $G_i(\eta)$ to the factorization of the two matrices of order two $G_{t1}(\eta)$ and $G_{t2}(\eta)$. These matrices are of the Daniele-Khrapkov type and can be factorized according to the equations in chapter 4, section 4.8.4.

To complete the formulation we need the primary field that we assume due to be a plane wave with direction $\varphi = \varphi_o$ (Fig. 8). Taking into account the considerations of section 7.5, we assume as primary field the geometrical optical field (superscript g). To this end in the regions where the reflected plane wave is present we have (the factor $e^{-j\alpha_o z}$ is omitted)

$$E_z^g(\rho,\varphi) = e^{j\tau_o \cos(\varphi-\varphi_o)} E_o + e^{j\tau_o \cos(\varphi-\varphi_r)} E_r$$

$$H_z^g(\rho,\varphi) = e^{j\tau_o \cos(\varphi-\varphi_o)} H_o + e^{j\tau_o \cos(\varphi-\varphi_r)} H_r$$

where E_r and H_r are the intensity of the reflected fields, and φ_r is the reflection angle. Snell's law yields $\varphi_r = -\varphi_o$.

To obtain E_r and H_r we introduce the radial components:

$$E_\rho^g(\rho,\varphi) = -j\frac{k}{\tau_o^2}\left[\frac{\alpha_o}{k}\frac{\partial}{\partial\rho}E_z^g(\rho,\varphi) + \frac{Z_o}{\rho}\frac{\partial}{\partial\varphi}H_z^g(\rho,\varphi)\right]$$

$$H_\rho^g(\rho,\varphi) = -j\frac{k}{\tau_o^2}\left[\frac{\alpha_o}{k}\frac{\partial}{\partial\rho}H_z^g(\rho,\varphi) - \frac{Y_o}{\rho}\frac{\partial}{\partial\varphi}E_z^g(\rho,\varphi)\right]$$

where $Z_o = \sqrt{\frac{\mu}{\varepsilon}}$, $Y_o = \sqrt{\frac{\varepsilon}{\mu}}$, and force the boundary conditions on $\varphi = \Phi = \pi$:

$$E_z^g(\rho,\varphi) = Z_a H_\rho^g(\rho,\varphi), \quad H_z^g(\rho,\varphi) = -\frac{1}{Z_a}E_\rho^g(\rho,\varphi)$$

This provides the values of E_r and H_r, that are not explicitly reported here for the sake of simplicity. According to these values and using the Laplace transforms, we obtain

$$X_-^g(\eta) = \begin{vmatrix} V_{\rho+}^g(-\eta,\Phi) - V_{\rho+}^g(-\eta,-\Phi) \\ V_{z+}^g(-\eta,\Phi) - V_{z+}^g(-\eta,-\Phi) \end{vmatrix} = \begin{vmatrix} V_{\rho+}^g(-\eta,\Phi) \\ V_{z+}^g(-\eta,\Phi) \end{vmatrix} = \frac{1}{\eta - \tau_o\cos(\Phi-\varphi_o)}\begin{vmatrix} E_{\rho g} \\ E_{zg} \end{vmatrix}$$

$$Y_-^g(\eta) = \begin{vmatrix} I_{z+}^g(-\eta,\Phi) - I_{z+}^g(-\eta,-\Phi) \\ -I_{\rho+}^g(-\eta,\Phi) + V_{\rho+}^g(-\eta,-\Phi) \end{vmatrix} = \frac{1}{Z_a}\frac{1}{\eta - \tau_o\cos(\Phi-\varphi_o)}\begin{vmatrix} -E_{\rho g} \\ E_{zg} \end{vmatrix}$$

where the constants $E_{\rho g}$ and E_{zg} are given by

$$E_{zg} = \frac{-H_o Z_o^2 \alpha_o \sin 2\varphi_o + kZ_a 2E_o \sin\varphi_o (Z_a\tau_o + kZ_o\sin\varphi_o)}{j[Z_a Z_o \alpha_o^2 \cos^2\varphi_o + (Z_o\tau_o + kZ_a\sin\varphi_o)(Z_a\tau_o + kZ_o\sin\varphi_o)]}$$

$$E_{\rho g} = -\frac{kZ_a 2\sin\varphi_o[E_o Z_a \alpha_o \cos\varphi_o + H_o Z_a(Z_o\tau_o + kZ_a\sin\varphi_o)]}{j[Z_a Z_o \alpha_o^2 \cos^2\varphi_o + (Z_o\tau_o + kZ_a\sin\varphi_o)(Z_a\tau_o + kZ_o\sin\varphi_o)]}$$

The solution of eq. (80) is reported in chapter 2, section 2.4.2:

$$\begin{vmatrix} E_{a+}(\eta) \\ E_{b+}(\eta) \end{vmatrix} = G_+^{-1}(\eta)\cdot G_-^{-1}(\eta_o)\frac{1}{\eta-\eta_o}\begin{vmatrix} X_g \\ Z_o Y_g \end{vmatrix}$$

where

$$\eta_o = -\tau_o \cos\varphi_o, \quad X_g = \begin{vmatrix} E_{\rho g} \\ E_{zg} \end{vmatrix}, \quad Y_g = \begin{vmatrix} -E_{\rho g} \\ E_{zg} \end{vmatrix}$$

7.7 The general problem of factorization

In the following we will show that the kernel matrix involved in arbitrary half-planes immersed in an isotropic medium commutes with a polynomial matrix. Whence, we can use the factorization technique described in chapter 4. Algebraic manipulations provide different expressions of the matrix kernel. We will use the two forms described in this section.

- First form of the matrix kernel $G_o(\eta)$

Premultiplying by $\begin{vmatrix} \mathbf{1} & -\mathbf{1} \\ \mathbf{1} & \mathbf{1} \end{vmatrix}$ reduces the factorization of $G_i(\eta)$ in (83) to that of

$$G_o(\eta) = \begin{vmatrix} \mathbf{1} + \frac{1}{2}(\mathbf{Z}_a + \mathbf{Z}_b)[\mathbf{Z}_c(\eta)]^{-1} & \frac{1}{2}\{\mathbf{Z}_a - \mathbf{Z}_b\} \\ \frac{1}{2}\{\mathbf{Z}_a - \mathbf{Z}_b\}[\mathbf{Z}_c(\eta)]^{-1} & \frac{1}{2}(\mathbf{Z}_a + \mathbf{Z}_b) + \mathbf{Z}_c(\eta) \end{vmatrix}$$

$$= \begin{vmatrix} \mathbf{1} & \frac{1}{2}\{\mathbf{Z}_a - \mathbf{Z}_b\} \\ 0 & \frac{1}{2}(\mathbf{Z}_a + \mathbf{Z}_b) \end{vmatrix} + \begin{vmatrix} \frac{1}{2}(\mathbf{Z}_a + \mathbf{Z}_b)[\mathbf{Z}_c(\eta)]^{-1} & 0 \\ \frac{1}{2}\{\mathbf{Z}_a - \mathbf{Z}_b\}[\mathbf{Z}_c(\eta)]^{-1} & \mathbf{Z}_c(\eta) \end{vmatrix}$$

Premultiplying by $\begin{vmatrix} \mathbf{1} & \frac{1}{2}\{\mathbf{Z}_a - \mathbf{Z}_b\} \\ 0 & \frac{1}{2}(\mathbf{Z}_a + \mathbf{Z}_b) \end{vmatrix}^{-1} = \begin{vmatrix} \mathbf{1} & -\{\mathbf{Z}_a - \mathbf{Z}_b\}(\mathbf{Z}_a + \mathbf{Z}_b)^{-1} \\ 0 & 2(\mathbf{Z}_a + \mathbf{Z}_b)^{-1} \end{vmatrix}$ yields the matrix

$$W(\eta) = \mathbf{1} + \frac{1}{\xi}P(\eta) \tag{84}$$

where $P(\eta)$ is a polynomial matrix of order four and degree two:

$$P(\eta) = \begin{vmatrix} \frac{1}{2}[(\mathbf{Z}_a + \mathbf{Z}_b) - (\mathbf{Z}_a - \mathbf{Z}_b)(\mathbf{Z}_a + \mathbf{Z}_b)^{-1}(\mathbf{Z}_a - \mathbf{Z}_b)]P_{y2}(\eta) & -(\mathbf{Z}_a - \mathbf{Z}_b)(\mathbf{Z}_a + \mathbf{Z}_b)^{-1}P_{z2}(\eta) \\ P_{y2}(\eta) & 2(\mathbf{Z}_a + \mathbf{Z}_b)^{-1}P_{z2}(\eta) \end{vmatrix}$$

with the polynomial matrices of order two and degree two given by

$$P_{z2}(\eta) = \frac{1}{\omega\varepsilon}\begin{vmatrix} k^2 - \eta^2 & \eta\,\alpha_o \\ \eta\,\alpha_o & \tau_o^2 \end{vmatrix}$$

$$P_{y2}(\eta) = \frac{1}{\omega\mu}\begin{vmatrix} \tau_o^2 & -\eta\,\alpha_o \\ -\eta\,\alpha_o & k^2 - \eta^2 \end{vmatrix}$$

The factorization of the matrix $W(\eta)$ commuting with the polynomial matrix $P(\eta)$ is very cumbersome and involves the solution of a nonlinear system of five unknowns (section 4.9.6). In the following, we will try to reduce the complexity of this problem by assuming particular values of the surface impedances.

- Second form of the matrix kernel $G_o(\eta)$.

We have

$$G_o(\eta) = \begin{vmatrix} \mathbf{1} + \frac{1}{2}(\mathbf{Z}_a + \mathbf{Z}_b)[\mathbf{Z}_c(\eta)]^{-1} & \frac{1}{2}\{\mathbf{Z}_a - \mathbf{Z}_b\} \\ \frac{1}{2}\{\mathbf{Z}_a - \mathbf{Z}_b\}[\mathbf{Z}_c(\eta)]^{-1} & \frac{1}{2}(\mathbf{Z}_a + \mathbf{Z}_b) + \mathbf{Z}_c(\eta) \end{vmatrix}$$
$$= \begin{vmatrix} \mathbf{1} & \frac{1}{2}\{\mathbf{Z}_a - \mathbf{Z}_b\} \\ 0 & \frac{1}{2}(\mathbf{Z}_a + \mathbf{Z}_b) \end{vmatrix} + \begin{vmatrix} \frac{1}{2}(\mathbf{Z}_a + \mathbf{Z}_b)[\mathbf{Z}_c(\eta)]^{-1} & 0 \\ \frac{1}{2}\{\mathbf{Z}_a - \mathbf{Z}_b\}[\mathbf{Z}_c(\eta)]^{-1} & \mathbf{Z}_c(\eta) \end{vmatrix}$$

Algebraic manipulation yields

$$G_o(\eta) = \begin{vmatrix} \frac{1}{2}(\mathbf{Z}_a + \mathbf{Z}_b + 2\mathbf{Z}_c(\eta)) & \frac{1}{2}\{\mathbf{Z}_a - \mathbf{Z}_b\} \\ \frac{1}{2}\{\mathbf{Z}_a - \mathbf{Z}_b\} & \frac{1}{2}(\mathbf{Z}_a + \mathbf{Z}_b + 2\mathbf{Z}_c(\eta)) \end{vmatrix} \begin{vmatrix} \frac{1}{\xi_+} & 0 \\ 0 & 1 \end{vmatrix} \begin{vmatrix} \frac{1}{\xi_-} & 0 \\ 0 & 1 \end{vmatrix} \begin{vmatrix} P_{y2}(\eta) & 0 \\ 0 & 1 \end{vmatrix}$$

$$G_o(\eta) = \begin{vmatrix} \frac{1}{2\xi_+}(\mathbf{Z}_a + \mathbf{Z}_b + 2\mathbf{Z}_c(\eta)) & \frac{1}{2}\{\mathbf{Z}_a - \mathbf{Z}_b\} \\ \frac{1}{2\xi_+}\{\mathbf{Z}_a - \mathbf{Z}_b\} & \frac{1}{2}(\mathbf{Z}_a + \mathbf{Z}_b + 2\mathbf{Z}_c(\eta)) \end{vmatrix} \begin{vmatrix} \frac{1}{\xi_-} & 0 \\ 0 & 1 \end{vmatrix} \begin{vmatrix} P_{y2}(\eta) & 0 \\ 0 & 1 \end{vmatrix}$$

$$G_o(\eta) = \begin{vmatrix} \frac{1}{\xi_+} & 0 \\ 0 & 1 \end{vmatrix} \begin{vmatrix} \frac{1}{2}(\mathbf{Z}_a + \mathbf{Z}_b + 2\mathbf{Z}_c(\eta)) & \frac{1}{2}\xi_+\{\mathbf{Z}_a - \mathbf{Z}_b\} \\ \frac{1}{2\xi_+}\{\mathbf{Z}_a - \mathbf{Z}_b\} & \frac{1}{2}(\mathbf{Z}_a + \mathbf{Z}_b + 2\mathbf{Z}_c(\eta)) \end{vmatrix} \begin{vmatrix} \frac{1}{\xi_-} & 0 \\ 0 & 1 \end{vmatrix} \begin{vmatrix} P_{y2}(\eta) & 0 \\ 0 & 1 \end{vmatrix}$$

where $P_{y2}(\eta)$ is the matrix polynomial defined by $P_{y2}(\eta) = \xi\, \mathbf{Y}_c$.

Since $\begin{vmatrix} \frac{1}{\xi_-} & 0 \\ 0 & 1 \end{vmatrix} \begin{vmatrix} P_{y2}(\eta) & 0 \\ 0 & 1 \end{vmatrix}$ is a (weak) minus factorized matrix and $\begin{vmatrix} \frac{1}{\xi_+} & 0 \\ 0 & 1 \end{vmatrix}$ is a plus factorized matrix, the factorization of $G_o = G_{o+}G_{o-}$ reduce to the factorization of the central matrix:

$$G_e(\eta) = \frac{1}{2} \begin{vmatrix} (\mathbf{Z}_a + \mathbf{Z}_b + 2\mathbf{Z}_c(\eta)) & \xi_+\{\mathbf{Z}_a - \mathbf{Z}_b\} \\ \frac{1}{\xi_+}\{\mathbf{Z}_a - \mathbf{Z}_b\} & (\mathbf{Z}_a + \mathbf{Z}_b + 2\mathbf{Z}_c(\eta)) \end{vmatrix} = G_{e+}(\eta)G_{e-}(\eta) \qquad (85)$$

or, alternatively,

$$\begin{vmatrix} \frac{1}{2}(\mathbf{Z}_a + \mathbf{Z}_b + 2\mathbf{Z}_c(\eta)) & \frac{1}{2}\xi_+\{\mathbf{Z}_a - \mathbf{Z}_b\} \\ \frac{1}{2\xi_+}\{\mathbf{Z}_a - \mathbf{Z}_b\} & \frac{1}{2}(\mathbf{Z}_a + \mathbf{Z}_b + 2\mathbf{Z}_c(\eta)) \end{vmatrix}$$
$$= \begin{vmatrix} 0 & \xi_+ \\ \frac{1}{\xi_+} & 0 \end{vmatrix} \begin{vmatrix} \frac{1}{2}\{\mathbf{Z}_a - \mathbf{Z}_b\} & \frac{1}{2}\xi_+(\mathbf{Z}_a + \mathbf{Z}_b + 2\mathbf{Z}_c(\eta)) \\ \frac{1}{2\xi_+}(\mathbf{Z}_a + \mathbf{Z}_b + 2\mathbf{Z}_c(\eta)) & \frac{1}{2}\{\mathbf{Z}_a - \mathbf{Z}_b\} \end{vmatrix}$$

7.7.1 The case of symmetric half-plane

We found that the form (85) is the more suitable to simplify the factorization in half-plane problems. For instance, if $\mathbf{Z}_a$ and $\mathbf{Z}_b$ are isotropic, then the solution process is simpler than the one considered in section 7.6.2.

In this section, we consider the explicit factorization for arbitrary symmetric half-plane $\mathbf{Z}_a = \mathbf{Z}_b$. Looking at (85), the factorization of the matrix of order four is reduced to the factorization of the matrix $\mathbf{Z}_a + \mathbf{Z}_c(\boldsymbol{\eta})$ of order two. This problem was faced by many authors and solved for the first time in Hurd and Luneburg (1985) when $Z_a = Z_b$ are diagonal. When $\mathbf{Z}_b = \mathbf{Z}_a = Z_o \begin{vmatrix} z_{11} & z_{12} \\ z_{21} & z_{22} \end{vmatrix} = Z_o \mathbf{z}_a$ are arbitrary, the factorization of $\mathbf{Z}_a + \mathbf{Z}_c(\boldsymbol{\eta})$ can be accomplished by observing that $1 + \mathbf{Z}_a^{-1}\mathbf{Z}_c(\boldsymbol{\eta})$ commutes with the polynomial matrix $P_2(\eta) = \mathbf{Z}_a^{-1}(\xi\, \mathbf{Z}_c(\boldsymbol{\eta}))$. The general procedure indicated in section 4.9.2 applies, and since the polynomial matrix is of order two the exponential behavior of the factorized matrices can be eliminated using the technique indicated in sections 4.8.5 and 4.9.2. In this case, we introduce the rational matrix $R(\eta) = 1 + xP_2(\eta)$, and to eliminate an offending exponential behavior we force x to satisfy the following nonlinear equation (section 4.9.2):

$$f(x) = \int_{-\infty}^{\infty} \left(\frac{\log[1 + x\lambda_1]}{\lambda_1 - \lambda_2} + \frac{\log[1 + x\,\lambda_2]}{\lambda_2 - \lambda_1} \right) d\eta$$

$$= \int_{-\infty}^{\infty} \left(\frac{\log\left[1 + \dfrac{1}{\xi(\eta)}\lambda_1\right]}{\lambda_1 - \lambda_2} + \frac{\log\left[1 + \dfrac{1}{\xi(\eta)}\lambda_2\right]}{\lambda_2 - \lambda_1} \right) d\eta = n$$

where λ_1 and λ_2 are the two eigenvalues of $P_2(\eta)$, and the integral n is known.

We have

$$\frac{df(x)}{dx} = \int_{-\infty}^{\infty} \left(\frac{1}{1 + x(\lambda_1 + \lambda_2) + x^2\lambda_2\lambda_1} \right) d\eta$$

The integrand function is rational in η, and the residue theorem yields

$$\frac{df(x)}{dx} = 2\pi j \frac{k^2}{\sqrt{x(ax^3 + bx^2 + cx + d)}}$$

where the parameters a, b, c, d were evaluated with MATHEMATICA:

$$a = 4 \det^2[\mathbf{z}_a]\tau_o^2$$
$$b = 4 \det[\mathbf{z}_a][k^2 z_{11} + (z_{11} + z_{22})\tau_o^2]$$
$$c = 4k^2(z_{11}^2 + \det[\mathbf{z}_a]) + (z_{12} + z_{21})^2\alpha_o^2 + 4z_{11}z_{22}\tau_o^2$$
$$d = 4k^2 z_{11}$$

Taking into account that $f(0) = 0$, $f(x)$ can be expressed in terms of elliptic integrals. In fact, putting

$$ax^3 + bx^2 + cx + d = a(x - x_1)(x - x_2)(x - x_3)$$

we get

$$f(x) = EllipticF\left[\arcsin[\sqrt{\frac{x(x_3 - x_1)}{(x - x_1)x_3}}, \frac{x_1(x_1 - x_2)}{(x_1 - x_3)x_2}\right]\frac{1}{\sqrt{x_2 x_1(x_1 - x_3)}} = n$$

By inverting this equation, we express x as a Jacobian elliptic function:

$$x = \frac{x_1 x_3\, JacobiSN\left[\frac{1}{2}(n x_1^2\sqrt{x_2} - n x_1 x_3\sqrt{x_2}, \frac{x_3(x_1 - x_2)}{(x_1 - x_3)x_2}\right]^2}{x_1 - x_3 + x_3\, JacobiSN\left[\frac{1}{2}(n x_1^2\sqrt{x_2} - n x_1 x_3\sqrt{x_2}, \frac{x_3(x_1 - x_2)}{(x_1 - x_3)x_2}\right]^2}$$

Factorization equations follow from

$$W = 1 + \frac{1}{\xi}\mathbf{Z}_a(\xi\,\mathbf{Z}_c(\boldsymbol{\eta})) = \tilde{W}_-\tilde{R}_- R^{-1}\tilde{R}_+\tilde{W}_+ = \tilde{W}_-\tilde{R}_-[R^{-1}]_-[R^{-1}]_+\tilde{R}_+\tilde{W}_+$$

where the logarithmic (nonstandard) factorization of $W = \tilde{W}_- \cdot \tilde{W}_+$ and $R = \tilde{R}_-\tilde{R}_+$ can be obtained with the general method to factorize matrices commuting with a polynomial matrix (section 4.9) and the standard factorization of $R^{-1}(\eta) = (1 + xP_2(\eta))^{-1}$ can be obtained by the method of factorization of rational matrices (see, e.g., section 4.4.2).

It is interesting to observe that in this case the Sommerfeld-Malyuzhintes (S-M) method yields a difference equation of order two for which there is no known method of solution. However, the progress reported recently (Antipov & Silvestrov, 2004) has overcome the difficulties encountered with the S-M approach.

7.7.2 The case of opposite diagonal impedances $\mathbf{Z}_b = -\mathbf{Z}_a$

In this case the matrix kernel reduces to

$$M = \begin{vmatrix} \mathbf{Z}_a & \xi_+\mathbf{Z}_c(\eta) \\ \frac{1}{\xi_+}\mathbf{Z}_c(\eta) & \mathbf{Z}_a \end{vmatrix} = \begin{vmatrix} \mathbf{Z}_a & \frac{1}{\xi_-}\mathbf{P}(\eta) \\ \frac{1}{\xi_+^2\xi_-}\mathbf{P}(\eta) & \mathbf{Z}_a \end{vmatrix} = M_+M_-$$

where the matrix polynomial $\mathbf{P}(\eta) = \xi\,\mathbf{Z}_c(\eta)$ is of order two. This matrix is a weak minus matrix where the offending pole arises from $\frac{1}{\xi_+^2} = \frac{1}{\tau_o - \eta}$. Consequently, the weak factorization is trivial since we can assume $M_{w+} = 1$, $M_{w-} = M$.

7.8 The jump or penetrable half-plane problem

Many problems that concern penetrable half-planes were solved by Senior (1959), Volakis and Senior (1989), and Senior and Volakis (1995). In this case the half-plane presents boundary conditions having the form

$$V_{a-}(\eta) = V_{b-}(\eta) = Z_R[I_{a-}(\eta) - I_{b-}(\eta)]$$

where Z_R is the matrix impedance of the penetrable half-plane. Taking into account that

$$V_{a+}(\eta) = V_{b+}(\eta) = V_+(\eta), \quad I_{a+}(\eta) = I_{b+}(\eta) = I_+(\eta)$$

the W-H equations of the problem can be derived as follows:

$$V_{a-}(\eta) + V_{a+}(\eta) = Z_R[I_{a-}(\eta) - I_{b-}(\eta)] + V_+(\eta) = Z_C I_{a-}(\eta) + Z_C I_+(\eta)$$
$$V_{b-}(\eta) + V_{b+}(\eta) = Z_R[I_{a-}(\eta) - I_{b-}(\eta)] + V_+(\eta) = -Z_C I_{b-}(\eta) - Z_C I_+(\eta)$$

Summing and subtracting the last two equations yields

$$2Z_R[I_{a-}(\eta) - I_{b-}(\eta)] + 2V_+(\eta) = Z_C[I_{a-}(\eta) - I_{b-}(\eta)]$$
$$0 = Z_C[I_{a-}(\eta) + I_{b-}(\eta)] + 2Z_C I_+(\eta)$$

or

$$2V_+(\eta) = (Z_C - 2Z_R)[I_{a-}(\eta) - I_{b-}(\eta)] \tag{86}$$
$$2I_+(\eta) = -[I_{a-}(\eta) + I_{b-}(\eta)] \tag{87}$$

Equation (87) may be solved immediately and yields the incident contribution

$$I_+(\eta) = \frac{I_o^i}{\eta - \eta_o}, \quad I_{a-}(\eta) + I_{b-}(\eta) = -2\frac{I_o^i}{\eta - \eta_o}$$

In the general case, eq. (86) involves the factorization of the kernel $Z_C - 2Z_R$. This kernel has been studied in the case of symmetric half-planes. The factorization is scalar when Z_R is a scalar. In fact, the matrix $t(\eta) = \begin{vmatrix} \eta & -\alpha_o \\ \alpha_o & \eta \end{vmatrix}$ renders diagonal the matrix $t \cdot (Z_C - 2Z_R) \cdot t^{-1}$.

7.9 Full-plane junction at skew incidence

With the same notations of the previous section, we have (Fig. 9)

$$V(\eta, 0_+) = V_+(\eta) + V_-(\eta) = Z_C I(\eta, 0_+) = Z_C(I_+(\eta) + I_-(\eta))$$
$$V_+(\eta) = Z_r I_+(\eta), \quad V_-(\eta) = -Z_a I_-(\eta)$$

This yields the W-H equation

$$(Z_r - Z_C)I_+(\eta) = (Z_a + Z_C)I_-(\eta)$$

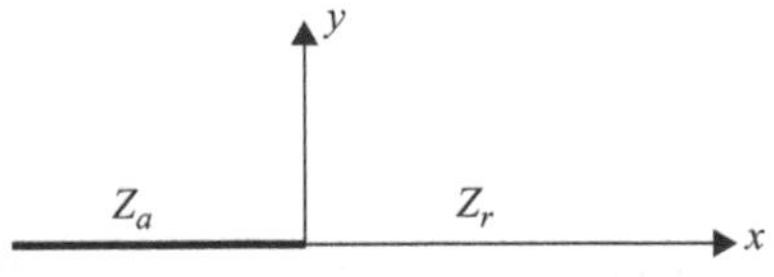

Fig. 9: Junction between two coplanar half-planes

or

$$(1 - Z_C Z_r^{-1}) Z_r I_+(\eta) = (1 - Z_C Z_a^{-1}) Z_a I_-(\eta)$$

If the impedances Z_r and Z_a are isotropic, the matrix $t(\eta) = \begin{vmatrix} \eta & -\alpha_o \\ \alpha_o & \eta \end{vmatrix}$ renders diagonal the matrix $t \cdot (1 - Z_C Z_a^{-1})^{-1}(1 - Z_C Z_r^{-1}) \cdot t^{-1}$, and the factorization is scalar. We deal with a matrix factorization similar to that of the symmetric half-plane Z_r if and Z_a are related by $Z_a = aZ_r$ with a scalar. This case also includes the perfectly conducting half-plane.

The problem considered in this section is important for the study of propagation of radio waves across a coastline for both a flat-earth model (Clemmow, 1953; Bazer & Karp, 1962; Weinstein, 1969, p. 317) and a curved-earth model (Thompson, 1962; Chang, 1969; Wait, 1970).

7.10 Diffraction by an half plane immersed in arbitrary linear medium

7.10.1 Transverse equation in an indefinite medium

We consider only time harmonic electromagnetic fields with a time dependence specified by the factor $e^{j\omega t}$, which is omitted. In every layer, the medium presents constitutive relations having the form

$$\mathbf{D} = \boldsymbol{\varepsilon} \cdot \mathbf{E} + \boldsymbol{\xi} \cdot \mathbf{H}$$
$$\mathbf{B} = \boldsymbol{\zeta} \cdot \mathbf{E} + \boldsymbol{\mu} \cdot \mathbf{H}$$

where **D, B** define the induction fields, **E, H** the electromagnetic fields, and the electromagnetic dyadics $\boldsymbol{\varepsilon}$, $\boldsymbol{\xi}$, $\boldsymbol{\zeta}$, $\boldsymbol{\mu}$ are known. In particular for lossless media, we have the following lossless conditions (Kong, 1975):

$$\boldsymbol{\varepsilon} = \boldsymbol{\varepsilon}^+, \quad \boldsymbol{\mu} = \boldsymbol{\mu}^+, \quad \boldsymbol{\zeta} = \boldsymbol{\xi}^+$$

where the superscript $+$ denotes a transpose and complex conjugate operation.

Without loss of generality we assume the z dependence of the electromagnetic fields **E** and **H** to be specified by the factor $e^{-j\alpha_o z}$, which is omitted.

By using the Bresler-Marcuvitz formalism (Bresler & Marcuvitz, 1956; Daniele, 1971) we can obtain the transverse field equations having as unknowns the transverse fields $\mathbf{E}_t = \hat{z}E_z + \hat{x}E_x$ and $\mathbf{H}_t = \hat{z}H_z + \hat{x}H_x$ where $\hat{z}, \hat{x}$ and $\hat{y}$ are the unit vectors of the Cartesian system (z, x, y). To this end, Maxwell's equations can be rewritten in an abstract form:

$$\Gamma_\nabla \cdot \psi = \theta \tag{88}$$

where

$$\Gamma_\nabla = \begin{vmatrix} 0 & \nabla x1 \\ \nabla x1 & 0 \end{vmatrix}, \quad \psi = \begin{vmatrix} \mathbf{E} \\ \mathbf{H} \end{vmatrix}, \quad \theta = j\omega \begin{vmatrix} \mathbf{D} \\ -\mathbf{B} \end{vmatrix} = W \cdot \psi \tag{89}$$

and where **1** is the unity dyadic in the Euclidean space and

$$W = j\omega \begin{vmatrix} \boldsymbol{\varepsilon} & \boldsymbol{\xi} \\ -\boldsymbol{\zeta} & -\boldsymbol{\mu} \end{vmatrix}$$

The decomposition

$$\nabla = \nabla_t + \hat{y}\frac{\partial}{\partial y}, \quad \nabla_t = \hat{z}\frac{\partial}{\partial z} + \hat{x}\frac{\partial}{\partial x} \tag{90}$$

yields

$$\Gamma_\nabla = \Gamma_t + \Gamma_y \frac{\partial}{\partial y} \tag{91}$$

where

$$\Gamma_t = \begin{vmatrix} 0 & \nabla_t x 1 \\ \nabla_t x 1 & 0 \end{vmatrix}, \quad \Gamma_y = \begin{vmatrix} 0 & \hat{y}x1 \\ \hat{y}x1 & 0 \end{vmatrix}$$

The following equations hold:

$$I_t \cdot \Gamma_t = \Gamma_t \cdot I_y, \quad I_t \cdot \Gamma_y = \Gamma_y \cdot I_t = \Gamma_y, \quad I_y \cdot \Gamma_t = \Gamma_t \cdot I_t, \quad I_y \cdot \Gamma_y = \Gamma_y \cdot I_y = 0 \tag{92}$$

where

$$I_t = \begin{vmatrix} 1_t & 0 \\ 0 & 1_t \end{vmatrix}, \quad I_y = \begin{vmatrix} 1_y & 0 \\ 0 & 1_y \end{vmatrix}, \quad 1_t = \hat{z}\hat{z} + \hat{x}\hat{x}, \quad 1_y = \hat{y}\hat{y} \tag{93}$$

Taking these equations into account, the first member of (88) becomes

$$\Gamma_\nabla \cdot \psi = \left(\Gamma_t + \Gamma_y \frac{\partial}{\partial y}\right)\psi = \Gamma_t \cdot \psi_t + \Gamma_y \frac{\partial}{\partial y}\psi_t + \Gamma_t \cdot \psi_y \tag{94}$$

where $\psi_t = \begin{vmatrix} \mathbf{E}_t \\ \mathbf{H}_t \end{vmatrix}$ and $\psi_y = \begin{vmatrix} E_y\hat{y} \\ H_y\hat{y} \end{vmatrix}$ with $\mathbf{E}_t = \hat{z}E_z + \hat{x}E_x$, $\mathbf{H}_t = \hat{z}H_z + \hat{x}H_x$.

Using the decomposition

$$W = W_{tt} + W_{ty} + W_{yt} + W_{yy} \tag{95}$$

where $W_{tt} = I_t \cdot W \cdot I_t$, $W_{ty} = I_t \cdot W \cdot I_y$, $W_{yt} = I_y \cdot W \cdot I_t$, $W_{yy} = I_y \cdot W \cdot I_y$ yields the following decomposition in transversal and longitudinal components of the Maxwell eq. (88):

$$I_y \cdot \Gamma_t \cdot \psi_t = W_{yt} \cdot \psi_t + W_{yy} \cdot \psi_y \tag{96}$$

$$I_t \cdot \frac{\partial}{\partial y}\Gamma_y \cdot \psi_t + I_t \cdot \Gamma_t \cdot \psi_y = W_{tt} \cdot \psi_t + W_{ty} \cdot \psi_y \tag{97}$$

Introduction in (96) of the matrix $\hat{W}_y$ defined by

$$\hat{W}_y \cdot W_{yy} = W_{yy} \cdot \hat{W}_y = I_y \tag{98}$$

yields the equation that expresses the longitudinal field ψ_y in terms of the transversal field ψ_t:

$$\psi_y = \hat{W}_y \cdot (I_y \cdot \Gamma_t - W_{yt}) \cdot \psi_t \tag{99}$$

Equation (99) leads to

$$\hat{W}_y = \frac{1}{j\omega(\varepsilon_y \mu_y - \xi_y \zeta_y)} \begin{vmatrix} \mu_y 1_y & \xi_y 1_y \\ -\zeta_y 1_y & -\varepsilon_y 1_y \end{vmatrix} \tag{100}$$

Taking into account that $\Gamma_y^2 = -I_t$, substitution of (99) into (97) yields the transverse Maxwell equations:

$$-\frac{\partial}{\partial y}\psi_t = M\left(j\frac{\partial}{\partial z}, j\frac{\partial}{\partial x}\right) \cdot \psi_t \tag{101}$$

where the matrix of order four $M\left(j\frac{\partial}{\partial z}, j\frac{\partial}{\partial x}\right)$ is given by

$$M\left(j\frac{\partial}{\partial z}, j\frac{\partial}{\partial x}\right) = -\Gamma_y \cdot [I_t \cdot (\Gamma_t - W_{ty}) \cdot \hat{W}_y \cdot (I_y \cdot \Gamma_t - W_{yt}) - W_{tt}] \tag{102}$$

This operator matrix $M\left(j\frac{\partial}{\partial z}, j\frac{\partial}{\partial x}\right)$ depends on ω and the parameters $\boldsymbol{\varepsilon}$, $\boldsymbol{\xi}$, $\boldsymbol{\zeta}$, $\boldsymbol{\mu}$ of the medium. It is of second order in the operators $j\frac{\partial}{\partial z}$ and $j\frac{\partial}{\partial x}$ and in a general can be evaluated using MATHEMATICA.

7.10.2 Field equations in the Fourier domain

Without loss of generality we assume the z dependence of the electromagnetic field **E** and **H** to be specified by the factor $e^{-j\alpha_o z}$, which is omitted. Let us introduce the Fourier transform:

$$\begin{aligned} \mathbf{V}(\eta, y) &= \int_{-\infty}^{\infty} \hat{y} \times \mathbf{E}_t(\alpha_o, y, x) e^{j\eta x} dx \\ \mathbf{I}(\eta, y) &= \int_{-\infty}^{\infty} \mathbf{H}_t(\alpha_o, y, x) e^{j\eta x} dx \end{aligned} \tag{103}$$

Equation (101) yield

$$-\frac{d}{dy}\begin{vmatrix} \mathbf{V} \\ \mathbf{I} \end{vmatrix} = \begin{vmatrix} \mathbf{T}_e(\eta) & \mathbf{Z}(\eta) \\ \mathbf{Y}(\eta) & \mathbf{T}_h(\eta) \end{vmatrix} \begin{vmatrix} \mathbf{V} \\ \mathbf{I} \end{vmatrix} \tag{104}$$

where the polynomial matrix of order four $\mathbf{P}(\eta) = \begin{vmatrix} \mathbf{T}_e(\eta) & \mathbf{Z}(\eta) \\ \mathbf{Y}(\eta) & \mathbf{T}_h(\eta) \end{vmatrix}$ involves the polynomial matrices of order two: $\mathbf{T}_e(\eta), \mathbf{Z}(\eta), \mathbf{Y}(\eta), \mathbf{T}_h(\eta)$ having degree two.

Let us consider a homogeneous slab defined between $y = 0$ and $y = d$. Solution of (104) yields

$$\begin{vmatrix} \mathbf{V}_1 \\ \mathbf{I}_1 \end{vmatrix} = e^{\mathbf{P}d}\begin{vmatrix} \mathbf{V}_2 \\ \mathbf{I}_2 \end{vmatrix} = \mathbf{T}\begin{vmatrix} \mathbf{V}_2 \\ \mathbf{I}_2 \end{vmatrix} \quad \text{or} \quad \begin{matrix} \mathbf{V}_1 = \mathbf{A}\cdot\mathbf{V}_2 + \mathbf{B}\cdot\mathbf{I}_2 \\ \mathbf{I}_1 = \mathbf{C}\cdot\mathbf{V}_2 + \mathbf{D}\cdot\mathbf{I}_2 \end{matrix} \tag{105}$$

where

$$\begin{matrix} \mathbf{V}_1 = \mathbf{V}|_{y=0}, & \mathbf{I}_1 = \mathbf{I}|_{y=0} \\ \mathbf{V}_2 = \mathbf{V}|_{y=d}, & \mathbf{I}_1 = \mathbf{I}|_{y=d} \end{matrix} \tag{106}$$

and

$$\mathbf{T} = \mathbf{T}(\alpha_o, \eta) = \begin{vmatrix} \mathbf{A} & \mathbf{B} \\ \mathbf{C} & \mathbf{D} \end{vmatrix} \tag{107}$$

is the transmission matrix of the slab. It is convenient to introduce a circuit model representing the slab as a two-port (Fig. 10). This representation is valid also for more slabs, in which case the transmission matrix is the product of the transmission matrices relevant to the different slabs (Fig. 11).

If the slab is indefinite in the positive (or negative) y-direction, the vanishing of V_2 and I_2 as $d \to \infty$ is compensated by the increase in the matrix $\mathbf{T}$, so that it is necessary to introduce a limit process to model a homogeneous semi-infinite medium. This process can be complex. It is better to introduce the eigenvalues and the eigenvectors of the matrix $\mathbf{P}$ according to the following considerations.

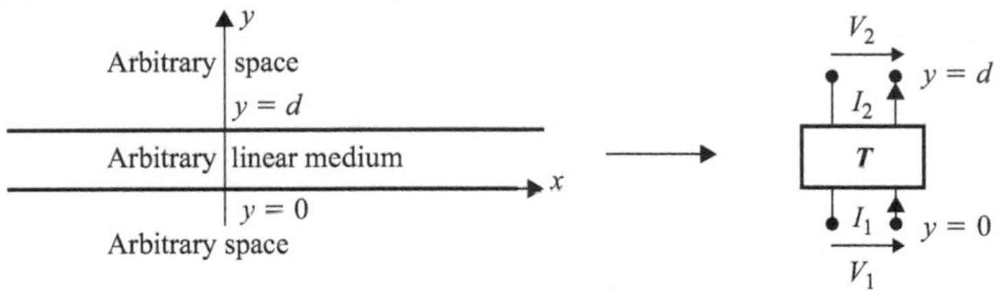

Fig. 10: Behavior in the Fourier domain of a slab filled by arbitrary linear medium

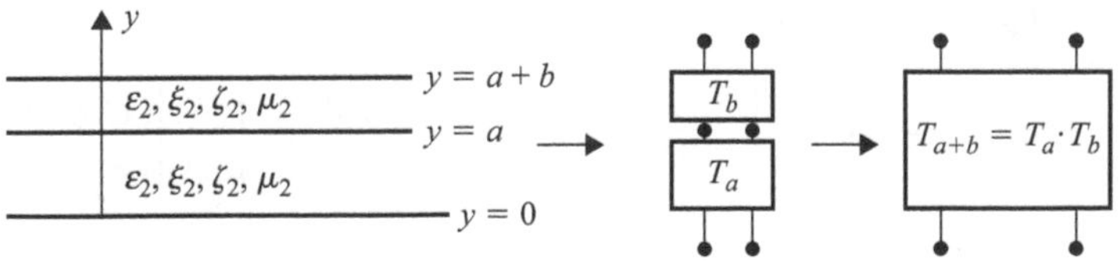

Fig. 11: Transmission matrix of two contiguous slabs

By indicating with γ_i, $i = 1, 2, 3, 4$ and $\boldsymbol{\psi}_i = \begin{vmatrix} \mathbf{V}_i \\ \mathbf{I}_i \end{vmatrix}$, $i = 1, 2, 3, 4$ the four eigenvalues and eigenvectors of $\mathbf{P}$, the more general solution of (3.2) assumes the form

$$\begin{vmatrix} \mathbf{V} \\ \mathbf{I} \end{vmatrix} = C_1 e^{-\gamma_1 y} \boldsymbol{\psi}_1 + C_2 e^{-\gamma_2 y} \boldsymbol{\psi}_2 + C_3 e^{-\gamma_3 y} \boldsymbol{\psi}_3 + C_4 e^{-\gamma_4 y} \boldsymbol{\psi}_4 \tag{108}$$

or

$$\begin{cases} \mathbf{V} = C_1 e^{-\gamma_1 y} \mathbf{V}_1 + C_2 e^{-\gamma_2 y} \mathbf{V}_2 + C_3 e^{-\gamma_3 y} \mathbf{V}_3 + C_4 e^{-\gamma_4 y} \mathbf{V}_4 \\ \mathbf{I} = C_1 e^{-\gamma_1 y} \mathbf{I}_1 + C_2 e^{-\gamma_2 y} \mathbf{I}_2 + C_3 e^{-\gamma_3 y} \mathbf{I}_3 + C_4 e^{-\gamma_4 y} \mathbf{I}_4 \end{cases} \tag{109}$$

where C_i, $i = 1, 2, 3, 4$ are four arbitrary scalars that do not depend on y.

In presence of a passive medium, we conjecture that two eigenvalues (e.g., γ_1 and γ_2) present no negative real part and the other two (e.g., γ_3 and γ_4) present no positive real part. The authors were unable to make a direct proof of this conjecture. However, this property holds for all the media considered in the following. Moreover, at the end of this section we will consider a reflection problem for an arbitrary linear medium. Reasoning based on the existence and uniqueness of the reflected field should constitute an indirect proof of the conjecture.

It follows that in an indefinite upper medium the scalars C_3 and C_4 must be zero and the solution must have the form

$$\begin{cases} \mathbf{V} = C_1 e^{-\gamma_1 y} \mathbf{V}_1 + C_2 e^{-\gamma_2 y} \mathbf{V}_2 \\ \mathbf{I} = C_1 e^{-\gamma_1 y} \mathbf{I}_1 + C_2 e^{-\gamma_2 y} \mathbf{I}_2 \end{cases} \quad y > 0 \tag{110}$$

Conversely, in the presence of an indefinite lower medium the scalars C_1 and C_2 must be zero and the solution must have the form

$$\begin{cases} \mathbf{V} = C_3 e^{-\gamma_3 y} \mathbf{V}_3 + C_4 e^{-\gamma_4 y} \mathbf{V}_4 \\ \mathbf{I} = C_3 e^{-\gamma_3 y} \mathbf{I}_3 + C_4 e^{-\gamma_4 y} \mathbf{I}_4 \end{cases} \quad y < 0 \tag{111}$$

Let us write the components of the eigenvectors in the form

$$\mathbf{V}_i = \begin{vmatrix} V_z^{(i)} \\ V_x^{(i)} \end{vmatrix}, \quad \mathbf{I}_i = \begin{vmatrix} I_z^{(i)} \\ I_x^{(i)} \end{vmatrix} \quad i = 1, 2, 3, 4 \tag{112}$$

It follows that for $y > 0$

$$\mathbf{V} = \begin{vmatrix} V_z \\ V_x \end{vmatrix} = C_1 e^{-\gamma_1 y} \begin{vmatrix} V_z^{(1)} \\ V_x^{(1)} \end{vmatrix} + C_2 e^{-\gamma_2 y} \begin{vmatrix} V_z^{(2)} \\ V_x^{(2)} \end{vmatrix} = \begin{vmatrix} V_z^{(1)} & V_z^{(2)} \\ V_x^{(1)} & V_x^{(2)} \end{vmatrix} \begin{vmatrix} C_1 e^{-\gamma_1 y} \\ C_2 e^{-\gamma_2 y} \end{vmatrix} \tag{113}$$

$$\mathbf{I} = \begin{vmatrix} I_z \\ I_x \end{vmatrix} = C_1 e^{-\gamma_1 y} \begin{vmatrix} I_z^{(1)} \\ I_x^{(1)} \end{vmatrix} + C_2 e^{-\gamma_2 y} \begin{vmatrix} I_z^{(2)} \\ I_x^{(2)} \end{vmatrix} = \begin{vmatrix} I_z^{(1)} & I_z^{(2)} \\ I_x^{(1)} & I_x^{(2)} \end{vmatrix} \begin{vmatrix} C_1 e^{-\gamma_1 y} \\ C_2 e^{-\gamma_2 y} \end{vmatrix} \tag{114}$$

By eliminating the vector $\begin{vmatrix} C_1 e^{-\gamma_1 y} \\ C_2 e^{-\gamma_2 y} \end{vmatrix}$, eqs. (113) and (114) yield

$$\mathbf{V} = \vec{\mathbf{Z}}_c \, \mathbf{I} \quad y > 0 \tag{115}$$

where the impedance $\vec{\mathbf{Z}}_k$ is defined by

$$\vec{\mathbf{Z}}_c = \begin{vmatrix} V_z^{(1)} & V_z^{(2)} \\ V_x^{(1)} & V_x^{(2)} \end{vmatrix} \cdot \begin{vmatrix} I_z^{(1)} & I_z^{(2)} \\ I_x^{(1)} & I_x^{(2)} \end{vmatrix}^{-1} \tag{116}$$

Equations (115) and (116) extend the fundamental concept of characteristic impedance to an indefinite arbitrary linear medium. Equation (116) provides the characteristic impedance for an indefinite upper medium ($y > 0$). Similarly, for an indefinite lower medium ($y < 0$) we have

$$\mathbf{V} = -\overleftarrow{\mathbf{Z}}_c \, \mathbf{I} \quad y < 0 \tag{117}$$

where the characteristic impedance $\overleftarrow{\mathbf{Z}}_k$ is defined by

$$\overleftarrow{\mathbf{Z}}_c = -\begin{vmatrix} V_z^{(3)} & V_z^{(4)} \\ V_x^{(3)} & V_x^{(4)} \end{vmatrix} \cdot \begin{vmatrix} I_z^{(3)} & I_z^{(4)} \\ I_x^{(3)} & I_x^{(4)} \end{vmatrix}^{-1} \tag{118}$$

In general as defined by (116) and (118), the characteristic impedances are rather cumbersome and do not allow to discuss and ascertain the analytical properties of the characteristic impedances in the complex $\eta -$ plane. For the purpose of obtaining more convenient expressions of $\vec{\mathbf{Z}}$ and $\overleftarrow{\mathbf{Z}}$, we consider the following equations obtained by the elimination of either **I** or **V** in (104):

$$\begin{aligned} &\frac{d^2 V}{dy^2} + (T_e + Z T_h Z^{-1}) \frac{dV}{dy} - Z(Y - T_h Z^{-1} T_e) V = 0 \\ &\frac{d^2 I}{dy^2} + (T_h + Y T_e Y^{-1}) \frac{dI}{dy} - Y(Z - T_e Y^{-1} T_h) I = 0 \end{aligned} \tag{119}$$

If we are looking for solutions having the form

$$\mathbf{V} = e^{-\gamma_v y} \, \mathbf{V}_o, \quad \mathbf{I} = e^{-\gamma_I y} \, \mathbf{I}_o \tag{120}$$

these equations yield

$$\begin{aligned} &\gamma_V^2 - (T_e + Z T_h Z^{-1}) \gamma_V - Z(Y - T_h Z^{-1} T_e) = 0 \\ &\gamma_I^2 - (T_h + Y T_e Y^{-1}) \gamma_I - Y(Z - T_e Y^{-1} T_h) = 0 \end{aligned} \tag{121}$$

In general, there are several solutions of the previous equations. We choose them so that the matrix propagation constants γ_V and γ_I have the same eigenvalues. Thus, they are chosen with a real part positive or negative, respectively, in the half-spaces $y > 0$ and $y < 0$.

From eq. (104) we have

$$\begin{aligned} -\gamma_V V &= -T_e \cdot V - Z \cdot I \\ -\gamma_I I &= -Y \cdot V - T_h \cdot I \end{aligned} \tag{122}$$

Taking into account that $I = Y_c V$, $V = Z_c I$, this yields the following expressions for the characteristic impedance and admittance:

$$\begin{aligned} Y_c &= Z^{-1} \cdot (\gamma_V - T_e) \\ Z_c &= Y^{-1} \cdot (\gamma_I - T_h) \end{aligned} \tag{123}$$

We let

$$A_V = -(T_e + ZT_h Z^{-1}), \quad B_V = -Z(Y - T_h Z^{-1} T_e)$$

or

$$A_I = -(T_h + YT_e Y^{-1}), \quad B_I = -Y(Z - T_e Y^{-1} T_h)$$

In the following for the sake of simplicity we will omit the subscript V or I. The obtained results apply to both eq. (121).

The problem of solving (121) explicitly is difficult when A and B do not commute. However, we are dealing with matrices of order two, which means that the following characteristic equation holds:

$$\gamma^2 - t_\gamma \gamma + \Delta_\gamma 1 = 0 \tag{124}$$

where $t_\gamma = -(\gamma_{1,3} + \gamma_{2,4})$ and $\Delta_\gamma = \gamma_{1,3}\gamma_{2,4}$ are the trace and the determinant of γ, respectively. By eliminating γ^2 between (124) and (121) the following expression is obtained:

$$\gamma = (\gamma_{1,3} + \gamma_{2,4})^{-1} \cdot (-B + \Delta_\gamma 1)$$

The inversion of $(A + t_\gamma 1)$ yields

$$(A + t_\gamma 1)^{-1} = x\, 1 + y_\gamma A$$

where

$$x = \frac{t_\gamma + t_A}{t_\gamma(t_\gamma + t_A) + \Delta_A}, \quad y = -\frac{1}{t_\gamma^2 + t_\gamma t_A + \Delta_A}$$

where t_A and Δ_A are the trace and the determinant of A, respectively. It follows that

$$\gamma = x\Delta_\gamma 1 - xB + y\, \Delta_\gamma A - yA \cdot B$$

This equation provides the matrices γ involved in eq. (123) that evaluate Z_c and Y_c. It requires only the knowledge of the trace t_γ and the determinant Δ_γ. The obtained expression apparently differs from (116) and (118). However, in the all worked examples we showed numerically that the two different expressions lead to identical results.

It is important to observe that the characteristic impedances do not depend on the particular section considered. Taking this property into account, we are able to solve the elementary and fundamental problem illustrated in Fig. 12. An incident wave with transversal wave numbers α_o and η_o impinges on the arbitrary stratified medium located in $y \geq 0$. In the circuit model of the problem, $\vec{\mathbf{Z}}_c$ represents the characteristic impedance of the indefinite medium for

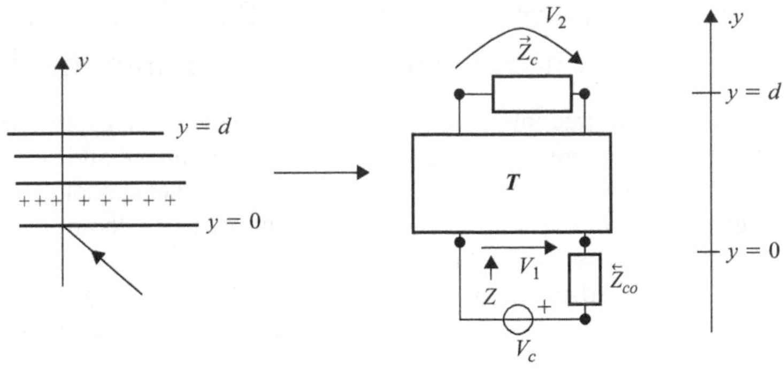

Fig. 12: The reflection problem from a linear stratified medium

$y > d$, $\overleftarrow{\mathbf{Z}}_{co}$ represents the characteristic impedance of the indefinite medium for $y < 0$, $\mathbf{T}$ the transmission matrix of the stratified medium located in $0 \leq y \leq d$, and $\mathbf{V}_o$ the open voltage on the section $y = 0$. Circuit theory provides the following expression for $\mathbf{V}_o$:

$$\mathbf{V}_o = (\overleftarrow{\mathbf{Z}}_{co} + \vec{\mathbf{Z}}_{co}) \cdot \vec{\mathbf{Z}}_{co}^{-1} \cdot \mathbf{V}^i \tag{125}$$

where $\mathbf{V}^i$ is the Fourier transform of the transverse field $\mathbf{E}_t^i$ of the incident plane wave evaluated at $y = 0$, and

$$\mathbf{V}^i = 2\pi\hat{y} \times \mathbf{E}_t^i \delta(\eta - \eta_o) \tag{126}$$

Circuit analysis yields the solution

$$\mathbf{V}_1 = \mathbf{Z}(0) \cdot [\overleftarrow{\mathbf{Z}}_c + \mathbf{Z}(0)]^{-1} \cdot \mathbf{V}_0 \tag{127}$$

where $\mathbf{Z}(0) = (\mathbf{A} \cdot \vec{\mathbf{Z}}_c + \mathbf{B}) \cdot (\mathbf{C} \cdot \vec{\mathbf{Z}}_c + \mathbf{D})^{-1}$. Once $\mathbf{V}_1$ is known, we may evaluate the voltage and current for every value of y by using the theory of stratified media. The inverse Fourier transforms

$$\mathbf{E}_t(\alpha_o, y, x) = \frac{1}{2\pi} \int_{-\infty}^{\infty} \mathbf{V}(\eta, y) \times \hat{y}\, e^{-j\eta x} d\eta$$

$$\mathbf{H}_t(\alpha_o, y, x) = \frac{1}{2\pi} \int_{-\infty}^{\infty} \mathbf{I}(\eta, y) e^{-j\eta x} d\eta$$

provide the transverse electromagnetic field everywhere. The presence of $\delta(\eta - \eta_o)$[5] renders the integrals very easy to evaluate.

---

[5] The delta functions means that there is a single line $\eta = \eta_o$ in the spatial spectrum.

7.10.3 The W-H equation for a PEC or a PMC half-plane immersed in a homogeneous linear arbitrary medium

Figure 13 illustrates the diffraction of an incident plane wave by an imperfect half-plane immersed in a linear homogenous medium. This problem was studied and solved in Daniele and Graglia (2007). In general the factorization of the matrix kernel cannot be done in closed form. Consequently in the presence of an arbitrary medium, the Fredholm factorization has been used (Daniele & Graglia, 2007). In this section we consider the particular cases where either $Z_a = 0 = Z_b$ (PEC half-plane) or $Z_a = \infty = Z_b$ (PMC half-plane).

The previous circuit considerations yield immediately the W-H equation of the problem. We get

$$(\overrightarrow{\mathbf{Y}}_c + \overleftarrow{\mathbf{Y}}_c)\mathbf{V}_+ = \mathbf{A}_- \text{ (PEC half-plane)} \tag{128}$$

$$(\overrightarrow{\mathbf{Z}}_c + \overleftarrow{\mathbf{Z}}_c)\mathbf{I}_+ = \mathbf{M}_- \text{ (PMC half-plane)} \tag{129}$$

where $\overrightarrow{\mathbf{Y}}_c = \overrightarrow{\mathbf{Z}}_c^{-1}$, $\overleftarrow{\mathbf{Y}}_c = \overleftarrow{\mathbf{Z}}_c^{-1}$

$$\mathbf{V}_+(\eta) = \mathbf{V}(\eta, 0), \quad \mathbf{A}_-(\eta) = \mathbf{I}(\eta, 0_+) - \mathbf{I}(\eta, 0_-)$$

$$\mathbf{I}_+(\eta) = \mathbf{I}(\eta, 0), \quad \mathbf{M}_-(\eta) = \mathbf{V}(\eta, 0_+) - \mathbf{V}(\eta, 0_-)$$

These equations are homogeneous since we used the Laplace domain (see section 7.3). To take the source into account, we must study the possible plane waves that propagate in the linear bianisotropic medium surrounding the half-plane. Here we assume the source to be an incident plane wave:

$$\mathbf{E}_t^i(x, 0, z) = E_o \hat{\mathbf{e}}_{ot} e^{-j\eta_o x} e^{-j\alpha_o z}, \quad \mathbf{H}_t^i(x, 0, z) = E_o \mathbf{h}_{ot} e^{-j\eta_o x} e^{-j\alpha_o z}$$

where η_o is the propagation constant in the x-direction. By denoting with $\hat{\mathbf{e}}_\mathbf{o}$, $\hat{\mathbf{h}}_\mathbf{o}$ the polarization for the electric and magnetic fields of the incident wave, we have

$$\hat{\mathbf{e}}_{ot} = \hat{\mathbf{e}}_\mathbf{o} - \hat{\mathbf{e}}_\mathbf{o} \cdot \hat{y}\hat{y}, \quad \hat{\mathbf{h}}_{ot} = \hat{\mathbf{h}}_\mathbf{o} - \hat{\mathbf{h}}_\mathbf{o} \cdot \hat{y}\hat{y}$$

By separating the nonconventional part in the Laplace transform we get

$$\mathbf{V}_+^s + j\frac{1}{\eta - \eta_o} E_o \hat{\mathbf{e}}_{ot} = (\overrightarrow{\mathbf{Y}}_c + \overleftarrow{\mathbf{Y}}_c)^{-1} \cdot \mathbf{A}_- \text{ (PEC half-plane)}$$

$$\mathbf{I}_+^s + j\frac{1}{\eta - \eta_o} E_o \hat{\mathbf{h}}_{ot} = (\overrightarrow{\mathbf{Z}}_c + \overleftarrow{\mathbf{Z}}_c)^{-1} \cdot \mathbf{M}_- \text{ (PMC half-plane)}$$

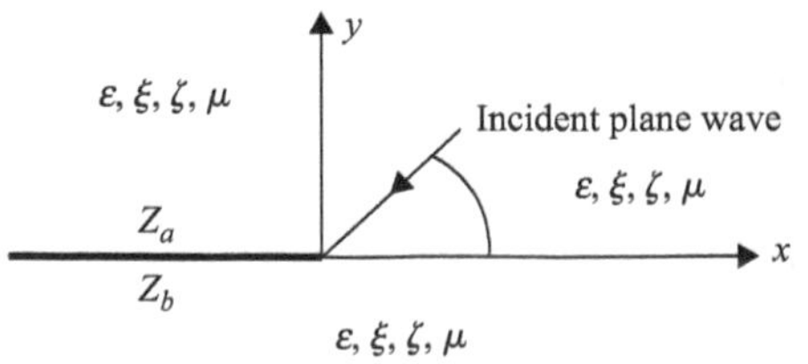

Fig. 13: Geometry of the problem

The above equations constitute the W-H equations of the problem. Their solution requires the factorization of either $(\overrightarrow{\mathbf{Y}}_c + \overleftarrow{\mathbf{Y}}_c)^{-1}$ or $(\overrightarrow{\mathbf{Z}}_c + \overleftarrow{\mathbf{Z}}_c)^{-1}$.

7.10.3.1 Some cases of closed form factorization

Generally, the expressions of the characteristic impedances obtained via eqs. (123), (116), or (118) are very cumbersome. However, they simplify considerably for simple media. For instance, for an isotropic free space defined by the parameters $\boldsymbol{\varepsilon} = \varepsilon 1$, $\boldsymbol{\mu} = \mu 1$, $\boldsymbol{\xi} = 0$, $\boldsymbol{\zeta} = 0$, we obtain the following expected result:

$$\overrightarrow{\mathbf{Z}}_c = \overleftarrow{\mathbf{Z}}_c = \mathbf{Z}_c = \frac{Z_o}{k\sqrt{\tau_o^2 - \eta^2}} \begin{vmatrix} k^2 - \eta^2 & \eta\,\alpha_o \\ \eta\,\alpha_o & \tau_o^2 \end{vmatrix} \tag{130}$$

where

$$Z_o = \sqrt{\frac{\mu}{\varepsilon}}, \quad k^2 = \omega^2 \varepsilon\mu, \quad \tau_o^2 = k^2 - \alpha_o^2$$

Another simple medium is constituted by a scalar chiral medium. It is defined by the scalar parameters $\boldsymbol{\varepsilon} = \varepsilon 1$, $\boldsymbol{\mu} = \mu 1$, $\boldsymbol{\xi} = -j\kappa 1$, $\boldsymbol{\zeta} = j\kappa 1$.

In this case eqs. (116) and (118) yield

$$\overrightarrow{\mathbf{Z}}_c = Z_o \frac{k_1 + k_2}{k(j(\tau_{o1}^2 - \tau_{o2}^2)\alpha_o\eta + k_1\tau_{o2}^2\chi_1 + k_2\tau_{o1}^2\chi_2)}$$

$$\bullet \begin{vmatrix} (k_1\chi_1 - j\alpha_o\eta)(k_2\chi_2 + j\alpha_o\eta) & \dfrac{(\tau_{o1}^2 + \tau_{o2}^2)\alpha_o\eta + jk_1\tau_{o2}^2\chi_1 - jk_2\tau_{o1}^2\chi_2}{2} \\ \dfrac{(\tau_{o1}^2 + \tau_{o2}^2)\alpha_o\eta + jk_1\tau_{o2}^2\chi_1 - jk_2\tau_{o1}^2\chi_2}{2} & \tau_{o1}^2\tau_{o2}^2 \end{vmatrix}$$

$$= \begin{vmatrix} \overrightarrow{Z}_{c11}(\alpha_o, \eta) & \overrightarrow{Z}_{c12}(\alpha_o, \eta) \\ \overrightarrow{Z}_{c21}(\alpha_o, \eta) & \overrightarrow{Z}_{c22}(\alpha_o, \eta) \end{vmatrix} \tag{131}$$

$$\overleftarrow{\mathbf{Z}}_c = \begin{vmatrix} \overrightarrow{Z}_{c11}(-\alpha_o, \eta) & -\overrightarrow{Z}_{c12}(-\alpha_o, \eta) \\ -\overrightarrow{Z}_{c21}(-\alpha_o, \eta) & \overrightarrow{Z}_{c22}(-\alpha_o, \eta) \end{vmatrix} \tag{132}$$

where

$$Z_o = \sqrt{\frac{\mu}{\varepsilon}}, \quad k = \omega\sqrt{\mu\,\varepsilon}, \quad k_{1,2} = \omega(\sqrt{\mu\,\varepsilon} \pm \kappa)$$

$$\tau_{o1,2} = \sqrt{k_{1,2}^2 - \alpha_o^2}, \quad \chi_{1,2} = \sqrt{\tau_{o1,2}^2 - \eta^2}$$

Note that the matrix representing $\overleftarrow{\mathbf{Z}}_c$ can be obtained from the matrix representing $\overrightarrow{\mathbf{Z}}_c$ by changing α_o into $-\alpha_o$. In addition, both these matrices are symmetric:

$$\overrightarrow{Z}_{c21}(\alpha_o, \eta) = \overrightarrow{Z}_{c12}(\alpha_o, \eta), \quad \overleftarrow{Z}_{c21}(\alpha_o, \eta) = \overleftarrow{Z}_{c12}(\alpha_o, \eta) \tag{133}$$

Of course, when the chirality factor κ is zero, the two characteristic impedances $\overrightarrow{\mathbf{Z}}_c$ and $\overleftarrow{\mathbf{Z}}_c$ are equal and reduce to that of free space.

The expressions of $\vec{\mathbf{Z}}_c = \vec{\mathbf{Z}}$ and $\overleftarrow{\mathbf{Z}}_c = \overleftarrow{\mathbf{Z}}$ are very important to ascertain the possibility of obtaining closed form factorization of the matrix kernel introduced in W-H eqs. (128) and (129). In particular, the closed-form solutions obtained in the literature concern only the case of PEC or PCM that involves factorization of a matrix or order two.

To simplify the factorization, it must be observed that for skew incidence ($\alpha_o \neq 0$) it is important to consider the transformation introduced in Senior (1978) and Lüneburg and Serbest (2000):

$$t = \begin{pmatrix} \eta & -\alpha_o \\ \alpha_o & \eta \end{pmatrix}, \quad t_a = \begin{pmatrix} \eta & \alpha_o \\ -\alpha_o & \eta \end{pmatrix} \tag{134}$$

For instance, the PMC half-plane immersed in a chiral medium requires the factorization of the matrix $\vec{\mathbf{Z}} + \overleftarrow{\mathbf{Z}}$ (129). Taking into account that

$$t \cdot (\vec{\mathbf{Z}} + \overleftarrow{\mathbf{Z}}) \cdot t_a = Z_o \cdot \begin{pmatrix} \dfrac{2(k_1+k_2)(\alpha_o^2+\eta^2)\chi_1\chi_2}{k(k_2\chi_1+k_1\chi_2)} & 0 \\ 0 & \dfrac{2k_1k_2(k_1+k_2)(\alpha_o^2+\eta^2)}{k(k_2\chi_1+k_1\chi_2)} \end{pmatrix} \tag{135}$$

the factorization of the matrix of order two $\vec{\mathbf{Z}} + \overleftarrow{\mathbf{Z}}$ is accomplished by the factorization of the scalars χ_1, χ_2, $k_2\chi_1 + k_1\chi_2$. In fact, it reduces to the diagonal matrix present in the second member. Similar considerations apply for the case of a PEC. This last problem is also discussed in a paper by Przezdziecki (2000).

We remark that for suitable values of the electromagnetic parameters, $\mathbf{T}_e(\eta)$ and $\mathbf{T}_h(\eta)$ may vanish. For instance, this happens when $\boldsymbol{\xi} = 0$, $\boldsymbol{\zeta} = 0$ and the permittivity $\boldsymbol{\varepsilon}$ and the permeability $\boldsymbol{\mu}$ have the form

$$\begin{aligned} \boldsymbol{\varepsilon} &= \varepsilon_{zz}\hat{z}\hat{z} + \varepsilon_{zx}\hat{z}\hat{x} + \varepsilon_{xz}\hat{x}\hat{z} + \varepsilon_{xx}\hat{x}\hat{x} + \varepsilon_{yy}\hat{y}\hat{y} \Rightarrow \begin{vmatrix} \varepsilon_{zz} & \varepsilon_{zx} & 0 \\ \varepsilon_{xz} & \varepsilon_{xx} & 0 \\ 0 & 0 & \varepsilon_{yy} \end{vmatrix} \\ \boldsymbol{\mu} &= \mu_{zz}\hat{z}\hat{z} + \mu_{zx}\hat{z}\hat{x} + \mu_{xz}\hat{x}\hat{z} + \mu_{xx}\hat{x}\hat{x} + \mu_{yy}\hat{y}\hat{y} \Rightarrow \begin{vmatrix} \mu_{zz} & \mu_{zx} & 0 \\ \mu_{xz} & \mu_{xx} & 0 \\ 0 & 0 & \mu_{yy} \end{vmatrix} \end{aligned} \tag{136}$$

In this case, eq. (121) simplify:

$$\gamma_V^2 - Z \cdot Y = 0, \quad \gamma_I^2 - Y \cdot Z = 0$$

and we are dealing with equations similar to the classical transmission line equations well studied in the literature. For instance, we have the following characteristic impedances (Paul, 1975):

$$\vec{\mathbf{Z}}_c = \overleftarrow{\mathbf{Z}}_c = \mathbf{Z}_c = \boldsymbol{\gamma}_V^{-1}\mathbf{Z} = \boldsymbol{\gamma}_V\mathbf{Y}^{-1} = \mathbf{Y}^{-1}\boldsymbol{\gamma}_I = \mathbf{Z}\boldsymbol{\gamma}_I^{-1} \tag{137}$$

where $\gamma_V = \sqrt{ZY}$ and $\gamma_I = \sqrt{YZ}$. Taking into account that $\gamma_V = \sqrt{ZY}$ commutes with the polynomial matrix ZY, or that $\gamma_I = \sqrt{YZ}$ commutes with the polynomial matrix YZ, we can express the kernels $\vec{\mathbf{Y}} + \overleftarrow{\mathbf{Y}}$ and $\vec{\mathbf{Z}} + \overleftarrow{\mathbf{Z}}$ by a polynomial matrix multiplied by a matrix that commutes with a polynomial matrix. This is a very remarkable fact because we are able to factorize in closed form matrices that commute with a polynomial matrix (chapter 4).

These particular cases were addressed by Hurd and Przezdziecki (1981), who also simplified the factorization problem by reducing it to a Hilbert problem (Hurd, 1976). Other explicit solutions of (121) can be obtained when the matrices

$$A_V = -(T_e + ZT_hZ^{-1}), \quad B_V = -Z(Y - T_hZ^{-1}T_e)$$

or

$$A_I = -(T_h + YT_eY^{-1}) \quad B_I = -Y(Z - T_eY^{-1}T_h)$$

commute. For instance, this happens if $\mathbf{T}_e(\eta)$ and $\mathbf{T}_h(\eta)$ are scalars. In this case for eq. (104) we obtain the solution

$$V = e^{\frac{-(T_e+T_h)-\sqrt{(T_e-T_h)^2+4(\mathbf{ZY}-T_eT_h)}}{2}y}V_1 + e^{\frac{-(T_e+T_h)+\sqrt{(T_e-T_h)^2+4(\mathbf{ZY}-T_eT_h)}}{2}y}V_2, \tag{138}$$

which yields the following expressions of the characteristic impedances:

$$\begin{aligned}\vec{\mathbf{Z}}_c &= \left[\frac{(T_e+T_h)+\sqrt{(T_e-T_h)^2+4(\mathbf{ZY}-T_eT_h)}}{2}\right]^{-1}\mathbf{Z}\\ \overleftarrow{\mathbf{Z}}_c &= \left[\frac{-(T_e+T_h)+\sqrt{(T_e-T_h)^2+4(\mathbf{ZY}-T_eT_h)}}{2}\right]^{-1}\mathbf{Z}\end{aligned} \tag{139}$$

To discover all the cases where we have scalar expressions of $\mathbf{T}_e(\eta)$ and $\mathbf{T}_h(\eta)$ is not a simple task. For instance, this is the case if

$$\boldsymbol{\varepsilon} \Rightarrow \begin{vmatrix}\varepsilon_{zz} & \varepsilon_{zx} & 0\\ \varepsilon_{xz} & \varepsilon_{xx} & 0\\ 0 & 0 & \varepsilon_{yy}\end{vmatrix}, \quad \boldsymbol{\mu} \Rightarrow \begin{vmatrix}\mu_{zz} & \mu_{zx} & 0\\ \mu_{xz} & \mu_{xx} & 0\\ 0 & 0 & \mu_{yy}\end{vmatrix}, \quad \boldsymbol{\xi} \Rightarrow \begin{vmatrix}0 & \xi_{zx} & 0\\ -\xi_{zx} & 0 & 0\\ 0 & 0 & 0\end{vmatrix}, \quad \boldsymbol{\zeta} \Rightarrow \begin{vmatrix}0 & \zeta_{zx} & 0\\ -\zeta_{zx} & 0 & 0\\ 0 & 0 & 0\end{vmatrix} \tag{140}$$

Again, expressions (139) show that the kernel $\vec{\mathbf{Y}} + \overleftarrow{\mathbf{Y}}$ and $\vec{\mathbf{Z}} + \overleftarrow{\mathbf{Z}}$ reduce to matrices that commute with polynomials matrices. This means that closed-form solutions can be obtained for PEC or PCM half-planes immersed in a bianisotropic medium defined by (140).

Other cases that can been solved in closed form involve more general gyrotropic media. For instance, let us consider

$$\boldsymbol{\varepsilon} \Rightarrow \begin{vmatrix}\varepsilon_{zz} & 0 & 0\\ 0 & \varepsilon_{xx} & \varepsilon_{xy}\\ 0 & -\varepsilon_{xy} & \varepsilon_{yy}\end{vmatrix}, \quad \boldsymbol{\varepsilon} = \varepsilon 1, \quad \boldsymbol{\mu} = \mu 1, \quad \boldsymbol{\xi} = 0, \quad \boldsymbol{\zeta} = 0$$

This involves in eq. (104) the matrix polynomial

$$\mathbf{P} = \begin{vmatrix}\frac{j\varepsilon_{xy}\eta}{\varepsilon_{xx}} & 0 & -\frac{j(\eta^2 - \varepsilon_{xx}\mu_o{}^2\omega^2)}{\varepsilon_{xx}\mu_o\omega} & \frac{j\alpha_o\eta}{\varepsilon_{xx}\omega}\\ -\frac{j\alpha_o\varepsilon_{xy}}{\varepsilon_{xx}} & 0 & \frac{j\alpha_o\eta}{\varepsilon_{xx}\omega} & -\frac{j(\alpha_o{}^2\mu_o - \varepsilon_{xx}\mu_o{}^2\omega^2)}{\varepsilon_{xx}\mu_o\omega}\\ j\frac{(\varepsilon_{xx}^2+\varepsilon_{xy}^2)\mu_o\omega^2 - \alpha_o{}^2\varepsilon_{xx}}{\varepsilon_{xx}\mu_o\omega} & -\frac{j\alpha_o\eta}{\mu_o\omega} & -\frac{j\varepsilon_{xy}\eta}{\varepsilon_{xx}} & \frac{j\alpha_o\varepsilon_{xy}}{\varepsilon_{xx}}\\ -\frac{j\alpha_o\eta}{\mu_o\omega} & -j\frac{\varepsilon_{xx}\eta^2 - \varepsilon_{xx}\varepsilon_{zz}\mu_o\omega^2}{\varepsilon_{xx}\mu_o\omega} & 0 & 0\end{vmatrix}$$

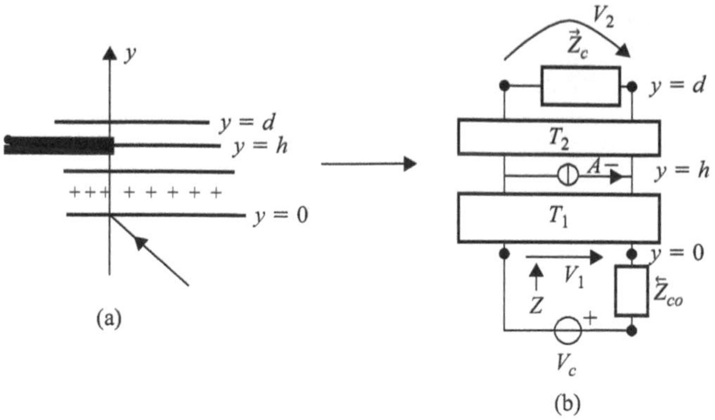

Fig. 14: A PEC half-plane immersed in an arbitrary stratified medium

It is possible to show that the matrix $t \cdot (\vec{\mathbf{Z}} + \overleftarrow{\mathbf{Z}}) \cdot t_a$ in this case presents the form

$$t \cdot (\vec{\mathbf{Z}} + \overleftarrow{\mathbf{Z}}) \cdot t_a = f(\eta) \begin{pmatrix} a_o(\eta) + a_1(\eta)\varphi(\eta) & \eta \\ \eta & b_o(\eta) + b_1(\eta)\varphi(\eta) \end{pmatrix} \tag{141}$$

where the functions $a_o(\eta), a_1(\eta), b_o(\eta), b_1(\eta)$ are rational function of η not reported here.

To rewrite this matrix in the form

$$\begin{pmatrix} a_o(\eta) + a_1(\eta)\varphi(\eta) & \eta \\ \eta & b_o(\eta) + b_1(\eta)\varphi(\eta) \end{pmatrix} = \mathbf{R}_o(\eta) + \mathbf{R}_1(\eta)\varphi(\eta)$$

where the rational matrices $\mathbf{R}_o(\eta)$ and $\mathbf{R}_1(\eta)$ are defined by

$$\mathbf{R}_o(\eta) = \begin{pmatrix} a_o(\eta) & \eta \\ \eta & b_o(\eta) \end{pmatrix}, \quad \mathbf{R}_1(\eta) = \begin{pmatrix} a_1(\eta) & 0 \\ 0 & b_1(\eta) \end{pmatrix}$$

we again must deal with matrices commuting with polynomial matrices. Similar considerations apply for the PEC case involving the matrix $t \cdot (\vec{\mathbf{Y}} + \overleftarrow{\mathbf{Y}}) \cdot t_a$. This last case was solved for the first time by Hurd and Przezdziecki (1985).

7.11 The half-plane immersed in an arbitrary planar stratified medium

Figure 14a illustrates the geometry involving a PEC half-plane[6] immersed in an arbitrary stratified medium. The solution of the circuit model shown in Fig. 14b yields the W-H equation having as unknowns the minus function A_- (Laplace transform of the total current

[6] We consider a PEC half-plane for the sake of simplicity. The presence of an arbitrary impedance half-plane requires a slight modification of the deduction of the W-H equations, as is indicated in section 7.6.

induced on the half-plane) and $V_+ = V_+(\eta, h)$ (Laplace transform of the transverse electric field in the aperture $(x > 0, y = h)$).

The two-port transmission matrices T_1 and T_2 represent the slab $y = h, y = d$ and the stratification between $y = h$ and $y = d$, respectively. The other elements in the circuit model are defined in section 7.10.2.

CHAPTER 8

Planar discontinuities in stratified media

8.1 The planar waveguide problem

8.1.1 The E-polarization case

Let us consider the planar waveguide shown in Fig. 1A. This structure can be studied via the circuit representation indicated in Fig. 1B. In particular, the PEC walls of the waveguide are simulated by the current generators A_{1-}, A_{2-}, and the slab $-d \leq y \leq 0$ in the physical structure is equivalent to the Π two-port introduced in Fig. 3a of chapter 7.

Writing the node equations on the circuit 1B yields directly the W-H equations of the problem:

$$\begin{aligned} &\text{node 1} \quad (Y_1 + Y_3 + Y_c)V_{1+} - Y_3 V_{2+} = A_{1-} \\ &\text{node 2} \quad -Y_3 V_{1+} + (Y_2 + Y_3 + Y_c)V_{2+} = A_{2-} \end{aligned} \tag{1}$$

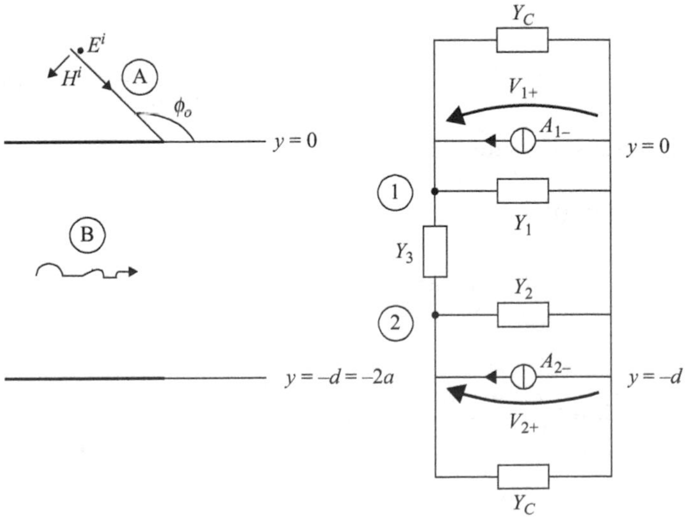

Fig. 1: Scattering by a planar waveguide (A) source: plane wave (B) source: incident mode

where $\tau = \tau(\eta) = \sqrt{k^2 - \eta^2}$, $Y_c = \frac{1}{Z_c} = \frac{\tau}{\omega\mu_o}$, $Y_1 = Y_2 = jY_c \tan\frac{\tau d}{2}$, $Y_3 = -jY_c \frac{1}{\sin \tau d}$, and $-A_{1-}$ and $-A_{2-}$ are the Fourier transforms of the total currents induced on the two half-planes.

Since $Y_2 = Y_1$, by summing and subtracting the two equations, system (1) decouples into two scalar equations:

$$\begin{aligned} (Y_1 + Y_c)(V_{1+} + V_{2+}) &= A_{1-} + A_{2-} \\ (Y_1 + 2Y_3 + Y_c)(V_{1+} - V_{2+}) &= A_{1-} - A_{2-} \end{aligned} \tag{2}$$

which is equivalent to rewriting the matrix kernel $G(\eta)$ of the W-H eq. (1):

$$G(\eta) \cdot \begin{vmatrix} V_{1+} \\ V_{2+} \end{vmatrix} = \begin{vmatrix} A_{1-} \\ A_{2-} \end{vmatrix}$$

in the form:

$$\begin{aligned} G(\eta) &= \begin{pmatrix} Y_1 + Y_3 + Y_c & -Y_3 \\ -Y_3 & Y_1 + Y_3 + Y_c \end{pmatrix} \\ &= \frac{1}{2}\begin{pmatrix} 1 & 1 \\ 1 & -1 \end{pmatrix} \cdot \begin{pmatrix} Y_1 + Y_c & 0 \\ 0 & Y_1 + 2Y_3 + Y_c \end{pmatrix} \cdot \begin{pmatrix} 1 & 1 \\ 1 & -1 \end{pmatrix} \end{aligned} \tag{3}$$

whence the factorization problem of $G(\eta)$ is reduced to the factorization of the diagonal matrix:

$$M(\eta) = \begin{pmatrix} Y_1 + Y_c & 0 \\ 0 & Y_1 + 2Y_3 + Y_c \end{pmatrix} = \begin{pmatrix} \dfrac{\tau e^{j\tau a}}{\omega\mu \cos(\tau a)} & 0 \\ 0 & -j\dfrac{\tau e^{j\tau a}}{\omega\mu \sin(\tau a)} \end{pmatrix} = \begin{pmatrix} m_1 & 0 \\ 0 & m_2 \end{pmatrix} \tag{4}$$

where

$$m_1 = \frac{\tau e^{j\tau a}}{\omega\mu \cos(\tau a)} = m_{1-}(\eta) m_{1+}(\eta), \quad m_2(\eta) = -j\frac{\tau e^{j\tau a}}{\omega\mu \sin(\tau a)} = m_{2-}(\eta) m_{2+}(\eta) \tag{5}$$

According to the results of sections 3.2.6.1 and 3.2.6.2 we get

(a) factorization of $m_1(\eta)$

$$m_{1+}(\eta) = \frac{1}{\omega\mu} m_{c+}(\eta), \quad m_{1-}(\eta) = m_{c-}(\eta) \tag{6}$$

where

$$m_{c+}(\alpha) = \frac{\sqrt{k - \eta}\, \Gamma\left(-\dfrac{\eta}{A} + \dfrac{B}{A} + 1\right) \exp\left[\dfrac{\tau a}{\pi} \log\dfrac{j\tau - \eta}{k} - q\eta\right]}{\sqrt{\cos(ka)}\, e^{\gamma\frac{\eta}{A}}\, \Gamma\left(\dfrac{B}{A} + 1\right)} \prod_{n=1}^{\infty} \frac{\left(1 - \dfrac{\eta}{An + B}\right)}{\left(1 - \dfrac{\eta}{\alpha_n}\right)}$$

$$m_{c-}(\eta) = m_{c+}(-\eta)$$

$$\alpha_n = \sqrt{k^2 - \left(\frac{(n-1/2)\pi}{a}\right)^2}, \quad A = -j\frac{\pi}{a}, \quad B = -\frac{1}{2}A, \quad n = 1, 2, \ldots$$

$$q = j\frac{a}{\pi}\left[\log\left(-\frac{j2\pi}{ka}\right) + 1 - \gamma\right]$$

(b) factorization of $m_2(\eta)$

$$m_{2+}(\eta) = -j\frac{1}{\omega\mu a}m_{s+}(\eta), \quad m_{2-}(\eta) = m_{s-}(\eta) \tag{7}$$

where

$$m_{s+}(\eta) = \left[\frac{e^{j\tau a}}{\frac{\sin(\tau a)}{\tau a}}\right]_+ = \sqrt{\frac{ka}{\sin(ka)}}\frac{e^{\frac{\tau a}{\pi}\log\frac{j\tau-\eta}{k}}\prod_{n=1}^{\infty}\frac{1-\frac{\eta a}{-j\pi n}}{1-\frac{\eta}{\eta_{an}}}}{e^{\gamma\frac{\eta a}{-j\pi}}\Gamma(1)\, e^{q\eta}}\Gamma\left(1-\frac{\eta a}{-j\pi}\right)$$

$$m_{s-}(\eta) = \left[\frac{e^{j\tau a}}{\frac{\sin(\tau a)}{\tau a}}\right]_- = m_{s+}(-\eta)$$

$$\eta_{an} = \sqrt{k^2 - \left(\frac{n\pi}{a}\right)^2} \quad n = 1, 2, \ldots, \quad \text{Im}[\eta_{an}] < 0$$

$$q = j\frac{a}{\pi}\left[\log\left(-\frac{j2\pi}{ka}\right) + 1 - \gamma\right]$$

To conclude, to accomplish the solution of $G(\eta)\cdot\begin{vmatrix} V_{1+} \\ V_{2+} \end{vmatrix} = \begin{vmatrix} A_{1-} \\ A_{2-} \end{vmatrix}$, we must take the source into account. This source can be an external plane wave, or a mode that propagates in the planar waveguide.

8.1.2 Source constituted by plane wave

Similarly to the half-plane problem, in the presence of a plane wave $E_z^i(x, y) = E_o e^{jk\,\rho\cos(\varphi-\varphi_o)}$, the primary contribution is present only in the terms A_{1-} and has the value

$$A_{1-}^p = \int_{-\infty}^{0} H_x^g(x, 0_+)e^{j\eta x}dx = 2j\frac{E_o\sin\varphi_o}{Z_o(\eta-\eta_o)}$$

where $\eta_o = -k\cos\varphi_o$. This yields the nonhomogeneous W-H equation

$$G(\eta)\cdot\begin{vmatrix} V_{1+} \\ V_{2+} \end{vmatrix} = \begin{vmatrix} A_{1-}^d \\ A_{2-} \end{vmatrix} + \begin{vmatrix} 2j\dfrac{E_o\sin\varphi_o}{Z_o} \\ 0 \end{vmatrix}\frac{1}{\eta-\eta_o} \tag{8}$$

Alternatively, by letting $V_{1+} = V_{1+}^s + V_{1+}^i$, $V_{2+} = V_{2+}^s + V_{2+}^i$, the contributions V_{1+}^i and V_{2+}^i of the incident field $E_z^i(x,y)$ on the two apertures $y = 0, x \geq 0$ and $y = -d, x \geq 0$ are

$$V_{1+}^i = j\frac{E_o}{\eta-\eta_o}, \quad V_{2+}^i = j\frac{E_o e^{-jk\sin\varphi_o d}}{\eta-\eta_o} \tag{9}$$

which confirms that the characteristic part of $\begin{vmatrix} A_{1-} \\ A_{2-} \end{vmatrix}$ in the pole $\eta_o = -k\cos\varphi_o$ is identical to its primary contribution: $\begin{vmatrix} 2j\frac{E_o\sin\varphi_o}{Z_o} \\ 0 \end{vmatrix}\frac{1}{\eta-\eta_o}$. In fact, we have

$$G(\eta_o)\begin{vmatrix} V_{1+}^i \\ V_{2+}^i \end{vmatrix} = \begin{vmatrix} 2j\dfrac{E_o\sin\varphi_o}{Z_o} \\ 0 \end{vmatrix}\frac{1}{\eta-\eta_o} \tag{10}$$

A third way to evaluate the source term is to set $V_{1+} = V_{1+}^d + V_{1+}^g$, $V_{2+} = V_{2+}^d + V_{2+}^g$ where V_{1+}^g and V_{2+}^g represent the geometrical optics contribution of the field on the two apertures. Assuming $\varphi_o > \frac{\pi}{2}$, since the aperture $y = -d, x \geq 0$ is illuminated only in the region $-d\cot\varphi_o < x < \infty$, it follows that V_{2+}^g differs from V_{2+}^i by the contribution

$$V_+^i(\eta) - V_+^g(\eta) = \int_0^{-d\cot\varphi_o} E_o e^{jk\rho\cos\varphi_o} e^{j\eta x} dx \tag{11}$$

However, this contribution is regular in $\eta_o = -k\cos\varphi_o$, and, consequently, we have again that the characteristic part of $\begin{vmatrix} A_{1-} \\ A_{2-} \end{vmatrix}$ is $\begin{vmatrix} 2j\frac{E_o\sin\varphi_o}{Z_o} \\ 0 \end{vmatrix}\frac{1}{\eta-\eta_o} = R_o\frac{1}{\eta-\eta_o}$.

Using the solution of the W-H equations obtained in section 2.4.2:

$$F_+(\eta) = G_+^{-1}(\eta)\cdot G_-^{-1}(\eta_o)\cdot\frac{R_o}{\alpha-\eta_o}, \quad F_-(\alpha) = G_-(\alpha)G_-^{-1}(\alpha_o)\frac{R_o}{\alpha-\alpha_o}$$

one obtains the explicit solution

$$\begin{aligned} \begin{vmatrix} V_{1+} \\ V_{2+} \end{vmatrix} &= \begin{vmatrix} \dfrac{1}{m_{1-}(\eta_o)m_{1+}(\eta)} + \dfrac{1}{m_{2-}(\eta_o)m_{2+}(\eta)} \\ \dfrac{1}{m_{1-}(\eta_o)m_{1+}(\eta)} - \dfrac{1}{m_{2-}(\eta_o)m_{2+}(\eta)} \end{vmatrix}\frac{jE_o\sin\varphi_o}{Z_o(\eta-\eta_o)} \\ \begin{vmatrix} A_{1-} \\ A_{2-} \end{vmatrix} &= \begin{vmatrix} \dfrac{m_{1-}(\eta)}{m_{1-}(\eta_o)} + \dfrac{m_{2-}(\eta)}{m_{2-}(\eta_o)} \\ \dfrac{m_{1-}(\eta)}{m_{1-}(\eta_o)} - \dfrac{m_{2-}(\eta)}{m_{2-}(\eta_o)} \end{vmatrix}\frac{jE_o\sin\varphi_o}{Z_o(\eta-\eta_o)} \end{aligned} \tag{12}$$

8.1.3 Source constituted by an incident mode

The modes in the planar waveguide are defined by the vanishing of $\cos \tau a$ (present in the denominator of $m_1(\eta)$) and by the vanishing of $\frac{\sin \tau a}{\tau}$ (present in the denominator of $m_2(\eta)$). They are also obtained via the zeroes of the function

$$\frac{\sin \tau d}{\tau} = 2\frac{\sin \tau a}{\tau}\cos \tau a = 0 \tag{13}$$

that yields

$$\eta_{dn} = \sqrt{k^2 - \left(\frac{n\pi}{d}\right)^2} \quad n = 1,2,\ldots, \quad \mathrm{Im}[\eta_{an}] < 0 \tag{14}$$

The first mode occurs for $n = 1$:

$$\eta_{d1} = \alpha_1 = \sqrt{k^2 - \left(\frac{\pi}{d}\right)^2} \tag{15}$$

For this mode we have

$$H_x^i(x,y) = H_o \cos\frac{\pi}{d}y\, e^{-j\alpha_1 x} \tag{16}$$

As indicated with A_{1-}^s and A_{2-}^s the conventional parts of the Laplace transforms of the currents present on the walls of the waveguide, we have

$$A_{1-} = A_{1-}^s + j\frac{H_o}{\eta - \alpha_1}, \quad A_{2-} = A_{2-}^s + j\frac{H_o}{\eta - \alpha_1} \tag{17}$$

Thus, the normal form of the W-H equation is

$$G(\eta)\cdot\begin{vmatrix} V_{1+} \\ V_{2+} \end{vmatrix} = \begin{vmatrix} A_{1-}^s \\ A_{2-}^s \end{vmatrix} + \begin{vmatrix} jH_o \\ jH_o \end{vmatrix}\frac{1}{\eta - \alpha_1} \tag{18}$$

which yields the solution

$$\begin{vmatrix} V_{1+} \\ V_{2+} \end{vmatrix} = \begin{vmatrix} \dfrac{1}{m_{1-}(\alpha_1)m_{1+}(\eta)} \\ \dfrac{1}{m_{1-}(\alpha_1)m_{1+}(\eta)} \end{vmatrix}\frac{jH_o}{\eta - \alpha_1} \tag{19}$$

$$\begin{vmatrix} A_{1-} \\ A_{2-} \end{vmatrix} = \begin{vmatrix} \dfrac{m_1(\eta)}{m_{2-}(\alpha_1)m_{2+}(\eta)} \\ \dfrac{m_1(\eta)}{m_{2-}(\alpha_1)m_{2+}(\eta)} \end{vmatrix}\frac{jH_o}{\eta - \alpha_1} \tag{20}$$

Note that in this case $\begin{vmatrix} V_{1+} \\ V_{2+} \end{vmatrix}$ presents only a continuous spectrum and, in particular, it does not contain the pole contribution at $\eta = \alpha_1$.

8.1.4 The skew plane wave case

Let us consider an incident plane wave impinging at a skew incidence angle β:

$$E_z^i = E_o e^{j\tau_o\rho\cos(\varphi-\varphi_o)} e^{-j\alpha_o z}, \quad H_z^i = H_o e^{j\tau_o\rho\cos(\varphi-\varphi_o)} e^{-j\alpha_o z} \tag{21}$$

where $\alpha_o = k\cos\beta$, $\tau_o = k\sin\beta = \sqrt{k^2-\alpha_o^2}$, and $\rho\cos(\varphi-\varphi_o) = x\cos\varphi_o + y\sin\varphi_o$.

Using Maxwell's equations yields the following other components of the incident plane wave:

$$\begin{aligned}
E_x^i &= e^{j\tau_o\rho\cos(\varphi-\varphi_o)} e^{-j\alpha_o z}\frac{E_o\alpha_o\cos\varphi_o + Z_o H_o k\sin\varphi_o}{\tau_o}\\
E_y^i &= e^{j\tau_o\rho\cos(\varphi-\varphi_o)} e^{-j\alpha_o z}\frac{E_o\alpha_o\sin\varphi_o - Z_o H_o k\cos\varphi_o}{\tau_o}\\
H_x^i &= e^{j\tau_o\rho\cos(\varphi-\varphi_o)} e^{-j\alpha_o z}\frac{-E_o k Y_o\sin\varphi_o + \alpha_o H_o\cos\varphi_o}{\tau_o}\\
H_y^i &= e^{j\tau_o\rho\cos(\varphi-\varphi_o)} e^{-j\alpha_o z}\frac{E_o k Y_o\cos\varphi_o + H_o\alpha_o\sin\varphi_o}{\tau_o}
\end{aligned} \tag{22}$$

where Z_o and $Y_o = Z_o^{-1}$ are the impedance and admittance of free space.

In the spectral domain we introduce the Fourier transforms:

$$\begin{aligned}
\mathbf{V}(\eta, y) &= e^{j\alpha_o z}\int_{-\infty}^{\infty}\hat{y}\times\mathbf{E}_t(x,y,z)e^{j\eta x}dx\\
\mathbf{I}(\eta, y) &= e^{j\alpha_o z}\int_{-\infty}^{\infty}\mathbf{H}_t(x,y,z)e^{j\eta x}dx
\end{aligned} \tag{23}$$

where

$$\mathbf{E}_t = \hat{z}E_z + \hat{x}E_x, \quad \mathbf{H}_t = \hat{z}H_z + \hat{x}H_x$$

The previous formulation again yields

$$\mathbf{G}(\eta)\cdot\begin{vmatrix}\mathbf{V}_{1+}\\ \mathbf{V}_{2+}\end{vmatrix} = \begin{vmatrix}\mathbf{A}_{1-}\\ \mathbf{A}_{2-}\end{vmatrix} \tag{24}$$

However, now the unknowns are vectors rather than scalars:

$$\mathbf{V}_{1+}(\eta) = \mathbf{V}(\eta, 0), \quad \mathbf{V}_{2+}(\eta) = \mathbf{V}(\eta, -d), \quad \mathbf{A}_{1-}(\eta) = \mathbf{I}(\eta, 0_+) - \mathbf{I}(\eta, 0_-)$$

$$\mathbf{A}_{2-}(\eta) = \mathbf{I}(\eta, -d_+) - \mathbf{I}(\eta, -d_-)$$

$$\begin{aligned}
\mathbf{G}(\eta) &= \begin{pmatrix}\mathbf{Y}_1+\mathbf{Y}_3+\mathbf{Y}_c & -\mathbf{Y}_3\\ -\mathbf{Y}_3 & \mathbf{Y}_1+\mathbf{Y}_3+\mathbf{Y}_c\end{pmatrix}\\
\xi = \xi(\eta) &= \sqrt{\tau_o^2-\eta^2}, \quad \mathbf{Y}_1 = \mathbf{Y}_2 = j\mathbf{Y}_c\tan\frac{\xi d}{2}, \quad \mathbf{Y}_3 = -j\mathbf{Y}_c\frac{1}{\sin\xi d}\\
\mathbf{Y}_c &= \frac{Y_o}{k\sqrt{k^2-\eta^2-\alpha_o^2}}\begin{vmatrix}\tau_o^2 & -\eta\,\alpha_o\\ -\eta\,\alpha_o & k^2-\eta^2\end{vmatrix}
\end{aligned} \tag{25}$$

The factorization of $\mathbf{G}(\eta)$ can be accomplished by putting the 4 × 4 matrix in the form

$$\mathbf{G}(\eta) = \begin{pmatrix} \mathbf{Y}_1 + \mathbf{Y}_3 + \mathbf{Y}_c & -\mathbf{Y}_3 \\ -\mathbf{Y}_3 & \mathbf{Y}_1 + \mathbf{Y}_3 + \mathbf{Y}_c \end{pmatrix} = G(\eta) \otimes R(\eta)$$

$$\begin{pmatrix} \mathbf{Y}_1+\mathbf{Y}_3+\mathbf{Y}_c & -\mathbf{Y}_3 \\ -\mathbf{Y}_3 & \mathbf{Y}_1+\mathbf{Y}_3+\mathbf{Y}_c \end{pmatrix} = \begin{pmatrix} Y_1+Y_3+Y_c & -Y_3 \\ -Y_3 & Y_1+Y_3+Y_c \end{pmatrix} \otimes \left(\frac{1}{\xi^2} \begin{vmatrix} \tau_o^2 & -\eta\,\alpha_o \\ -\eta\,\alpha_o & k^2-\eta^2 \end{vmatrix} \right) \tag{26}$$

where the first 2 × 2 matrix $G(\eta)$ is the matrix considered in the E-polarization case, $\otimes$ is the Kronecker product symbol, and the rational matrix $R(\eta)$ is defined by

$$R(\eta) = \frac{1}{\xi^2} \begin{vmatrix} \tau_o^2 & -\eta\,\alpha_o \\ -\eta\,\alpha_o & k^2-\eta^2 \end{vmatrix} = \frac{k}{\xi Y_o} \mathbf{Y}_c(\eta) \tag{27}$$

Taking into account that (see also section 4.2)

$$\begin{aligned} \mathbf{G}(\eta) &= G(\eta) \otimes R(\eta) = [G_-(\eta) \cdot G_+(\eta)] \otimes [R_-(\eta) \cdot R_+(\eta)] \\ &= [G_-(\eta) \otimes R_-(\eta)] \cdot [G_+(\eta) \otimes R_+(\eta)] = \mathbf{G}_-(\eta) \cdot \mathbf{G}_+(\eta) \end{aligned} \tag{28}$$

we obtain

$$\mathbf{G}_-(\eta) = [G_-(\eta) \otimes R_-(\eta)], \quad \mathbf{G}_+(\eta) = [G_+(\eta) \otimes R_+(\eta)] \tag{29}$$

Consequently, the factorization of $\mathbf{G}(\eta)$ is reduced to the factorization of $G(\eta)$ and $R(\eta)$ obtained in the previous sections.

To take into account the source we set

$$\mathbf{G}(\eta) \cdot \begin{vmatrix} \mathbf{V}_{1+} \\ \mathbf{V}_{2+} \end{vmatrix} = \begin{vmatrix} \mathbf{A}_{1-}^s \\ \mathbf{A}_{2-}^s \end{vmatrix} + \frac{\mathbf{R}_o}{\eta - \eta_o} \tag{30}$$

with $\eta_o = -\tau_o \cos\varphi_o$ and where, according to the previous considerations, $\mathbf{R}_o$ can be evaluated either as the characteristic part of the minus function $\begin{vmatrix} \mathbf{A}_{1-} \\ \mathbf{A}_{2-} \end{vmatrix}$ or the characteristic part of $\mathbf{G}(\eta) \cdot \begin{vmatrix} \mathbf{V}_{1+} \\ \mathbf{V}_{2+} \end{vmatrix}$.

We observe that geometrical optics arises from the following contributions:

wall $y = 0$, $x < 0$: incident wave plus reflected wave due to the PEC plane $y = 0$,
wall $y = -d$, $x < 0$: no geometrical optics contribution,
aperture $y = 0$, $x > 0$: incident wave,
aperture $y = -d$, $x > 0$: the incident wave is present provided that $x > -d \cot\varphi_o$.

The evaluation of the contribution by the walls is very simple since in this case for $\mathbf{A}_{1-}$ we have the same characteristic part considered in the skew incidence on a single PEC half-plane, whereas the characteristic part of $\mathbf{A}_{2-}$vanishes. It follows that

$$\mathbf{R}_o = \begin{vmatrix} -2j\,H_o \\ -\dfrac{2j\alpha_o \cos\varphi_o}{\tau_o} H_o + \dfrac{2jk \sin\varphi_o}{\tau_o Z_o} E_o \\ 0 \\ 0 \end{vmatrix} \tag{31}$$

Conversely, the evaluation the characteristic part of $\begin{vmatrix} \mathbf{V}_{1+} \\ \mathbf{V}_{2+} \end{vmatrix}$ is given by

$$\frac{\mathbf{T}_t}{\eta - \eta_o} \tag{32}$$

where, taking into account the considerations relevant to eq. (11),

$$\mathbf{T}_t = \begin{vmatrix} -j\dfrac{E_o \alpha_o \cos\varphi_o + Z_o H_o k \sin\varphi_o}{\tau_o} \\ jE_o \\ -j\, e^{-j\tau_o d \sin\varphi_o} \dfrac{E_o \alpha_o \cos\varphi_o + Z_o H_o k \sin\varphi_o}{\tau_o} \\ j\, e^{-j\tau_o d \sin\varphi_o} E_o \end{vmatrix}$$

Of course the following equation holds:

$$\mathbf{R}_o = \mathbf{G}(\eta_o) \cdot \mathbf{T}_t \tag{33}$$

8.2 The reversed half-planes problem

8.2.1 The E-polarization case

Let us consider the two reversed half-planes shown in Fig. 2a. This structure can be studied by the circuit representation indicated in Fig. 2b. In particular, the PEC walls of the

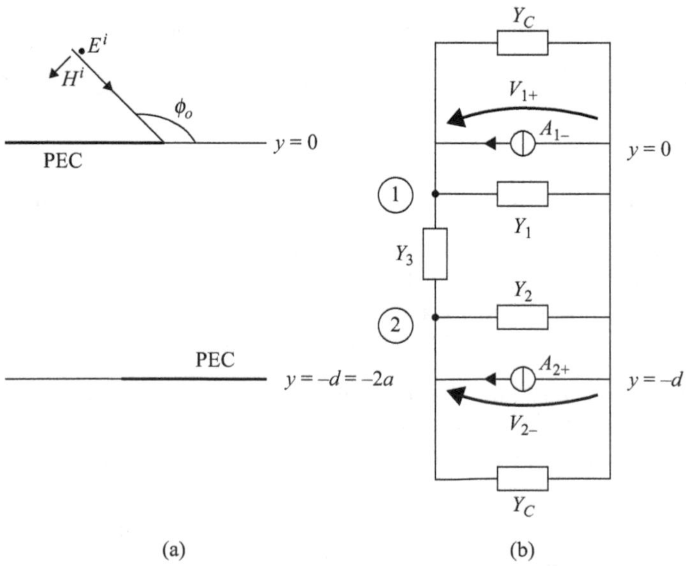

Fig. 2: Scattering by two inverted half-planes

waveguide are simulated by the current generators A_{1-}, A_{2+} and the slab $-d \leq y \leq 0$ in the physical structure is equivalent to the Π two-port introduced in chapter 7, Fig. 3a.

Writing the node equations on the circuit 2b yields directly the W-H equations[1] of the following problem:

$$\begin{aligned} &\text{node 1} \quad (Y_1 + Y_3 + Y_c)V_{1+} - Y_3 V_{2-} = A_{1-} \\ &\text{node 2} \quad -Y_3 V_{1+} + (Y_2 + Y_3 + Y_c)V_{2-} = A_{2+} \end{aligned} \tag{34}$$

where $Y_1 = Y_2 = jY_c \tan\frac{\tau d}{2}$, $Y_3 = -jY_c \frac{1}{\sin \tau d}$ and $-A_{1-}$ and $-A_{2+}$ are the Fourier transforms of the total currents induced on the two half-planes.

With respect to the problem considered in Fig. 1, here we deal with different plus and minus functions. In fact, now the plus functions are V_{1+} and A_{2+} and the minus functions are V_{2-} and A_{1-}. Algebraic manipulations yields the W-H equations written in normal form:

$$\begin{pmatrix} \dfrac{(Y_1 + Y_c)\cdot(Y_1 + 2Y_3 + Y)}{Y_1 + Y_3 + Y_c} & -\dfrac{Y_3}{Y_1 + Y_3 + Y_c} \\ \dfrac{Y_3}{Y_1 + Y_3 + Y_c} & \dfrac{1}{Y_1 + Y_3 + Y_c} \end{pmatrix} \cdot \begin{vmatrix} V_{1+} \\ A_{2+} \end{vmatrix} = \begin{vmatrix} A_{1-} \\ V_{2-} \end{vmatrix} \tag{35}$$

Substituting the values of Y_1, Y_2, Y_3, Y_c and taking into account the primary field contribution, for the E-polarization case (normal incidence) we get

$$G(\eta)\begin{vmatrix} V_{1+} \\ Z_o A_{2+} \end{vmatrix} = \begin{vmatrix} Z_o A^d_{1-} \\ V_{2-} \end{vmatrix} + \begin{vmatrix} 2j\dfrac{E_o \sin\varphi_o}{Z_o} \\ 0 \end{vmatrix} \dfrac{1}{\eta - \eta_o} \tag{36}$$

where the matrix kernel is given by

$$G(\eta) = \begin{pmatrix} \dfrac{2\tau}{k} & -e^{-j\tau d} \\ e^{-j\tau d} & \dfrac{1 - e^{-2j\tau d}}{2\tau} k \end{pmatrix} \tag{37}$$

and $\eta_o = -k\cos\varphi_o$.

Despite many attempts, up to now this matrix has not been factorized in closed form. A week factorizaction has been obtained in Abrahams and Wickham (1991) and Buyukaksoy and Serbest (1993). Alternatively, in the following we will use the Fredholm factorizaction described in chapter 5.

8.2.2 Qualitative characteristics of the solution

Taking into account that $\det[G(\eta)] = 1$, we observe that the structural singularities of the W-H unknowns are only the branch points $\eta = \pm k$. After obtaining the W-H unknowns

[1] Equation (34) are not written in normal form.

V_{1+} and A_{2+}, and V_{2-} and A_{1-}, we can use the following expressions to evaluate $V(\eta, y)$ and $I(\eta, y)$ and the corresponding transverse components of the electromagnetic field everywhere:

for $y > 0$

$$E_z(x,y) = E_z^i(x,y) + E_z^r(x,y) + \frac{1}{2\pi}\int_{-\infty}^{\infty} V(\eta,0)e^{-j\eta x}e^{-j\tau y}d\eta \tag{38}$$

$$H_x(x,y) = H_x^i(x,y) + H_x^r(x,y) + \frac{1}{2\pi}\int_{-\infty}^{\infty} I(\eta,0)e^{-j\eta x}e^{-j\tau y}d\eta$$
$$V(\eta,y) = V_{1+}e^{-j\tau y}, \quad I(\eta,y) = \frac{V_{1+}}{Z_c}e^{-j\tau y} \tag{39}$$

for $-d < y < 0$

$$V(\eta,y) = \frac{V_{1+}\sin(\tau(d+y)) - V_{2-}\sin(\tau y)}{\sin(\tau d)}$$
$$I(\eta,y) = j\frac{V_{1+}\cos(\tau(d+y)) - V_{2-}\cos(\tau y)}{Z_c\sin(\tau d)} \tag{40}$$

for $y < -d$

$$V(\eta,y) = V_{2-}e^{-j\tau(y+d)}, \quad I(\eta,y) = -\frac{V_{2-}}{Z_c}e^{-j\tau(y+d)}$$

These expressions show that $V(\eta, y)$ and $I(\eta, y)$ contain as singularities only the branch points $\eta = \pm k$.

It seems that $V(\eta, y)$ and $I(\eta, y)$ in eqs. (40) are exponentially unbounded as $\eta \to \pm\infty$. The following alternative expressions are more suitable for the evaluation in the region $(-d < y < 0)$ through an inverse Fourier transform:

$$V(\eta,y) = \omega\mu\frac{A_{1-}e^{j\tau y} + e^{-j\tau(d+y)})A_{2+}}{2\tau}$$

$$I(\eta,y) = \frac{\tau(e^{-j\tau y} + e^{j\tau(2d+y)})V_{1+}}{(1-e^{2j\tau d})\omega\mu} - \frac{\tau(e^{j\tau(d+y)} + e^{j\tau(d-y)})V_{2-}}{(1-e^{2j\tau d})\omega\mu} \quad (-d < y < 0)$$

8.2.3 Numerical evaluation of the electromagnetic field

The electromagnetic field can be evaluated through the inverse Fourier transforms

$$E_z(x,y) = \frac{1}{2\pi}\int_{-\infty}^{\infty} V(\eta,y)e^{-j\eta x}d\eta \tag{41}$$

$$H_x(x,y) = \frac{1}{2\pi} \int_{-\infty}^{\infty} I(\eta, y) e^{-j\eta x} d\eta \tag{42}$$

In the region $(-d < y < 0)$, it is sufficient to use the inverse discrete Fourier transform. In the open region, the far field can be evaluated with the saddle point method (section 2.9.2).

8.2.4 Numerical solution of the W-H equations

The factorization of the matrix $G(\eta)$ has been considered by Büyükaksoy and Serbest (1993). Alternatively an efficient approximate factorization can be obtained using the Fredholm integral equation technique (chapter 5). First we normalize $G(\eta)$ in the form

$$G(\eta) = \begin{pmatrix} \frac{2\tau}{k} & -e^{-j\tau d} \\ e^{-j\tau d} & \frac{1-e^{-2j\tau d}}{2\tau} k \end{pmatrix} = \begin{pmatrix} \frac{\sqrt{k+\eta}}{\sqrt{k}} & 0 \\ 0 & \frac{\sqrt{k}}{\sqrt{k+\eta}} \end{pmatrix} \cdot M(\eta) \cdot \begin{pmatrix} \frac{\sqrt{k-\eta}}{\sqrt{k}} & 0 \\ 0 & \frac{\sqrt{k}}{\sqrt{k-\eta}} \end{pmatrix} \tag{43}$$

where

$$M(\eta) = \begin{pmatrix} 2 & -\frac{\sqrt{k-\eta}}{\sqrt{k+\eta}} e^{-j\tau d} \\ \frac{\sqrt{k+\eta}}{\sqrt{k-\eta}} e^{-j\tau d} & \frac{1-e^{-2j\tau d}}{2} \end{pmatrix} \tag{44}$$

We observe that

$$M^{-1}(\eta) = \begin{pmatrix} \frac{1-e^{-2j\tau d}}{2} & \frac{\sqrt{k-\eta}}{\sqrt{k+\eta}} e^{-j\tau d} \\ -\frac{\sqrt{k+\eta}}{\sqrt{k-\eta}} e^{-j\tau d} & 2 \end{pmatrix} \tag{45}$$

is bounded. Hence, the Fredholm equation

$$M(\eta) F_+(\eta) + \frac{1}{2\pi j} \cdot \int_{-\infty}^{+\infty} \frac{[M(t) - M(\eta)] F_+(t)}{t-\eta} dt = \frac{R}{\eta - \eta_o} \tag{46}$$

involves a Fredholm kernel $M^{-1}(\eta) \frac{[M(t)-M(\eta)]}{t-\eta}$ that is compact. Next we use the approximate techniques indicated in section 5.1.5. Since the original integration line (real axis) is very near the branch points $\pm k$ and the pole η_o, it is convenient to warp the integration path. According to the indications of section 5.1.5.1, in the following we use three paths.

Path G:

$$\eta(y) = -k\cos\left[-\frac{\pi}{2} + \frac{1}{2}gd(-y) - jy\right], \quad \text{(gudermann line)} \tag{47}$$

with $gd(x) = arc\cos\frac{1}{\cosh x}\operatorname{sgn}(x)$.

Path B:

$$\eta(y) = e^{j\frac{\pi}{4}}y, \quad -\infty < y < \infty, \quad \text{(bisector straight line)} \tag{48}$$

Path C:

$$\eta(x) = k\left[x + j\frac{\arctan x}{10(1+x^2)}\right], \quad \text{(slight deformed real axis)} \tag{49}$$

8.2.4.1 Numerical simulations

Due to the compactness of the Fredholm kernel $M^{-1}(\eta)\frac{[M(t)-M(\eta)]}{t-\eta}$, we are sure that by increasing A and decreasing h in the numerical scheme indicated in section 5.1.5, the approximate solution converges to the exact solution (Kantorovich & Krylov, 1967).

Consequently, the problem is to find the best technique to minimize the computer time.

Previously we have indicated three possible ways based on the warping of the original real axis along the paths G, B, and C.

Working in the w – plane (path G) implies the possibility to select small values of A, since the kernel $m(\eta, t)$ behaves as $\frac{1}{\cosh y}$ instead of $\frac{1}{y}$ as $y \to \infty$. Conversely, working in the η – plane (paths B and C), we observe that the discretization is uniform in the η – plane and assures that the contribution of the integrand is well considered everywhere in the integral.

Since a comparison with an exact solution is not available, to have a criterion on the accuracy of the approximate solution we changed the values of A and h and compared the different solutions obtained with the three paths G, B, and C. For the sake of simplicity we report only the numerical results obtained with path G when $A = 20$ and $h = 0.04$.

We have considered the unknowns:

$$\tilde{V}_{1+}(\eta) = \sqrt{\frac{k-\eta}{k}}V_{1+}(\eta), \quad \tilde{V}_{2-}(\eta) = \sqrt{\frac{k+\eta}{k}}V_{2-}(\eta) \tag{50}$$

Figure 3 shows the value of the function $\left|\hat{V}_{2-}(\varphi)\right| = \tilde{V}_{2-}(-k\cos\varphi)$, whereas Fig. 4 reports the value of the function $\left|\hat{V}_{1+}(\varphi)\right| = \tilde{V}_{1+}(-k\cos\varphi)$.

The numerical evaluation of $\hat{V}_{1+}(\varphi)$ is very accurate. Changing the path of the integration and/or the values of A and h provides plots that are indistinguishable. The relative error from the different evaluations is less than 2–3%. This does not happen for the function $\hat{V}_{2-}(\varphi)$, which is related to the electric field in an aperture located in a shadow region. The plot obtained in Fig. 3 should be accurate, since it was obtained from high values of A and small values of h.

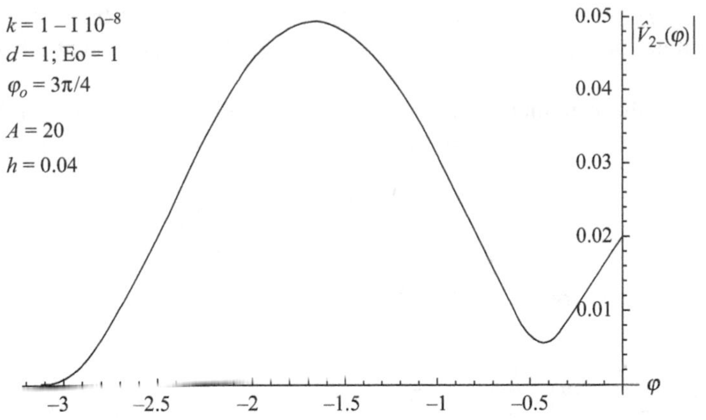

Fig. 3: Plot of $\left|\hat{V}_{2-}(\varphi)\right|$

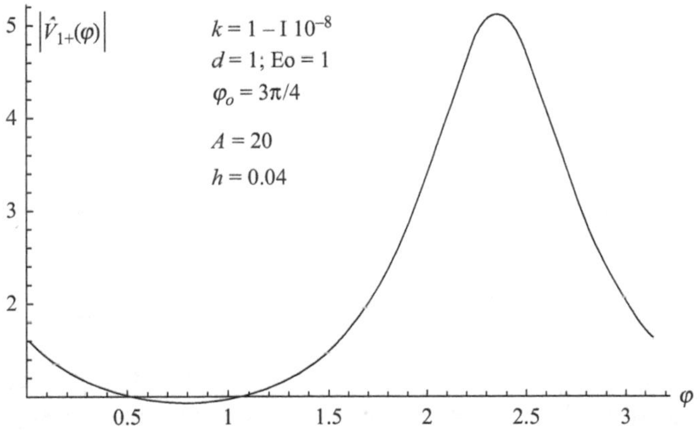

Fig. 4: Plot of $\left|\hat{V}_{1+}(\varphi)\right|$

8.2.4.2 A method for increasing the accuracy or the function $V_{2-}(\eta)$

From one of the W-H equations we have

$$V_{2-}(\eta) = e^{-j\tau d} V_{1+}(\eta) + jkd \frac{e^{-j\tau d} \sin(\tau d)}{\tau d} Z_o A_{2+}(\eta) \tag{51}$$

We observed that whereas the accuracy of the approximate evaluations of $V_{1+}(\eta)$ and $A_{2+}(\eta)$ is good, this does not happen for the function $V_{2-}(\eta)$. For instance, in its inverse Fourier transform $E_z(x, -d)$ presents significant values for $x > 0$ in contrast with the fact that $E_z(x, -d)$ must vanish on the PEC half-plane located in $x > 0, y = -d$.

The reason for the discrepancy is because especially if $k\,d$ is small the values of $E_z(x, y)$ on the aperture $x < 0, y = -d$ are very small with respect to those on the aperture $x > 0, y = 0$. Consequently, we need a high accuracy in the approximate expressions of

$V_{1+}(\eta)$ and $A_{2+}(\eta)$ to have a sufficient accuracy of $V_{2-}(\eta)$ obtained by eq. (51); for instance, this accuracy must be high near $\eta = k$, since $V_{1+}(\eta)$ and $A_{2+}(\eta)$ present a branch point but $V_{2-}(\eta)$ is regular.

To overcome these numerical problems, we consider (51) as a scalar W-H equation having $V_{2-}(\eta)$ and $A_{2+}(\eta)$ as unknowns. The factorization of the function

$$M(\alpha) = e^{-j\tau d} \frac{\sin\left[\sqrt{k^2-\alpha^2}\, d\right]}{\sqrt{k^2-\alpha^2} d} \tag{52}$$

was accomplished in chapter 3, section 3.2.6.2, example 6, yielding

$$M_+(\alpha) = \frac{\sqrt{\frac{\sin(kd)}{kd}} \exp\left[-\frac{\tau\, d}{\pi}\log\frac{j\tau-\alpha}{k} + q\,\alpha\right] e^{\gamma\frac{\alpha}{A}}\Gamma\left(\frac{B}{A}+1\right)}{\Gamma\left(-\frac{\alpha}{A}+\frac{B}{A}+1\right)} \prod_{n=1}^{\infty} \frac{\left(1-\frac{\alpha}{\alpha_n}\right)}{\left(1-\frac{\alpha}{An+B}\right)} \tag{53}$$

$$M_-(\alpha) = M_+(-\alpha) \tag{54}$$

where

$$\begin{aligned} &\tau = \sqrt{k^2-\alpha^2} \\ &A = -j\frac{\pi}{d}, \quad B = 0, \quad q = j\frac{d}{\pi}\left[\log\left(-\frac{j2\pi}{kd}\right) + 1 - \gamma\right] \\ &\alpha_n = \sqrt{k^2 - \left(\frac{n\pi}{d}\right)^2}, \quad n = 1, 2, \ldots, \quad \mathrm{Im}[\alpha_n] < 0 \end{aligned} \tag{55}$$

By assuming $\lambda = 1$, $k = \frac{2\pi}{\lambda}(1 - j10^{-6})$, $d = 1.1\frac{\lambda}{2}$ and a truncated product with $N_b = 200$, we obtain an error $M_-(\alpha)M_+(\alpha) - M(\alpha)$ that vanishes in the range $-100 \le \alpha \le 100$.

We also ascertained that with these approximations, the functions $\sqrt{k-\alpha}\, M_+(\alpha)$ and its inverse $(\sqrt{k-\alpha}\, M_+(\alpha))^{-1}$ behave as a bounded not vanishing constant as $\alpha \to \infty$.

The W-H technique yields the following expression for $V_{2-}(\eta)$:

$$V_{2-}(\eta) = M_-(\eta) Y_-(\eta) \tag{56}$$

where $Y_-(\eta) = \left\{M_-^{-1}(\eta) e^{-j\sqrt{k^2-\eta^2}\, d} V_{1+}(\eta)\right\}_-$ is the minus decomposed function of $M_-^{-1}(\eta) e^{-j\sqrt{k^2-\eta^2}\, d} V_{1+}(\eta)$ that is expressed by (see chapter 3, section 3.1, eq. (3)):

$$Y_-(\eta) = -\frac{1}{2\pi j} \int_{\gamma_2} \frac{M_-^{-1}(u) e^{-j\tau(u)d} V_{1+}(u)}{u-\eta}\, du \tag{57}$$

where $\tau(\eta) = \sqrt{k^2-\eta^2}$. By decomposing $\tau = \tau_- + \tau_+$ (chapter 3, section 3.1.1, eq. (13)):

$$\tau_+(\eta) = \frac{\tau(\eta)}{j\pi} \log\frac{j\tau(\eta)-\eta}{k}$$

$$\tau_-(\eta) = \frac{\tau(\eta)}{j\pi}\log\frac{j\tau(\eta)+\eta}{k} \tag{58}$$

we obtain

$$\begin{aligned} Y_-(\eta) &= -\frac{1}{2\pi j}\int_{\gamma_2} \frac{M_-^{-1}(u)e^{-j\tau_-(u)d}e^{-j\tau_+(u)d}V_{1+}(u)}{u-\eta}du \\ &= -\frac{1}{2\pi j}\int_{\gamma_2} \frac{\left[M_-^{-1}(u)e^{-j\tau_-(u)d} - M_-^{-1}(\eta)e^{-j\tau_-(\eta)d}\right]e^{-j\tau_+(u)d}V_{1+}(u)}{u-\eta}du \\ &\quad + \frac{M_-^{-1}(\eta)e^{-j\tau_-(\eta)d}}{2\pi j}\int_{\gamma_2} \frac{e^{-j\tau_+(u)d}V_{1+}(u)}{u-\eta}du \end{aligned} \tag{59}$$

The last integral vanishes since it represents the minus part of a plus function. This yields

$$M_-^{-1}(\eta)V_{2-}(\eta) = -\frac{1}{2\pi j}\int_{-\infty}^{\infty} \frac{\left[M_-^{-1}(u)e^{-j\tau_-(u)d} - M_-^{-1}(\eta)e^{-j\tau_-(\eta)d}\right]e^{-j\tau_+(u)d}V_{1+}(u)}{u-\eta}du \tag{60}$$

A numerical quadrature gives

$$M_-^{-1}(\eta)V_{2-}(\eta) = -\frac{h}{2\pi j}\sum_{i=-A/h}^{A/h} \theta(\eta, hi)e^{-j\tau_+(hi)d}V_{1+}(hi) \tag{61}$$

where

$$\theta(\eta, u) = \mathrm{If}\left[u=\eta,\ \frac{d}{d\eta}[\lambda_-(\eta)],\ \frac{\lambda_-(u)-\lambda_-(\eta)}{u-\eta}\right] \tag{62}$$

with

$$\lambda_-(\eta) = M_-^{-1}(\eta)e^{-j\tau_-(\eta)d} = \frac{\exp[-q\,\eta]e^{-\gamma\frac{\eta}{A}}\Gamma\left(-\frac{\eta}{A}+\frac{B}{A}+1\right)}{\sqrt{\frac{\sin(kd)}{kd}}\Gamma\left(\frac{B}{A}+1\right)}\prod_{n=1}^{\infty}\frac{\left(1-\frac{\eta}{\alpha_n}\right)}{\left(1-\frac{\eta}{An+B}\right)} \tag{63}$$

8.2.5 Source constituted by a skew plane wave

Consider an incident plane wave impinging at a skew incidence angle β on the two reversed planes shown in Fig. 2a:

$$E_z^i = E_o e^{j\tau_o\rho\cos(\varphi-\varphi_o)}e^{-j\alpha_o z}, \quad H_z^i = H_o e^{j\tau_o\rho\cos(\varphi-\varphi_o)}e^{-j\alpha_o z} \tag{64}$$

where $\alpha_o = k\cos\beta$, $\tau_o = k\sin\beta = \sqrt{k^2-\alpha_o^2}$, $\rho\cos(\varphi-\varphi_o) = x\cos\varphi_o + y\sin\varphi_o$

Using Maxwell's equations yields the following additional components of the incident plane wave:

$$\begin{aligned}
E_x^i &= e^{j\tau_o \rho \cos(\varphi-\varphi_o)} e^{-j\alpha_o z} \frac{E_o\alpha_o \cos\varphi_o + Z_o H_o k \sin\varphi_o}{\tau_o} \\
E_y^i &= e^{j\tau_o \rho \cos(\varphi-\varphi_o)} e^{-j\alpha_o z} \frac{E_o\alpha_o \sin\varphi_o - Z_o H_o k \cos\varphi_o}{\tau_o} \\
H_x^i &= e^{j\tau_o \rho \cos(\varphi-\varphi_o)} e^{-j\alpha_o z} \frac{-E_o k Y_o \sin\varphi_o + \alpha_o H_o \cos\varphi_o}{\tau_o} \\
H_y^i &= e^{j\tau_o \rho \cos(\varphi-\varphi_o)} e^{-j\alpha_o z} \frac{E_o k Y_o \cos\varphi_o + H_o \alpha_o \sin\varphi_o}{\tau_o}
\end{aligned} \tag{65}$$

where Z_o and $Y_o = Z_o^{-1}$ are the impedance and admittance of free space. In the spectral domain we introduce the following Fourier transforms:

$$\begin{aligned}
\mathbf{V}(\eta, y) &= e^{j\alpha_o z} \int_{-\infty}^{\infty} \hat{y} \times \mathbf{E}_t(x, y, z) e^{j\eta x} dx \\
\mathbf{I}(\eta, y) &= e^{j\alpha_o z} \int_{-\infty}^{\infty} \mathbf{H}_t(x, y, z) e^{j\eta x} dx
\end{aligned} \tag{66}$$

where $\mathbf{E}_t = \hat{z}E_z + \hat{x}E_x$, $\mathbf{H}_t = \hat{z}H_z + \hat{x}H_x$.

The previous formulation yield again

$$\mathbf{G}(\eta) \cdot \begin{vmatrix} \mathbf{V}_{1+} \\ \mathbf{A}_{2+} \end{vmatrix} = \begin{vmatrix} \mathbf{A}_{1-} \\ \mathbf{V}_{2-} \end{vmatrix} = \begin{vmatrix} \mathbf{A}_{1-}^s \\ \mathbf{V}_{2-}^s \end{vmatrix} + \frac{\mathbf{R}_o}{\eta - \eta_o} \tag{67}$$

where

$$\begin{aligned}
&\eta_o = -\tau_o \cos\varphi_o \\
&\mathbf{V}_{1+}(\eta) = \mathbf{V}(\eta, 0), \quad \mathbf{V}_{2-}(\eta) = \mathbf{V}(\eta, -d), \quad \mathbf{A}_{1-}(\eta) = \mathbf{I}(\eta, 0_+) - \mathbf{I}(\eta, 0_-) \\
&\mathbf{A}_{2+}(\eta) = \mathbf{I}(\eta, -d_+) - \mathbf{I}(\eta, -d_-)^2
\end{aligned}$$

$$\mathbf{G}(\eta) = \begin{pmatrix} \dfrac{(\mathbf{Y_1}+\mathbf{Y_c})\cdot(\mathbf{Y_1}+2\mathbf{Y_3}+\mathbf{Y})}{\mathbf{Y_1}+\mathbf{Y_3}+\mathbf{Y_c}} & -\dfrac{\mathbf{Y_3}}{\mathbf{Y_1}+\mathbf{Y_3}+\mathbf{Y_c}} \\ \dfrac{\mathbf{Y_3}}{\mathbf{Y_1}+\mathbf{Y_3}+\mathbf{Y_c}} & \dfrac{\mathbf{1}}{\mathbf{Y_1}+\mathbf{Y_3}+\mathbf{Y_c}} \end{pmatrix} = \begin{pmatrix} 2\,\mathbf{Y}_c & -e^{-j\xi d}\mathbf{1}_2 \\ e^{-j\xi d}\mathbf{1}_2 & je^{-j\xi d}\sin(\xi d)\mathbf{Z}_c \end{pmatrix} \tag{68}$$

with

$$\xi = \xi(\eta) = \sqrt{\tau_o^2 - \eta^2}, \quad \mathbf{Y_1} = \mathbf{Y_2} = j\mathbf{Y_c}\tan\frac{\xi d}{2}, \quad \mathbf{Y_3} = -j\mathbf{Y}_c\frac{1}{\sin\xi d}$$

$$\mathbf{Y_c} = \frac{Y_o}{k\sqrt{k^2-\eta^2-\alpha_o^2}} \begin{vmatrix} \tau_o^2 & -\eta\,\alpha_o \\ -\eta\,\alpha_o & k^2-\eta^2 \end{vmatrix}, \quad \mathbf{Z_c} = \frac{Z_o}{k\sqrt{k^2-\alpha_o^2-\eta^2}} \begin{vmatrix} k^2-\eta^2 & \eta\,\alpha_o \\ \eta\,\alpha_o & \tau_o^2 \end{vmatrix}$$

---

[2] The expressions used in the components of $\mathbf{G}(\eta)$ make sense since all the introduced matrices commute.

and $\mathbf{1}_2$ is the 2×2 identity matrix. According to the previous considerations $\mathbf{R}_o$ can be evaluated either as the characteristic part of the minus function $\begin{vmatrix} \mathbf{A}_{1-} \\ \mathbf{V}_{2-} \end{vmatrix}$ or the characteristic part of $\mathbf{G}(\eta) \cdot \begin{vmatrix} \mathbf{V}_{1+} \\ \mathbf{A}_{2+} \end{vmatrix}$.

We observe that geometrical optics arises from the following contributions:

Wall: $y = 0,\ x < 0$: incident wave plus reflected wave due to the PEC plane,
Aperture: $y = -d,\ x < 0$: no geometrical optics contribution,
Aperture : $y = 0,\ x > 0$: incident wave plus reflected wave due to the PEC plane $y = -d$,
Wall: $y = -d,\ x > 0$: incident wave plus reflected wave due to a PEC plane $y = -d$.

The first evaluation is very simple since in this case for $\mathbf{A}_{1-}$ we have the same characteristic part considered in the skew incidence on a single PEC half-plane, whereas the characteristic part of $\mathbf{V}_{2-}$vanishes. It follows that

$$\mathbf{R}_o = \begin{vmatrix} -2j\,H_o \\ -\dfrac{2j\alpha_o \cos\varphi_o}{\tau_o} H_o + \dfrac{2jk\sin\varphi_o}{\tau_o\, Z_o} E_o \\ 0 \\ 0 \end{vmatrix} \tag{69}$$

Conversely, in the alternative evaluation the characteristic part of $\begin{vmatrix} \mathbf{V}_{1+} \\ \mathbf{V}_{2+} \end{vmatrix}$ is given by

$$\frac{\mathbf{T}_t}{\eta - \eta_o} \tag{70}$$

where, taking into account the considerations relevant to eq. (11),

$$\mathbf{T}_t = \begin{vmatrix} \dfrac{(-j + je^{-2jd\tau_o \sin\varphi_o})(E_o\alpha_o \cos\varphi_o + H_o k Z_o \sin\varphi_o)}{\tau_o} \\ (j - je^{-2jd\tau_o \sin\varphi_o})E_o \\ 2je^{-jd\tau_o \sin\varphi_o} H_o \\ \dfrac{2je^{-jd\tau_o \sin\varphi_o}(H_o Z_o \alpha_o \cos\varphi_o - E_o k \sin\varphi_o)}{\tau_o Z_o} \end{vmatrix}$$

Of course the following equation holds:

$$\mathbf{R}_o = \mathbf{G}(\eta_o) \cdot \mathbf{T}_t \tag{71}$$

Taking into account that $\mathbf{Z_c} = \mathbf{Y}_c^{-1}$ commutes with $\mathbf{Y}_c$, the factorization of $\mathbf{G}(\eta)$ can be accomplished by putting the 4×4 matrix in the form:

$$\mathbf{G}(\eta) = \mathbf{A}(\eta) \otimes \mathbf{1}_2 + \mathbf{B}(\eta) \otimes \mathbf{Y}_c(\eta)$$

where the matrices $\mathbf{A}(\eta)$ and $\mathbf{B}(\eta)$ are 2×2 matrices.

The key point to reduce the order of matrices to be factorized is the observation that the polynomial matrix $t(\eta) = \begin{vmatrix} \eta & -\alpha_o \\ \alpha_o & \eta \end{vmatrix}$ makes $\mathbf{Y_c}$ and $\mathbf{Z_c}$ diagonal. In fact,

$$\mathbf{Z}_c = t^{-1}(\eta) \cdot \begin{vmatrix} \dfrac{\xi}{\omega\varepsilon} & 0 \\ 0 & \dfrac{k^2}{\omega\varepsilon\xi} \end{vmatrix} \cdot t(\eta), \quad \mathbf{Z}_c = t^{-1}(\eta) \cdot \begin{vmatrix} \dfrac{\omega\varepsilon}{\xi} & 0 \\ 0 & \dfrac{\xi}{\omega\mu} \end{vmatrix} \cdot t(\eta) \tag{72}$$

or

$$\begin{vmatrix} k^2 - \eta^2 & \eta\,\alpha_o \\ \eta\,\alpha_o & \tau_o^2 \end{vmatrix} = t^{-1}(\eta) \cdot d(\eta) \cdot t(\eta) \tag{73}$$

where

$$d(\eta) = \begin{vmatrix} \xi^2 & 0 \\ 0 & k^2 \end{vmatrix}$$

It follows that putting $T(\eta) = \begin{vmatrix} t(\eta) & 0 \\ 0 & t(\eta) \end{vmatrix} = 1_2 \otimes t(\eta)$, the matrix $T(\eta) \cdot G(\eta) \cdot T^{-1}(\eta)$ has the form

$$T(\eta) \cdot G(\eta) \cdot T^{-1}(\eta) = \begin{vmatrix} x & 0 & x & 0 \\ 0 & x & 0 & x \\ x & 0 & x & 0 \\ 0 & x & 0 & x \end{vmatrix} \tag{74}$$

where x means a nonzero element. Introducing the permutation matrix $P = P^{-1}$ defined by

$$P = \begin{vmatrix} 1 & 0 & 0 & 0 \\ 0 & 0 & 1 & 0 \\ 0 & 1 & 0 & 0 \\ 0 & 0 & 0 & 1 \end{vmatrix} \tag{75}$$

we get the following quasi-diagonal matrix:

$$P \cdot T(\eta) \cdot G(\eta) \cdot T^{-1}(\eta) \cdot P = \begin{vmatrix} G_{d1}(\eta) & 0 \\ 0 & G_{d2}(\eta) \end{vmatrix} \tag{76}$$

where the matrices $G_{d1}(\eta)$ and $G_{d2}(\eta)$ are of order two.

To have matrices that with their inverses exist and are bounded as $\eta \to \pm\infty$, we operate a normalization by introducing the matrix $\mathbf{G}_t(\eta)$ defined by

$$\mathbf{G}_t(\eta) = n_- \cdot diag[1, Y_o, Z_o, 1] \cdot P \cdot T \cdot \mathbf{G}(\eta) \cdot T^{-1} \cdot P \cdot diag[Z_o, 1, 1, Y_o] \cdot n_+ = \begin{pmatrix} \mathbf{G}_{th}(\eta) & \mathbf{0} \\ \mathbf{0} & \mathbf{G}_{te}(\eta) \end{pmatrix} \tag{77}$$

where the normalization matrices $n_\pm$ are given by

$$n_\mp = diag\left[\frac{\sqrt{\tau_o \pm \eta}}{\sqrt{\tau_o}}, \frac{\sqrt{\tau_o}}{\sqrt{\tau_o \pm \eta}}, \frac{\sqrt{\tau_o}}{\sqrt{\tau_o \pm \eta}}, \frac{\sqrt{\tau_o \pm \eta}}{\sqrt{\tau_o}}\right] \tag{78}$$

The 2 × 2 matrices $G_{th}(\eta)$ and $G_{te}(\eta)$ are suitable for the Fredholm factorization since they are bounded with their inverses as $\alpha \to \infty$. They have the following explicit expressions:

$$\mathbf{G}_{th}(\eta) = \begin{pmatrix} 2 & -e^{-j\xi d}\dfrac{\sqrt{\tau_o+\eta}}{\sqrt{\tau_o-\eta}} \\ e^{-j\xi d}\dfrac{\sqrt{\tau_o-\eta}}{\sqrt{\tau_o+\eta}} & je^{-j\xi d}\sin(\xi d) \end{pmatrix}, \quad \mathbf{G}_{te}(\eta) = \begin{pmatrix} 2 & -e^{-j\xi d}\dfrac{\sqrt{\tau_o-\eta}}{\sqrt{\tau_o+\eta}} \\ e^{-j\xi d}\dfrac{\sqrt{\tau_o+\eta}}{\sqrt{\tau_o-\eta}} & je^{-j\xi d}\sin(\xi d) \end{pmatrix} \tag{79}$$

In the previous section we have factorized $G_{te}(\eta)$ by the Fredholm method. Now we show that the factorization of $G_{th}(\eta)$ reduces to that of $G_{te}(\eta) = G_{te-}(\eta) \cdot G_{te+}(\eta)$. In fact, the following equation holds:

$$G_{th}(\eta) = \begin{pmatrix} 1 & 0 \\ 0 & \dfrac{\tau_o - \eta}{\tau_o + \eta} \end{pmatrix} \cdot G_{te}(\eta) \cdot \begin{pmatrix} 1 & 0 \\ 0 & \dfrac{\tau_o + \eta}{\tau_o - \eta} \end{pmatrix} \tag{80}$$

which provides the weak factorization of $G_{th}(\eta) = G_{thw-}(\eta) \cdot G_{thw+}(\eta)$, where

$$G_{thw-}(\eta) = \begin{pmatrix} 1 & 0 \\ 0 & \dfrac{\tau_o - \eta}{\tau_o + \eta} \end{pmatrix} \cdot G_{te-}(\eta), \quad G_{thw+}(\eta) = G_{te+}(\eta) \cdot \begin{pmatrix} 1 & 0 \\ 0 & \dfrac{\tau_o + \eta}{\tau_o - \eta} \end{pmatrix} \tag{81}$$

Starting from the weak factorizations (81), we can accomplish the standard factorizations of $G_{th}(\eta) = G_{th-}(\eta) \cdot G_{th+}(\eta)$ and $G_{te}(\eta) = G_{te-}(\eta) \cdot G_{te+}(\eta)$ by using known techniques (section 2.7). With the standard factorized matrices, taking into account eq. (78), we get the following factorization of $\mathbf{G}(\eta) = \mathbf{G}_{w-}(\eta) \cdot \mathbf{G}_{w+}(\eta)$:

$$\mathbf{G}_{w-}(\eta) = [n_- \cdot diag[1, Y_o, Z_o, 1] \cdot P \cdot T]^{-1} \cdot \begin{pmatrix} \mathbf{G}_{th-}(\eta) & \mathbf{0} \\ \mathbf{0} & \mathbf{G}_{te-}(\eta) \end{pmatrix} \tag{82}$$

$$\mathbf{G}_{w+}(\eta) = \begin{pmatrix} \mathbf{G}_{th+}(\eta) & \mathbf{0} \\ \mathbf{0} & \mathbf{G}_{te+}(\eta) \end{pmatrix} \cdot \left[T^{-1} \cdot P \cdot diag[Z_o, 1, 1, Y_o] \cdot n_+\right]^{-1} \tag{83}$$

This factorization is again weak for the presence of the polynomial matrix T, which introduces a nonvanishing entire function in the W-H solution technique (section 2.4.1). To evaluate this entire function we will use the following result:

$$t^{-1}(\eta) = \frac{1}{\eta^2 + \alpha_o^2}\begin{vmatrix} \eta & \alpha_o \\ -\alpha_o & \eta \end{vmatrix} \quad \text{or} \quad T^{-1} = \frac{T_a(\eta)}{\eta^2 + \alpha_o^2} \tag{84}$$

where

$$T_a(\eta) = \begin{vmatrix} \eta & \alpha_o & 0 & 0 \\ -\alpha_o & \eta & 0 & 0 \\ 0 & 0 & \eta & \alpha_o \\ 0 & 0 & -\alpha_o & \eta \end{vmatrix}$$

8.2.5.1 Formal solution

Rewrite the W-H equation in the form

$$\mathbf{G}(\eta) \cdot \mathbf{F}_+(\eta) = \mathbf{F}_-(\eta) = \mathbf{X}_-(\eta) + \frac{\mathbf{R}_o}{\eta - \eta_o} \tag{85}$$

where

$$\mathbf{F}_+(\eta) = \begin{vmatrix} \mathbf{V}_{1+} \\ \mathbf{A}_{2+} \end{vmatrix}, \quad \mathbf{X}_-(\eta) = \begin{vmatrix} \mathbf{A}^s_{1-} \\ \mathbf{V}^s_{2-} \end{vmatrix}, \quad \mathbf{F}_-(\eta) = \begin{vmatrix} \mathbf{A}_{1-} \\ \mathbf{V}_{2-} \end{vmatrix}$$

After operating the factorization $\mathbf{G}(\eta) = \mathbf{G}_{w-}(\eta) \cdot \mathbf{G}_{w+}(\eta)$, the W-H technique yields

$$\mathbf{G}_{w+}(\eta) \cdot \mathbf{F}_+(\eta) = \mathbf{G}^{-1}_{w-}(\eta) \cdot \mathbf{X}_-(\eta) + \mathbf{G}^{-1}_{w-}(\eta) \cdot \frac{\mathbf{R}_o}{\eta - \eta_o} \tag{86}$$

or

$$\mathbf{G}_{w+}(\eta) \cdot \mathbf{F}_+(\eta) - \mathbf{G}^{-1}_{w-}(\eta_o) \cdot \frac{\mathbf{R}_o}{\eta - \eta_o} = \mathbf{G}^{-1}_{w-}(\eta) \cdot \mathbf{F}_-(\eta) - \mathbf{G}^{-1}_{w-}(\eta_o) \cdot \frac{\mathbf{R}_o}{\eta - \eta_o} = \mathbf{w} \tag{87}$$

Taking into account the form of the factorized matrices $\mathbf{G}_{w-}(\eta)$ and $\mathbf{G}_{w+}(\eta)$, it follows that the vector $\mathbf{w}$ is constant. To obtain this unknown vector, we resort to the same procedure used to factorize $\mathbf{Y}_c(\eta)$ in chapter 7. For instance, taking into account (85), $\mathbf{w}$ must be chosen in order to ensure that the solutions

$$\mathbf{F}_+(\eta) = \mathbf{G}^{-1}_{w+}(\eta) \cdot \left[\mathbf{w} + \mathbf{G}^{-1}_{w-}(\eta_o) \cdot \frac{\mathbf{R}_o}{\eta - \eta_o}\right] \tag{88}$$

$$\mathbf{F}_-(\eta) = \mathbf{G}_{w-}(\eta) \cdot \left[\mathbf{w} + \mathbf{G}^{-1}_{w-}(\eta_o) \cdot \frac{\mathbf{R}_o}{\eta - \eta_o}\right] \tag{89}$$

do not present the offending poles $\eta = j\alpha_o$ and $\eta = -j\alpha_o$, respectively.

In the following the null spaces of $T_a(j\alpha_o)$ and $T_a(-j\alpha_o)$ will be important. By using MATHEMATICA we get

$$NullSpace[T_a(-j\alpha_o)] = c_1\,\mathbf{u}_1 + c_2\,\mathbf{u}_2$$
$$NullSpace[T_a(j\alpha_o)] = -c_3\,\mathbf{u}_3 - c_4\,\mathbf{u}_4$$

where the constant c_i, $i = 1,2,3,4$ are arbitrary, and the vectors $\mathbf{u}_i$, $i = 1,2,3,4$ are defined by

$$\mathbf{u}_1 = \begin{vmatrix} 0 \\ 0 \\ -j \\ 1 \end{vmatrix}, \quad \mathbf{u}_2 = \begin{vmatrix} -j \\ 1 \\ 0 \\ 0 \end{vmatrix}, \quad \mathbf{u}_3 = \begin{vmatrix} 0 \\ 0 \\ j \\ 1 \end{vmatrix}, \quad \mathbf{u}_4 = \begin{vmatrix} j \\ 1 \\ 0 \\ 0 \end{vmatrix}$$

Looking at eq. (88), the absence of the offending pole $\eta = j\alpha_o$ in $\mathbf{F}_+(\eta)$ implies that the vector

$$P \cdot diag[Z_o, 1, 1, Y_o] \cdot n_+(j\alpha_o) \cdot \begin{pmatrix} \mathbf{G}_{th+}(j\alpha_o) & \mathbf{0} \\ \mathbf{0} & \mathbf{G}_{te+}(j\alpha_o) \end{pmatrix}^{-1} \cdot \left[\mathbf{w} + \mathbf{G}_{w-}^{-1}(\eta_o) \cdot \frac{\mathbf{R}_o}{j\alpha_o - \eta_o}\right]$$

must be in the null space of $T_a(j\alpha_o)$:

$$P \cdot diag[Z_o, 1, 1, Y_o] \cdot n_+(j\alpha_o) \cdot \begin{pmatrix} \mathbf{G}_{th+}(j\alpha_o) & \mathbf{0} \\ \mathbf{0} & \mathbf{G}_{te+}(j\alpha_o) \end{pmatrix}^{-1} \cdot \left[\mathbf{w} + \mathbf{G}_{w-}^{-1}(\eta_o) \cdot \frac{\mathbf{R}_o}{j\alpha_o - \eta_o}\right]$$
$$= -c_3\,\mathbf{u}_3 - c_4\,\mathbf{u}_4 \tag{90}$$

Similarly, the absence of the offending pole $\eta = -j\alpha_o$ in $\mathbf{F}_-(\eta)$ implies

$$P \cdot diag[1, Z_o, Y_o, 1] \cdot n_-^{-1}(-j\alpha_o) \cdot \begin{pmatrix} \mathbf{G}_{th-}(-j\alpha_o) & \mathbf{0} \\ \mathbf{0} & \mathbf{G}_{te-}(-j\alpha_o) \end{pmatrix} \cdot \left[\mathbf{w} + \mathbf{G}_{w-}^{-1}(\eta_o) \cdot \frac{\mathbf{R}_o}{-j\alpha_o - \eta_o}\right]$$
$$= c_1\,\mathbf{u}_1 + c_2\,\mathbf{u}_2 \tag{91}$$

The previous eight scalar equations allow us to determine the eight unknowns c_i, $i = 1,2,3,4$ and the four components of $\mathbf{w}$. We get

$$\begin{vmatrix} c_1 \\ c_2 \\ c_3 \\ c_4 \end{vmatrix} = 2j\,\mathbf{M}^{-1} \cdot \mathbf{G}_{w-}^{-1}(\eta_o) \cdot \frac{\mathbf{R}_o \alpha_o}{\alpha_o^2 + \eta_o^2} \tag{92}$$

$$\mathbf{w} = \left[P \cdot diag[1, Z_o, Y_o, 1] \cdot n_-^{-1}(-j\alpha_o) \cdot \begin{pmatrix} \mathbf{G}_{th-}(-j\alpha_o) & \mathbf{0} \\ \mathbf{0} & \mathbf{G}_{te-}(-j\alpha_o) \end{pmatrix}\right]^{-1} \cdot (c_1\,\mathbf{u}_1 + c_2\,\mathbf{u}_2)$$
$$- \mathbf{G}_{w-}^{-1}(\eta_o) \cdot \frac{\mathbf{R}_o}{-j\alpha_o - \eta_o} \tag{93}$$

where the 4×4 matrix $\mathbf{M}$ is defined by

$$M = |\mathbf{U}_1, \mathbf{U}_2, \mathbf{U}_3, \mathbf{U}_4| \tag{94}$$

with the vector $\mathbf{U}_i$, $i = 1,2,3,4$ defined by

$$
\begin{aligned}
\mathbf{U}_1 &= \left[P \cdot diag[1, Z_o, Y_o, 1] \cdot n_-^{-1}(-j\alpha_o) \cdot \begin{pmatrix} \mathbf{G}_{th-}(-j\alpha_o) & 0 \\ 0 & \mathbf{G}_{te-}(-j\alpha_o) \end{pmatrix} \right]^{-1} \cdot \mathbf{u}_1 \\
\mathbf{U}_2 &= \left[P \cdot diag[1, Z_o, Y_o, 1] \cdot n_-^{-1}(-j\alpha_o) \cdot \begin{pmatrix} \mathbf{G}_{th-}(-j\alpha_o) & 0 \\ 0 & \mathbf{G}_{te-}(-j\alpha_o) \end{pmatrix} \right]^{-1} \cdot \mathbf{u}_2 \\
\mathbf{U}_3 &= \left[P \cdot diag[Z_o, 1, 1, Y_o] \cdot n_+(j\alpha_o) \cdot \begin{pmatrix} \mathbf{G}_{th+}(j\alpha_o) & 0 \\ 0 & \mathbf{G}_{te+}(j\alpha_o) \end{pmatrix}^{-1} \right]^{-1} \cdot \mathbf{u}_3 \\
\mathbf{U}_4 &= \left[P \cdot diag[Z_o, 1, 1, Y_o] \cdot n_+(j\alpha_o) \cdot \begin{pmatrix} \mathbf{G}_{th+}(j\alpha_o) & 0 \\ 0 & \mathbf{G}_{te+}(j\alpha_o) \end{pmatrix}^{-1} \right]^{-1} \cdot \mathbf{u}_4
\end{aligned} \tag{95}
$$

8.3 The three half-planes problem

8.3.1 The E-polarization case (normal incidence case)

Figure 5 illustrates the geometry of the problem. In the spectral domain, this geometry can be studied with a circuit representation, as shown in Fig. 6. In particular, the three PEC walls are

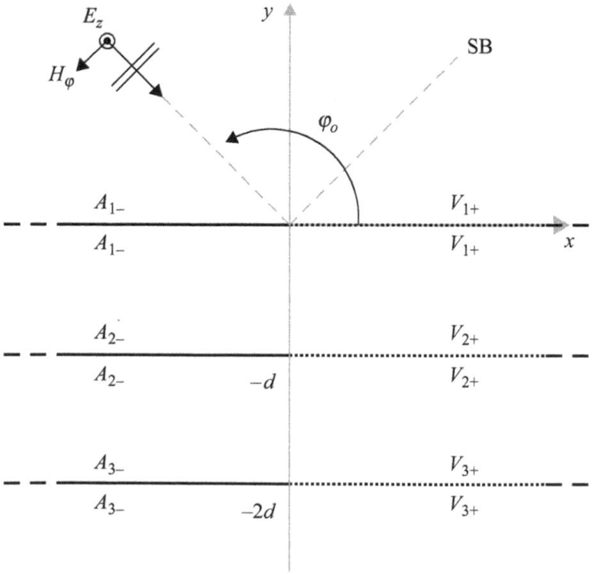

Fig. 5: Diffraction on three equally spaced semi-infinite PEC planes: E-polarization; Wiener-Hopf unknowns also shown

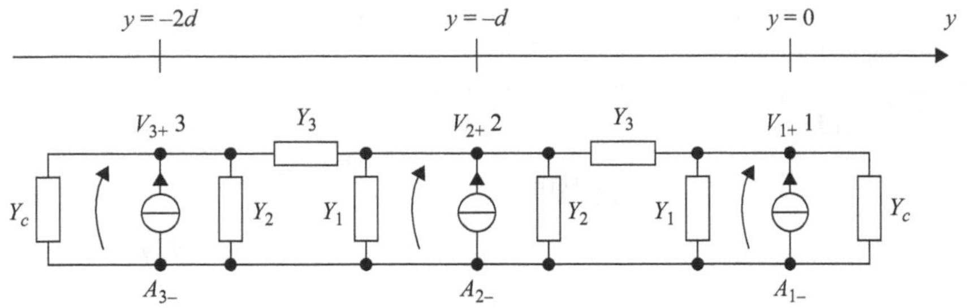

Fig. 6: Equivalent circuit model of the problem under study

simulated by the current generators A_{1-}, A_{2-} and A_{3-}; the two slabs $-d \leq y \leq 0$ and $-2d \leq y \leq -d$ in the physical structure are equivalent to the two Π two-ports indicated in figure.

Writing the node equations on circuit 6 leads directly the W-H equations of the following problem:

$$\begin{aligned} &\text{node1} \Rightarrow (Y_c + Y_1 + Y_3)V_{1+} - Y_3 V_{2+} = A_{1-} \\ &\text{node3} \Rightarrow (Y_c + Y_1 + Y_3)V_{3+} - Y_3 V_{2+} = A_{3-} \\ &\text{node2} \Rightarrow 2(Y_1 + Y_3)V_{2+} - Y_3 V_{1+} - Y_3 V_{3+} = A_{2-} \end{aligned} \tag{96}$$

where

$$Y_1 = Y_2 = jY_c \tan\frac{\tau d}{2}, \quad Y_3 = -jY_c \frac{1}{\sin \tau d}, \quad Y_c = \frac{\tau}{\omega\, \mu}, \quad \tau = \sqrt{k^2 - \eta^2}$$

V_{1+}, V_{2+} and V_{3+} and $-A_{1-}$, $-A_{2-}$ and $-A_{3+}$ are the Fourier transforms of the electric field E_z on the three apertures $y = 0$, $y = -d$, $y = -2d$, respectively, and the total currents induced on the three half-planes.

The previous equations constitute a W-H system of third order. The symmetry of the geometry allows us to reduce the order of this system. By summing the first two equations we get

$$G(\eta) \cdot \left| \begin{array}{c} \sqrt{k_o - \eta}(V_{1+} + V_{3+}) \\ 2\sqrt{k_o - \eta} V_{2+} \end{array} \right| = \left| \begin{array}{c} Z_o \dfrac{k}{\sqrt{k_o + \eta}}(A_{1-} + A_{3-}) \\ Z_o \dfrac{k}{\sqrt{k_o + \eta}} A_{2-} \end{array} \right| \tag{97}$$

where

$$\begin{aligned} G(\eta) &= \frac{2}{(1 - e^{-2j\tau d})} \begin{pmatrix} 1 & -e^{-j\tau d} \\ -e^{-j\tau d} & \dfrac{1 + e^{-2j\tau d}}{2} \end{pmatrix} \\ G^{-1}(\eta) &= \begin{pmatrix} \dfrac{1 + e^{-2j\tau d}}{2} & e^{-j\tau d} \\ e^{-j\tau d} & 1 \end{pmatrix} = e^{-j\tau d} \begin{pmatrix} \cos(\tau d) & 1 \\ 1 & e^{j\tau d} \end{pmatrix} \end{aligned} \tag{98}$$

The factorization of $G(\eta)$ was studied by Jones (1986), who obtained a nonstandard factorization where the factorized matrices present an offending behavior at ∞. The possibility of obtaining a nonstandard factorization is because the kernel matrix commutes with an entire matrix. In fact,

$$\begin{pmatrix} \cos(\tau d) & 1 \\ 1 & e^{j\tau d} \end{pmatrix} = \begin{pmatrix} \cos(\tau d) & 1 \\ 1 & \cos(\tau d) \end{pmatrix} + j \begin{pmatrix} 0 & 0 \\ 0 & \sin(\tau d) \end{pmatrix}$$

$$= P_- \cdot \left[\mathbf{1} + j\tau d \, P_-^{-1} \cdot \begin{pmatrix} 0 & 0 \\ 0 & \dfrac{\sin(\tau d)}{\tau d} \end{pmatrix} \cdot P_+^{-1} \right] \cdot P_+$$

where

$$P = \begin{pmatrix} \cos(\tau d) & 1 \\ 1 & \cos(\tau d) \end{pmatrix} = P_- \cdot P_+$$

The factorization of P is simple and yields entire factorized matrices P_- and P_+ expressed by infinite products. The previous equations show that we reduced the factorization of $G(\eta)$ to the factorization of the matrix

$$\mathbf{1} + j\tau d \, P_-^{-1} \cdot \begin{pmatrix} 0 & 0 \\ 0 & \dfrac{\sin(\tau d)}{\tau d} \end{pmatrix} \cdot P_+^{-1}$$

This matrix commutes with the entire matrix

$$W(\eta) = P_-^{-1} \cdot \begin{pmatrix} 0 & 0 \\ 0 & \dfrac{\sin(\tau d)}{\tau d} \end{pmatrix} \cdot P_+^{-1}$$

The logarithmic decomposition of $\mathbf{1} + j\tau d \, P_-^{-1} \cdot \begin{pmatrix} 0 & 0 \\ 0 & \frac{\sin(\tau d)}{\tau d} \end{pmatrix} \cdot P_+^{-1}$ introduces offending behavior that can be eliminated by multiplying the nonstandard factorized matrix by a suitable entire matrix (chapter 4). The complete specification of this entire matrix requires the solution of an infinite set of linear equations.

Accurate calculations are very cumbersome, so again it is better to use the Fredholm factorization. It applies immediately, since $G(\eta)$ and its inverse $G^{-1}(\eta)$ are bounded on the real axis as $\eta \to \pm\infty$. Taking into account that

$$A_{1-} = A_{1-}^s + 2j \frac{E_o \sin \varphi_o}{Z_o} \frac{1}{\eta - \eta_o} \tag{99}$$

we can rewrite the W-H system in the normal form:

$$G(\eta) \cdot F_+(\eta) = X_-(\eta) + \frac{R_o}{\eta - \eta_o} \tag{100}$$

where

$$F_+(\eta) = \begin{vmatrix} \sqrt{k_o - \eta}(V_{1+} + V_{3+}) \\ 2\sqrt{k_o - \eta}\, V_{2+} \end{vmatrix}, \quad R_o = \begin{vmatrix} \dfrac{k}{\sqrt{k_o + \eta_o}} 2jE_o \sin \varphi_o \\ 0 \end{vmatrix} \tag{101}$$

By using the expedient of section 5.1.5 of warping the real axis in the integration line $\eta(y) = e^{j\frac{\pi}{4}}y$, $-\infty < y < \infty$, we obtain the resulted indicated in Fig. 7 and Fig. 8.

8.3.2 The skew incidence case

As it happens for the half-plane problem (section 7.5) or for the planar waveguide problem (section 8.1.4), in the skew incidence case we are dealing with a matrix kernel that has the form

$$\mathbf{G}(\eta) = G(\eta) \otimes R(\eta)$$

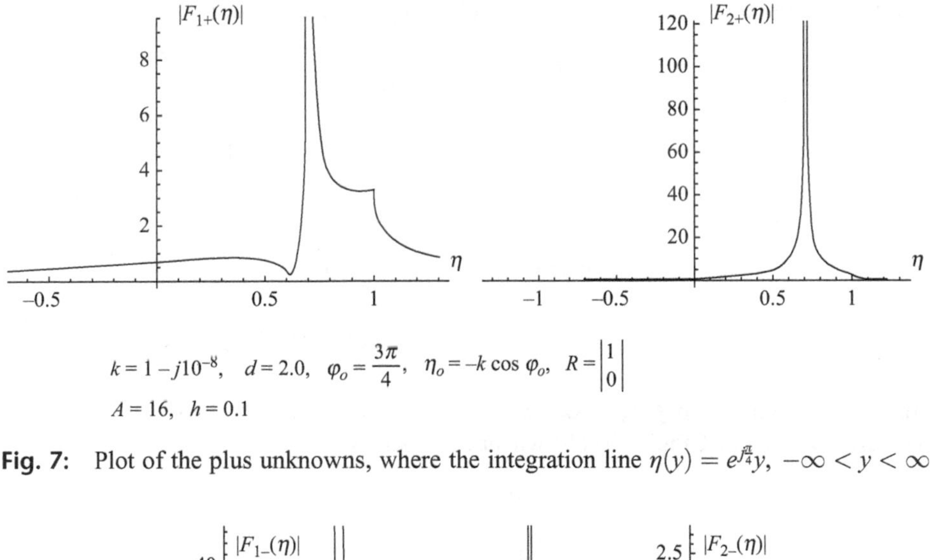

$$k = 1 - j10^{-8}, \quad d = 2.0, \quad \varphi_o = \frac{3\pi}{4}, \quad \eta_o = -k\cos\varphi_o, \quad R = \begin{vmatrix} 1 \\ 0 \end{vmatrix}$$

$$A = 16, \quad h = 0.1$$

Fig. 7: Plot of the plus unknowns, where the integration line $\eta(y) = e^{j\frac{\pi}{4}}y$, $-\infty < y < \infty$

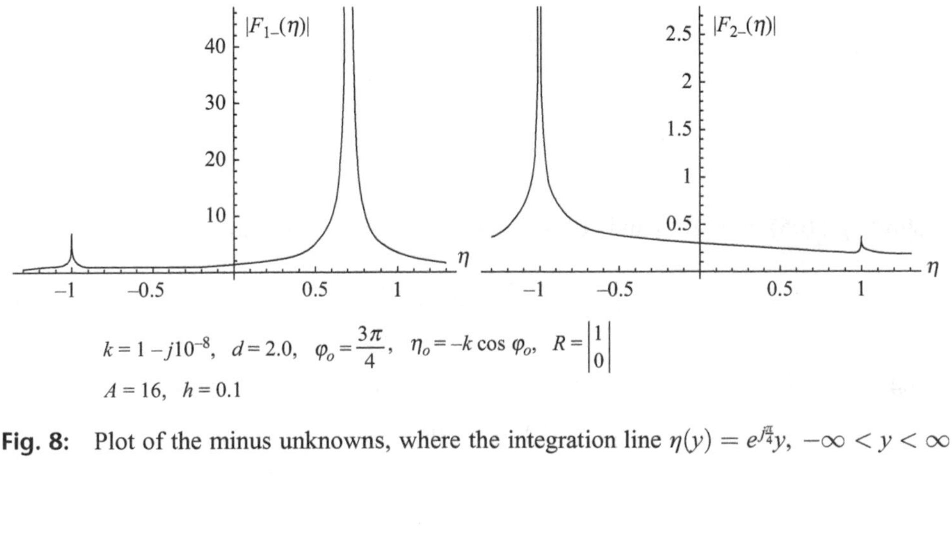

$$k = 1 - j10^{-8}, \quad d = 2.0, \quad \varphi_o = \frac{3\pi}{4}, \quad \eta_o = -k\cos\varphi_o, \quad R = \begin{vmatrix} 1 \\ 0 \end{vmatrix}$$

$$A = 16, \quad h = 0.1$$

Fig. 8: Plot of the minus unknowns, where the integration line $\eta(y) = e^{j\frac{\pi}{4}}y$, $-\infty < y < \infty$

where $G(\eta)$ is defined by eq. (98) and

$$R(\eta) = \frac{1}{\xi^2}\begin{vmatrix} \tau_o^2 & -\eta\,\alpha_o \\ -\eta\,\alpha_o & k^2-\eta^2 \end{vmatrix}$$

We achieve the factorization with the well-known equations

$$\begin{aligned}\mathbf{G}(\eta) &= G(\eta)\otimes R(\eta) = [G_-(\eta)\cdot G_+(\eta)]\otimes[R_-(\eta)\cdot R_+(\eta)] \\ &= [G_-(\eta)\otimes R_-(\eta)]\cdot[G_+(\eta)\otimes R_+(\eta)] = \mathbf{G}_-(\eta)\cdot\mathbf{G}_+(\eta)\end{aligned}$$

8.4 Arrays of parallel wire antennas in stratified media

8.4.1 The single antenna case

Consider Fig. 13 in chapter 6 and assume that the radius of the wire is a. In the spectral domain the component $\tilde{E}_z$ must satisfy the following wave equation:

$$\left(\frac{\partial^2}{\partial x^2}+\frac{\partial^2}{\partial y^2}+\tau^2\right)\tilde{E}_z = 0 \tag{102}$$

where $\tau = \sqrt{k^2-\alpha^2}$. The solution of (102) that satisfies the radiation condition is given by

$$\tilde{E}_z = C(\alpha)H_o^{(2)}(\tau\rho) \tag{103}$$

To evaluate $C(\alpha)$, we consider the azimuthal component H_φ:

$$\tilde{H}_\varphi(\rho) = -j\frac{k}{Z_o\tau^2}\frac{\partial\tilde{E}_z}{\partial\rho} = -j\frac{k}{Z_o\tau}C(\alpha)H_0^{(2)\prime}(\tau\rho) \tag{104}$$

In the whole cylinder $\rho \le a$, $-\infty < z < \infty$ we can neglect the displacement currents (Canavero, Daniele & Graglia, 1988) whence Ampere's law yields

$$2\pi a\tilde{H}_\varphi(a) = \tilde{I}(\alpha) \tag{105}$$

where $\tilde{I}(\alpha)$ is the Fourier transform of the current $I(z)$:

$$\tilde{I}(\alpha) = \int_{-L/2}^{L/2} I(z)e^{j\alpha z}dz.$$

Substituting (105) into (104) and (103) yields the radiated field $\tilde{E}_z^s(\rho)$

$$\tilde{E}_z^s(\rho) = \frac{j\tau Z_o H_o^{(2)}(\tau\rho)}{2\pi k a H_o^{(2)\prime}(\tau a)}\tilde{I}(\alpha) \tag{106}$$

Taking into account that $\tilde{E}_z(a)$ vanishes on the surface of the wire we have

$$\tilde{E}_z^s(a) = V_\odot(\alpha) - V_o$$

where $V_{\odot}(\alpha)$ is the Fourier transform of a function that vanishes in the range $-L/2 < z < L/2$. Substituting the previous equation in (106) we get the longitudinal modified W-H equation:

$$\frac{j\tau Z_o H_o^{(2)}(\tau a)}{2\pi k a H_o^{(2)\prime}(\tau a)} \tilde{I}(\alpha) = V_{\odot}(\alpha) - V_o \tag{107}$$

To obtain the impedance of the wire antenna, we do not solve this equation with the factorization method but prefer to use the moment method. In particular, by using the symbols in section 6.4 we observe that

$$A \rightarrow \frac{j\tau Z_o H_o^{(2)}(\tau a)}{2\pi k a H_o^{(2)\prime}(\tau a)}$$

$$x \rightarrow i(z) \Leftrightarrow \tilde{i}(\alpha) = I(\alpha)$$

$$y \rightarrow -V_o \delta(z) \Leftrightarrow \tilde{y}(\alpha) = -V_o$$

$$h \rightarrow \frac{\delta(z)}{V_o} \Leftrightarrow \tilde{h}(\alpha) = \frac{1}{V_o}$$

Considering $L = \frac{\lambda}{2}$ and assuming that a sinusoidal distribution of $i(z)$ is physically acceptable, we use only one moment:

$$\psi_1(z) = \varphi_1(z) = \cos \frac{2\pi}{\lambda} z \tag{108}$$

which yields

$$\tilde{\psi}_1(\alpha) = \tilde{\varphi}_1(\alpha) = \int_{-L/2}^{L/2} \cos\left(\frac{2\pi}{\lambda} z\right) e^{j\alpha z} dz = \frac{4\pi}{\lambda} \frac{\cos(\alpha\lambda/4)}{(2\pi/\lambda)^2 - \alpha^2}$$

Hence, from (6.111) the admittance of the half-wave wire antenna is expressed by

$$Y_a = h \cdot x = \frac{\varphi_1 \cdot y}{\varphi_1 \cdot A \cdot \psi_1} h \cdot \psi_1 = \frac{1}{2\pi} \frac{\int_{-\infty}^{\infty} \tilde{\varphi}_1(-\alpha)\tilde{y}(\alpha) d\alpha \int_{-\infty}^{\infty} \tilde{h}(-\alpha)\tilde{\psi}_1(\alpha) d\alpha}{\int_{-\infty}^{\infty} \tilde{\varphi}_1(-\alpha)\tilde{A}(\alpha)\tilde{\psi}_1(\alpha) d\alpha}$$

The us of Parseval's theorem for the integrals in the numerators yields

$$Y_a = -2\pi \frac{\varphi_1(0)\psi_1(0)}{\int_{-\infty}^{\infty} \tilde{\varphi}_1(-\alpha)\tilde{A}(\alpha)\tilde{\psi}_1(\alpha) d\alpha}$$

To evaluate this integral we can take into account that for a small wire radius ($\frac{a}{\lambda} \ll 1$)

$$\frac{j\tau Z_o H_o^{(2)}(\tau\, a)}{2\pi k a H_o^{(2)\prime}(\tau\, a)} \approx j\frac{\tau^2 Z_o}{2\pi k}\log(\tau\, a)$$

After algebraic manipulations, this yields the impedance

$$Z_a = \frac{1}{Y_a} = j\frac{2Z_o}{\pi}\int_0^\infty \log\left[\left((2\pi)^2 - u^2\right)\left(\frac{a}{\lambda}\right)^2\right]\frac{\cos^2(u/4)}{u^2-(2\pi)^2}du$$

$$= j\frac{2Z_o}{\pi}\int_0^\infty \log\left[(2\pi)^2 - u^2\right]\frac{\cos^2(u/4)}{u^2-(2\pi)^2}du + j\frac{4Z_o}{\pi}\log\left(\frac{a}{\lambda}\right)\int_0^\infty \frac{\cos^2(u/4)}{u^2-(2\pi)^2}du$$

The last integral is null. Taking into account that $Z_o = 377\ \Omega$, we find numerically that

$$Z_a = 73.13 + j42.55\ \Omega$$

8.4.2 The W-H equations of an array of wire antennas

Figure 9 illustrates an array of n wire antennas that radiate in an arbitrary stratified medium. If the wire r with radius a_r is not perfectly conducting, it presents an impedance per unit length:

$$Z_{wr} = \frac{J_o(k_{wr}a_r)k_{wr}}{2\pi\sigma_{wr}a_r J_1(k_{wr}a_r)}$$

where k_{wr} and σ_{wr} are the propagation constant and the conductivity of the r-wire.

The spectral theory of a bundle of wires immersed in a stratified medium generalizes the results of section 8.4.1 and is presented in Canavero, Daniele, and Graglia (1988). We obtain the following system of modified W-H equations:

$$\sum_{q=1}^{n}\tilde{B}_q(\alpha)\left[H_0^{(2)}(\tau\, d_{qr}) + \tilde{e}_q(\alpha, x_r, y_r)\right]\tilde{I}_q(\alpha) - Z_{wr}\tilde{I}_r(\alpha)$$
$$= -V_r + E_{zr\odot}(\alpha, x_r, y_r) + E_{zr\oplus}(\alpha, x_r, y_r) \quad (r = 1, 2, \ldots, n) \qquad (109)$$

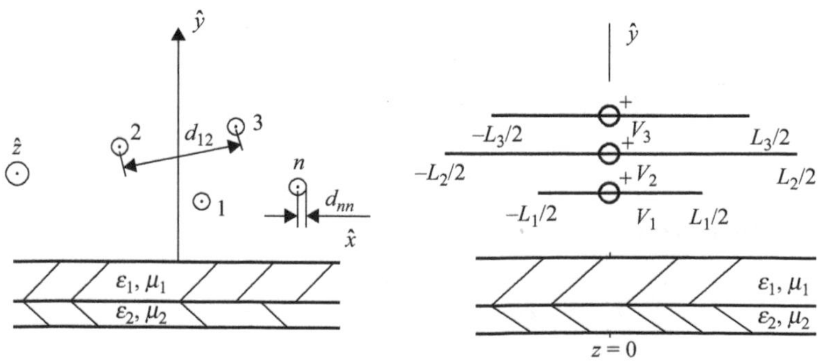

Fig. 9: Array of wire antennas

where the $\odot$ type functions $E_{zr\odot}(\alpha, x_r, y_r)$ are regular in the half-plane $\mathrm{Im}[\alpha] \leq 0$, and the $\oplus$ type functions $E_{zr\oplus}(\alpha, x_r, y_r)$ are regular in the half-plane $\mathrm{Im}[\alpha] \geq 0$,

$$d_{qq} = a_q$$

$$\tilde{B}_q(\alpha) = \frac{j\tau Z_o}{2\pi k a H_o^{(2)\prime}(\tau\, d_{qq})}$$

$$\tilde{e}_q(\alpha, x, y) = \frac{1}{\pi\tau^2}\int_{-\infty}^{\infty} \frac{1}{\alpha^2+\eta^2} \frac{\alpha^2(\tau^2-\eta^2)\tilde{\Gamma}^{TM}(\eta,\alpha) + \eta^2 k_o^2 \tilde{\Gamma}^{TE}(\eta,\alpha)}{\sqrt{\tau^2-\eta^2}} \cdot \exp\left[-j\eta(x-x_q) - j\sqrt{\tau^2-\eta^2}(y+y_q)\right] d\eta \tag{110}$$

The reflection coefficients $\tilde{\Gamma}^{TM}(\eta,\alpha)$ and $\tilde{\Gamma}^{TE}(\eta,\alpha)$ are considered in the basis $\hat{\boldsymbol{\sigma}}$, $\hat{\boldsymbol{\beta}}$ where the characteristic impedance of the isotropic media is diagonal (section 7.5); the propagation constant χ and the characteristic impedance for an isotropic medium are given by

$$\chi = \sqrt{k^2-\alpha^2-\eta^2}, \quad \mathbf{Z_c} = \mathbf{Z}_c^{TE}\hat{\boldsymbol{\sigma}}\hat{\boldsymbol{\sigma}} + \mathbf{Z}_c^{TM}\hat{\boldsymbol{\beta}}\hat{\boldsymbol{\beta}}$$

where $\mathbf{Z}_c^{TE} = \frac{\omega\mu}{\chi}$, $\mathbf{Z}_c^{TM} = \frac{\chi}{\omega\varepsilon}$, $y = -d$.

Example

In the basis $\hat{\boldsymbol{\sigma}}$, $\hat{\boldsymbol{\beta}}$ we have two scalar circuits modeled with transmission lines.

We want to evaluate the reflections coefficients $\tilde{\Gamma}^{TM}(\eta,\alpha)$ and $\tilde{\Gamma}^{TE}(\eta,\alpha)$ for the stratification indicated in Fig. 10. The transmission line relevant to the free space has

$$\chi = \sqrt{k^2-\alpha^2-\eta^2}, \quad \mathbf{Z}_c^{TE} = \frac{\omega\mu}{\chi}, \quad \mathbf{Z}_c^{TM} = \frac{\chi}{\omega\varepsilon}$$

whereas the transmission line relevant to the medium defined by ε_1, μ_1 has

$$\chi_1 = \sqrt{k_1^2-\alpha^2-\eta^2}, \quad \mathbf{Z}_{c1}^{TE} = \frac{\omega\mu_1}{\chi_1}, \quad \mathbf{Z}_{c1}^{TM} = \frac{\chi_1}{\omega\varepsilon_1}$$

The reflection coefficients $\tilde{\Gamma}^{TM}(\eta,\alpha)$ and $\tilde{\Gamma}^{TE}(\eta,\alpha)$ are given by

$$\tilde{\Gamma}^{TE}(\eta,\alpha) = \frac{Z^{TE}-Z_c^{TE}}{Z^{TE}+Z_c^{TE}}, \quad \tilde{\Gamma}^{TM}(\eta,\alpha) = \frac{Z^{TM}-Z_c^{TM}}{Z^{TM}+Z_c^{TM}}$$

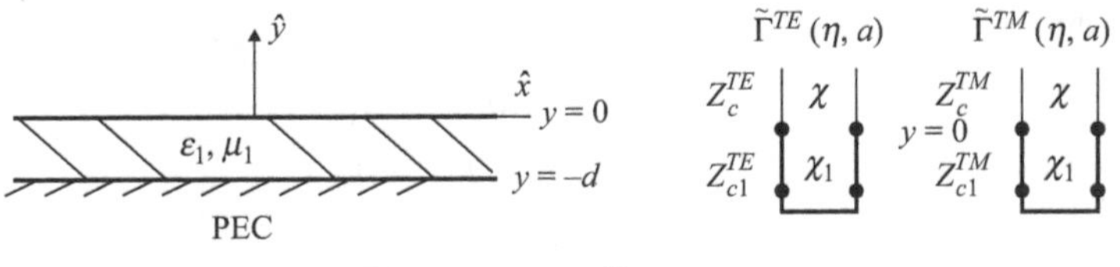

Fig. 10: Evaluation of $\tilde{\Gamma}^{TM}(\eta,\alpha)$ and $\tilde{\Gamma}^{TE}(\eta,\alpha)$ for a particular stratification

where $Z^{TM}(\eta, \alpha)$ and $Z^{TE}(\eta, \alpha)$ are the impedances in the input ($y = 0$) of the transmission line and are evaluated by

$$Z^{TM} = jZ_{c1}^{TM} \tan(\chi_1 d), \quad Z^{TE} = jZ_{c1}^{TE} \tan(\chi_1 d)$$

The matrix kernel

$$G_{rq}(\alpha) = \tilde{B}_q(\alpha)\left[H_0^{(2)}(\tau\, d_{qr}) + \tilde{e}_q(\alpha, x_r, y_r)\right] \to \mathbf{G}(\alpha) \tag{111}$$

in eq. (111) contains an integral that makes its factorization very difficult. The integral is zero if the array is in the free space, and sometimes it can be evaluated explicitly. For instance, if $d = 0$ we have $\tilde{\Gamma}^{TM}(\eta, \alpha) = -1\; \tilde{\Gamma}^{TE}(\eta, \alpha) = -1$ and we get

$$\tilde{e}_q(\alpha, x, y) = -H_0^{(2)}(\tau\, r_{iq})$$

where r_{iq} is the distance from the point x, y to the image of the wire q.

8.4.2.1 The admittance matrix

To get the admittance matrix $Y_{rq} \to \mathbf{Y}_a$ of the array defined by

$$\begin{aligned} I_1(0) &= Y_{11}V_1 + Y_{12}V_2 + \cdots + Y_{1n}V_n \\ I_2(0) &= Y_{21}V_1 + Y_{22}V_2 + \cdots + Y_{2n}V_n \\ &\dots\ \dots\ \dots\ \dots\ \dots\ \dots\ \dots\ \dots\ \dots \\ I_n(0) &= Y_{n1}V_1 + Y_{n2}V_2 + \cdots + Y_{nn}V_n \end{aligned}$$

it is convenient to resort to the moment method, as it has been used in section 8.4.1.

For the sake of simplicity, let us assume that the voltages V_i are all vanishing except for $V_1 = 1$. If follows that $Y_{r1} = I_r(0)$.

Let us introduce the space $\mathfrak{A} \otimes E_n$, where $\mathfrak{A}$ is the Hilbert space, and E_n the Euclidean space on n dimensions. In this space, a vector x has the following representations in the natural space and in the spectral space, respectively:

$$x = \begin{vmatrix} x^{(1)}(z) \\ x^{(2)}(z) \\ \dots \\ x^{(n)}(z) \end{vmatrix}, \quad \tilde{x} = \begin{vmatrix} \tilde{x}^{(1)}(\alpha) \\ \tilde{x}^{(2)}(\alpha) \\ \dots \\ \tilde{x}^{(n)}(\alpha) \end{vmatrix}$$

In our case the vector x has the components

$$x^{(r)}(z) = I_r(z) \quad \text{or} \quad \tilde{x}^{(r)}(\alpha) = \tilde{I}_r(\alpha)$$

while the vector y has component $y^{(r)} = \delta_{r1}\delta(z)$. Introducing the vector $\mathbf{h}_s$ that in the natural space has component

$$h_s^{(r)} = \delta_{rs}\delta(z)$$

we get that $I_r(0)$ is a functional in the space $\mathfrak{A} \otimes E_n$. In fact,

$$I_r(0) = \mathbf{h}_r \cdot \mathbf{x}$$

According to the results of section 6.4.2, this means that if we evaluate $\mathbf{x}$ with the moment method the quantity $I_r(0) = Y_{r1}$ is stationary.

To evaluate **x** with the moment method we introduce the following approximate representation of the current $I_r(z)/I_r(0)$:

$$\frac{I_r(z)}{I_r(0)} = f_r(z) = \sin\left[k\left(\frac{L_r}{2} - |z|\right)\right]$$

or

$$\frac{\tilde{I}_r(\alpha)}{I_r} = \tilde{f}_r(\alpha) = \frac{k\, e^{j\alpha L_r/2} - ke^{-j\alpha L_r/2}\cos(kL_r) - j\alpha e^{-j\alpha L_r/2}\sin(kL_r)}{k^2 - \alpha^2} \tag{112}$$

and introduce the n moment ($r = 1, 2, \dots, n$):

$$\tilde{\psi}_r(\alpha) = \tilde{\varphi}_r(\alpha) = \left|0, 0, \frac{\tilde{f}_r(\alpha)}{\sin\left(k\frac{L_r}{2}\right)}, 0, 0\right|^t \tag{113}$$

In this way from the representation

$$x = x_1\psi_1 + x_2\psi_2 + \cdots + x_n\psi_n$$

we get

$$Y_{r1} = I_r(0) = x_r.$$

The values of x_r are given by the solution of the system

$$\begin{aligned}
\varphi_1 \cdot G \cdot \psi_1\, x_1 + \varphi_1 \cdot G \cdot \psi_2\, x_2 + \cdots \varphi_1 \cdot G \cdot \psi_n\, x_n &= \varphi_1 \cdot y = 1\\
\varphi_2 \cdot G \cdot \psi_1\, x_1 + \varphi_2 \cdot G \cdot \psi_2\, x_2 + \cdots \varphi_2 \cdot G \cdot \psi_n\, x_n &= \varphi_2 \cdot y = 0\\
\cdots\ \cdots\ \cdots\ \cdots\ \cdots\ \cdots\ \cdots\ \cdots\ \cdots\ \cdots\ \cdots\ \cdots\ \cdots\ \cdots\\
\varphi_n \cdot G \cdot \psi_1\, x_1 + \varphi_n \cdot G \cdot \psi_2\, x_2 + \cdots \varphi_n \cdot G \cdot \psi_n\, x_n &= \varphi_n \cdot y = 0
\end{aligned}$$

In the spectral domain the coefficient $\varphi_r \cdot G \cdot \psi_q$ are expressed by

$$\varphi_r \cdot G \cdot \psi_q = \frac{1}{2\pi}\int_{-\infty}^{\infty} \tilde{\varphi}_r(-\alpha) \cdot \mathbf{G}(\alpha) \cdot \tilde{\psi}_r(\alpha)d\alpha$$

where the entries of $\mathbf{G}(\alpha)$ are defined by (111).

By repeating the same reasoning, when only the voltage V_q is not vanishing, we get the following system that provides the stationary values of $Y_{rq} = I_r(0) = x_r$:

$$\begin{aligned}
\varphi_1 \cdot G \cdot \psi_1\, x_1 + \varphi_1 \cdot G \cdot \psi_2\, x_2 + \cdots \varphi_1 \cdot G \cdot \psi_n\, x_n &= \varphi_1 \cdot y = 0\\
\varphi_2 \cdot G \cdot \psi_1\, x_1 + \varphi_2 \cdot G \cdot \psi_2\, x_2 + \cdots \varphi_2 \cdot G \cdot \psi_n\, x_n &= \varphi_2 \cdot y = 0\\
\cdots\ \cdots\ \cdots\ \cdots\ \cdots\ \cdots\ \cdots\ \cdots\ \cdots\ \cdots\ \cdots\ \cdots\ \cdots\ \cdots\\
\varphi_q \cdot G \cdot \psi_1\, x_1 + \varphi_q \cdot G \cdot \psi_2\, x_2 + \cdots \varphi_q \cdot G \cdot \psi_n\, x_n &= \varphi_n \cdot y = 1\\
\cdots\ \cdots\ \cdots\ \cdots\ \cdots\ \cdots\ \cdots\ \cdots\ \cdots\ \cdots\ \cdots\ \cdots\ \cdots\ \cdots\\
\varphi_n \cdot G \cdot \psi_1\, x_1 + \varphi_n \cdot G \cdot \psi_2\, x_2 + \cdots \varphi_n \cdot G \cdot \psi_n\, x_n &= \varphi_n \cdot y = 0
\end{aligned}$$

whence we can verify that the admittance $\mathbf{Y}_a$ of the array is given by

$$-\mathbf{Y}_a = \begin{vmatrix} \varphi_1 \cdot G \cdot \psi_1 & \varphi_1 \cdot G \cdot \psi_2 & \cdots & \varphi_1 \cdot G \cdot \psi_n \\ \varphi_2 \cdot G \cdot \psi_1 & \varphi_2 \cdot G \cdot \psi_2 & \cdots & \varphi_2 \cdot G \cdot \psi_n \\ \cdots & \cdots & \cdots & \cdots \\ \varphi_n \cdot G \cdot \psi_1 & \varphi_n \cdot G \cdot \psi_2 & \cdots & \varphi_n \cdot G \cdot \psi_n \end{vmatrix}^{-1}$$

The elements Z_{rq} of the matrix impedance $\mathbf{Z}_a = \mathbf{Y}_a^{-1}$ of the array present the simple form

$$-Z_{rq} = \varphi_r \cdot G \cdot \psi_q = \frac{1}{2\pi} \int_{-\infty}^{\infty} \tilde{\varphi}_r(-\alpha) \cdot \mathbf{G}(\alpha) \cdot \tilde{\psi}_r(\alpha) d\alpha \tag{114}$$

The hypothesis that the voltage sources are all located at $z = 0$ is not mandatory. Equations (114) hold for an arbitrary location of the parallel wires, provided that eqs. (113) are slightly modified.

8.4.3 Spectral theory of transmission lines constituted by bundles of wires

The theory considered in the previous section applies also for the study of a transmission line constituted by a bundle of wires (Canavero, Daniele & Graglia, 1988). Using this theory, it is possible to ascertain the validity limits of the telegrapher equations (Daniele, Gilli & Pignari, 1996).

8.5 Spectral theory of microstrip and coplanar transmission lines

8.5.1 Coplanar line with two strips

Figure 11 shows a coplanar line with two strips located in the plane $y = 0$.

The two strips have the same width w and are equidistant from the origin$x = 0$. The abscissa of the center of strip 1 is $x_o = (s + w)/2$, and that of the strip 2 is $-x_o$. By assuming a negligible thickness of the strips in the y-direction and a width w small with respect to the

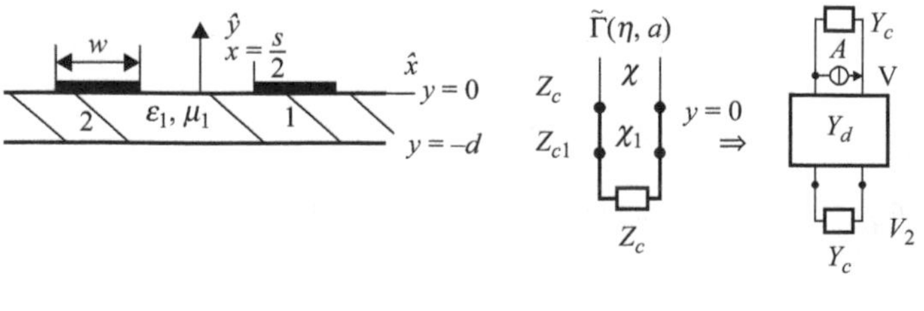

Fig. 11: An ungrounded coplanar line

wavelength λ, we can ignore the component j_x of the current and assume only longitudinal current distributions on the strips having the form

$$j_z = I(z)f(x)\delta(y)$$

In the following we assume a longitudinal dependence of the type $I\,e^{-j\alpha z}$, which is omitted.

We remember that the modal voltage and current are defined by eqs. (40) and (41) of chapter 7:

$$V = \begin{vmatrix} V_x(\eta,0) \\ \\ V_z(\eta,0) \end{vmatrix} = e^{j\alpha z} \int_{-\infty}^{\infty} \begin{vmatrix} -E_x(x,0) \\ \\ E_z(x,0) \end{vmatrix} e^{j\eta x} dx$$

$$I = \begin{vmatrix} I_z(\eta,0) \\ \\ I_x(\eta,0) \end{vmatrix} = e^{j\alpha z} \int_{-\infty}^{\infty} \begin{vmatrix} H_z(x,0) \\ \\ H_x(x,0) \end{vmatrix} e^{j\eta x} dx$$

Taking into account that assuming a vanishing j_x yields vanishing values of H_z, in the circuit model the two strips are modeled by a the current generator:

$$A = \begin{vmatrix} 0 \\ \\ A_2(\eta) \end{vmatrix} \tag{115}$$

The nodal equations of the circuit of Fig. 11 yields

$$\begin{aligned} (\mathbf{Y}_{11}^d + \mathbf{Y}_c)\cdot\mathbf{V} + \mathbf{Y}_{12}^d\cdot\mathbf{V}_2 &= \mathbf{A} \\ \mathbf{Y}_{21}^d\cdot\mathbf{V} + (\mathbf{Y}_{22}^d + \mathbf{Y}_c)\cdot\mathbf{V}_2 &= 0 \end{aligned} \tag{116}$$

where k and k_d are the propagation constant of free space and the dielectric substrate, respectively, we have

$$\chi = \sqrt{k^2 - \eta^2 - \alpha^2}, \quad \chi_d = \sqrt{k_d^2 - \eta^2 - \alpha^2}, \quad \tau_o^2 = \sqrt{k^2 - \alpha^2}, \quad \tau_{od} = \sqrt{k_d^2 - \alpha^2}$$

$$\mathbf{Y_c} = \frac{Y_o}{k\sqrt{k^2 - \eta^2 - \alpha^2}} \begin{vmatrix} \tau_o^2 & -\eta\,\alpha \\ -\eta\,\alpha & k^2 - \eta^2 \end{vmatrix}, \quad \mathbf{Y_{cd}} = \frac{Y_{od}}{k_d\sqrt{k_d^2 - \eta^2 - \alpha^2}} \begin{vmatrix} \tau_{od}^2 & -\eta\,\alpha \\ -\eta\,\alpha & k_d^2 - \eta^2 \end{vmatrix}$$

$$\mathbf{Y}_{11}^d = \mathbf{Y}_{22}^d = j\mathbf{Y}_c^d \cot(\chi_d\,d), \quad \mathbf{Y}_{12}^d = \mathbf{Y}_{21}^d = j\mathbf{Y}_c^d \frac{1}{\sin(\chi_d\,d)}$$

Elimination of $\mathbf{V}_2$ in (116) yields the following homogeneous equation:

$$Z_e(\eta,\alpha)A(\eta) = \begin{vmatrix} V_x(\eta,0) \\ V_z(\eta,0) \end{vmatrix}$$

or

$$\begin{aligned} Z_{e12}(\eta,\alpha)A_2(\eta) &= V_x(\eta,0) \\ Z_{e22}(\eta,\alpha)A_2(\eta) &= V_z(\eta,0) \end{aligned} \tag{117}$$

where

$$Z_e(\eta,\alpha) = \left[(\mathbf{Y}_{11}^d + \mathbf{Y}_c) - \mathbf{Y}_{12}^d \cdot (\mathbf{Y}_{22}^d + \mathbf{Y}_c)^{-1} \cdot \mathbf{Y}_{21}^d\right]^{-1}$$

In particular, the second of eqs. (117) is a modified W-H equation having as unknowns the function $A_2(\eta)$, which is the Fourier transform of a function that is not vanishing in the range

$$in_+ = \left(-x_o - \frac{s}{2} < x < -x_o + \frac{s}{2}\right) \bigcup \left(x_o - \frac{s}{2} < x < x_o + \frac{s}{2}\right)$$

and the Fourier transform $V_z(\eta) = V_z(\eta,0)$

$$V_z(\eta) = \int_{-\infty}^{-x_o-\frac{s}{2}} E_z(x)e^{j\eta x}dx + \int_{-x_o+\frac{s}{2}}^{x_o-\frac{s}{2}} E_z(x)e^{j\eta x}dx + \int_{x_o+\frac{s}{2}}^{\infty} E_z(x)e^{j\eta x}dx$$

that is nonvanishing in the complementary interval of in_+:

$$in_- = \left(-\infty < x < -x_o - \frac{s}{2}\right) \bigcup \left(-x_o + \frac{s}{2} < x < x_o - \frac{s}{2}\right) \bigcup \left(x_o + \frac{s}{2} < x < \infty\right)$$

With this equation we can solve different problems. An important problem is that of the excitation. It consists in evaluating the currents present on the strips when an external source is present. Considering the spectral representation of the incident field:

$$\tilde{E}_z^i(\eta,z) = \int_{-\infty}^{\infty} V_z^i(\eta)e^{-j\alpha z}d\alpha$$

we must solve a W-H equation where the kernel is given by

$$Z_{e22}(\eta,\alpha)$$

where α is to be considered as a parameter. Another important problem that we can solve is to find the modes of the coplanar transmission line. This requires the evaluation of the parameter α that produces not vanishing solution of the homogeneous equation. In the following we use the Garlekin method to solve this last problem.

Consider the following expansion:

$$A_2 = I_2\tilde{f}_2(\eta) + I_1\tilde{f}_1(\eta)$$

When we introduce the expansion (or test) functions, we assume the same distribution of the currents on the strips:

$$f_1(x) = s(x - x_o), \quad f_2(x) = s(x + x_o)$$

where the function $s(x)$ is zero for $|x| > w/2$. Hence,

$$\tilde{f}_1(\eta) = S(\eta)e^{j\eta x_o}, \quad \tilde{f}_2(\eta) = S(\eta)e^{-j\eta x_o}$$

We have

$$f_1 \cdot f_2 = \int_{-\infty}^{\infty} f_1(x)f_2(x)dx = 0$$

$$f_1 \cdot E_z = \int_{-\infty}^{\infty} f_1(x)E_z(x)dx = 0 = \frac{1}{2\pi}\int_{-\infty}^{\infty} S(-\eta)e^{-j\eta x_o}V_z(\eta)d\eta)$$

$$f_2 \cdot E_z = \int_{-\infty}^{\infty} f_2(x)E_z(x)dx = 0 = \frac{1}{2\pi}\int_{-\infty}^{\infty} S(-\eta)e^{+j\eta x_o}V_z(\eta)d\eta)$$

The Garlekin method yields the homogeneous equations

$$\begin{aligned} G_{11}(\alpha)I_1 + G_{12}(\alpha)I_2 = 0 \\ G_{21}(\alpha)I_1 + G_{22}(\alpha)I_2 = 0 \end{aligned} \tag{119}$$

where

$$G_{11}(\alpha) = G_{22}(\alpha) = \frac{1}{2\pi}\int_{-\infty}^{\infty} S(-\eta)Z_{e22}(\alpha,\eta)S(\eta)d\eta$$

$$G_{12}(\alpha) = \frac{1}{2\pi}\int_{-\infty}^{\infty} S(-\eta)Z_{e22}(\alpha,\eta)S(\eta)e^{-2j\eta x_o}d\eta$$

$$G_{21}(\alpha) = \frac{1}{2\pi}\int_{-\infty}^{\infty} S(-\eta)Z_{e22}(\alpha,\eta)S(\eta)e^{2j\eta x_o}d\eta$$

In isotropic media, the function $Z_{e22}(\alpha,\eta)$ is an even function of both α and η, whence $G_{21}(\alpha) = G_{12}(\alpha)$.

The propagation constants α_o of the modes are given by those values of α that provide a non-zero solution of the homogeneous equations, yielding

$$\det\begin{vmatrix} G_{11}(\alpha) & G_{12}(\alpha) \\ G_{21}(\alpha) & G_{22}(\alpha) \end{vmatrix} = G_{11}^2(\alpha) - G_{12}^2(\alpha) \rightarrow G_{12}(\alpha_o) = \pm G_{11}(\alpha_o) \tag{120}$$

Due to the presence of even function of α, if α_o[3] is a solution of the above equation, $-\alpha_o$ is also a solution. In applications, the quasi-TEM modes that are characterized by negligible values of the longitudinal components of the electromagnetic fields are very important.

[3] We assume α_o positive or having negative imaginary part.

For the TEM modes, the sum of the currents on the strips must be zero[4] when the dispersion equation is

$$G_{12}(\alpha_o) = G_{11}(\alpha_o)$$

which yields $I_2 = -I_1$.

To obtain the characteristic impedance of a quasi-TEM mode we must evaluate the voltage of the progressive quasi-TEM mode. For the considered geometry V_{TEM} can be defined by

$$V_{TEM} = -\int_{-x_o+s/2}^{x_o-s/2} E_x(x,0)dx \tag{121}$$

The evaluation of $E_x(x,0)$ requires the inverse Fourier transform of the spectral voltage $V_x(\eta) = \hat{x} \cdot \mathbf{V}(\eta)$. It can be obtained via

$$V_x(\eta) = Z_{e12}(\eta, \alpha)A_2(\eta) \tag{122}$$

8.5.1.1 *The absence of the dielectric substrate*

For illustrative purposes, in this section we consider the evaluation of the propagation constant and the TEM mode of the coplanar line shown in Fig. 11 when the dielectric substrate is absent.

In general, if the conductors of a transmission line are immersed in a homogeneous isotropic medium, we have ideal TEM modes having propagation constants defined by the propagation constants of the medium surrounding the conductors. This happens in the case considered herein. First we observe that the kernel matrix of the W-H equation simplifies considerably, and we get

$$Z_{e22}(\alpha, \eta) = \frac{1}{2}Z_{c22} = \frac{Z_o(k^2 - \alpha^2)}{2k\sqrt{k^2 - \eta^2 - \alpha^2}} \tag{123}$$

The correct evaluation of the parameters of the transmission line requires an appropriate choice of the expansion function $s(x)$. To satisfy the edge behavior at $x = \pm w/2$, we assume

$$s(x) = \frac{u(x + w/2) - u(x - w/2)}{\sqrt{(w/2)^2 - x^2}} \tag{124}$$

where $u(x)$ is the step function. With this choice we get

$$S(\eta) = \int_{-w/2}^{w/2} \frac{e^{j\eta x}}{\sqrt{(w/2)^2 - x^2}}dx = J_o\left(\frac{\eta w}{2}\right)$$

[4] Otherwise the TEM mode radiates infinite power in every transverse section of the waveguide.

The dispersion eq. (120) becomes

$$(k^2-\alpha_o^2)\int_{-\infty}^{\infty} J_o^2\left(\frac{\eta w}{2}\right)\frac{1}{\sqrt{k^2-\eta^2-\alpha_o^2}}d\eta = (k^2-\alpha_o^2)\int_{-\infty}^{\infty} J_o^2\left(\frac{\eta w}{2}\right)\frac{\cos(2\eta\alpha_0)}{\sqrt{k^2-\eta^2-\alpha_o^2}}d\eta$$

Taking into account that the integral have different values we get the expected result:

$$\alpha_o = k \tag{125}$$

To evaluate the characteristic impedance of the coplanar line in the absence of a dielectric substrate, we observe that

$$A_2(\eta) = I_2\tilde{f}_2(\eta) + I_1\tilde{f}_1(\eta) = I(\tilde{f}_1(\eta) - \tilde{f}_2(\eta)) = 2jI\ S(\eta)\sin(x_o\eta)$$

Equation (122) yields

$$V_x(\eta) = \frac{Z_o\alpha_o\eta}{k\sqrt{k^2-\eta^2-\alpha_o^2}} jJ_o\left(\frac{\eta w}{2}\right)\sin(x_o\eta)I = -Z_o sign(\eta)J_o\left(\frac{\eta w}{2}\right)\sin(x_o\eta)I$$

where according to the convention for the propagation constants, we set

$$\left.\sqrt{k^2-\eta^2-\alpha_o^2}\right|_{\alpha_o=k} = -j|\eta|$$

The inverse transform yields the component $E_x(x,0)$

$$E_x(x,0) = \frac{Z_oI}{2\pi}\int_{-\infty}^{\infty} sign(\eta)J_o\left(\frac{\eta w}{2}\right)\sin(x_o\eta)e^{-j\eta x}d\eta = \frac{Z_oI}{\pi}\int_0^{\infty} J_o\left(\frac{\eta w}{2}\right)\sin(x_o\eta)\cos(x\eta)d\eta$$

$$= \frac{Z_oI}{2\pi}\int_0^{\infty} J_o\left(\frac{\eta w}{2}\right)\sin\left[(x_o-x)\eta\right]\cos(x\eta)d\eta + \frac{Z_oI}{2\pi}\int_0^{\infty} J_o\left(\frac{\eta w}{2}\right)\sin\left[(x_o+x)\eta\right]\cos(x\eta)d\eta$$

The integrals in the last member can be evaluated in closed form (Gradshteyn & Ryzhik, 1965, p. 731, formula 9). We get

$$\int_0^{\infty} J_o\left(\frac{\eta w}{2}\right)\sin\left[(x_o\pm x)\eta\right]\cos(x\eta)d\eta = \frac{u(x_o\pm x - w/2)}{\sqrt{(x_o\pm x)^2 - w^2/4}}$$

and then

$$Z_\infty = \frac{V_{TEM}}{I} = \frac{Z_o}{\pi}\int_{-x_o+w/2}^{x_o-w/2}\int_0^{\infty} J_o\left(\frac{\eta w}{2}\right)\sin(x_o\eta)\cos(x\eta)d\eta dx$$

$$= \frac{Z_o}{\pi}\log\frac{w}{-w+4x_o-4\sqrt{x_o\left(x_o-\frac{w}{2}\right)}}$$

where we have taken into account using MATHEMATICA that

$$\int_{-x_o+w/2}^{x_o-w/2} \frac{u(x_o \pm x - w/2)}{\sqrt{(x_o \pm x)^2 - w^2/4}} dx = \log \frac{w}{-w + 4x_o - 4\sqrt{x_o\left(x_o - \frac{w}{2}\right)}}$$

The exact formula obtained with the evaluation of the capacitance between the two coplanar strips through the conformal mapping method is (Paul, 1992)

$$Z_{e\infty} = Z_o \frac{\text{EllipticK}[m]}{\text{EllipticK}[1-m]}$$

where

$$m = \left(\frac{2x_o - w}{2x_o - w}\right)^2$$

To compare the two formulas, assuming $x_o = w$ we get $Z_\infty = 211\ \Omega$ and $Z_{e\infty} = 241\ \Omega$.

To improve the approximation, we must choose more accurate expansion functions that take into account the proximity of the two strips.

8.5.2 The shielded microstrip transmission line

Figure 12 illustrates the cross section of a shielded transmission line. The problem of propagation of the quasi-TEM mode can be tackled with the W-H technique. In the limit $c \to 0$ this problem can be formulated in terms of the following modified W-H equation (Mittra & Lee, 1971, p. 283):

$$G(\eta)F_1(\eta) = \Phi_1(\eta) + e^{-j\eta L}\Phi_+(\eta) + e^{j\eta L}\Phi_+(-\eta)$$

where

$$F_1(\eta) = \int_{-L}^{L} \rho_S(x)e^{j\eta x}dx, \quad \rho_S(x) = -\left[\varepsilon_o \frac{\partial E_z(x,y)}{\partial y}\bigg|_{y=(b+c)_+} - \frac{\partial E_z(x,y)}{\partial y}\bigg|_{y=(b+c)_-}\right]$$

$$\Phi_+(\eta) = \int_{L}^{\infty} E_z(x, b+c)e^{j\eta x}dx, \quad \Phi_1(\eta) = \pi \frac{\sin(\eta L)}{\eta}$$

$$G(\eta) = \frac{\sinh(\eta b)\sinh[\eta(a-b)]}{\varepsilon_o \eta[\varepsilon_r \sinh[\eta(a-b)\cosh(\eta b) + \cosh[\eta(a-b)\sinh(\eta b)]}$$

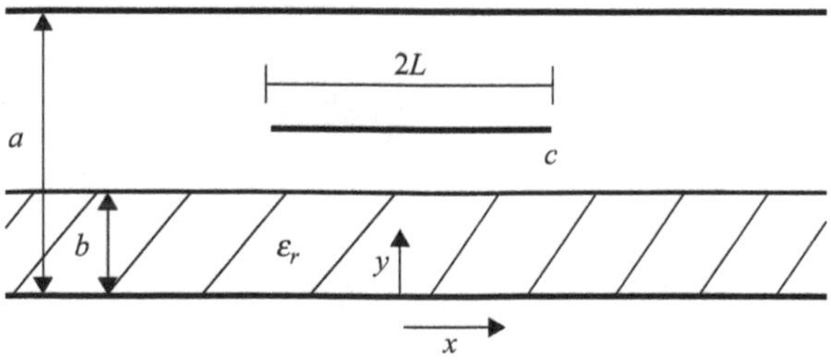

Fig. 12: Geometry of the problem

In the limit $a \to \infty$, that is, when the top shield is absent, $G(\eta)$reduces to

$$G(\eta) = \frac{1}{\varepsilon_o|\eta|\left[1 + \varepsilon_r \coth\left(|\eta|b\right)\right]}$$

The presence of a nonanalytic function $G(\eta)$ requires us to work within the framework of a Hilbert-Riemann problem.

8.6 General W-H formulation of planar discontinuity problems in arbitrary stratified media

A planar discontinuity in an arbitrary stratified medium is defined by patches or apertures located in particular sections of the stratification. For example, Fig. 13 illustrates the presence of two patches in the section $y = y_a$ and $y = y_b$.

In the section where the patch is located, the plane is divided into two regions. The region + is the region where the patch is present. The region – is an aperture where the transverse components $\mathbf{E}_t$ and $\mathbf{H}_t$ of the electromagnetic field are continuous.

The boundary conditions on the patches depend on the constituent materials. These conditions are studied in Senior and Volakis (1995). For the sake of simplicity, in the following we assume PEC patches whose boundary conditions are

$$\mathbf{E}_t(z,x,y_{A-}) = \mathbf{E}_t(z,x,y_{A+}) = 0, \quad \mathbf{E}_t(z,x,y_{B-}) = \mathbf{E}_t(z,x,y_{B+}) = 0$$

The dependence on z,x, requires the introduction of two-dimensional Fourier transforms instead the one-dimensional transforms introduced in chapter 7, eqs. (103):

$$\mathbf{V}(\alpha,\eta,y) = \int_{-\infty}^{\infty}\int_{-\infty}^{\infty} \hat{y} \times \mathbf{E}_t(z,x,y)e^{j\alpha z}e^{j\eta x}dzdx$$

$$\mathbf{I}(\alpha,\eta,y) = \int_{-\infty}^{\infty}\int_{-\infty}^{\infty} \mathbf{H}_t(z,x,y)e^{j\alpha z}e^{j\eta x}dzdx$$

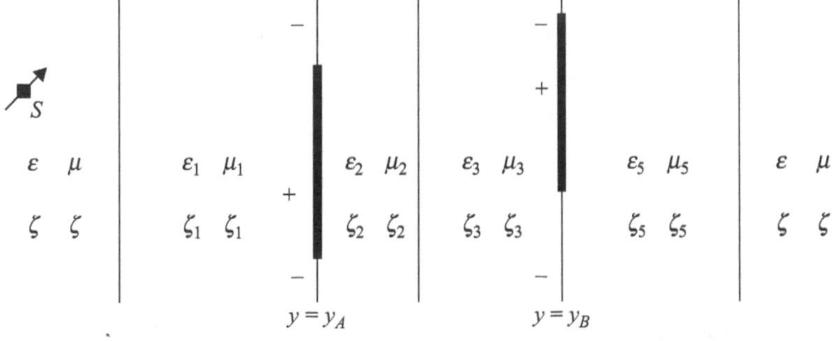

Fig. 13: A planar discontinuity problem a in stratified medium

In the following we use $\boldsymbol{\sigma} = \alpha\hat{z} + \eta\hat{x}$. All the consideration in chapter 7 hold, and the only difference is that α is a spectral variable and not a parameter. In particular, circuit models as those indicated in chapter 7 are very useful in deriving the equations of the problems. By assuming that the source of the electromagnetic field is located in $y < y_a$, a Norton representation of the circuit modeling the region $y < y_a$ is indicated in Fig. 14a.

In this representation, the current $\mathbf{I}_p$ represents the known Fourier transform of $\mathbf{H}_t^p(z, x, y_{A-})$, which is the transverse magnetic field in the section $y = y_a$ when the whole plane $y = y_a$ is a PEC plane. The admittance $\mathbf{Y}_a$ is the matrix that relates $-I(y_{A-})$ to V_A when the source is absent. Similarly, the region $y > y_b$ is modeled by an admittance $\mathbf{Y}_b$ that relates $I(y_{B+})$ to V_B.

Since the patches are PEC, they are modeled by plus current generators that are defined by

$$A_{A+} = I(y_{A+}) - I(y_{A-}), \quad A_{B+} = I(y_{B+}) - I(y_{B-})$$

The subscript $+$ means that the support of A_{A+} and A_{B+} in the natural domain is the region $+$.

Finally, the region between the section $y = y_{A+}$ and $y = y_{B-}$ can be modeled by a two-port having an admittance $\mathbf{Y}_e$. With these considerations in mind, for the problem indicated in Fig. 13 we get the circuit representation of Fig. 15:

$$\mathbf{Y}_e = \begin{vmatrix} Y_{e11} & Y_{e12} \\ Y_{e21} & Y_{e22} \end{vmatrix}$$

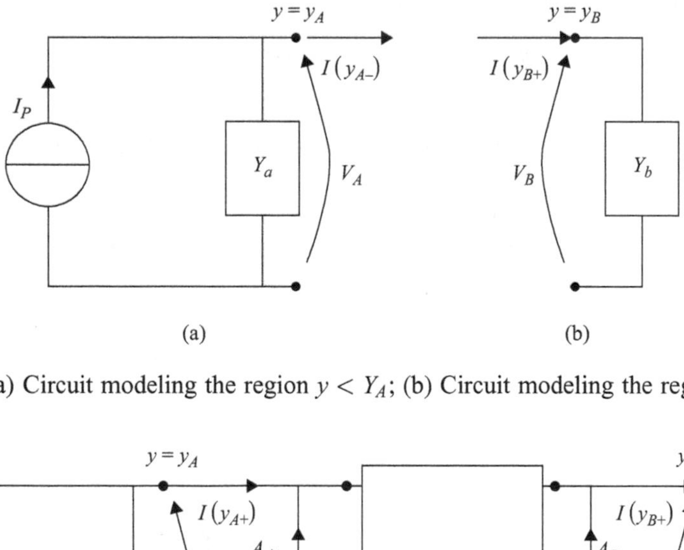

Fig. 14: (a) Circuit modeling the region $y < Y_A$; (b) Circuit modeling the region $y > Y_B$

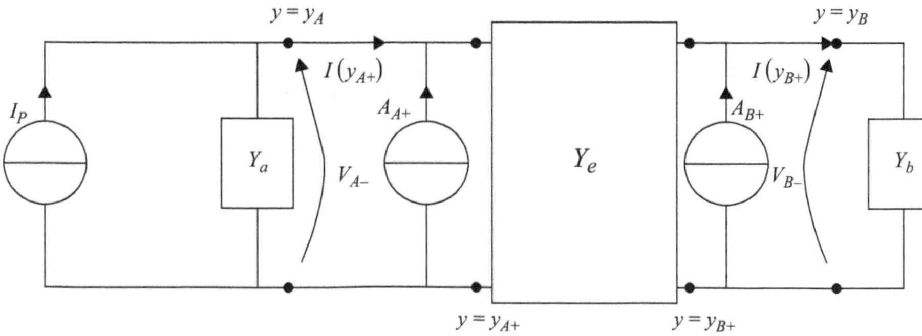

Fig. 15: Circuit representation of the geometry of Fig. 13

In the representation of Fig. 15 we use the subscript – for the voltage V_{A-} and V_{B-} since they are Fourier transforms of functions that are vanishing in the support +. Using the nodal analysis in Fig. 15 we get

$$Z_t(\boldsymbol{\sigma})A_+(\boldsymbol{\sigma}) = V_-(\boldsymbol{\sigma}) - Z_t(\boldsymbol{\sigma})I_s(\boldsymbol{\sigma}) \tag{126}$$

where

$$Y_t(\boldsymbol{\sigma}) = \begin{vmatrix} Y_A(\boldsymbol{\sigma}) + Y_{e11}(\boldsymbol{\sigma}) & Y_{e12}(\boldsymbol{\sigma}) \\ Y_{e21}(\boldsymbol{\sigma}) & Y_B(\boldsymbol{\sigma}) + Y_{e22}(\boldsymbol{\sigma}) \end{vmatrix}$$

$$Z_t(\boldsymbol{\sigma}) = Y_t^{-1}(\boldsymbol{\sigma}), \quad A_+(\boldsymbol{\sigma}) = \begin{vmatrix} A_{A+}(\boldsymbol{\sigma}) \\ A_{B+}(\boldsymbol{\sigma}) \end{vmatrix}, \quad V_-(\boldsymbol{\sigma}) = \begin{vmatrix} V_{A-}(\boldsymbol{\sigma}) \\ V_{B-}(\boldsymbol{\sigma}) \end{vmatrix}, \quad I_s(\boldsymbol{\sigma}) = \begin{vmatrix} I_p(\boldsymbol{\sigma}) \\ 0 \end{vmatrix}$$

Equations (126) are vector multidimensional W-H equations. The matrix kernel $Z_t(\boldsymbol{\sigma}) = Z_t(\alpha, \beta)$ is a function of two complex variables α, β. Radlow (1961, 1965) attempted to factorize the matrix kernel in a particular problem,[5] but his solution is incorrect (Albani, 2007). Conversely, the moment method can be applied successfully provided that a good choice is made of the two-dimensional expansion and test functions.

The applications of the aforementioned theory are numerous and important and have been presented in many papers and books. In particular, the book by Munk (2000) is very complete on the theory and applications of the method of moments in the presence of patches having a periodic distribution in discontinuity planes.

8.6.1 Formal solution with the factorization method

To obtain the formal solution of the W-H eq. (126), let us consider the presence of a patch of arbitrary form located in the section $y = 0$. For the sake of simplicity let us assume that the plane $y = 0$ is a symmetry plane for the geometric region under study. Hence, Fig. 15 is modified as indicated in Fig. 16.

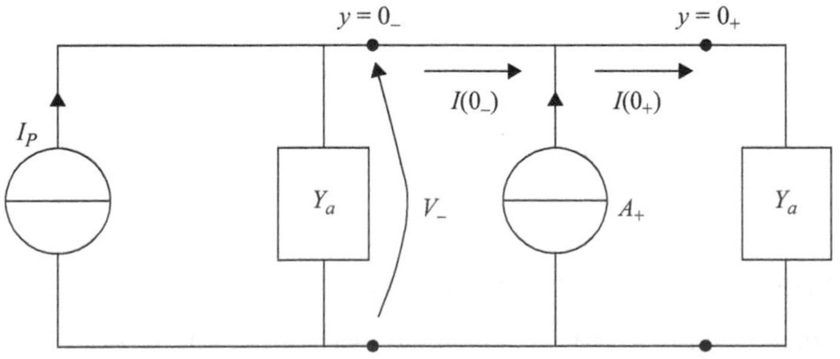

Fig. 16: Presence of one only discontinuity plane that separates two identical geometrical regions

[5] In the problem considered by Radlow, the patch is a PEC quarter-plane immersed in free space.

The W-H equation is

$$\frac{1}{2}Z_a(\boldsymbol{\sigma})A_+(\boldsymbol{\sigma}) = V_-(\boldsymbol{\sigma}) - V^p(\boldsymbol{\sigma}) \tag{127}$$

where

$$Z_a = Y_a^{-1}, \quad V^p(\boldsymbol{\sigma}) = \frac{1}{2}Z_a(\boldsymbol{\sigma})I_p(\boldsymbol{\sigma})$$

For multidimensional functions, we define decomposition and factorization. In the natural space the function $+(-)$ is the function having as support the region $+(-)$ and vanishing in the complementary region $-(+)$. We indicate with P the operator that expresses the part $+$ of an arbitrary function f. In the spatial domain P is defined by

$$P(\boldsymbol{\rho}) = 1 \text{ if } \boldsymbol{\rho} \in S_+, \quad P(\boldsymbol{\rho}) = 0 \text{ if } \boldsymbol{\rho} \in S_-$$

and we get $f_+(\boldsymbol{\rho}) = P(\boldsymbol{\rho})f(\boldsymbol{\rho})$. In the spectral domain this equation becomes

$$\tilde{f}_+(\boldsymbol{\sigma}) = \frac{1}{(2\pi)^2}\tilde{P}(\boldsymbol{\sigma}) * \tilde{f}(\boldsymbol{\sigma}) = \frac{1}{(2\pi)^2}\int_{\boldsymbol{\sigma}'} \tilde{P}(\boldsymbol{\sigma} - \boldsymbol{\sigma}')\tilde{f}(\boldsymbol{\sigma}')d\boldsymbol{\sigma}'$$

where $*$ means convolution and $\tilde{f}(\boldsymbol{\sigma}) = \tilde{f}(\alpha, \eta)$ is the double Fourier transform of $f(\boldsymbol{\rho}) = f(z, x)$. We let $m(\boldsymbol{\sigma}) = \frac{1}{2}Z_a(\boldsymbol{\sigma})$.

The problem of the factorization of (127), or of multidimensional W-H equations, consists in the factorization of a function m into two functions m_- and m_+ that with their inverses are functions minus and plus, respectively. However, while for the classical W-H equations the properties of plus and minus functions can be quickly ascertained by inspecting the singularities, this does not happen in the multidimensional W-H equations. This constitutes also the reason why we cannot resort to the logarithmic decomposition.

The formal solution of eq. (127) is

$$\begin{aligned} A_+(\boldsymbol{\sigma}) &= [m_+(\boldsymbol{\sigma})]^{-1} \cdot S_+(\boldsymbol{\sigma}) \\ V_-(\boldsymbol{\sigma}) &= m_-(\boldsymbol{\sigma}) \cdot S_-(\boldsymbol{\sigma}) \end{aligned} \tag{128}$$

where $S_+(\boldsymbol{\sigma})$ and $S_-(\boldsymbol{\sigma})$ arise from the decomposition of the function

$$-[m_-(\boldsymbol{\sigma})]^{-1} \cdot V^p(\boldsymbol{\sigma}) = S_-(\boldsymbol{\sigma}) + S_+(\boldsymbol{\sigma})$$

8.6.1.1 Some properties derived from the formal solution

(a) Exact evaluation of the transverse magnetic field on the apertures

Even though we are unable to obtain the factorization of m, the concept of decomposition can produce interesting results. For instance, from Fig. 16 we get

$$I(0_+) = \frac{I_p}{2} + \frac{A_+}{2}, \quad I(0_-) = I(0_+) - A_+$$

Decomposing the previous equations into plus and minus functions yields

$$I_-(0_+) = \frac{I_{p-}}{2}, \quad I_-(0_-) = I_-(0_+) = \frac{I_{p-}}{2}$$

Hence, we know the exact transverse magnetic field on the aperture (minus function) that is given by

$$\mathbf{H}_t(\boldsymbol{\rho}, 0_-) = \mathbf{H}_t(\boldsymbol{\rho}, 0_+) = \frac{1}{2}\mathbf{H}_t^p(\boldsymbol{\rho}, 0), \quad \boldsymbol{\rho} \in S_-$$

(b) Impossibility to define an equivalent admittance for the discontinuity

The circuit model of Fig. 16 introduces the following question. Is it possible to model the discontinuity section $y = 0$ with a lumped admittance? To answer this question we observe that (128) can be rewritten as

$$V_-(\boldsymbol{\sigma}) = T \cdot V^p(\boldsymbol{\sigma})$$

where

$$T = -m_-(\boldsymbol{\sigma}) \cdot \overline{P} \cdot m_-^{-1}(\boldsymbol{\sigma})$$

and $\overline{P}$ is the operator that expresses the part $-$ of a function. We observe that

$$T^2 = m_- \cdot \overline{P} \cdot m_-^{-1} \cdot m_- \cdot \overline{P} \cdot m_-^{-1} = -T$$

from where we deduce that T is not invertible. For the sake of simplicity let us assume that the discontinuities occur in free space.[6] In this case $V^p(\boldsymbol{\sigma})$ has the meaning of incident voltage on the section $y = 0$, and we observe that $T = 1 + \Gamma$ where Γ is the reflection coefficient due to the impedance Z_{eq}.

$$V_-(0_-) = V(0_-) = Z_{eq} \cdot I(0_-)$$

where

$$Z_{eq} = (1 + \Gamma)(1 - \Gamma)^{-1} Z_c$$

or

$$Y_{eq} = Y_c(1 - \Gamma)(1 + \Gamma)^{-1}$$

This means that it is not possible to define Y_{eq} because the operator $T = 1 + \Gamma$ is not invertible.

(c) Babinet's principle

We consider only a planar discontinuity in free space. We call complementary two discontinuity planes S_1 and S_2 where

$$S_{1-} \equiv S_{2+} \quad \text{and} \quad S_{1+} \equiv S_{2-}$$

---

[6] In this case we have $Y_a = Y_c$ where Y_c is the admittance of free space.

Let us consider in free space the two problems P_1 and P_2:

- Problem P_1
  To solve the problem when the discontinuity plane is S_1.
- Problem P_2
  To solve the problem when the discontinuity plane is S_2.
  The problem P_1 yields the Wiener-Hopf equation

$$V_-^{(1)}(\boldsymbol{\sigma}) = V^{(1)i}(\boldsymbol{\sigma}) - \frac{1}{2} Z_c(\boldsymbol{\sigma}) \cdot A_+^{(1)}(\boldsymbol{\sigma}) \tag{129}$$

whereas problem P_2 yields the Wiener-Hopf equation

$$A_+^{(2)}(\boldsymbol{\sigma}) = 2I^{(2)i}(\boldsymbol{\sigma}) - 2Y_c(\boldsymbol{\sigma}) \cdot V_-^{(2)}(\boldsymbol{\sigma}) \tag{130}$$

The possibility of relating the solutions of problems P_1 and P_2 is because of

$$Y_c = Z_c^{-1} = -Y_o^2 \, \hat{y} \times Z_c \times \hat{y}$$

where $Y_o = \frac{1}{Z_o}$ is the admittance of free space.

Substituting the previous equation into (130) and multiplying by $\frac{1}{2} Z_o \hat{y} \times$ yields

$$\frac{1}{2} Z_o \hat{y} \times A_+^{(2)}(\boldsymbol{\sigma}) = Z_o \hat{y} \times I^{(2)i}(\boldsymbol{\sigma}) - Y_o Z_c(\boldsymbol{\sigma}) \cdot \hat{y} \times V_-^{(2)}(\boldsymbol{\sigma})$$

If $Z_o \hat{y} \times I^{(2)i}(\boldsymbol{\sigma}) = V^{(1)i}(\boldsymbol{\sigma})$, we obtain Babinet's principle that relates the solution of problem P_1 to the solution of problem P_2:

$$\frac{1}{2} Z_o \hat{y} \times A_+^{(2)}(\boldsymbol{\sigma}) = V_-^{(1)}(\boldsymbol{\sigma})$$
$$Y_o \hat{y} \times V_-^{(2)}(\boldsymbol{\sigma}) = \frac{1}{2} A_+^{(1)}(\boldsymbol{\sigma})$$

(d) The physical optics approximation

Let us consider the solution of

$$A_+(\boldsymbol{\sigma}) = I^s(\boldsymbol{\sigma}) - Y_t(\boldsymbol{\sigma}) \cdot V_-(\boldsymbol{\sigma})$$

If the source is constituted by a plane wave having the spatial pulsation $\boldsymbol{\sigma}_o$, the spectrum is concentrated on the line $\boldsymbol{\sigma}_o$ and we can use the approximation

$$Y_t(\boldsymbol{\sigma}) \approx Y_t(\boldsymbol{\sigma}_o)$$

This approximation is very accurate for very high values of the propagation constant k.[7] The W-H equation can be rewritten as

$$A_+(\boldsymbol{\sigma}) \approx I^s(\boldsymbol{\sigma}) - Y_t(\boldsymbol{\sigma}_o) \cdot V_-(\boldsymbol{\sigma})$$

[7] Taking into account that in free space if the propagation constant k assumes very high values, we have $\sqrt{k^2 - \sigma^2} \approx \sqrt{k^2 - \sigma_o^2}$.

Taking into account that $\overline{P}A_+(\boldsymbol{\sigma}) = 0$, the projection with the operator $\overline{P}$ yields

$$V_-(\boldsymbol{\sigma}) \approx Y_t^{-1}(\boldsymbol{\sigma}_o) \cdot \overline{P} \cdot I^s(\boldsymbol{\sigma})$$

For instance, if S_- is an aperture in the PEC plane $z = 0$ located in free space, we get

$$\mathbf{E}_t(\boldsymbol{\rho}, z) = \mathbf{E}_t^p(\boldsymbol{\rho}, z) + \frac{1}{(2\pi)^2} \int_{\boldsymbol{\sigma}} V(\boldsymbol{\sigma}, z) e^{-j\boldsymbol{\sigma}\cdot\boldsymbol{\rho}} d\boldsymbol{\sigma} \tag{131}$$

where

$$V(\boldsymbol{\sigma}, z) = V_-(\boldsymbol{\sigma}) e^{-j\sqrt{k^2 - \boldsymbol{\sigma}^2}|z|}, \quad V_-(\boldsymbol{\sigma}) \approx Y_c^{-1}(\boldsymbol{\sigma}_o) \cdot \overline{P} \cdot I^i(\boldsymbol{\sigma})$$

and $\mathbf{E}_t^p(\boldsymbol{\rho}, z)$ is the field when the aperture is absent, that is, when the plane $z = 0$ is a PEC plane.

8.6.2 The method of stationary phase for multiple integrals

In many instances, one requires the evaluation of multiple integral having the form

$$\int f(\mathbf{x}) \exp(-j\Omega q(\mathbf{x})) d\mathbf{x}$$

where $\mathbf{x} = (x_1, x_2, \ldots, x_n)$ represents a point of the euclidean space E_n. For $\Omega \gg 1$ and under certain limitations, we get (Jones, 1964)

$$\int f(\mathbf{x}) \exp(-j\Omega q(\mathbf{x})) d\mathbf{x} = \left(\frac{2\pi}{\Omega}\right)^{n/2} \frac{1}{d} f(\mathbf{x}_s) e^{-j[\Omega\, q(\mathbf{x}_s) + \ell\pi/4]} \tag{132}$$

where $\mathbf{x}_s$ is the point of stationary phase defined by

$$\frac{\partial q}{\partial x_i} = 0, \quad (i = 1, 2, \ldots, n)$$

d is the determinant of the Hessian evaluated in $\mathbf{x}_s$:

$$H_s = \left| \begin{matrix} \frac{\partial^2 q}{\partial x_1 \partial x_1} & \frac{\partial^2 q}{\partial x_1 \partial x_2} & \cdots & \frac{\partial^2 q}{\partial x_1 \partial x_n} \\ \frac{\partial^2 q}{\partial x_2 \partial x_1} & \frac{\partial^2 q}{\partial x_2 \partial x_2} & \cdots & \frac{\partial^2 q}{\partial x_2 \partial x_n} \\ \cdots & \cdots & \cdots & \cdots \\ \frac{\partial^2 q}{\partial x_n \partial x_1} & \frac{\partial^2 q}{\partial x_n \partial x_2} & \cdots & \frac{\partial^2 q}{\partial x_n \partial x_n} \end{matrix} \right|_{\mathbf{x} = \mathbf{x}_s}$$

and the indicator ℓ is given by

$$\ell = \sum_{i=1}^{n} sign(h_i)$$

where h_i are the eigenvalues of the Hessian H_s.

8.6.2.1 Far field radiated by apertures

As an example of the development in section 8.6.2, we evaluate the far field in the double integral of eq. (132). We let

$$\mathbf{x} = \boldsymbol{\sigma}, \quad f(\mathbf{x}) = \frac{1}{(2\pi)^2} V_-(\boldsymbol{\sigma}), \quad \Omega = r$$

$$q(\mathbf{x}) = \frac{\sqrt{k^2 - \alpha^2 - \eta^2}\, z + \alpha x + \eta y}{r}$$
$$= \sqrt{k^2 - \alpha^2 - \eta^2} \cos\theta + \alpha \cos\varphi \sin\theta + \eta \sin\varphi \sin\theta$$

The point $\boldsymbol{\sigma}_s = (\alpha_s, \eta_s)$ of stationary phase is defined by

$$\frac{\partial q}{\partial \alpha_s} = -\frac{\alpha_s}{\sqrt{k^2 - \alpha_s^2 - \eta_s^2}} \cos\theta + \cos\varphi \sin\theta = 0$$

$$\frac{\partial q}{\partial \alpha_s} = -\frac{\eta_s}{\sqrt{k^2 - \alpha_s^2 - \eta_s^2}} \cos\theta + \sin\varphi \sin\theta = 0$$

The solution yields $\alpha_s = k \cos\varphi \sin\theta$, $\eta_s = k \sin\varphi \sin\theta$, $q(\boldsymbol{\sigma}_s) = k$. The Hessian at the point $\boldsymbol{\sigma}_s = (\alpha_s, \eta_s)$ is

$$H_s = \begin{vmatrix} 1 + \tan^2\theta \cos^2\varphi & \tan^2\theta \sin\varphi \cos\varphi \\ \tan^2\theta \sin\varphi \cos\varphi & 1 + \tan^2\theta \sin^2\varphi \end{vmatrix}$$

whence

$$d = \frac{1}{(k \cos\theta)^2}, \quad d_1 = -\frac{1}{k \cos^2\theta}, \quad d_2 = -\frac{1}{k}, \quad \ell = -1 - 1 = -2$$

Substituting in (132) we get

$$\mathbf{E}_t(\boldsymbol{\rho}, z) = j \frac{k \cos\theta}{2\pi} V_-(\boldsymbol{\sigma}_s) \frac{e^{-jkr}}{r} \tag{133}$$

8.6.3 The circular aperture

In this section we apply the previous theory to the solution of the problem of a circular aperture located in a PEC plane (Fig. 17). Since we introduce a system of cartesian coordinates that is a different from the one used in chapter 7, we introduce the following Fourier transforms:

$$\mathbf{V}(\alpha, \eta, z) = \int_{-\infty}^{\infty} \int_{-\infty}^{\infty} \mathbf{E}_t(x, y, z) e^{j\alpha x} e^{j\eta y} dx dy$$

$$\mathbf{I}(\alpha, \eta, z) = \int_{-\infty}^{\infty} \int_{-\infty}^{\infty} \mathbf{H}_t(x, y, z) \times \hat{z}\, e^{j\alpha x} e^{j\eta y} dx dy$$

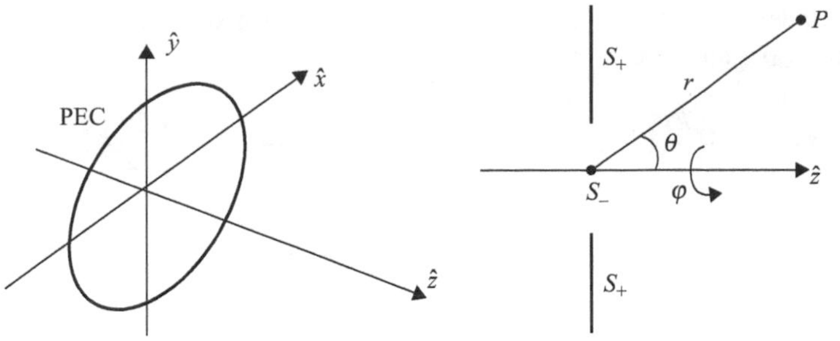

Fig. 17: Aperture S_- in a PEC plane

or, being that $\boldsymbol{\sigma} = \alpha\,\hat{x} + \eta\,\hat{y}$, $\boldsymbol{\rho} = x\,\hat{x} + y\,\hat{y}$,

$$\mathbf{V}(\boldsymbol{\sigma}, z) = \int_{\boldsymbol{\rho}} \mathbf{E}_t(\boldsymbol{\rho}, z) e^{j\,\boldsymbol{\sigma}\times\boldsymbol{\rho}} d\boldsymbol{\rho}$$

$$\mathbf{I}(\boldsymbol{\sigma}, z) = \int_{\boldsymbol{\rho}} \mathbf{H}_t(\boldsymbol{\rho}, z) \times \hat{z}\, e^{j\,\boldsymbol{\sigma}\times\boldsymbol{\rho}} d\boldsymbol{\rho}$$

For the aperture problem shown in Fig. 17, we get the W-H equation

$$A_+(\boldsymbol{\sigma}) = I^p(\boldsymbol{\sigma}) - Y_e(\boldsymbol{\sigma}) \cdot V_-(\boldsymbol{\sigma})$$

where

$$A_+(\boldsymbol{\sigma}) = I(\boldsymbol{\sigma}, 0_-) - I(\boldsymbol{\sigma}, 0_+), \quad V_-(\boldsymbol{\sigma}) = V(\boldsymbol{\sigma}, 0_-) = V(\boldsymbol{\sigma}, 0_+)$$

$$Y_e(\boldsymbol{\sigma}) = 2Z_c^{-1}(\boldsymbol{\sigma}), \quad Z_c(\boldsymbol{\sigma}) = \frac{Z_o}{k\sqrt{k^2 - \alpha^2 - \eta^2}} \begin{vmatrix} k^2 - \alpha^2 & -\alpha\,\eta \\ -\alpha\,\eta & k^2 - \eta^2 \end{vmatrix}$$

The source term $I^s(\boldsymbol{\sigma})$ depends on the plane wave

$$\mathbf{E}^i(\boldsymbol{\rho}, z) = \mathbf{E}_o^i e^{-j\boldsymbol{\sigma}_o\cdot\boldsymbol{\rho} - j\sqrt{k^2-\boldsymbol{\sigma}_o^2}\,z}, \quad \mathbf{H}^i(\boldsymbol{\rho}, z) = \mathbf{H}_o^i e^{-j\boldsymbol{\sigma}_o\cdot\boldsymbol{\rho} - j\sqrt{k^2-\boldsymbol{\sigma}_o^2}\,z}, \quad \mathbf{H}_o^i = Y_o \hat{k}_o \times \mathbf{E}_o^i$$

with $\mathbf{E}_o^i$ normal to the propagation direction $\mathbf{k}_o = \boldsymbol{\sigma}_o + \sqrt{k^2 - \boldsymbol{\sigma}_o^2}\,\hat{z} = k\,\hat{k}_o : \mathbf{E}_o^i \cdot \mathbf{k}_o = 0$.

With this source we obtain

$$I^p(\boldsymbol{\sigma}) = 4\pi^2 Y_e(\boldsymbol{\sigma}_o) \cdot \mathbf{E}_{ot}^i \delta(\boldsymbol{\sigma} - \boldsymbol{\sigma}_o)$$

where $\mathbf{E}_{ot}^i = \mathbf{E}_o^i - \hat{z}\hat{z} \cdot \mathbf{E}_o^i = \hat{x}\,E_{ox}^i + \hat{y}\,E_{oy}^i$.

The circular aperture problem is a fundamental problem that cannot be solved in closed form. Low-frequency solutions have been studied by many authors (see, e.g., Bouwkamp,

1954; Van Bladel, 2007). In the case of normal incidence $\boldsymbol{\sigma}_o = 0$, we find the following electric field on the aperture of radius a:

$$E_\rho = \frac{2jk}{3\pi}\left(E^i_{ox}\cos\varphi + E^i_{oy}\sin\varphi\right)\left(4\sqrt{a^2-\rho^2} + \frac{2\rho^2}{\sqrt{a^2-\rho^2}}\right)$$

$$E_\varphi = \frac{8jk}{3\pi}\left(-E^i_{ox}\sin\varphi + E^i_{oy}\cos\varphi\right)\sqrt{a^2-\rho^2}$$

wherefrom, by supposing $E^i_{ox} = 0$:

$$V_-(\boldsymbol{\sigma}) = \frac{8jka^3}{3}\frac{\sin(\sigma a)}{\sigma a}E^i_{oy}\hat{y}$$

Conversely, working at high frequencies, the physical optics approximation considered in section 8.6.1.1 provides the solution

$$V_-(\boldsymbol{\sigma}) = E^i_{oy}\hat{y}\int_{\boldsymbol{\rho}} \overline{p}(\boldsymbol{\rho})e^{j\boldsymbol{\sigma}\cdot\boldsymbol{\rho}}d\boldsymbol{\rho} = E^i_{oy}\hat{y}\int_0^a \rho \int_0^{2\pi} e^{j\sigma\rho\cos(\varphi-\beta)}d\varphi d\rho$$

$$= 2\pi a^2\frac{J_1(\sigma a)}{\sigma a}E^i_{oy}\hat{y}$$

In the intermediate frequency band, we can obtain the solution through the method of moments. In the natural domain this method has been used by Levine and Schwinger (1949). These authors proved that the far field is stationary with respect to small variations of the aperture field. In the spectral domain we introduce a set of expansion functions $\tilde{\psi}_j(\boldsymbol{\sigma})$ and test functions $\tilde{\varphi}_j(\boldsymbol{\sigma})$ and let

$$V_-(\boldsymbol{\sigma}) = \sum_j V_j\,\tilde{\psi}_j(\boldsymbol{\sigma})$$

The moment method yields

$$\sum_j Y_{ij}\,V_j = I_i$$

where

$$Y_{ij} = \int_{\boldsymbol{\sigma}} \tilde{\varphi}_i(-\boldsymbol{\sigma})\cdot Y_e(\boldsymbol{\sigma})\cdot\tilde{\psi}_j(\boldsymbol{\sigma})d\boldsymbol{\sigma}, \quad I_i = \int_{\boldsymbol{\sigma}} \tilde{\varphi}_i(-\boldsymbol{\sigma})\cdot I^p(\boldsymbol{\sigma})d\boldsymbol{\sigma} \tag{134}$$

The accuracy of the solution depends on the choice of the functions $\tilde{\psi}_j(\boldsymbol{\sigma})$ and $\tilde{\varphi}_j(\boldsymbol{\sigma})$. In the following, we adopt the Garlekin scheme by assuming $\tilde{\psi}_j(\boldsymbol{\sigma}) = \tilde{\varphi}_j(\boldsymbol{\sigma})$.

A natural set of expansion functions is constituted by the modes of the circular waveguide of radius a. These modes are well described in the natural domain. We have

$$\psi_j(\boldsymbol{\rho}) = \varphi_j(\boldsymbol{\rho}) \rightarrow \mathbf{e}^{TE,TM}_{m\,n}$$

where

$$\mathbf{e}^{TE}_{m\,n} = \hat{z}\times\nabla_t\Psi_{m\,n}, \quad \mathbf{e}^{TM}_{m\,n} = \nabla_t\Phi_{m\,n}$$

and

$$\Psi_{mn} = J_m\left(\frac{\chi'_{mn}}{a}\rho\right)\begin{Bmatrix}\cos m\varphi\\ \sin m\varphi\end{Bmatrix}, \quad \Phi_{mn} = J_m\left(\frac{\chi_{mn}}{a}\rho\right)\begin{Bmatrix}\cos m\varphi\\ \sin m\varphi\end{Bmatrix}, \quad J'_m(\chi'_{mn}) = 0, \quad J_m(\chi_{mn}) = 0$$

In the spectral domain we have[8]

$$\tilde{\Phi}_{mn}(\boldsymbol{\sigma}) = \Phi_{mn}(\sigma,\beta) = \int_0^a \rho J_m\left(\frac{\chi_{mn}}{a}\rho\right)\int_0^{2\pi}\cos(m\varphi)e^{j\sigma\rho\cos(\varphi-\beta)}d\varphi d\rho$$

$$= 2\pi j^m\cos(m\beta)\int_0^a \rho J_m\left(\frac{\chi_{mn}}{a}\rho\right)J_m(\sigma\rho)d\rho$$

The last integral is a Lommel integral and can be evaluated in closed form. We get

$$\tilde{\Phi}_{mn}(\boldsymbol{\sigma}) = -2\pi j^m a^2\cos(m\beta)\frac{1}{\chi_{mn}^2 - (\sigma a)^2}J'_m(\chi_{mn})J_m(\sigma a)$$

and taking into account that Φ_{mn} is zero on the rim $\rho = a$:

$$\tilde{\mathbf{e}}_{mn}^{TM} = -j\boldsymbol{\sigma}\tilde{\Phi}_{mn}$$

Conversely, the function Ψ_{mn} is nonzero on $\rho = a$. This means that

$$\tilde{\mathbf{e}}_{mn}^{TE} = -j\hat{z}\times\boldsymbol{\sigma}\Psi_{mn}$$

is not valid. By using distribution theory we can write

$$\tilde{\mathbf{e}}_{mn}^{TE} = -j\hat{z}\times\boldsymbol{\sigma}\tilde{\Psi}_{mn} + \tilde{\mathbf{e}}_{c,mn}$$

where the corrective term arises from the Fourier transform of

$$\mathbf{e}_{c,mn} = \hat{z}\times(J_{mn}(\chi'_{mn})\cos(m\varphi)\delta(\rho - a)\hat{\rho}$$

With the procedure used for the TM modes, we get

$$-j\hat{z}\times\boldsymbol{\sigma}\tilde{\Psi}_{mn} = -2\pi j^{m-1}a\cos(m\beta)\frac{(\sigma a)^2}{\chi'_{mn}2 - (\sigma a)^2}J_m(\chi'_{mn})J'_m(\sigma a)\hat{\beta}$$

$$\tilde{\mathbf{e}}_{c,mn} = -2\pi j^{m+1}aJ_m(\chi'_{mn})\left[\sin(m\beta)\frac{mJ_m(\sigma a)}{\sigma a}\hat{\sigma} + \cos(m\beta)J'_m(\sigma a)\hat{\beta}\right]$$

After the evaluation of the expansion functions in the spectral domain, we must evaluate the integrals (134). These integrals cannot be calculated in closed form and have been evaluated numerically.

---

[8] We consider only the even modes.

Once the solution $V_-(\boldsymbol{\sigma})$is obtained, the far field for $z > 0$ is given by

$$\mathbf{E}_t(\boldsymbol{\rho}, z) = j\frac{k\cos\theta}{2\pi} V_-(\boldsymbol{\sigma}_s)\frac{e^{-jkr}}{r}$$

where $\boldsymbol{\sigma}_s = k(\hat{x}\cos\varphi\sin\theta + \hat{y}\sin\varphi\sin\theta)$. The longitudinal component $E_z(\boldsymbol{\rho}, z)$ satisfies

$$(\hat{z}E_z(\boldsymbol{\rho}, z) + \mathbf{E}_t(\boldsymbol{\rho}, z)) \cdot \hat{r} = 0$$

hence

$$E_z(\boldsymbol{\rho}, z) = -\frac{\mathbf{E}_t(\boldsymbol{\rho}, z)\cdot\hat{r}}{\cos\theta}$$

Some numerical simulations are shown in Fig. 18, Fig. 19, and Fig. 20.

8.6.4 The quarter plane problem

8.6.4.1 Introduction

For the sake of simplicity, in this presentation we consider only the case of scalar diffraction in the presence of a soft quarter-plane defined by $z = 0$, $x \geq 0$, $y \geq 0$ (Fig. 21).

The primary source is the incident plane wave:

$$\psi^i(x, y, z) = e^{-j\eta_o x - j\xi_o x - j\alpha_o z} \tag{135}$$

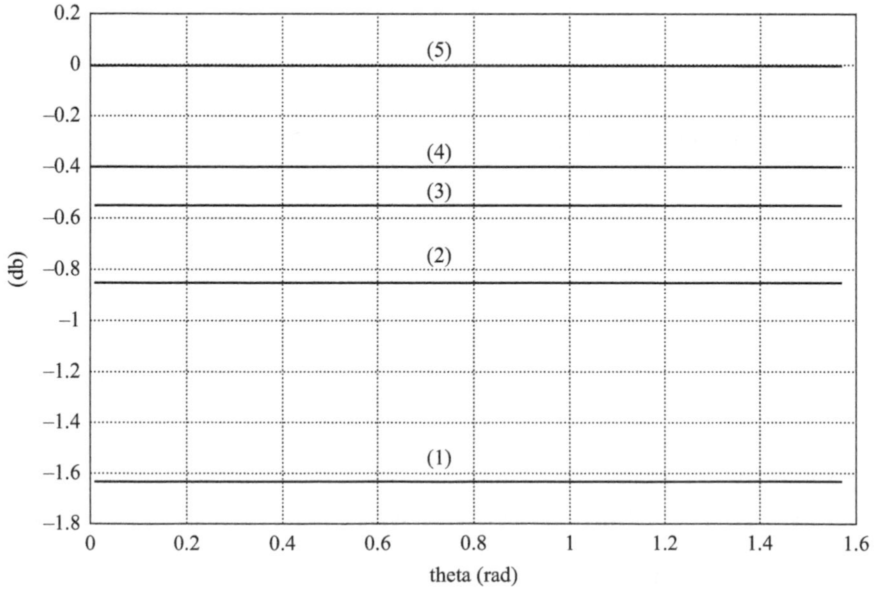

Fig. 18: Far field for small aperture ($k\,a = 1$): (1) used moment: TE_{11}; (2) used moments: TE_{11}, TE_{12}; (3) used moments: $TE_{11}, TE_{12}, TE_{13}$; (4) used moments: $TE_{11}, TE_{12}, TE_{13}, TE_{14}$; (5) far field evaluated with Bouwkamp's equation

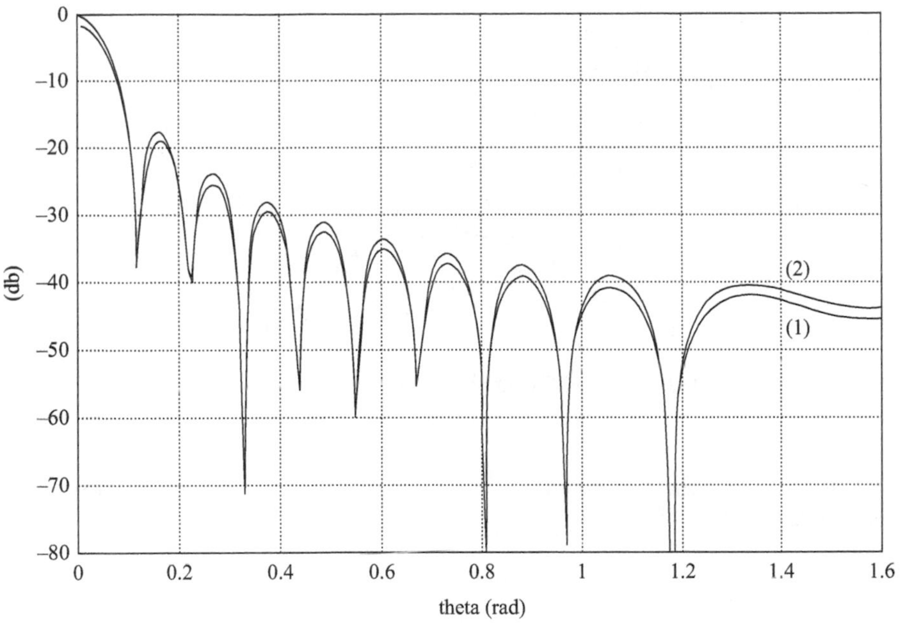

Fig. 19: Far field for large aperture ($k\,a = 31.4$), E – plane: (1) used moment: TE_{11}, (2) far field evaluated with physical optics

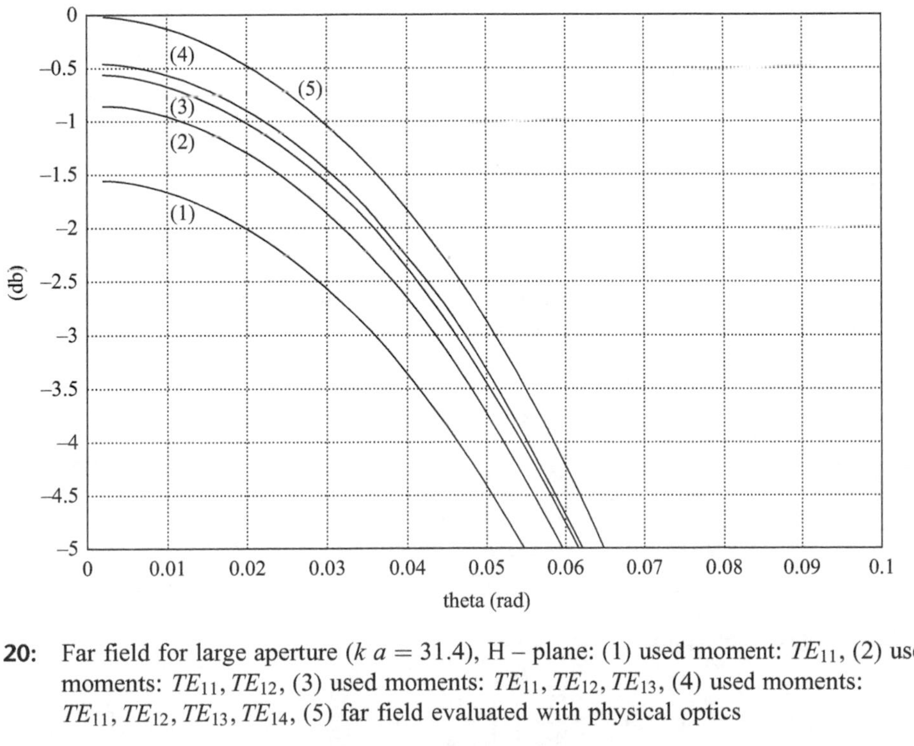

Fig. 20: Far field for large aperture ($k\,a = 31.4$), H – plane: (1) used moment: TE_{11}, (2) used moments: TE_{11}, TE_{12}, (3) used moments: $TE_{11}, TE_{12}, TE_{13}$, (4) used moments: $TE_{11}, TE_{12}, TE_{13}, TE_{14}$, (5) far field evaluated with physical optics

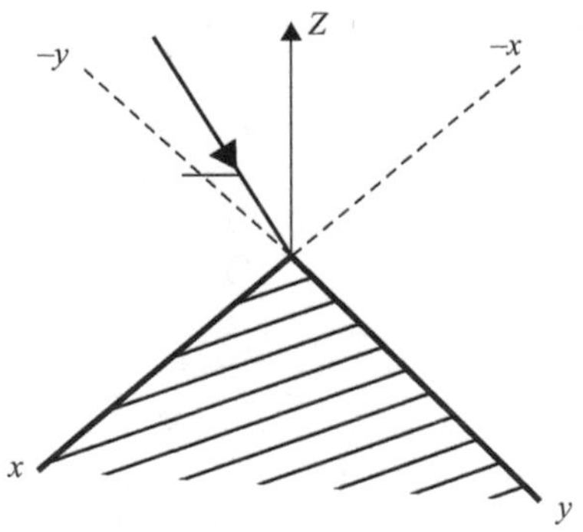

Fig. 21: Geometry of the quarter-plane problem

where k is the propagation constant in free space, and θ_o and φ_o are the azimuthal and zenithal angles of the incident wave:

$$\eta_o = k\sin\theta_o\cos\varphi_o, \quad \xi_o = k\sin\vartheta_o\sin\varphi_o$$

$$\alpha_o = \sqrt{k^2 - \eta_o^2 - \xi_o^2} = k\cos\vartheta_o$$

By indicating with $\psi(x, y, z)$ the total field, we must solve

$$\nabla^2\psi + k^2\psi = 0 \tag{136}$$

with the boundary condition

$$\psi(x, y, 0) = 0 \quad x \geq 0, \quad y \geq 0x \geq 0$$

The scattered field $\psi^s(x, y, z)$ satisfies

$$\nabla^2\psi^s + k^2\psi^s = j(x, y)\delta(z) \tag{137}$$

where $j(x, y)$ is the equivalent source induced on the quarter-plane.

To obtain the meaning of $j(x, y)$, we apply $\int_{0_-}^{0_+} \ldots \; dz$ to the previous equation:

$$\left.\frac{\partial\psi(x, y, z)}{\partial z}\right|_{z=0_+} - \left.\frac{\partial\psi(x, y, z)}{\partial z}\right|_{z=0_-} = j(x, y)$$

Taking into account the relationship

$$(\nabla^2 + k^2)\frac{e^{-jk\sqrt{(x-x')^2+(y-y')^2+z^2}}}{\sqrt{(x-x')^2 + (y-y')^2 + z^2}} = -4\pi\delta(x - x')\delta(y - y')\delta(z) \tag{138}$$

the following Green representation holds:

$$\psi(x, y, z) = \psi^i(x, y, z) + -\frac{1}{4\pi}\int_0^\infty\int_0^\infty \frac{e^{-jk\sqrt{(x-x')^2+(y-y')^2+z^2}}}{\sqrt{(x-x')^2 + (y-y')^2 + z^2}} j(x', y')dx'dy'$$

At the points $z = 0, x \geq 0, y \geq 0$ where $\psi(x, y, z) = 0$, the two-dimensional W-H equation follows:

$$\int_0^\infty \int_0^\infty g(x - x', y - y')j(x', y')dx'dy' = \psi^i(x, y, 0), \quad x \geq 0, \quad y \geq 0 \tag{139}$$

where

$$g(x, y) = \frac{1}{4\pi} \frac{e^{-jk\sqrt{x^2+y^2}}}{\sqrt{x^2 + y^2}}$$

The double Fourier transform of $g(x, y)$ is given by

$$G(\eta, \xi) = \int_{-\infty}^{\infty} \int_{\infty}^{\infty} g(x, y)e^{j\eta x}e^{j\xi y}dxdy = \frac{1}{4\pi} \int_{-\infty}^{\infty} \int_{\infty}^{\infty} \frac{e^{-jk\sqrt{x^2+y^2}}}{\sqrt{x^2 + y^2}} e^{j\eta x}e^{j\xi y}dxdy$$

$$= \frac{1}{4\pi} \int_{-\pi}^{\pi} \int_0^\infty \frac{e^{-jk\rho}}{\rho} e^{j\eta\rho\cos\varphi}e^{j\xi\rho\sin\varphi}\rho d\rho d\varphi = -j\frac{1}{4\pi} \int_{-\pi}^{\pi} \frac{1}{k - \eta\cos\varphi - \xi\sin\varphi} d\varphi$$

$$= -j\frac{1}{2} \frac{1}{\sqrt{k^2 - \eta^2 - \xi^2}}$$

This yields the following equation in the spectral domain:

$$G(\eta, \xi)F_+(\eta, \xi) = F_+^i(\eta, \xi) + F_-(\eta, \xi) \tag{140}$$

where

$$F_+(\eta, \xi) = J_{++}(\eta, \xi) = \int_0^\infty \int_0^\infty j(x, y)e^{j\eta x}e^{j\xi y}dxdy$$

$$F_+^i(\eta, \xi) = \Psi_{++}^i(\eta, \xi) = \int_0^\infty \int_0^\infty e^{-j\eta_o x - j\xi_o x}e^{j\eta x - j\xi x}dxdy$$

$$= -\frac{1}{\eta - \eta_o} \frac{1}{\xi - \xi_o}$$

$$F_-(\eta, \xi) = V_{--}(\eta, \xi) + Y_{-+}(\eta, \xi) + Z_{+-}(\eta, \xi)$$

In the previous formulas, the subscript $++$ means that the considered function is (independently of the value of the complex parameter ξ) regular in an upper half-plane of the complex variable η and (independently of the value of the complex parameter η) regular in an upper half-plane of the complex variable ξ.

Similarly:

The subscript $+-$ means that the considered function is (independently of the value of the complex parameter ξ) regular in an upper half-plane of the complex variable η and (independently of the value of the complex parameter η) regular in a lower half-plane of the complex variable ξ.

The subscript $-+$ means that the considered function is (independently of the value of the complex parameter ξ) regular in a lower half-plane of the complex variable η and (independently of the value of the complex parameter η) regular in an upper half-plane of the complex variable ξ.

The subscript $--$ means that the considered function is (independently of the value of the complex parameter ξ) regular in a lower half-plane of the complex variable η and (independently of the value of the complex parameter η) regular in a lower half-plane of the complex variable ξ.

The inverse Fourier transform of a function with subscript $++$ produces a function that in the spatial domain vanishes at $x < 0$ and $y < 0$. Similarly considerations apply for the other subscripts. For instance, the subscript $-+$ produces a function that in the spatial domain vanishes at $x > 0$ and $y < 0$. Of course the order of the subscripts is mandatory. For example, the product of the two functions $f_{+-}g_{-+}$ is neither a function $+-$ nor a function $-+$. Consequently, it is no guaranteed that its inverse Fourier transform vanishes in the quadrant $x > 0, y > 0$.

By using the logarithmic decomposition we can write

$$\begin{aligned} G(\eta,\xi) &= e^{\log[G(\eta,\xi)]} \\ &= e^{\{\log[G(\eta,\xi)]\}_{++}+\{\log[G(\eta,\xi)]\}_{+-}+\{\log[G(\eta,\xi)]\}_{-+}+\{\log[G(\eta,\xi)]\}_{--}} \\ &= G_{++}(\eta,\xi)G_{+-}(\eta,\xi)G_{-+}(\eta,\xi)G_{--}(\eta,\xi) \end{aligned} \tag{141}$$

where the decomposed functions of $\log[G(\eta,\xi)]$ are given by

$$\{\log[G(\eta,\xi)]\}_{++} = \frac{1}{(2\pi j)^2}\int_{\gamma_1}\int_{\gamma_1}\frac{\log[G(\eta',\xi')]}{(\eta'-\eta)(\xi'-\xi)}d\eta' d\xi'$$

$$\{\log[G(\eta,\xi)]\}_{+-} = -\frac{1}{(2\pi j)^2}\int_{\gamma_1}\int_{\gamma_2}\frac{\log[G(\eta',\xi')]}{(\eta'-\eta)(\xi'-\xi)}d\eta' d\xi'$$

$$\{\log[G(\eta,\xi)]\}_{-+} = -\frac{1}{(2\pi j)^2}\int_{\gamma_2}\int_{\gamma_1}\frac{\log[G(\eta',\xi')]}{(\eta'-\eta)(\xi'-\xi)}d\eta' d\xi'$$

$$\{\log[G(\eta,\xi)]\}_{--} = \frac{1}{(2\pi j)^2}\int_{\gamma_2}\int_{\gamma_2}\frac{\log[G(\eta',\xi')]}{(\eta'-\eta)(\xi'-\xi)}d\eta' d\xi'$$

and γ_1, γ_2 are the smile and frown real axes, respectively (chapter 1, Fig. 7).

The previous integrals have been evaluated explicitly by Radlow (Albani, 2007). From these considerations, eq. (141) does not constitute a W-H factorization. However, by using the factors present in the second member of (141), Radlow proposed an explicit factorization of $G(\eta,\xi)$. Even though some authors gave credit to this solution (Albertsen, 1997), now it definitely appears Radlow's factorization is incorrect (Albani, 2007).

8.6.4.2 Reduction to Fredholm equations

In the natural domain, the W-H eq. (139) is an integral equations defined by a convolution kernel. In particular, we can extend to the multidimensional case the method discussed in chapter 5 (see also Daniele, 2004) to reduce the W-H equation to a Fredholm equation. For this task, we introduce the step function $u(x)$ and the projection operators:

$$P \to p(x,y) = u(x)u(y)$$

$$\overline{P} \to 1 - u(x)u(y)$$

$$= \overline{p}(x,y) = u(x)u(-y) + u(-x)u(y) + u(-x)u(-y)$$

We observe that

$$PF_+ = F_+, \quad \overline{P}F_+ = (1-P)F_+ = 0 \tag{142}$$

$$PF_- = 0, \quad \overline{P}F_- = F_- \tag{143}$$

Multiplying (140) by P and taking into account the (142), we get

$$PGF_+ = PF_+^i \to \frac{1}{(2\pi j)^2}\int_{\gamma_1}\int_{\gamma_1}\frac{G(\eta',\xi')}{(\eta'-\eta)(\xi'-\xi)}F_+(\eta',\xi')d\eta' d\xi'$$
$$= -\frac{1}{\eta-\eta_o}\frac{1}{\xi-\xi_o} \tag{144}$$

Multiplying the second of (144) by G:

$$\frac{1}{(2\pi j)^2}\int_{\gamma_1}\int_{\gamma_1}\frac{G(\eta,\xi)}{(\eta'-\eta)(\xi'-\xi)}F_+(\eta',\xi')d\eta' d\xi' - G(\eta,\xi)F_+(\eta,\xi) = 0 \tag{145}$$

Subtracting (145) from (144),

$$G(\eta,\xi)F_+(\eta,\xi) + \frac{1}{(2\pi j)^2}\int_{-\infty}^{\infty}\int_{-\infty}^{\infty}\frac{G(\eta',\xi') - G(\eta,\xi)}{(\eta'-\eta)(\xi'-\xi)}F_+(\eta',\xi')d\eta' d\xi' = -\frac{1}{\eta-\eta_o}\frac{1}{\xi-\xi_o} \tag{146}$$

Equation (146) is a two-dimensional Fredholm integral equation of the second kind. Its solution provides the factorization of the two-dimensional kernel $G(\eta,\xi)$.

8.6.4.3 Numerical quadrature

As happens in the one-dimensional case, the accuracy for obtaining numerical solutions of (146) considerably increases if one deforms the contour path constituted by the real axis of the η – plane (ξ – plane) into the straight line $\lambda_\eta(\lambda_\xi)$ that joins the points $-jk$ and $+jk$. Furthermore, for facilitating function-theoretic manipulations that may be obscure in the η, ξ – planes, we also introduce the complex planes w_η, w_ξ defined by

$$\eta = -k\cos w_\eta, \quad \xi = -k\cos w_\xi \tag{147}$$

In these planes the straight lines λ_η and λ_ξ are represented by the vertical lines

$$w_\eta = -\frac{\pi}{2} + j\,u_\eta, \quad w_\xi = -\frac{\pi}{2} + j\,u_\xi$$

where u_η and u_ξ are real. With these new variables eq. (146) becomes

$$\hat{G}(u_\eta, u_\xi)\hat{F}_+(u_\eta, u_\xi) + \frac{1}{(2\pi)^2}\int_{-\infty}^{\infty}\int_{-\infty}^{\infty}\frac{M(u_\eta, u_\xi, u'_\eta, u'_\xi)\operatorname{ch}u'_\eta \operatorname{ch}u'_\xi}{(\operatorname{sh}u'_\eta - \operatorname{sh}u_\eta)(\operatorname{sh}u'_\xi - \operatorname{sh}u_\xi)}\hat{F}_+(u'_\eta, u'_\xi)du'_\eta\, du'_\xi$$
$$= -\frac{1}{j\operatorname{sh}u_\eta - \eta_o}\frac{1}{j\operatorname{sh}u_\xi - \xi_o} \tag{148}$$

where

$$\hat{G}(u_\eta, u_\xi) = G(\eta, \xi), \quad \hat{F}_+(u_\eta, u_\xi) = F_+(\eta, \xi), \quad M(u_\eta, u_\xi, u'_\eta, u'_\xi) = G(\eta', \xi') - G(\eta, \xi)$$

For the one-dimensional factorization, usually (148) is solved by numerical quadrature. For the two-dimensional case, the problem of the numerical quadrature of the two folded integral remains cumbersome. To reduce the number of numerical unknowns, the following representation of the function $\hat{F}_+(u_\eta, u_\xi)$ is proposed:

$$\hat{F}_+(u_\eta, u_\xi) = \frac{p_n(u_\eta, u_\xi)}{(j\operatorname{sh}u_\eta - \eta_o)(j\operatorname{sh}u_\xi - \xi_o)} \tag{149}$$

where $p_n(u_\eta, u_\xi)$ is an unknown polynomial of order n whose coefficients can be obtained by applying the collocation method on (148). Using representation (149), we experienced accurate approximate solutions even if we used moderate values of the order n of the polynomial $p_n(u_\eta, u_\xi)$.

8.6.4.4 Moment method

As an alternative to the quadrature method, is perhaps preferable to use the method of moments. For example, one could evaluate the integral in (148) to a first approximation by assuming

$$\hat{F}_+(u'_\eta, u'_\xi) = \hat{F}^i_+(u'_\eta, u'_\xi)$$

CHAPTER 9

Wiener-Hopf analysis of waveguide discontinuities

9.1 Marcuvitz-Schwinger formalism

The Marcuvitz-Schwinger formalism (Felsen & Marcuvitz, 1973) provides the general representation of electromagnetic fields in an arbitrary waveguide filled with a homogeneous isotropic medium. By assuming that the longitudinal axis of the waveguide is the axis $\hat{z}$, we have the following representation of the transverse electromagnetic field in the TM and TE modes:

$$\begin{aligned}\mathbf{E}_t(\boldsymbol{\rho}, z) &= \sum_i V_i'(z)\mathbf{e}_i'(\boldsymbol{\rho}) + \sum_i V_i''(z)\mathbf{e}_i''(\boldsymbol{\rho}) \\ \mathbf{H}_t(\boldsymbol{\rho}, z) &= \sum_i I_i'(z)\mathbf{h}_i'(\boldsymbol{\rho}) + \sum_i V_i''(z)\mathbf{h}_i''(\boldsymbol{\rho})\end{aligned} \tag{1}$$

where the modal voltages and currents V_i, I_i specify the z-dependence and $\mathbf{e}_i(\boldsymbol{\rho})$ and $\mathbf{h}_i(\boldsymbol{\rho}) = \hat{z} \times \mathbf{e}_i(\boldsymbol{\rho})$ are the transverse eigenvectors of the guide.

The index i is twofold and can be discrete or continuous depending as to whether the transverse cross section of the guide is bounded or unbounded. The transverse eigenvectors are related to the transverse eigenfunctions through

$$\mathbf{e}_i'(\boldsymbol{\rho}) = -\frac{1}{\tau_i'}\nabla_t\phi_i(\boldsymbol{\rho}), \quad \mathbf{h}_i''(\boldsymbol{\rho}) = -\frac{1}{\tau_i''}\nabla_t\psi_i(\boldsymbol{\rho}) \tag{2}$$

where

$$\nabla_t^2\phi_i(\boldsymbol{\rho}) + (\tau_i')^2\phi_i(\boldsymbol{\rho}) = 0, \quad \nabla_t^2\psi_i(\boldsymbol{\rho}) + (\tau_i'')^2\psi_i(\boldsymbol{\rho}) = 0 \tag{3}$$

The functions $\phi_i(\boldsymbol{\rho})$ and $\psi_i(\boldsymbol{\rho})$ provide the TM and TE modes. On the contour γ that limits the transverse section of the waveguide it must be

$$\phi_i(\boldsymbol{\rho})|_\gamma = 0, \quad \left.\frac{\partial\psi_i(\boldsymbol{\rho})}{\partial\nu}\right|_\gamma = 0$$

where ν is the normal to the contour γ. If $\tau_i' = 0$, the TM mode becomes a TEM mode and a different normalization in the first of eq. (2) must be used. We recall that the TEM mode can occur only if the transverse section of the waveguide is not simply connected.

9.1.1 Example 1

Let us consider a rectangular waveguide oriented in the z-direction (Fig. 1b), and let us assume that the guide has longitudinal direction y and cross section consisting of two parallel planes separated by a distance A (Fig. 1a).

The unnormalized eigenfunctions are

$$\phi_i(\mathbf{\rho}) = \sin\left(\frac{m\pi}{A}x\right)\frac{1}{2\pi}e^{-j\alpha z}, \quad \psi_i(\mathbf{\rho}) = \cos\left(\frac{m\pi}{A}x\right)\frac{1}{2\pi}e^{-j\alpha z}$$

The modal index m is discrete ($m = 0, 1, 2, \ldots$), while the modal index α is continuous ($-\infty < \alpha < \infty$).

Let us indicate with $V_m'(y,\alpha)$, $V_m''(y,\alpha)$ and $I_m'(y,\alpha)$, $I_m''(y,\alpha)$ the $TM_{m\alpha}$ and $TE_{m\alpha}$ voltages and the currents of the transmission line relevant to the waveguide with cross section in the plane x, z (Fig. 1a). If the existing modes all have the same index m, we have

$$E_x(x,y,z) = e_x(y,z)\cos\left(m\frac{\pi}{A}x\right) = \cos\left(m\frac{\pi}{A}x\right)\frac{1}{2\pi}\int_{-\infty}^{\infty}\left[m\frac{\pi}{A}V_m'(y,\alpha) + j\alpha V_m''(y,\alpha)\right]e^{-j\alpha z}d\alpha$$

$$E_z(x,y,z) = e_z(y,z)\sin\left(m\frac{\pi}{A}x\right) = \sin\left(m\frac{\pi}{A}x\right)\frac{1}{2\pi}\int_{-\infty}^{\infty}\left[-j\alpha V_m'(y,\alpha) - m\frac{\pi}{A}V_m''(y,\alpha)\right]e^{-j\alpha z}d\alpha$$

$$H_x(x,y,z) = h_x(y,z)\sin\left(m\frac{\pi}{A}x\right) = \sin\left(m\frac{\pi}{A}x\right)\frac{1}{2\pi}\int_{-\infty}^{\infty}\left[-j\alpha I_m'(y,\alpha) - m\frac{\pi}{A}I_m''(y,\alpha)\right]e^{-j\alpha z}d\alpha$$

$$H_z(x,y,z) = h_z(y,z)\cos\left(m\frac{\pi}{A}x\right) = \cos\left(m\frac{\pi}{A}x\right)\frac{1}{2\pi}\int_{-\infty}^{\infty}\left[-m\frac{\pi}{A}I_m'(y,\alpha) - j\alpha I_m''(y,\alpha)\right]e^{-j\alpha z}d\alpha$$

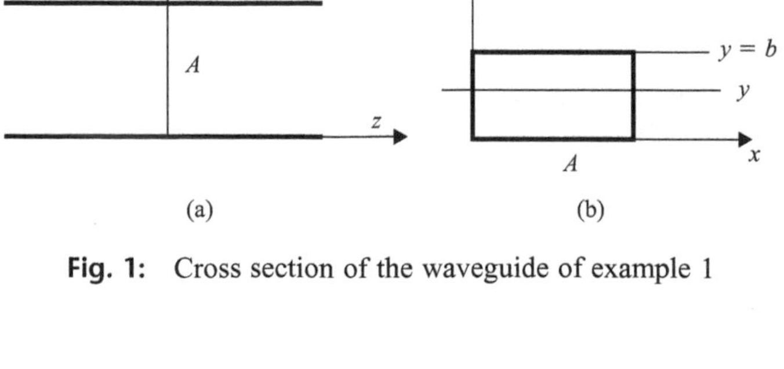

Fig. 1: Cross section of the waveguide of example 1

The inversion in the Fourier domain yields

$$m\frac{\pi}{A}V'_m(y,\alpha)+j\alpha V''_m(y,\alpha)] = \int_{-\infty}^{\infty} e_x(y,z)e^{j\alpha z}dz = \tilde{E}_1(y,\alpha)$$

$$j\alpha V'_m(y,\alpha)+m\frac{\pi}{A}V''_m(y,\alpha)] = \int_{-\infty}^{\infty} -e_z(y,z)e^{j\alpha z}dz = \tilde{E}_2(y,\alpha)$$

$$m\frac{\pi}{A}I'_m(y,\alpha)+j\alpha I''_m(y,\alpha)] = \int_{-\infty}^{\infty} -h_z(y,z)e^{j\alpha z}dz = -\tilde{H}_z(y,\alpha)$$

$$j\alpha I'_m(y,\alpha)+m\frac{\pi}{A}I''_m(y,\alpha)] = \int_{-\infty}^{\infty} -h_x(y,z)e^{j\alpha z}dz = -\tilde{H}_x(y,\alpha)$$

or in matrix form

$$\begin{pmatrix} m\frac{\pi}{A} & j\alpha \\ j\alpha & m\frac{\pi}{A} \end{pmatrix} \cdot \begin{pmatrix} V'_m(y,\alpha) \\ V''_m(y,\alpha) \end{pmatrix} = \begin{pmatrix} \tilde{E}_1(y,\alpha) \\ \tilde{E}_2(y,\alpha) \end{pmatrix} \tag{4}$$

$$\begin{pmatrix} m\frac{\pi}{A} & j\alpha \\ j\alpha & m\frac{\pi}{A} \end{pmatrix} \cdot \begin{pmatrix} I'_m(y,\alpha) \\ I''_m(y,\alpha) \end{pmatrix} = -\begin{pmatrix} \tilde{H}_z(y,\alpha) \\ \tilde{H}_x(y,\alpha) \end{pmatrix} \tag{5}$$

Taking into account the loads of the transmission lines, the modal voltage and the modal current can be related. For instance, a PEC wall if in the section $y = 0$ (Fig. 1b) produces a short circuit on the transmission lines. In this case we get

$$I'_m(y,\alpha) = -\overleftarrow{Y}'_m V'_m(y,\alpha), \quad I''_m(y,\alpha) = -\overleftarrow{Y}''_m V''(y,\alpha) \tag{6}$$

where

$$\chi_o = \sqrt{k^2 - \left(m\frac{\pi}{A}\right)^2}, \quad \chi = \sqrt{\chi_o^2 - \alpha^2}$$

$$\overleftarrow{Y}'_m = -\frac{j\omega\varepsilon}{\chi}\cot(\chi y), \quad \overleftarrow{Y}''_m = -\frac{\chi}{j\omega\mu}\cot(\chi y)$$

Equation (6) allow for the elimination of the voltages and currents in (4) and (5). We get

$$\begin{pmatrix} m\frac{\pi}{A} & j\alpha \\ j\alpha & m\frac{\pi}{A} \end{pmatrix} \cdot \begin{pmatrix} -\overleftarrow{Y}'_m & 0 \\ 0 & -\overleftarrow{Y}''_m \end{pmatrix} \begin{pmatrix} m\frac{\pi}{A} & j\alpha \\ j\alpha & m\frac{\pi}{A} \end{pmatrix}^{-1} \begin{pmatrix} \tilde{E}_1(y,\alpha) \\ \tilde{E}_2(y,\alpha) \end{pmatrix} = -\begin{pmatrix} \tilde{H}_z(y,\alpha) \\ \tilde{H}_x(y,\alpha) \end{pmatrix}$$

Algebraic manipulations yield

$$\begin{pmatrix} m\frac{\pi}{A} & j\alpha \\ j\alpha & m\frac{\pi}{A} \end{pmatrix} \cdot \begin{pmatrix} Y'_m & 0 \\ 0 & Y''_m \end{pmatrix} \begin{pmatrix} m\frac{\pi}{A} & j\alpha \\ j\alpha & m\frac{\pi}{A} \end{pmatrix}^{-1}$$

$$= \frac{1}{\beta^2} \begin{vmatrix} \left(m\frac{\pi}{A}\right)^2 Y'_m + \alpha^2 Y''_m & -j\alpha m\frac{\pi}{A}(Y'_m - Y''_m) \\ j\alpha m\frac{\pi}{A}(Y'_m - Y''_m) & \left(m\frac{\pi}{A}\right)^2 Y''_m + \alpha^2 Y'_m \end{vmatrix}$$

where

$$\beta^2 = \left(m\frac{\pi}{A}\right)^2 + \alpha^2 = k^2 - \chi^2$$

We observe that

$$\left(m\frac{\pi}{A}\right)^2 Y'_m + \alpha^2 Y''_m = -j\cot(\chi y)\left(\frac{\omega\varepsilon}{\chi}\left(m\frac{\pi}{A}\right)^2 + \frac{\chi}{\omega\mu}\alpha^2\right) = -j\frac{\cot(\chi y)}{\omega\mu\chi}\beta^2\tau^2$$

$$j\alpha m\frac{\pi}{A}(Y'_m - Y''_m) = \alpha m\frac{\pi}{A}\frac{\cot(\chi y)}{\omega\mu\chi}\beta^2$$

$$\left(m\frac{\pi}{A}\right)^2 Y''_m + \alpha^2 Y'_m = -j\cot(\chi y)\left(\frac{\chi}{\omega\mu}\left(m\frac{\pi}{A}\right)^2 + \frac{\omega\varepsilon}{\chi}\alpha^2\right) = -j\frac{\cot(\chi y)}{\omega\mu\chi}\chi_o^2\beta^2$$

where $\tau^2 = k^2 - \alpha^2$. By taking these equations into account, (7) can be rewritten as

$$-g_y(\alpha)\mathbf{P}_m(\alpha) \cdot Y_o \begin{pmatrix} \tilde{E}_1(y,\alpha) \\ \tilde{E}_2(y,\alpha) \end{pmatrix} = -\begin{pmatrix} \tilde{H}_z(y,\alpha) \\ \tilde{H}_x(y,\alpha) \end{pmatrix} \tag{7}$$

where

$$g_y(\alpha) = -j\frac{k\cot(\chi y)}{\chi}, \quad \mathbf{P}_m(\alpha) = \frac{1}{k^2}\begin{vmatrix} \tau^2 & -j\alpha\frac{m\pi}{A} \\ j\alpha\frac{m\pi}{A} & \chi_o^2 \end{vmatrix}$$

Similarly, if the boundary $y = a$ is a PEC wall, there is a short circuit on the transmission lines at this boundary, where we get

$$I'_m(y,\alpha) = \vec{Y}'_m V'_m(y,\alpha), \quad I''_m(y,\alpha) = \vec{Y}''_m V''(y,\alpha)$$

where

$$\vec{Y}'_m = -\frac{j\omega\varepsilon}{\chi}\cot(\chi(a-y)), \quad \vec{Y}''_m = -\frac{\chi}{j\omega\mu}\cot(\chi(a-y))$$

Equation (7) becomes

$$g_{a-y}(\alpha)\mathbf{P}_m(\alpha) \cdot Y_o \begin{pmatrix} \tilde{E}_1(y,\alpha) \\ \tilde{E}_2(y,\alpha) \end{pmatrix} = -\begin{pmatrix} \tilde{H}_z(y,\alpha) \\ \tilde{H}_x(y,\alpha) \end{pmatrix} \tag{8}$$

The previous equations will be useful for obtaining Wiener-Hopf equations when a discontinuity is present in plane y (Fig. 1).

9.1.2 Example 2

Let us consider a waveguide having the transverse section indicated in Fig. 2. We can assume that

$$\phi_i(\boldsymbol{\rho}) = \sin\left(\frac{m\pi}{A}x\right)\frac{1}{\pi}\sin(\alpha z) \quad \psi_i(\boldsymbol{\rho}) = \cos\left(\frac{m\pi}{A}x\right)\frac{1}{\pi}\cos(\alpha z)$$

The modal index is discrete ($m = 0, 1, 2, \ldots$), and the modal index α ($0 \leq \alpha < \infty$) is continuous. Consider the section y, and let's indicate with $V_m'(y,\alpha)$, $V_m''(y,\alpha)$ and $I_m'(y,\alpha)$, $I_m''(y,\alpha)$ the $TM_{m\alpha}$ and $TE_{m\alpha}$ voltages and currents of the transmission line relevant to the waveguide with cross section in the plane x, z. If only modes with index m propagate, we have

$$E_x(x,y,z) = e_x(y,z)\cos\left(m\frac{\pi}{A}x\right) = \cos\left(m\frac{\pi}{A}x\right)\frac{1}{\pi}\int_0^\infty \left[m\frac{\pi}{A}\hat{V}_m'(y,\alpha) + \alpha\hat{V}_m''(y,\alpha)\right]\sin(\alpha z)d\alpha$$

$$E_z(x,y,z) = e_z(y,z)\sin\left(m\frac{\pi}{A}x\right) = \sin\left(m\frac{\pi}{A}x\right)\frac{1}{\pi}\int_0^\infty \left[\alpha\,\hat{V}_m'(y,\alpha) - m\frac{\pi}{A}\hat{V}_m''(y,\alpha)\right]\cos(\alpha z)d\alpha$$

$$H_x(x,y,z) = h_x(y,z)\sin\left(m\frac{\pi}{A}x\right) = \sin\left(m\frac{\pi}{A}x\right)\frac{1}{\pi}\int_0^\infty \left[\alpha\hat{I}_m'(y,\alpha) - m\frac{\pi}{A}\hat{I}_m''(y,\alpha)\right]\cos(\alpha z)d\alpha$$

$$H_z(x,y,z) = h_z(y,z)\cos\left(m\frac{\pi}{A}x\right) = \cos\left(m\frac{\pi}{A}x\right)\frac{1}{\pi}\int_0^\infty \left[-m\frac{\pi}{A}\hat{I}_m'(y,\alpha) - \alpha\hat{I}_m''(y,\alpha)\right]\sin(\alpha z)d\alpha$$

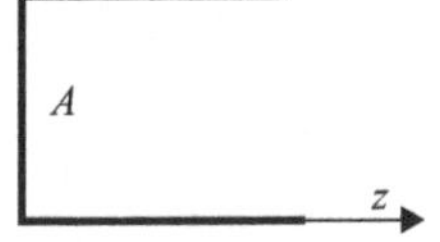

Fig. 2: Cross section of the waveguide of example 2

To obtain the inversion of these equations, we observe that both even and odd Fourier transforms are involved (see section 10.1). We get

$$m\frac{\pi}{A}j\hat{V}'_m(y,\alpha)+j\alpha\,\hat{V}''_m(y,\alpha)]=2j\int_0^\infty e_x(y,z)\sin(\alpha z)dz=\tilde{E}_{x+}(y,\alpha)-\tilde{E}_{x+}(y,-\alpha)$$

$$-\alpha\hat{V}'_m(y,\alpha)+m\frac{\pi}{A}\,\hat{V}''_m(y,\alpha)]=2\int_0^\infty -e_z(y,z)\cos(\alpha z)dz=-[\tilde{E}_{z+}(y,\alpha)+\tilde{E}_{z+}(y,-\alpha)]$$

$$m\frac{\pi}{A}j\,\hat{I}'_m(y,\alpha)+j\alpha\,\hat{I}''_m(y,\alpha)]=2j\int_0^\infty -h_z(y,z)\sin(\alpha z)dz=-[\tilde{H}_{z+}(y,\alpha)-\tilde{H}_{z+}(y,-\alpha)]$$

$$-\alpha\hat{I}'_m(y,\alpha)+m\frac{\pi}{A}\,\hat{I}''_m(y,\alpha)]=2\int_0^\infty -h_x(y,z)\cos(\alpha z)dz=-[\tilde{H}_{z+}(y,\alpha)+\tilde{H}_{z+}(y,-\alpha)]$$

where the plus functions $\tilde{F}_+(\alpha)=\int_0^\infty f(z)e^{j\alpha z}dz$ are involved.

In the following it is useful to introduce the operators Δ, Σ, and Γ defined by

$$\Delta\tilde{F}_+(\alpha)=\tilde{F}_+(\alpha)-\tilde{F}_+(-\alpha),\quad \Sigma\tilde{F}_+(\alpha)=\tilde{F}_+(\alpha)+\tilde{F}_+(-\alpha),\quad \Gamma=\begin{vmatrix}\Delta & 0\\ 0 & \Sigma\end{vmatrix}$$

In matrix form the previous equations can be rewritten as

$$\begin{pmatrix} m\frac{\pi}{A} & j\alpha \\ j\alpha & m\frac{\pi}{A}\end{pmatrix}\cdot\begin{pmatrix} j\hat{V}'_m(y,\alpha)\\ \hat{V}''_m(y,\alpha)\end{pmatrix}=\Gamma\cdot\begin{pmatrix}\tilde{E}_{x+}(y,\alpha)\\ -\tilde{E}_{z+}(y,\alpha)\end{pmatrix} \tag{9}$$

$$\begin{pmatrix} m\frac{\pi}{A} & j\alpha \\ j\alpha & m\frac{\pi}{A}\end{pmatrix}\cdot\begin{pmatrix} j\hat{I}'_m(y,\alpha)\\ \hat{I}''_m(y,\alpha)\end{pmatrix}=\Gamma\cdot\begin{pmatrix}-\tilde{H}_{z+}(y,\alpha)\\ -\tilde{H}_{x+}(y,\alpha)\end{pmatrix} \tag{10}$$

The voltage and modal currents can be related by taking into account the loads of the transmission lines. For instance, if in the section $y=c$ there is a PEC wall we have

$$\hat{I}'_m(y,\alpha)=Y'_m\hat{V}'_m(y,\alpha),\quad \hat{I}''_m(y,\alpha)=Y''_m\hat{V}''_m(y,\alpha) \tag{11}$$

where $(y<c)$

$$Y'_m=-\frac{j\omega\varepsilon}{\chi}\cot(\chi(c-y)),\quad Y''_m=-\frac{\chi}{j\omega\mu}\cot(\chi(c-y)) \tag{12}$$

Equation (12) allow for the elimination of voltages and currents in (10) and (11). We get

$$g_{c-y}(\alpha)\mathbf{P}_m(\alpha)\cdot\Gamma\cdot Y_o\begin{pmatrix}\tilde{E}_{x+}(y,\alpha)\\ -\tilde{E}_{z+}(y,\alpha)\end{pmatrix}=\Gamma\cdot\begin{pmatrix}-\tilde{H}_{z+}(y,\alpha)\\ -\tilde{H}_{x+}(y,\alpha)\end{pmatrix} \tag{13}$$

It is important to observe that eq. (13) is unchanged by replacing α with $-\alpha$.

9.2 Bifurcation in a rectangular waveguide

Let us consider a TE_{10} mode propagating in the waveguide b of Fig. 3. The discontinuity present in the section $z = 0$ produces modes in the three waveguides a, b, and c. More precisely, the index 1 is conserved, and the excited modes are $TE_{1n} + TM_{1n}$ in all three waveguides. To obtain the W-H equations of this problem, we consider a single waveguide having y as its longitudinal axis (Fig. 1). Equation (7) in the section $y = b_-$ and eq. (8) in $y = b_+$ apply. The PEC wall at $0 < x < A,\ y = b,\ z < 0$ implies that $-e_x(b_+, z) = e_x(b_-, z)$ and $e_z(b_+, z) = e_z(b_-, z)$ are zero on $0 < x < A,\ y = b,\ z < 0$ and are continuous in the aperture $0 < x < A,\ y = b,\ z > 0$: $e_x(b_+, z) = e_x(b_-, z)$. Consequently, the functions $\tilde{E}_{1,2}(b_\pm, \alpha) = \int_0^\infty \pm e_{x,z}(b_\pm, z) e^{j\alpha z} dz = \tilde{E}_{1,2+}(b_\pm, \alpha)$ are plus functions. Since the quantities $-h_z(b_+, z) = h_z(b_-, z)$ and $h_x(b_+, z) = h_x(b_-, z)$ are continuous on the aperture $0 < x < A,\ y = b,\ z > 0$, the functions

$$-\tilde{H}_z(b_+, \alpha) + \tilde{H}_z(b_-, \alpha) = \int_{-\infty}^{0} [-h_z(b_+, z) + h_z(b_-, z)] e^{j\alpha z} d\alpha = A_{1-}(\alpha)$$

$$-\tilde{H}_x(b_+, \alpha) + \tilde{H}_x(b_-, \alpha) = \int_{-\infty}^{0} [-h_x(b_+, z) + h_x(b_-, z)] e^{j\alpha z} d\alpha = A_{2-}(\alpha)$$

are minus functions. With these considerations in mind, subtracting eq. (7) from eq. (8) gives (the subscript $m = 1$ has been omitted)

$$[g_c(\alpha) + g_b(\alpha)]\mathbf{P}(\alpha) \cdot Y_o \begin{pmatrix} E_{1+}(\alpha) \\ E_{2+}(\alpha) \end{pmatrix} = \begin{pmatrix} A_{1-}(\alpha) \\ A_{2-}(\alpha) \end{pmatrix}$$

This equation can be rewritten as

$$\frac{j\,G(\alpha)}{\omega\mu} \begin{vmatrix} \tau^2 & j\dfrac{\alpha\pi}{A} \\ -j\dfrac{\alpha\pi}{A} & \chi_o \end{vmatrix} \cdot \begin{pmatrix} E_{1+}(\alpha) \\ -E_{2+}(\alpha) \end{pmatrix} = \begin{pmatrix} -A_{1-}(\alpha) \\ A_{2-}(\alpha) \end{pmatrix} \tag{14}$$

where

$$\tau^2 = k^2 - \alpha^2, \quad G(\alpha) = \frac{\sin(\chi a)}{\chi \sin(\chi b) \sin(\chi c)}$$

The factorization of $G(\alpha)$ was accomplished in example 4 of section 3.2.5.

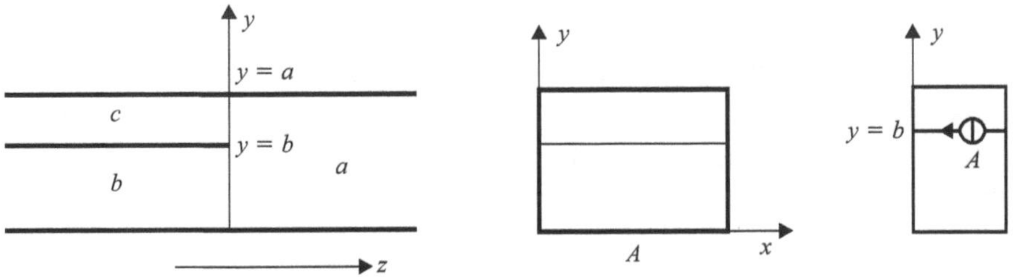

Fig. 3: Waveguide bifurcation problem: (a) longitudinal cross section; (b) transverse cross section; (c) transmission line in the y-direction

To put the W-H system in an inhomogeneous form, we consider the source term constituted by the incident TE_{10} mode that propagates in the b waveguide. This source produces in the minus functions $A_{1-}(\alpha)$ and $A_{2-}(\alpha)$ a nonstandard pole located at $\alpha = \chi_{10}^{b}{}'' = \sqrt{k^2 - \left(\frac{\pi}{A}\right)^2 - 0\left(\frac{\pi}{b}\right)^2} = \chi_o$. Taking into account the expressions of currents induced in the walls $y = 0_-$ and $y = 0_+$ that are due to the incident TE_{10} mode, we get

$$-A_{1-}(\alpha) = J^s_{x-}(\alpha) + \frac{H_o}{\chi_o A(\alpha - \chi_o)}, \quad A_{2-}(\alpha) = J^s_{z-}(\alpha) - j\frac{H_o}{(\alpha - \chi_o)}$$

The square matrix in (14) is a polynomial matrix and its factorization can be accomplished with a several techniques. However, it is more convenient to rewrite (14) in scalar form:

$$\frac{j\,G(\alpha)}{\omega\mu}\hat{X}_{1+}(\alpha) = J^s_{x-}(\alpha) + \frac{H_o}{\chi_o A(\alpha - \chi_o)} \tag{15}$$

$$-\frac{j\,G(\alpha)}{\omega\mu}[X_{2+}(\alpha)] = J^s_{z-}(\alpha) - j\frac{H_o}{(\alpha - \chi_o)} \tag{16}$$

where

$$\begin{aligned} \hat{X}_{1+}(\alpha) &= \tau^2 E_{1+}(\alpha) - j\frac{\alpha\pi}{A}E_{2+}(\alpha) \\ X_{2+}(\alpha) &= j\frac{\alpha\pi}{A}E_{1+}(\alpha) + \chi_o E_{2+}(\alpha) \end{aligned} \tag{17}$$

$X_{2+}(\alpha)$ is a plus function that vanishes as $\alpha \to \infty$. The factorization of $g(\alpha)$ yields

$$X_{2+}(\alpha) = \frac{\omega\mu}{G_+(\alpha)}\frac{G_-^{-1}(\chi_o)H_o}{(\alpha - \chi_o)}$$

Conversely, $\hat{X}_{1+}(\alpha)$ is a plus function but is not limited as $\alpha \to \infty$. We can resort to the consideration of section 2.5. It yields

$$\hat{X}_{1+}(\alpha) = \frac{\omega\mu}{jG_+(\alpha)}\left[\frac{G_-^{-1}(\chi_o)H_o}{\chi_o A(\alpha - \chi_o)} + c_o\right]$$

where c_o is a constant. To obtain c_o we invert eq. (17):

$$E_{1+}(\alpha) = \frac{\chi_o^2 \hat{X}_{1+}(\alpha) + j\frac{\alpha\pi}{A}X_{2+}(\alpha)}{k^2(\chi_o^2 - \alpha^2)}$$

$$E_{2+}(\alpha) = \frac{-j\frac{\alpha\pi}{A}\hat{X}_{1+}(\alpha) + (k^2 - \alpha^2)X_{2+}(\alpha)}{k^2(\chi_o^2 - \alpha^2)}$$

The pole $\alpha = -\chi_o$ is offending for the plus functions $E_{1+}(\alpha)$ and $E_{2+}(\alpha)$. Hence, we must force that

$$\chi_o^2 \hat{X}_{1+}(-\chi_o) - j\frac{\chi_o\pi}{A}X_{2+}(-\chi_o) = 0$$

$$j\frac{\chi_o\pi}{A}\hat{X}_{1+}(-\chi_o) + (k^2 - \chi_o{}^2)X_{2+}(-\chi_o) = 0$$

The previous equations constitute a single equation. In fact, they are equivalent to

$$\chi_o \hat{X}_{1+}(-\chi_o) - j\frac{\pi}{A} X_{2+}(-\chi_o) = 0 \tag{18}$$

which provides the evaluation of c_o.

By examining the obtained solutions, it can be verified that the spectrum of the functions $E_{1+}(\alpha)$, $E_{2+}(\alpha)$, $A_{1-}(\alpha)$, $A_{2-}(\alpha)$ is discrete. In particular, the poles of $A_{1-}(\alpha)$ and $A_{2-}(\alpha)$ are the propagation constants of the two waveguides a and b:

$$\chi_{10}^{b\;\prime\prime} = \pm\sqrt{k^2 - \left(\frac{\pi}{A}\right)^2}, \quad \chi_{1n}^{b\;\prime\prime} = \chi_{1n}^{b\;\prime} = -\sqrt{k^2 - \left(\frac{\pi}{A}\right)^2 - \left(n\frac{\pi}{b}\right)^2} \quad \text{and}$$

$$\chi_{1n}^{c\;\prime\prime} = \chi_{1n}^{c\;\prime} = -\sqrt{k^2 - \left(\frac{\pi}{A}\right)^2 - \left(n\frac{\pi}{c}\right)^2}$$

Conversely, the poles of $E_{1+}(\alpha)$ are $E_{2+}(\alpha)$ are $\chi_{1n}^{a\;\prime} = \chi_{1n}^{a\;\prime\prime} = \sqrt{k^2 - \left(\frac{\pi}{A}\right)^2 - \left(n\frac{\pi}{a}\right)^2}$.

To separate the contributions of the TE_{1n} modes from those of the TM_{1n} modes, we recall that that the longitudinal component of the electric field E_z is related to $E_{2+}(\alpha)$, whereas the longitudinal component of the magnetic field H_z is related to $A_{1-}(\alpha)$.

Consequently, the excitations coefficient of the mode TM_{1n} in waveguide a is related to the residue of $E_{2+}(\alpha)$ at the pole $\sqrt{k^2 - \left(\frac{\pi}{A}\right)^2 - \left(n\frac{\pi}{a}\right)^2}$.

The excitations coefficient of the modes TE_{1n} in waveguides b and c are related to the residue of $A_{1-}(\alpha)$ at the poles $\sqrt{k^2 - \left(\frac{\pi}{A}\right)^2 - \left(n\frac{\pi}{a}\right)^2}$ and $\sqrt{k^2 - \left(\frac{\pi}{A}\right)^2 - \left(n\frac{\pi}{c}\right)^2}$, respectively. To get the excitations coefficients of the other modes in waveguide a, we must consider $E_{1+}(\alpha)$ and, taking into account that the contributions of the TM_{1n} modes are known, we can evaluate the contributions of the modes TE_{1n} by difference.

Similarly, considerations apply to the modes TM_{1n} in the waveguides b and c.

9.3 The junction of two waveguides

Let us consider a TE_{10} mode propagating in the rectangular waveguide located in $z < 0$ (Fig. 4). The junction present in the section $z = 0$ produces modes in both waveguides. More precisely, the index 1 is conserved, and the excited modes are $TE_{1n} + TM_{1n}$ in the two waveguides.

Using eq. (8) in the section $y = b_-$ yields (the subscript $m = 1$ is omitted)

$$-g_b(\alpha)\mathbf{P}(\alpha) \cdot Y_o \begin{pmatrix} \tilde{E}_{x+}(b_-, \alpha) \\ -\tilde{E}_{z+}(b_-, \alpha) \end{pmatrix} = -\begin{pmatrix} \tilde{H}_{z+}(b_-, \alpha) + \tilde{H}_{z-}(b_-, \alpha) \\ \tilde{H}_{x+}(b_-, \alpha) + \tilde{H}_{x-}(b_-, \alpha) \end{pmatrix} \tag{19}$$

Similarly, using eq. (13) in the section $y = b_+$ yields (the subscript $m = 1$ is omitted and $c = a - b$)

$$g_c(\alpha)\mathbf{P}(\alpha) \cdot Y_o \begin{pmatrix} \tilde{E}_{x+}(b_+, \alpha) - \tilde{E}_{x+}(b_+, -\alpha) \\ -\tilde{E}_{z+}(b_+, \alpha) - \tilde{E}_{z+}(b_+, -\alpha) \end{pmatrix} = \begin{pmatrix} -\tilde{H}_{z+}(b_+, \alpha) + \tilde{H}_{z+}(b_+, -\alpha) \\ -\tilde{H}_{x+}(b_+, \alpha) - \tilde{H}_{x+}(b_+, -\alpha) \end{pmatrix} \tag{20}$$

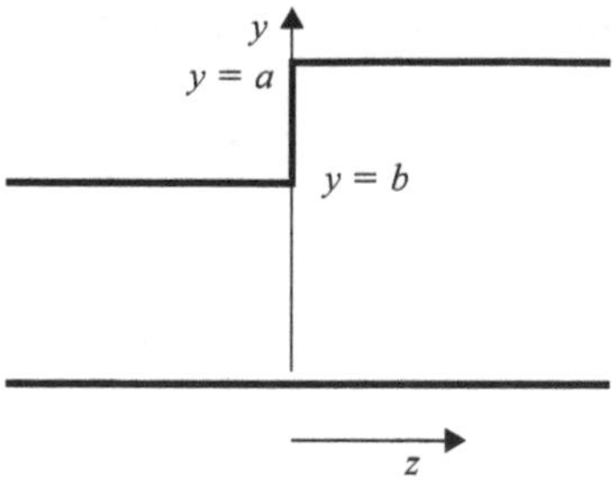

Fig. 4: Junction in the E – plane between two rectangular waveguides

Two additional functional equations can be obtained by changing α into $-\alpha$ in (19):

$$-g_b(\alpha)\mathbf{P}(-\alpha)\cdot Y_o\begin{pmatrix}\tilde{E}_{x+}(b_-,-\alpha)\\-\tilde{E}_{z+}(b_-,-\alpha)\end{pmatrix}=-\begin{pmatrix}\tilde{H}_{z+}(b_-,-\alpha)+\tilde{H}_{z-}(b_-,-\alpha)\\\tilde{H}_{x+}(b_-,-\alpha)+\tilde{H}_{x-}(b_-,-\alpha)\end{pmatrix} \tag{21}$$

Conversely, by interchanging α with $-\alpha$, eq. (20) remain unchanged.

The six eqs. (19), (20) and (21) contain as unknowns the six plus functions

$$\begin{gathered}\tilde{E}_{x+}(b_+,\alpha)=\tilde{E}_{x+}(b_-,\alpha)=\tilde{E}_{x+}(b,\alpha),\quad \tilde{E}_{z+}(b_+,\alpha)=\tilde{E}_{z+}(b_-,\alpha)=\tilde{E}_{z+}(b,\alpha),\\ \tilde{H}_{z+}(b_+,\alpha)=\tilde{H}_{z+}(b_-,\alpha)=\tilde{H}_{z+}(b,\alpha),\quad \tilde{H}_{x+}(b_+,\alpha)=\tilde{H}_{x+}(b_-,\alpha)=\tilde{H}_{x+}(b,\alpha),\\ \tilde{H}_{z-}(b_-,-\alpha),\quad \tilde{H}_{x-}(b_-,-\alpha)\end{gathered}$$

and the six minus functions

$$\begin{gathered}\tilde{E}_{x+}(b_+,-\alpha)=\tilde{E}_{x+}(b_-,-\alpha),\quad \tilde{E}_{z+}(b_+,-\alpha)=\tilde{E}_{z+}(b_-,-\alpha),\quad \tilde{H}_{z+}(b,-\alpha),\\ \tilde{H}_{x+}(b,-\alpha),\quad \tilde{H}_{x-}(b_-,\alpha),\quad \tilde{H}_{z-}(b_-,\alpha)\end{gathered}$$

By separating the plus functions from the minus functions, we obtain the W-H equations of the problem

$$\mathbf{G}(\alpha)\cdot X_+(\alpha)=X_+(-\alpha) \tag{22}$$

where

$$X_+(\alpha)=\left|\tilde{E}_{x+}(b,\alpha),\quad \tilde{E}_{z+}(b,\alpha),\quad \tilde{H}_{z+}(b,\alpha),\quad \tilde{H}_{x+}(b,\alpha),\quad \tilde{H}_{x-}(b_-,-\alpha),\quad \tilde{H}_{z-}(b_-,-\alpha)\right|^t$$

and $\mathbf{G}(\alpha)$ is obtained by algebraic manipulation on eqs. (19) through (21).

Equation (22) is consistent since $\mathbf{G}(\alpha)$ presents the property

$$\mathbf{G}(-\alpha)\cdot\mathbf{G}(\alpha)=\mathbf{1}$$

To take the source into account, we observe that the minus function $X_+(-\alpha)$ has the two components $\tilde{H}_{x-}(b_-,\alpha)$ and $\tilde{H}_{z-}(b_-,\alpha)$, which are not standard Laplace transforms because of the presence of the pole α_o located in the fourth quadrant. This pole is relevant to the mode TE_{10} propagating inside the waveguide at the left of the junction:

$$\tilde{H}_{x-}(b_-, \alpha) = \frac{R_x}{\alpha - \alpha_o} + \tilde{H}^s_{x-}(b_-, \alpha), \quad \tilde{H}_{z-}(b_-, \alpha) = \frac{R_z}{\alpha - \alpha_o} + \tilde{H}^s_{z-}(b_-, \alpha)$$

R_x and R_z are related to the amplitude of the incident TE_{10} mode.

This pole induces a nonstandard behavior to the last two components of $X_+(\alpha)$:

$$\tilde{H}_{x-}(b_-, -\alpha) = -\frac{R_x}{\alpha + \alpha_o} + \tilde{H}^s_{x-}(b_-, -\alpha), \quad \tilde{H}_{z-}(b_-, -\alpha) = -\frac{R_z}{\alpha + \alpha_o} + \tilde{H}^s_{z-}(b_-, -\alpha)$$

According to the general technique, we rewrite eq. (22) in the form

$$\mathbf{G}_+(\alpha) \cdot X_+(\alpha) - \mathbf{G}_-^{-1}(\alpha_o) \cdot \frac{R}{\alpha - \alpha_o} + \mathbf{G}_+^{-1}(-\alpha_o) \cdot \frac{R}{\alpha + \alpha_o}$$
$$= \mathbf{G}_-^{-1}(\alpha) \cdot X_+(-\alpha) - \mathbf{G}_-^{-1}(\alpha_o) \cdot \frac{R}{\alpha - \alpha_o} + \mathbf{G}_+^{-1}(-\alpha_o) \cdot \frac{R}{\alpha + \alpha_o} = 0$$

where

$$R = |0, 0, 0, 0, R_x, R_z|^t$$

From the previous equations we get the solution

$$X_+(\alpha) = \mathbf{G}_+^{-1}(\alpha) \cdot \mathbf{G}_-^{-1}(\alpha_o) \cdot \frac{R}{\alpha - \alpha_o} - \mathbf{G}_+^{-1}(\alpha) \cdot \mathbf{G}_+^{-1}(-\alpha_o) \cdot \frac{R}{\alpha + \alpha_o}$$

The factorization of $\mathbf{G}(\alpha) = \mathbf{G}_-(\alpha) \cdot \mathbf{G}_+(\alpha)$ can be accomplished in general by using the Fredholm factorization discussed in chapter 5. In some case, it is possible to achieve a closed form factorization with the methods indicated in chapter 4. For instance, this is the case if $b = c$.

9.4 A general discontinuity problem in a rectangular waveguide

Let us consider a TE_{10} mode propagating in the waveguide located in $z < 0$ (Fig. 5). The junction present in the section $z = 0$ produces modes in both waveguides. More precisely, the index 1 is conserved and the excited modes are $TE_{1n} + TM_{1n}$ in the two waveguides.

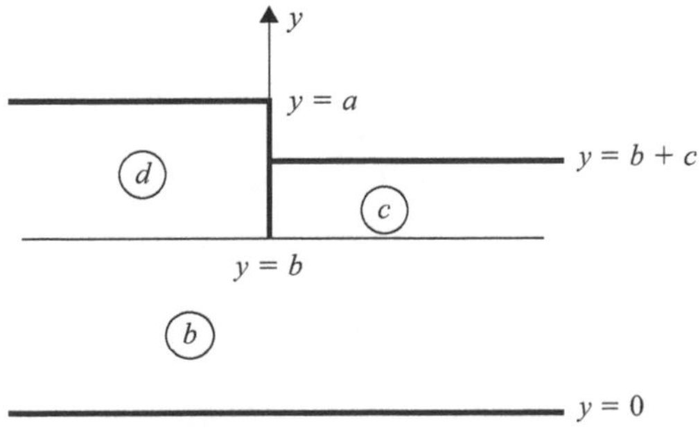

Fig. 5: A general discontinuity in a rectangular waveguide

To obtain the W-H equations for this problem, let us consider the three waveguides b, c, and d in the y-direction. For the waveguide b we get

$$-g_b(\alpha)\mathbf{P}(\alpha)\cdot Y_o\begin{pmatrix}\tilde{E}_{x-}(b_-,\alpha)+\tilde{E}_{x+}(b_-,\alpha)\\ -\tilde{E}_{z-}(b_-,\alpha)-\tilde{E}_{z+}(b_-,\alpha)\end{pmatrix}=-\begin{pmatrix}\tilde{H}_{z+}(b_-,\alpha)+\tilde{H}_{z-}(b_-,\alpha)\\ \tilde{H}_{x+}(b_-,\alpha)+\tilde{H}_{x-}(b_-,\alpha)\end{pmatrix} \quad (23)$$

Similarly, for the waveguide c we have

$$g_c(\alpha)\mathbf{P}(\alpha)\cdot Y_o\begin{pmatrix}\tilde{E}_{x+}(b_+,\alpha)-\tilde{E}_{x+}(b_+,-\alpha)\\ -\tilde{E}_{z+}(b_+,\alpha)--\tilde{E}_{z+}(b_+,-\alpha)\end{pmatrix}=\begin{pmatrix}-\tilde{H}_{z+}(b_+,\alpha)+\tilde{H}_{z+}(b_+,-\alpha)\\ -\tilde{H}_{x+}(b_+,\alpha)-\tilde{H}_{x+}(b_+,-\alpha)\end{pmatrix} \quad (24)$$

For the guide d, the same considerations of the waveguide c apply, and we get ($d = a - b$)

$$g_d(\alpha)\mathbf{P}(\alpha)\cdot Y_o\begin{pmatrix}\tilde{E}_{x-}(b_+,\alpha)-\tilde{E}_{x-}(b_+,-\alpha)\\ -\tilde{E}_{z-}(b_+,\alpha)--\tilde{E}_{z-}(b_+,-\alpha)\end{pmatrix}=\begin{pmatrix}-\tilde{H}_{z+}(b_-,\alpha)+\tilde{H}_{z+}(b_-,-\alpha)\\ -\tilde{H}_{x+}(b_-,\alpha)-\tilde{H}_{x+}(b_-,-\alpha)\end{pmatrix} \quad (25)$$

Two additional functional equations can be obtained by changing α into $-\alpha$ in (23):

$$-g_b(\alpha)\mathbf{P}(-\alpha)\cdot Y_o\begin{pmatrix}\tilde{E}_{x-}(b_-,-\alpha)+\tilde{E}_{x+}(b_-,-\alpha)\\ -\tilde{E}_{z-}(b_-,-\alpha)-\tilde{E}_{z+}(b_-,-\alpha)\end{pmatrix}=-\begin{pmatrix}\tilde{H}_{z+}(b_-,-\alpha)+\tilde{H}_{z-}(b_-,-\alpha)\\ \tilde{H}_{x+}(b_-,-\alpha)+\tilde{H}_{x-}(b_-,-\alpha)\end{pmatrix} \quad (26)$$

Conversely, by interchanging α with $-\alpha$, eqs. (24) and (25) remain unchanged.

We have the continuity of the transverse components:

$$\tilde{E}_{x-}(b_-,\alpha)=\tilde{E}_{x-}(b_+,\alpha)=\tilde{E}_{x-}(\alpha)$$
$$\tilde{E}_{x+}(b_-,\alpha)=\tilde{E}_{x+}(b_+,\alpha)=\tilde{E}_{x+}(\alpha)$$
$$\tilde{E}_{z-}(b_-,\alpha)=\tilde{E}_{z-}(b_+,\alpha)=\tilde{E}_{z-}(\alpha)$$
$$\tilde{E}_{z+}(b_-,\alpha)=\tilde{E}_{z+}(b_+,\alpha)=\tilde{E}_{z+}(\alpha)$$
$$\tilde{H}_{z-}(b_-,\alpha)=\tilde{H}_{z-}(b_+,\alpha)=\tilde{H}_{z-}(\alpha)$$
$$\tilde{H}_{z+}(b_-,\alpha)=\tilde{H}_{z+}(b_+,\alpha)=\tilde{H}_{z+}(\alpha)$$
$$\tilde{H}_{x-}(b_-,\alpha)=\tilde{H}_{x-}(b_+,\alpha)=\tilde{H}_{x-}(\alpha)$$
$$\tilde{H}_{x+}(b_-,\alpha)=\tilde{H}_{x+}(b_+,\alpha)=\tilde{H}_{x+}(\alpha)$$

The eight eqs. (23) through (26) contain unknown eight plus functions

$$X_+(\alpha)=\left|\tilde{E}_{x+}(\alpha),\ \tilde{E}_{z+}(\alpha),\ \tilde{H}_{z+}(\alpha),\ \tilde{H}_{x+}(\alpha),\ \tilde{E}_{x-}(-\alpha),\ \tilde{E}_{z-}(-\alpha),\ \tilde{H}_{z-}(-\alpha),\ \tilde{H}_{x-}(-\alpha)\right|^t$$

and eight minus functions

$$X_+(-\alpha) = \left|\tilde{E}_{x+}(-\alpha),\, \tilde{E}_{z+}(-\alpha),\, \tilde{H}_{z+}(-\alpha),\, \tilde{H}_{x+}(-\alpha),\, \tilde{E}_{x-}(\alpha),\, \tilde{E}_{z-}(\alpha),\, \tilde{H}_{z-}(\alpha),\, \tilde{H}_{x-}(\alpha)\right|^t$$

By separating the plus functions from the minus functions we obtain the W-H equations of the problem

$$\mathbf{G}(\alpha) \cdot X_+(\alpha) = X_+(-\alpha) \tag{27}$$

where $\mathbf{G}(\alpha)$ is obtained by algebraic manipulation on eqs. (23) through (26). Equation (27) is consistent since $\mathbf{G}(\alpha)$ presents the property $\mathbf{G}(-\alpha) \cdot \mathbf{G}(\alpha) = \mathbf{1}$.

To take the source into account, we observe that the minus function $X_+(-\alpha)$ has the last four components $\tilde{E}_{x-}(\alpha), \tilde{E}_{z-}(\alpha), \tilde{H}_{x-}(\alpha)$ and $\tilde{H}_{z-}(\alpha)$, which are nonstandard Laplace transforms because of the presence of the pole α_o located in the fourth quadrant. This pole is relevant to the mode TE_{10} propagating in the waveguide at the left of the junction:

$$\tilde{E}_{x-}(\alpha) = \frac{T_x}{\alpha - \alpha_o} + \tilde{E}^s_{x-}(\alpha), \quad \tilde{E}_{z-}(\alpha) = \frac{T_z}{\alpha - \alpha_o} + \tilde{E}^s_{z-}(\alpha)$$
$$\tilde{H}_{x-}(\alpha) = \frac{R_x}{\alpha - \alpha_o} + \tilde{H}^s_{x-}(\alpha), \quad \tilde{H}_{z-}(\alpha) = \frac{R_z}{\alpha - \alpha_o} + \tilde{H}^s_{z-}(\alpha)$$

T_x, T_z, R_x, and R_z are related to the amplitude of the incident TE_{10} mode.

This pole induces a nonstandard behavior in the last four components of $X_+(\alpha)$:

$$\tilde{E}_{x-}(-\alpha) = -\frac{T_x}{\alpha + \alpha_o} + \tilde{E}^s_{x-}(-\alpha), \quad \tilde{E}_{z-}(-\alpha) = -\frac{T_z}{\alpha + \alpha_o} + \tilde{E}^s_{z-}(-\alpha)$$
$$\tilde{H}_{x-}(-\alpha) = -\frac{R_x}{\alpha + \alpha_o} + \tilde{H}^s_{x-}(-\alpha), \quad \tilde{H}_{z-}(-\alpha) = -\frac{R_z}{\alpha + \alpha_o} + \tilde{H}^s_{z-}(-\alpha)$$

According to the general technique, we rewrite eq. (27) in the form

$$\mathbf{G}_+(\alpha) \cdot X_+(\alpha) - \mathbf{G}_-^{-1}(\alpha_o) \cdot \frac{R}{\alpha - \alpha_o} + \mathbf{G}_+^{-1}(-\alpha_o) \cdot \frac{R}{\alpha + \alpha_o}$$
$$= \mathbf{G}_-^{-1}(\alpha) \cdot X_+(-\alpha) - \mathbf{G}_-^{-1}(\alpha_o) \cdot \frac{R}{\alpha - \alpha_o} + \mathbf{G}_+^{-1}(-\alpha_o) \cdot \frac{R}{\alpha + \alpha_o} = 0$$

where

$$R = |0, 0, 0, 0, T_x, T_z, R_z, R_x|^t \tag{28}$$

From the previous equations we get the solution

$$X_+(\alpha) = \mathbf{G}_+^{-1}(\alpha) \cdot \mathbf{G}_-^{-1}(\alpha_o) \cdot \frac{R}{\alpha - \alpha_o} - \mathbf{G}_+^{-1}(\alpha) \cdot \mathbf{G}_+^{-1}(-\alpha_o) \cdot \frac{R}{\alpha + \alpha_o}$$

The factorization of $\mathbf{G}(\alpha) = \mathbf{G}_-(\alpha) \cdot \mathbf{G}_+(\alpha)$ can be accomplished in general using the Fredholm factorization indicated in chapter 5. In some cases it is possible to obtain a closed-form factorization with the methods indicated in chapter 4. For instance, this happens if $b = c = d$ (symmetric iris).

9.5 Radiation from truncated circular waveguides

This very important problem (Fig. 6) has been considered by Levine and Schwinger (1948) for the acoustic case and by Weinstein (1969) for the electromagnetic case. Let us consider a source constituted by a $TE_{mi} + TM_{mi}$ mode:

$$\tilde{E}_z^i = e_z^i(\rho,z)\sin(m\varphi+\varphi_o), \quad \tilde{H}_z^i = h_z^i(\rho,z)\cos(m\varphi+\varphi_o) \tag{29}$$

where

$$e_z^i(\rho,z) = J_m(\tau'_{mi}\rho)\hat{E}_o e^{-j\chi'_{mi}z}, \quad h_z^i(\rho,z) = \frac{J_m(\tau''_{mi}\rho)}{J_m(\tau''_{mi}a)} H_o e^{-j\chi''_{mi}z} = J_m(\tau''_{mi}\rho)\hat{H}_o e^{-j\chi''_{mi}z}$$

$$J_m(\tau'_{mi}a), \quad J'_m(\tau''_{mi}a) = 0, \quad \chi'_{mi} = \sqrt{k^2-\tau'^2_{mi}}, \quad \chi''_{mi} = \sqrt{k^2-\tau''^2_{mi}}$$

The field components that are of the interest for the solution of the problem are

$$\tilde{E}_z = e_z(\rho,z)\sin(m\varphi+\varphi_o), \quad \tilde{E}_\varphi = e_\varphi(\rho,z)\cos(m\varphi+\varphi_o)$$
$$\tilde{H}_z = h_z(\rho,z)\cos(m\varphi+\varphi_o), \quad \tilde{H}_\varphi = h_\varphi(\rho,z)\sin(m\varphi+\varphi_o)$$

On the wall of the waveguide exists a current given by

$$\tilde{J}_\varphi = j_\varphi(z)\cos(m\varphi+\varphi_o), \quad \tilde{J}_z = j_z(z)\sin(m\varphi+\varphi_o)$$

In the following, the Fourier transform will be defined as

$$E_z(\alpha,\rho) = \int_{-\infty}^{\infty} e_z(z,\rho)e^{j\alpha z}dz, \quad H_z(\alpha,\rho) = \int_{-\infty}^{\infty} h_z(z,\rho)e^{j\alpha z}dz$$

$$E_\varphi(\alpha,\rho) = \int_{-\infty}^{\infty} e_\varphi(z,\rho)e^{j\alpha z}dz, \quad H_\varphi(\alpha,\rho) = \int_{-\infty}^{\infty} h_\varphi(z,\rho)e^{j\alpha z}dz$$

$$J_\varphi(\alpha) = \int_{-\infty}^{\infty} j_\varphi(z)e^{j\alpha z}dz, \quad J_z(\alpha) = \int_{-\infty}^{\infty} j_z(z)e^{j\alpha z}dz$$

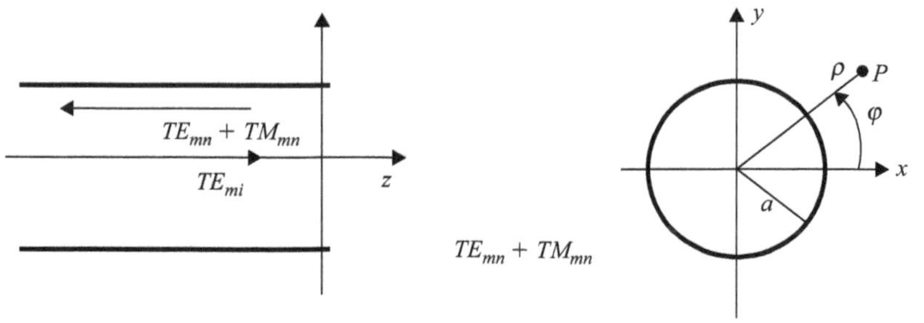

Fig. 6: Radiation from a circular waveguide

The wave equation for the Fourier transforms of the longitudinal components is the Bessel equation

$$\left(\frac{d^2}{d\rho^2}+\frac{1}{\rho}\frac{d}{d\rho}-\frac{m^2}{\rho^2}+\tau^2\right)\begin{vmatrix} E_z(\alpha,\rho) \\ H_z(\alpha,\rho) \end{vmatrix}=0$$

where $\tau^2 = k^2 - \alpha^2$.

We can relate the azimuthal components to the longitudinal components via

$$E_\varphi = -\frac{j}{\tau^2}\left(\frac{\alpha m}{\rho}E_z - \omega\mu\frac{\partial H_z}{\partial\rho}\right), \quad H_\varphi = -\frac{j}{\tau^2}\left(\omega\varepsilon\frac{\partial E_z}{\partial\rho} - \frac{\alpha m}{\rho}H_z\right) \tag{30}$$

The solutions of the Bessel equation that are regular at $\rho = 0$ and satisfy the radiation conditions as $\rho \to \infty$ are given by

$$\begin{aligned} E_z(\alpha,\rho) &= A(\alpha)J_m(\tau\rho) \\ H_z(\alpha,\rho) &= B(\alpha)J_m(\tau\rho) \end{aligned} \qquad \rho < a \tag{31}$$

$$\begin{aligned} E_z(\alpha,\rho) &= C(\alpha)H_m^{(2)}(\tau\rho) \\ H_z(\alpha,\rho) &= D(\alpha)H_m^{(2)}(\tau\rho) \end{aligned} \qquad \rho > a \tag{32}$$

It follows from (30) that

$$\begin{aligned} E_\varphi(\alpha,\rho) &= -\frac{jm\alpha}{\rho\,\tau^2}A(\alpha)J_m(\tau\rho) + \frac{j\omega\mu}{\tau}B(\alpha)J_m{}'(\tau\rho) \\ H_\varphi(\alpha,\rho) &= -\frac{j\omega\varepsilon}{\tau}A(\alpha)J_m'(\tau\rho) + \frac{jm\alpha}{\rho\,\tau^2}B(\alpha)J_m(\tau\rho) \end{aligned} \qquad \rho < a \tag{33}$$

$$\begin{aligned} E_\varphi(\alpha,\rho) &= -\frac{jm\alpha}{\rho\,\tau^2}C(\alpha)H_m^{(2)}(\tau\rho) + \frac{j\omega\mu}{\tau}D(\alpha)H_m^{(2)\prime}(\tau\rho) \\ H_\varphi(\alpha,\rho) &= -\frac{j\omega\varepsilon}{\tau}C(\alpha)H_m^{(2)\prime}(\tau\rho) + \frac{jm\alpha}{\rho\,\tau^2}D(\alpha)H_m^{(2)}(\tau\rho) \end{aligned} \qquad \rho > a \tag{34}$$

Let us consider the two functions

$$V_{z+}(\alpha) = E_z(\alpha, a_+) = E_z(\alpha, a_-), \quad V_{\varphi+}(\alpha) = E_\varphi(\alpha, a_+) = E_\varphi(\alpha, a_-)$$

which are plus functions since $E_z(\alpha, a_+) = E_z(\alpha, a_-)$ vanishes for $z < 0$.

From the first of (31) and the first of (32) evaluated at $\rho = a_- = a_+$, we get

$$A(\alpha) = \frac{V_{z+}(\alpha)}{J_m(\tau a)}, \quad C(\alpha) = \frac{V_{z+}(\alpha)}{H_m^{(2)}(\tau a)} \tag{35}$$

Substituting these equations into the first of (33) and the first of (34) evaluated at $\rho = a_- = a_+$, we can express $B(\alpha)$ and $D(\alpha)$ in terms of $V_{z+}(\alpha)$ and $V_{\varphi+}(\alpha)$:

$$\begin{aligned} B(\alpha) &= \frac{\tau}{j\omega\mu J_m'(\tau a)}\left[V_{\varphi+}(\alpha) + j\frac{m\alpha}{a\tau^2}V_{z+}(\alpha)\right] \\ D(\alpha) &= \frac{\tau}{j\omega\mu H_m^{(2)\prime}(\tau a)}\left[V_{\varphi+}(\alpha) + j\frac{m\alpha}{a\tau^2}V_{z+}(\alpha)\right] \end{aligned} \tag{36}$$

Introduce the two functions

$$J_{z-}(\alpha) = H_\varphi(\alpha, a_+) - H_\varphi(\alpha, a_-), \quad J_{\varphi-}(\alpha) = -H_z(\alpha, a_+) + H_z(\alpha, a_-)$$

which are minus functions since $H_{z,\varphi}(\alpha, a_+) = H_{z,\varphi}(\alpha, a_-)$ for $z > 0$. We can express $J_{z-}(\alpha)$ and $J_{\varphi-}(\alpha)$ by taking into account the second equations of (31) through (34) evaluated at $\rho = a_\pm$. We get

$$-J_{\varphi-}(\alpha) = D(\alpha)H_m^{(2)}(\tau a) - B(\alpha)J_m(\tau a) \tag{37}$$

$$J_{z-}(\alpha) = -\frac{j\omega\varepsilon}{\tau}[C(\alpha)H_m^{(2)\prime}(\tau a) - A(\alpha)J_m'(\tau a)] + \frac{jm\alpha}{a\tau^2}[D(\alpha)H_m^{(2)}(\tau a) - B(\alpha)J_m(\tau a)] \tag{38}$$

By substituting the expressions of $A(\alpha)$, $B(\alpha)$, $C(\alpha)$, and $D(\alpha)$ given by (35) and (36) into the previous equations, we obtain a W-H system having as unknowns the minus functions $J_{z-}(\alpha)$ and $J_{\varphi-}(\alpha)$ and the plus functions $V_{z+}(\alpha)$ and $V_{\varphi+}(\alpha)$:

$$\begin{aligned} -J_{\varphi-}(\alpha) &= \frac{\tau}{j\omega\mu}\left[V_{\varphi+}(\alpha) + j\frac{m\alpha}{a\tau^2}V_{z+}(\alpha)\right]\left[\frac{H_m^{(2)}(\tau a)}{H_m^{(2)\prime}(\tau a)} - \frac{J_m(\tau a)}{J_m'(\tau a)}\right] \\ J_{z-}(\alpha) &= -\frac{j\omega\varepsilon}{\tau}\left[\frac{H_m^{(2)\prime}(\tau a)}{H_m^{(2)}(\tau a)} - \frac{J_m'(\tau a)}{J_m(\tau a)}\right]V_{z+}(\alpha) - \frac{jm\alpha}{a\tau^2}J_{\varphi-}(\alpha) \end{aligned} \tag{39}$$

Taking into account that the Wronskian of $J_m(\tau a)$ and $H_m^{(2)}(\tau a)$ is given by

$$J_m(\tau a)H_m^{(2)\prime}(\tau a) - J_m'(\tau a)H_m^{(2)}(\tau a) = -\frac{2j}{\pi\tau a}$$

we can rewrite (39) in the form

$$\begin{aligned} J_{\varphi-}(\alpha) &= -\frac{2\tau}{\omega\mu\psi_m(\alpha)}\left[V_{\varphi+}(\alpha) + j\frac{m\alpha}{a\tau^2}V_{z+}(\alpha)\right] \\ J_{z-}(\alpha) + \frac{jm\alpha}{a\tau^2}J_{\varphi-}(\alpha) &= -\frac{2\omega\varepsilon}{\tau\phi_m(\alpha)}V_{z+}(\alpha) \end{aligned} \tag{40}$$

where

$$\phi_m(\alpha) = \pi\tau a H_m^{(2)}(\tau a)J_m(\tau a), \quad \psi_m(\alpha) = \pi\tau a H_m^{(2)\prime}(\tau a)J_m'(\tau a)$$

To take into account the source, we observe that the minus functions $J_{z-}(\alpha)$ and $J_{\varphi-}(\alpha)$ are nonstandard. We have

$$J_{\varphi-}(\alpha) = J_{\varphi-}^s(\alpha) + H_{z-}^i(\alpha, a_-), \quad J_{z-}(\alpha) = J_{z-}^s(\alpha) - H_{\varphi-}^i(\alpha, a_-)$$

where the nonstandard parts $H^i_{z-}(\alpha, a_-)$ and $H^i_{\varphi-}(\alpha, a_-)$ can be evaluated taking into account that

$$\tilde{E}^i_z = j\frac{J_m(\tau'_{mi}\rho)\tau'_{mi}}{J'_m(\tau'_{mi}a)k}e^{-j\chi'_{mi}z}\sin(m\varphi + \varphi_o)E_o, \quad \tilde{H}^i_z = \frac{J_m(\tau''_{mi}\rho)}{J_m(\tau''_{mi}a)}e^{-j\chi''_{mi}z}\cos(m\varphi + \varphi_o)H_o$$

$$\tilde{H}^i_\varphi = -\frac{j\omega\varepsilon}{\tau'^2_{mi}}\frac{\partial \tilde{E}^i_z}{\partial \rho} + j\frac{m\chi''_{mi}}{a\tau''^2_{mi}}\tilde{H}^i_z$$

We obtain

$$H^i_{z-}(\alpha, a_-) = -j\frac{H_o}{\alpha - \chi''_{mi}}$$

$$H^i_{\varphi-}(\alpha, a_-) = -\frac{jY_oE_o}{\alpha - \chi'_{mi}} - \frac{jm\chi''_{mi}}{a\chi''^2_{mi}}\frac{jH_o}{\alpha - \chi''_{mi}}$$

Thus, the W-H equations in normal form are

$$-\frac{2}{\omega\mu\tau\psi_m(\alpha)}\left[\tau^2 V_{\varphi+}(\alpha) + j\frac{m\alpha}{a}V_{z+}(\alpha)\right] = J^s_{\varphi-}(\alpha) - j\frac{H_o}{\alpha - \chi''_{mi}}$$

$$-\frac{2\omega\varepsilon\tau}{\phi_m(\alpha)}V_{z+}(\alpha) = \tau^2 J^s_{z-}(\alpha) + \frac{jm\alpha}{a}J^s_{\varphi-}(\alpha) + \frac{j\tau^2 Y_oE_o}{\alpha - \chi'_{mi}} - \frac{jm}{a}\left(\alpha - \frac{\tau^2\chi''_{mi}}{\tau''^2_{mi}}\right)\frac{jH_o}{\alpha - \chi''_{mi}}$$

These two W-H equations are decoupled if we introduce the unbounded plus and minus functions defined by (see section 2.5)

$$\hat{X}_+(\alpha) = \tau^2 V_{\varphi+}(\alpha) + j\frac{m\alpha}{a}V_{z+}(\alpha), \quad \hat{Y}_-(\alpha) = \tau^2 J^s_{z-}(\alpha) + \frac{jm\alpha}{a}J^s_{\varphi-}(\alpha)$$

We find

$$\begin{aligned}
-\frac{2}{\omega\mu\tau\psi_m(\alpha)}\hat{X}_+(\alpha) &= J^s_{\varphi-}(\alpha) - j\frac{H_o}{\alpha - \chi''_{mi}} \\
-\frac{2\omega\varepsilon\tau}{\phi_m(\alpha)}V_{z+}(\alpha) &= \hat{Y}_-(\alpha) + \frac{j\tau^2 Y_oE_o}{\alpha - \chi'_{mi}} - \frac{jm}{a}\left(\alpha - \frac{\tau^2\chi''_{mi}}{\tau''^2_{mi}}\right)\frac{jH_o}{\alpha - \chi''_{mi}}
\end{aligned} \tag{41}$$

The behavior of $\hat{X}_+(\alpha)$ and $\hat{Y}_-(\alpha)$ can be obtained by observing that

$$\tilde{E}_z(z, a, \varphi) \to z^{-1/2}, \quad \tilde{E}_\varphi(z, a, \varphi) \to z^{1/2} \quad \text{as } z \to 0_+$$

$$\tilde{J}_z(z) \to z^{1/2}, \quad \tilde{J}_\varphi(z) \to z^{-1/2} \quad \text{as } z \to 0_-$$

It follows that as $\alpha \to \infty$

$$V_{\varphi+}(\alpha) \to \alpha^{-3/2}, \ V_{z+}(\alpha) \to \alpha^{-1/2}, \ J_{z-}(\alpha) \to \alpha^{-3/2}, \ J_{\varphi-}(\alpha) \to \alpha^{-1/2}, \ \hat{X}_+(\alpha) \to \alpha^{1/2},$$

$$\hat{Y}_-(\alpha) \to \alpha$$

The solution of the two uncoupled eq. (41) requires the factorization of the functions $\psi_m(\alpha)$ and $\phi_m(\alpha)$. The asymptotic behavior of these functions follows from the asymptotic behavior of the Bessel functions. We have

$$J_m(\tau a) \approx \sqrt{\frac{2}{\pi\tau a}}\cos(\tau a - m\pi/2 - \pi/4) \approx \sqrt{\frac{2}{\pi\tau a}}e^{-j(\tau a - m\pi/2-\pi/4)}$$

$$H_m^{(2)}(\tau a) \approx \sqrt{\frac{2}{\pi\tau a}}e^{-j(\tau a - m\pi/2-\pi/4)}, \quad J_m'(\tau a) \approx j\sqrt{\frac{2}{\pi\tau a}}e^{-j(\tau a - m\pi/2-\pi/4)}$$

$$H_m^{(2)\prime}(\tau a) \approx -j\sqrt{\frac{2}{\pi\tau a}}e^{-j(\tau a - m\pi/2-\pi/4)}$$

It follows that $\psi_m(\alpha)$ and $\phi_m(\alpha)$ are constants as $\alpha \to \infty$.

The factorization of $\psi_m(\alpha)$ in the first of eq. (41) and the separation of the plus and the minus functions in two separate members yields the solution

$$J_{\varphi-}^s(\alpha) = \frac{A_o}{\sqrt{k+\alpha}\,\psi_{m-}(\alpha)} + j\frac{H_o[\sqrt{k+\alpha}\,\psi_{m-}(\alpha) - \sqrt{k+\chi_{mi}''}\,\psi_{m-}(\chi_{mi}'')]}{\sqrt{k+\alpha}\,\psi_{m-}(\alpha)(\alpha-\chi_{mi}'')} \tag{42}$$

$$\begin{aligned}\hat{X}_+(\alpha) &= \tau^2 V_{\varphi+}(\alpha) + j\frac{m\alpha}{a}V_{z+}(\alpha) \\ &= \left(-A_o + j\frac{\sqrt{k+\chi_{mi}''}}{\alpha-\chi_{mi}''}\psi_{m-}(\chi_{mi}'')H_o\right)\frac{\omega\mu\sqrt{k-\alpha}}{2}\psi_{m+}(\alpha)\end{aligned} \tag{43}$$

where A_o is the entire function that arises from the separations of the plus from the minus functions. According to the asymptotic behavior of the involved functions, A_o must be a constant.

By using the same reasoning in the second of eq. (41) we get

$$V_{z+}(\alpha) = -\frac{\phi_{m+}(\alpha)}{2\omega\varepsilon\sqrt{k-\alpha}}\left[B_o\alpha + C_o + j\frac{Y_oE_o\tau^2\phi_{m-}(\chi_{mi}')}{\sqrt{k+\chi_{mi}'}(\alpha-\chi_{mi}')}\right] \tag{44}$$

$$\begin{aligned}\hat{Y}_-(\alpha) &= \tau^2 J_{z-}^s(\alpha) + \frac{jm\alpha}{a}J_{\varphi-}^s(\alpha) = \frac{\sqrt{k+\alpha}}{\phi_{m-}(\alpha)}\left[B_o\alpha + C_o + \frac{jm}{a}\left(\alpha - \frac{\tau^2\chi_{mi}''}{\tau_{mi}''^2}\right)\frac{jH_o}{\alpha-\chi_{mi}''}\right] \\ &\quad + -j\tau^2\left[1 - \frac{\phi_{m-}(\chi_{mi}')\sqrt{k+\alpha}}{\sqrt{k+\chi_{mi}'}\phi_{m-}(\alpha)}\right]\frac{Y_oE_o}{\alpha-\chi_{mi}'}\end{aligned} \tag{45}$$

where $B_o\alpha + C_o$ is the entire function that arises from the separation of the plus from the minus functions. According to the asymptotic behavior of the involved functions, this must be a polynomial of first degree. The constant B_o is immediately found by taking into account the behavior of $\hat{Y}_-(\alpha)$ in (45). We get

$$B_o = j\frac{Y_oE_o\phi_{m-}(\chi_{mi}')}{\sqrt{k+\chi_{mi}'}} \tag{46}$$

To find A_o and C_o we observe that

$$V_{\varphi+}(\alpha) = \frac{\hat{X}_+(\alpha) - j\frac{m\alpha}{a} V_{z+}(\alpha)}{(k-\alpha)(k+\alpha)}$$

$$J^s_{z-}(\alpha) = \frac{\hat{Y}_-(\alpha) - \frac{jm\alpha}{a} J^s_{\varphi-}(\alpha)}{(k-\alpha)(k+\alpha)}$$

Since $V_{\varphi+}(\alpha)$ is regular $\alpha = -k$, and since $J^s_{z-}(\alpha)$ is regular $\alpha = k$. Hence, we get the following two equations:

$$\hat{X}_+(-k) + j\frac{mk}{a} V_{z+}(-k) = 0$$

$$\hat{Y}_-(k) - \frac{jmk}{a} J^s_{\varphi-}(k)$$

It can be shown that these two equations provide the evaluation of the two constants A_o and C_o.

9.6 Discontinuities in circular waveguides

The technique that we have elucidated for solving discontinuity problem in rectangular waveguides can be extended to circular waveguides. For instance, the following problems have been considered in the literature.

Problem (a): Radiation from a semi-infinite cylindrical waveguide into a larger cylindrical waveguide such that the two cylinders are coaxial (Fig. 7). This problem has been considered by many authors, including Mittra and Lee (1971) and Weinstein (1969).

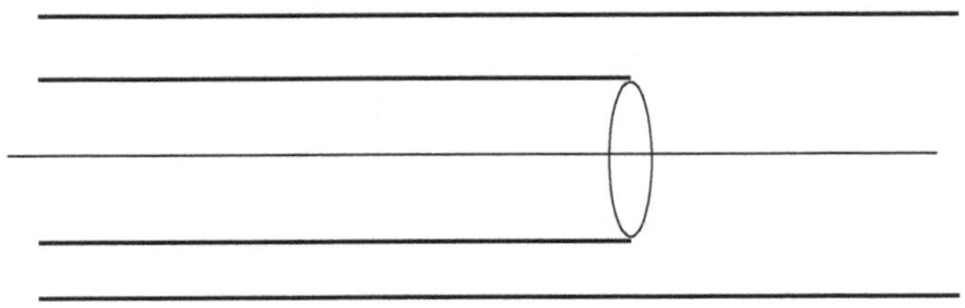

Fig. 7: Geometry of problem (a)

Problem (b): Scattering by a semi-infinite PEC rod of circular cross section (Fig. 8). This problem has been considered by Jones (1955).

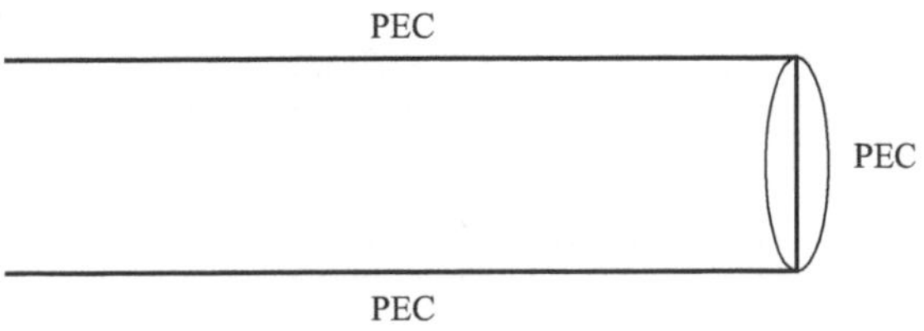

Fig. 8: Geometry of problem (b)

Problem (c): Scattering by a PEC cylinder of finite length (Fig. 9). This problem has been considered by Williams (1956).

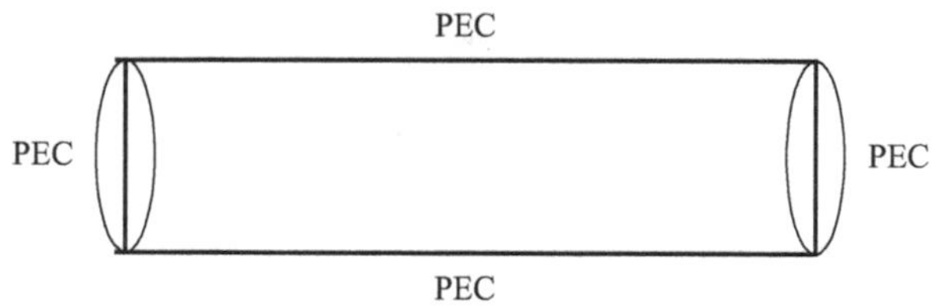

Fig. 9: Geometry of problem (c)

Problem (d): Junction between two circular waveguides (Fig. 10). This problem has been considered by Papadopoulos (1957).

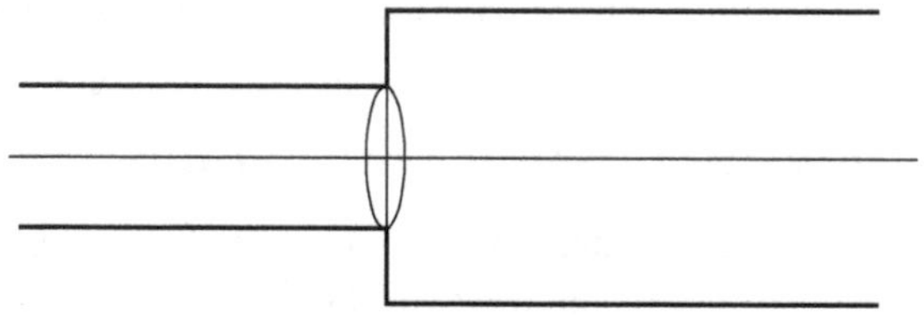

Fig. 10: Geometry of problem (d)

Problem (e): Junction between a coaxial cable and a circular waveguide (Fig. 11). This problem is considered by Weinstein (1969, p. 195).

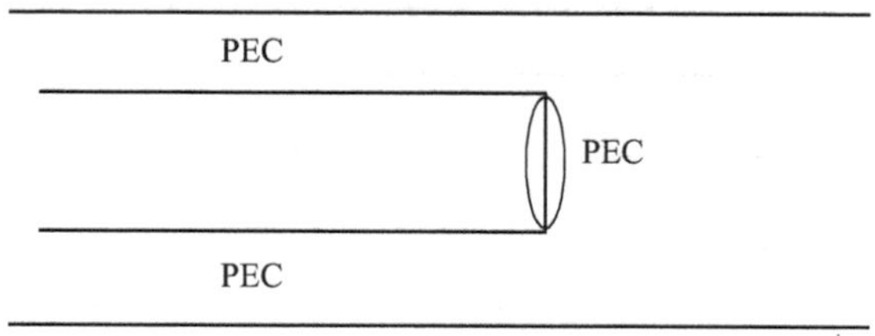

Fig. 11: Geometry of problem (e)

Problem (f): Semi-infinite transmission line (Fig. 12). This problem is considered by Weinstein (1969, p. 212)

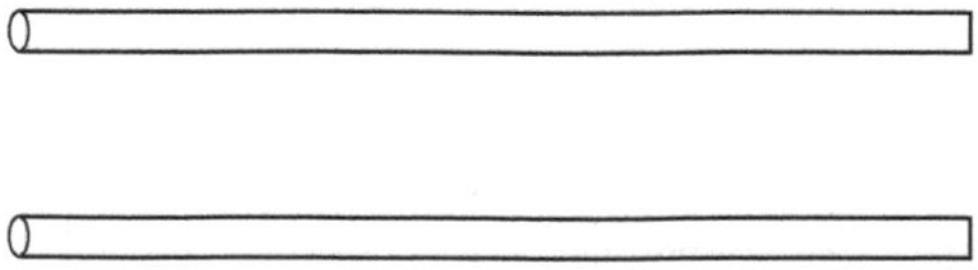

Fig. 12: Open-ended coaxial line

Problem (g): Semi-infinite sheath helix (Fig. 13). This problem is considered in Weinstein (1969, p. 345).

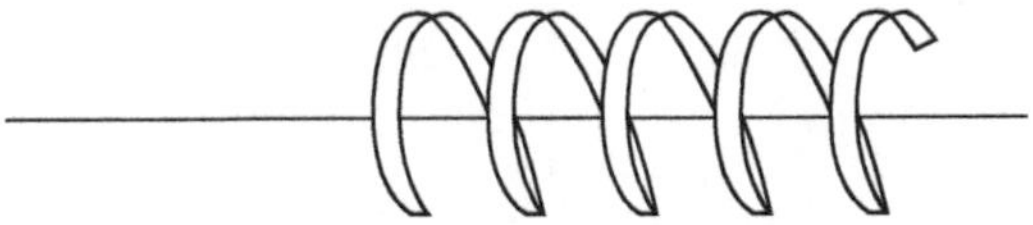

Fig. 13: Semi-infinite helical waveguide

CHAPTER 10

Further applications of the W-H technique

10.1 The step problem

10.1.1 Deduction of the transverse modified W-H equations (E-polarization case)

A simple transmission line model for the problem shown in Fig. 1 yields for $y > 0$

$$I(\alpha) = I_+(\alpha) + I_-(\alpha) = Y_c[V_+(\alpha) + V_-(\alpha)] = Y_c\, V_+(\alpha), \quad Y_c = \frac{\sqrt{k^2 - \alpha^2}}{k\, Z_o} \tag{1}$$

where

$$V_-(\alpha) = \int_{-\infty}^{0} E_z(x, 0) e^{j\alpha x} dx = 0, \quad I_-(\alpha) = \int_{-\infty}^{0} H_x(x, 0_+) e^{j\alpha x} dx$$

$$V_+(\alpha) = \int_{0}^{\infty} E_z(x, 0) e^{j\alpha x} dx, \quad I_+(\alpha) = \int_{0}^{\infty} H_x(x, 0) e^{j\alpha x} dx$$

For $y < 0$ we must introduce the odd and even Fourier transforms[1] defined by

$$F_o(\alpha) = 2j \int_{0}^{\infty} f(x) \sin(\alpha x) dx \quad F_e(\alpha) = 2 \int_{0}^{\infty} f(x) \cos(\alpha x) dx$$

[1] We can use different definitions of the odd and even Fourier transforms. In every case, the coefficients of the inverse transforms derive from

$$\int_{0}^{\infty} \cos(\alpha z) \cos(\alpha u) d\alpha = \frac{\pi}{2} \delta(z - u), \quad \int_{0}^{\infty} \sin(\alpha z) \sin(\alpha u) d\alpha = \frac{\pi}{2} \delta(z - u)$$

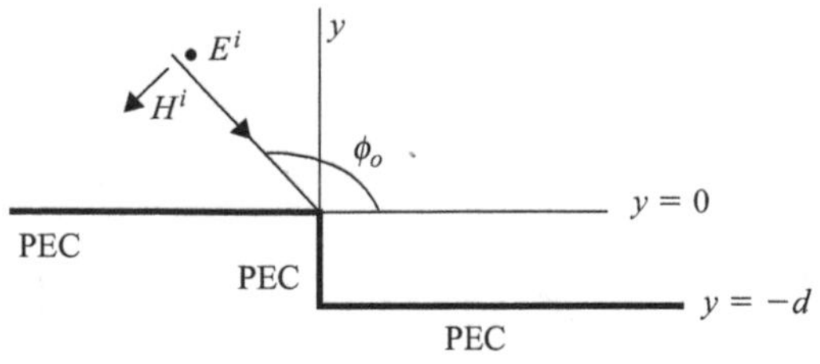

Fig. 1: The step problem

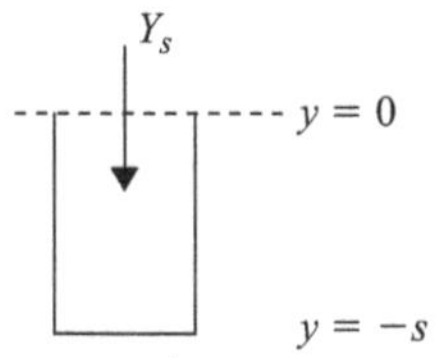

Fig. 2: Transmission line in the region $-s < y < 0$

$$f(x) = \frac{1}{j\pi}\int_0^\infty F_o(\alpha)\sin(\alpha x)d\alpha \quad f(x) = \frac{1}{\pi}\int_0^\infty F_e(\alpha)\cos(\alpha x)d\alpha$$

For these transforms the transmission line equations apply, and we get

$$\hat{I}(\alpha) = 2j\int_0^\infty H_x(x,0)\sin(\alpha x)dx = I_+(\alpha) - I_+(-\alpha)$$

$$= -Y_s\,\hat{V}(\alpha) = -Y_s\,2j\int_0^\infty E_z(x,0)\sin(\alpha x)dx = -Y_s[V_+(\alpha) - V_+(-\alpha)]$$

where from the transmission line theory we introduce Fig. 2.

By subtracting the two equations we get

$$G(\alpha)V_+(\alpha) + H(\alpha)V_+(-\alpha) = I_-(\alpha) + I_+(-\alpha)$$

where

$$G(\alpha) = Y_c + Y_s = -j\frac{\sin\left(\sqrt{k^2-\alpha^2}s\right)}{(k\,s)Z_o\sqrt{k^2-\alpha^2}s}e^{j\sqrt{k^2-\alpha^2}s}$$

$$H(\alpha) = Y_s = -j\frac{\sqrt{k^2-\alpha^2}}{k\,Z_o}\cot\left[\sqrt{k^2-\alpha^2}s\right]$$

To make explicit the source term, we observe that the plus functions $V_+(\alpha)$ and $I_+(\alpha)$ are standard since the pole $\alpha_o = -k\cos\varphi_o$ is in the fourth quadrant. Consequently, also the minus functions $V_+(-\alpha)$ and $I_+(-\alpha)$ are standard. This avoids the evaluation of the geometrical optics field in the aperture $y = 0,\ x > 0$. Conversely, for the minus function

$I_-(\alpha)$ the pole $\alpha_o = -k\cos\varphi_o$ is nonstandard. However, the nonstandard contribution is immediately evaluated because it is the primary contribution for the total current induced on the half-plane $y = 0,\ \ x < 0$. We have

$$I_-(\alpha) = I_-^s(\alpha) + 2j\frac{E_o\cos\varphi_o}{Z_o(\alpha - \alpha_o)}$$

where $I_-^s(\alpha)$ is standard. Thus,

$$G(\alpha)V_+(\alpha) + H(\alpha)V_+(-\alpha) = I_-^s(\alpha) + I_+(-\alpha) + 2j\frac{E_o\cos\varphi_o}{Z_o(\alpha - \alpha_o)}$$

involves only standard W-H unknowns.

The step problem is related to the thick semi-infinite plane problem. This important problem was studied by Jones (1953), who developed for its solution the very important method described in chapter 6, section 6.1.3.

10.1.2 Solution of the equations

The modified equations can be solved using Jones's method (section 6.1.3) (Daniele, Tascone & Zich, 1983). Alternatively, we can reduce them to a system of two standard W-H equations (section 1.5.1).

10.2 The strip problem

From the circuit representation of Fig. 3 we get

$$\frac{Z_c}{2}A_o(\alpha) = V\ (\alpha) + e^{j\alpha L}Y_+(\alpha)$$

where

$$A_o(\alpha) = \int_0^L [H_x(x,0_+) - H_x(x,0_-)]e^{j\alpha x}dx$$

$$Z_c = Y_c^{-1} = \frac{k\,Z_o}{\sqrt{k^2 - \alpha^2}} \qquad V_-(\alpha) = \int_{-\infty}^0 E_z(x,0)e^{j\alpha x}dx$$

$$Y_+(\alpha) = \int_0^\infty E_z(x+L,0)e^{j\alpha x}dx$$

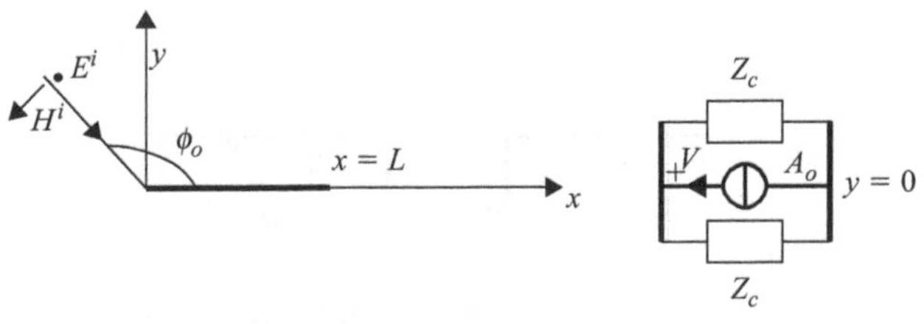

Fig. 3: The strip problem and its circuit model

To make explicit the source term, we observe that the only nonstandard unknown is

$$V_-(\alpha) = V_-^s(\alpha) - j\frac{E_o}{\alpha - \alpha_o}$$

Consequently,

$$\frac{Z_c}{2}A_o(\alpha) = V_-^s(\alpha) + e^{j\alpha L}Y_+(\alpha) - j\frac{E_o}{\alpha - \alpha_o}$$

is longitudinal modified W-H equation that contains only standard unknowns.

The solution of the previous equation can be accomplished according to the two methods presented in section 6.2 and the Jones method. In particular, the Jones method (see section 6.1.2 for details) is particularly appropriate for this problem (Serbest & Büyükaksoy, 1993).

10.2.1 Some longitudinally modified W-H geometries

Two parallel strips constitute a planar waveguide of finite length (Fig. 4). This important problem was solved for the first time by Jones (1952b). The Jones method was developed for solving the longitudinal W-H equations (section 6.1.2).

Some geometries considered in Noble (1958) and Mittra and Lee (1971) are indicated in the following sections.

10.3 The hole problem

By considering the transmission line for $y > 0$ (Fig. 5) we get the incomplete W-H equation:

$$I(\alpha) = I_-(\alpha) + I_o(\alpha) + \hat{I}_+(\alpha)e^{j\alpha L} = Y_c V_o(\alpha)$$

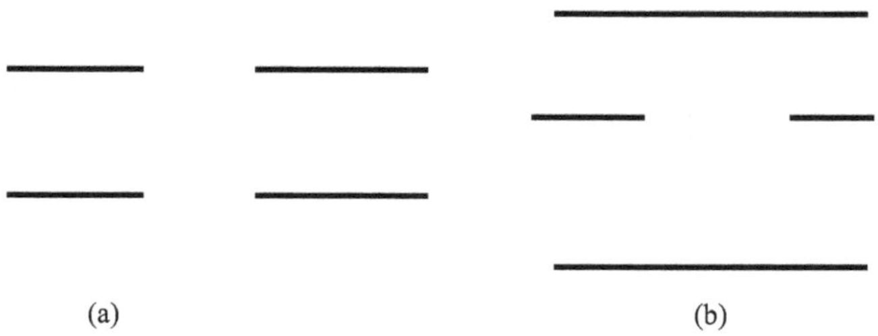

Fig. 4: Longitudinal modified geometries

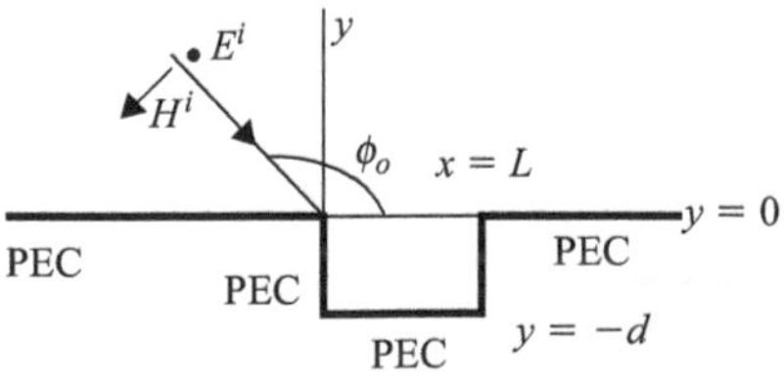

Fig. 5: The hole problem

where

$$Y_c = \frac{\sqrt{k^2 - \alpha^2}}{k\, Z_o}, \quad I_-(\alpha) = \int_{-\infty}^{0} H_x(x, 0_+) e^{j\alpha x} dx, \quad \hat{I}_+(\alpha) = \int_{0}^{\infty} H_x(x + L, 0) e^{j\alpha x} dx$$

$$V_o(\alpha) = \int_0^L E_z(x, 0) e^{j\alpha x} dx, \quad I_o(\alpha) = \int_0^L H_x(x, 0) e^{j\alpha x} dx$$

To complete the mathematical formulation of the problem, we observe that the hole constitutes a waveguide in the y-direction. We have

$$E_z(x, 0) = \sum_{n=1}^{\infty} A_n \sin\frac{n\pi}{L} x \quad H_x(x, 0) = -\sum_{n=1}^{\infty} Y_n A_n \sin\frac{n\pi}{L} x$$

where

$$Y_n = -j\frac{\sqrt{k^2 - \alpha_n^2}}{\omega\mu}, \quad \alpha_n = \frac{n\pi}{L}, \quad n = 1, 2, \ldots$$

It follows that

$$V_o(\alpha) = \sum_{n=1}^{\infty} A_n \frac{\alpha_n}{\alpha^2 - \alpha_n^2}(-1 + e^{j\alpha L} \cos(n\pi))$$

$$I_o(\alpha) = -\sum_{n=1}^{\infty} A_n Y_n \frac{\alpha_n}{\alpha^2 - \alpha_n^2}(-1 + e^{j\alpha L} \cos(n\pi))$$

Both functions $I_o(\alpha)$ and $V_o(\alpha)$ are entire and the previous equations constitute the sample representations of these functions (see eq. (6.54)). From these equations we get that the functional equation that relates $I_o(\alpha)$ to $V_o(\alpha)$ is constituted by the equation in the following samples:

$$I_o(\alpha_n) = -Y_n V_o(\alpha_n) \tag{2}$$

where $V_o(\alpha_n) = j\frac{L}{2} A_n$.

Keeping this fact in mind, we assume that $I_o(\alpha)$ is known and solve

$$I_-(\alpha) + I_o(\alpha) + \hat{I}_+(\alpha) e^{j\alpha L} = Y_c\, V_o(\alpha)$$

using the Jones method. In particular, this solution yields a functional equation that relates $V_o(\alpha)$ to $I_o(\alpha)$. By assuming $\alpha = \alpha_n = \frac{n\pi}{L}(n = 1, 2, \ldots)$ and taking (2) into account, we obtain a system of equations in the samples $V_o(\alpha_n)$.

10.4 The wall problem

Considering the transmission line for $y > 0$ we get (Fig. 6)

$$Y_c[V_-(\alpha) + \hat{V}_+(\alpha) e^{j\alpha L}] = I_-(\alpha) + I_o(\alpha) + \hat{I}_+(\alpha) e^{j\alpha L} \tag{3}$$

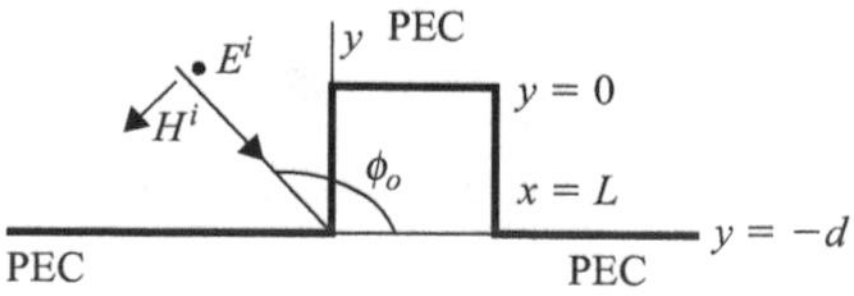

Fig. 6: The wall problem

where

$$Y_c = \frac{\sqrt{k^2 - \alpha^2}}{k\, Z_o}$$

Considering the transmission line $-d < y < 0, x < 0$, we get

$$-Y_d[V_-(\alpha) - V_-(-\alpha)] = I_-(\alpha) - I_-(-\alpha) \tag{4}$$

where

$$Y_d = -j\frac{\sqrt{k^2 - \alpha^2}}{k\, Z_o} \cot\left[\sqrt{k^2 - \alpha^2} d\right]$$

Considering the transmission line $-d < y < 0, L < x < \infty$, we get

$$-Y_d[\widehat{V}_+(\alpha)e^{j\alpha L} - \widehat{V}_+(-\alpha)e^{-j\alpha L}] = \widehat{I}_+(\alpha)e^{j\alpha L} - \widehat{I}_+(-\alpha)e^{-j\alpha L} \tag{5}$$

Another equation can be obtained by replacing α with $-\alpha$ in eq. (3):

$$Y_c[V_-(-\alpha) + \widehat{V}_+(-\alpha)e^{-j\alpha L}] = I_-(-\alpha) + I_o(-\alpha) + \widehat{I}_+(-\alpha)e^{-j\alpha L} \tag{6}$$

Conversely, with this substitution eqs. (4) and (5) remain unchanged.

Equations (3) through (6) constitute a system of W-H equations having as unknowns the plus function $\widehat{V}_+(\alpha)$, $\widehat{I}_+(\alpha)$, $V_-(-\alpha)$ and $I_-(-\alpha)$ and the minus functions $\widehat{V}_+(-\alpha)$, $\widehat{I}_+(-\alpha)$, $V_-(\alpha)$ and $I_-(\alpha)$. This system is incomplete, since also the unknowns $I_o(\pm\alpha)$ are present. We ignore this fact and solve the system using the Jones method. In this way we obtain a solution that depends on the functions $I_o(\pm\alpha)$. To complete the solution, we need the functional equations that relate $I_o(\pm\alpha)$ to the W-H unknowns. For this purpose we introduce the additional information that the function $I_o(\alpha)$ is entire; hence, it can be represented by the sampling theorem in the form

$$I_o(\alpha) = \sum_{m=-\infty}^{\infty} \frac{(e^{j\alpha L} - 1)I_o(\alpha_m)}{jL(\alpha - \alpha_m)} \tag{7}$$

where $\alpha_n = n\frac{2\pi}{L}$.

By substituting (7) into (3) and assuming $\alpha = \pm\alpha_n$, we get

$$Z_c(\pm\alpha_n)\left[I_-(\pm\alpha_n) + \sum_{m=-\infty}^{\infty} \frac{(e^{\pm j\alpha_n L} - 1)I_o(\alpha_m)}{jL(\pm\alpha_n - \alpha_m)} + \widehat{I}_+(\pm\alpha_n)e^{\pm j\alpha_n L}\right] - V_-(\pm\alpha_n)$$
$$- \widehat{V}_+(\pm\alpha_n)e^{\pm j\alpha_n L} = 0 \tag{8}$$

where $Z_c(\alpha) = (Y_c)^{-1}$. The W-H solutions relate the samples

$$I_-(\pm\alpha_n), \quad \hat{I}_+(\pm\alpha_n), \quad V_-(\pm\alpha_n), \quad \hat{V}_+(\pm\alpha_n)$$

to $I_o(\alpha_n)$. Substitution of these relations into (8) yields a system of infinite unknown in the samples $I_o(\alpha_n)$.

The wall problem can be related to the diffraction by a perfectly electrical conducting (PEC) rectangular cylinder. This problem has been tackled using the W-H technique (Aoki & Uchida, 1978; Kobayashi, 1982a, 1982b). In these papers the deduction of the incomplete W-H equations has been accomplished using an approach different from the one used in this section and based on the introduction of transmission lines.

10.5 The semi-infinite duct with a flange

Let us consider the geometry shown in Fig. 7. We can derive the Wiener-Hopf equations by introducing two alternative formulations. The first formulation is similar to that introduced in the hole problem[2] and is based on the introduction of the Fourier transform in the x-direction, leading to an incomplete W-H equation.

Alternatively, we can use Fourier transforms in the y-direction defined by

$$\tilde{\Phi}(x,\eta) = \int_{-\infty}^{\infty} \Phi(x,y)e^{j\eta y}dy$$

To simplify the deduction of the W-H equation, we reduce the problem to the study of the two simpler problems illustrated in Fig. 8.

In problem (a), in the region $x < b$, $-\infty < y < \infty$ we get

$$-Y_b[\tilde{E}_{z-}(b,\eta) + \tilde{E}_{z+}(b,\eta)] = -Y_b\tilde{E}_{z+}(b,\eta) = \tilde{H}_{x-}(b,\eta) + \tilde{H}_{x+}(b,\eta)$$

where

$$Y_b = -j\frac{\sqrt{k^2-\eta^2}}{k\,Z_o}\cot\left[\sqrt{k^2-\eta^2}\,b\right]$$

In the region $x > b$, $0 < y < \infty$, the use of the odd Fourier transform yields

$$Y_c[\tilde{E}_{z+}(b,\eta) - \tilde{E}_{z+}(b,-\eta)] = \tilde{H}_{x+}(b,\eta) - \tilde{H}_{x+}(b,-\eta)$$

where

$$Y_c = \frac{\sqrt{k^2-\eta^2}}{k\,Z_o}$$

By subtracting the two equations we obtain the transverse modified W-H equation:

$$(Y_b + Y_c)\tilde{E}_{z+}(b,\eta) - Y_c\tilde{E}_{z+}(b,-\eta) = -\tilde{H}_{x-}(b,\eta) - \tilde{H}_{x+}(b,-\eta) = X_-(\eta)$$

---

[2] In practice we have a hole in which $d = -\infty$.

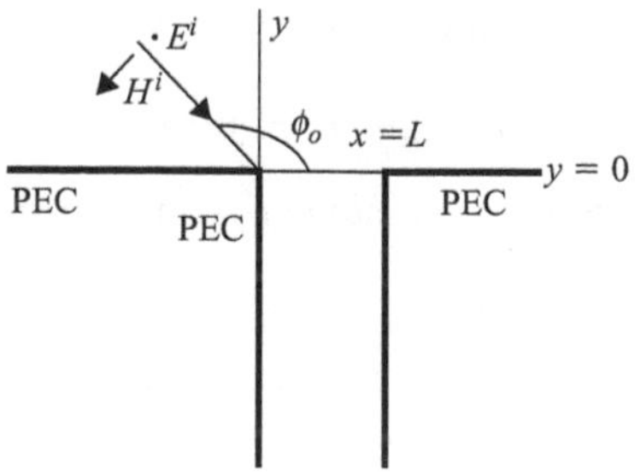

Fig. 7: Geometry of the problem

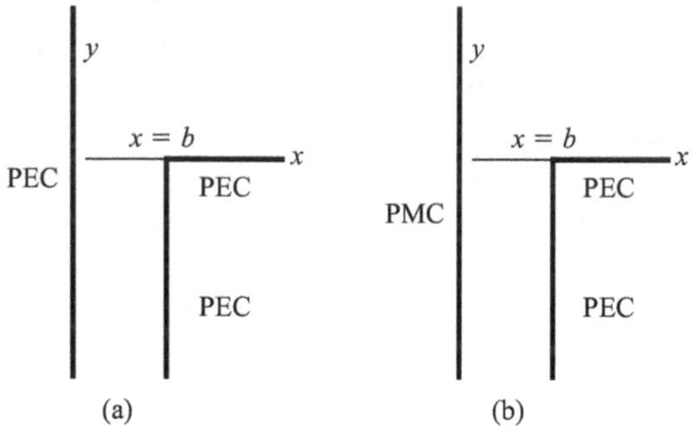

Fig. 8: Simplified geometry of the problem

Problem (b) can be studied in the same manner, and we get the same equations where

$$Y_b = j\frac{\sqrt{k^2 - \eta^2}}{k\, Z_o}\tan\left[\sqrt{k^2 - \eta^2}b\right]$$

10.6 Presence of dielectrics

In the previous examples we resorted to transmission line models to obtain the W-H equations for the problems. These transmission lines are relevant to waveguides with transverse sections filled by homogeneous media. Sometimes we encounter discontinuities due to the presence of dielectrics (Fig. 9). To study this structure, for the sake of simplicity, in the following we will consider only the presence of the component $E_z(x,y)$ of the electric field.

The wave equations in the two regions 1 and 2 are

$$\left(\frac{\partial^2}{\partial x^2} + \frac{\partial^2}{\partial y^2} + k_1^2\right)E_z(x,y) = 0, \quad \left(\frac{\partial^2}{\partial x^2} + \frac{\partial^2}{\partial y^2} + k_2^2\right)E_z(x,y) = 0$$

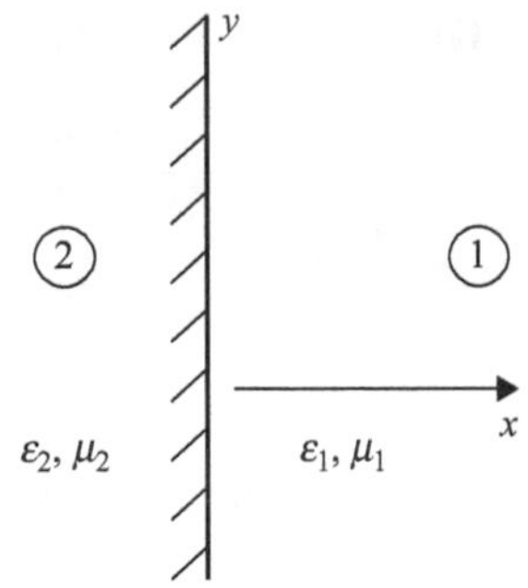

Fig. 9: Presence of a dielectric half-space

Applying the unilateral Fourier transforms

$$\tilde{E}_z(\alpha, y) = \int_0^\infty E_z(x, y) e^{j\alpha x} dx$$

to the previous equations we get

$$\left(\frac{d^2}{dy^2} + \tau_1^2\right)\tilde{E}_z(\alpha, y) = -j\alpha E_z(0_+, y) + \frac{\partial}{\partial y} E_z(0_+, y)$$

$$\left(\frac{d^2}{dy^2} + \tau_2^2\right)\tilde{E}_z(\alpha, y) = j\alpha E_z(0_-, y) - \frac{\partial}{\partial y} E_z(0_-, y)$$

where $\tau_{1,2} = \sqrt{k_{1,2}^2 - \alpha^2}$.

In the following it is important to evaluate the particular integral of the equation

$$\left(\frac{d^2}{dy^2} + \tau^2\right)\tilde{\Phi}(\alpha, y) = f_\alpha(y) \quad (a < y < b) \tag{9}$$

which is given by (Felsen & Marcuvitz, 1973, p. 278)

$$\tilde{\Phi}(\alpha, y) = \int_a^b g_\alpha(y, y') f_\alpha(y') dy' \tag{10}$$

where the Green function $g_\alpha(y, y')$ is

$$g_\alpha(y, y') = \frac{\overleftarrow{\varphi}_\alpha(y_<)\overrightarrow{\varphi}_\alpha(y_>)}{W_r[\overleftarrow{\varphi}_\alpha(y), \overrightarrow{\varphi}_\alpha(y)]} \tag{11}$$

In this equation, $\overleftarrow{\varphi}_\alpha(y)$, $\overrightarrow{\varphi}_\alpha(y)$ are solutions of eq. (9) that satisfy the boundary condition at the sections $y = a$ and $y = b$, respectively. $W_r[\overleftarrow{\varphi}_\alpha(y), \overrightarrow{\varphi}_\alpha(y)]$ is the Wronskian of the two functions $\overleftarrow{\varphi}_\alpha(y)$, $\overrightarrow{\varphi}_\alpha(y)$, which is independent on y, and $y_<$ and $y_>$ denote, respectively, the lesser and greater of the quantities y and y'.

10.7 A problem involving a dielectric slab

Figure 10 illustrates the geometry of the problem under study. In region 1 we define

$$\tilde{E}_{x+}(y,\alpha) = \int_0^\infty E_x(y,z)e^{j\alpha z}dz, \quad \tilde{H}_{z+}(y,\alpha) = \int_0^\infty H_z(y,z)e^{j\alpha z}dz$$

We have

$$\overleftarrow{\varphi}_\alpha(y) = \tilde{E}_x(0,\alpha)\cos(\tau y) = V_+(\alpha)\cos(\tau y)$$
$$\vec{\varphi}_\alpha(y) = \sin[\tau(y-s)]$$

where $\tau_{1,2} = \sqrt{k_{1,2}^2 - \alpha^2}$. This yields

$$g_\alpha(y,y') = \frac{\cos(\tau y_<)\cos(\tau(y_> - s))}{\tau\cos(\tau s)}$$

and provides the particular integral

$$\int_0^s \frac{\cos(\tau y_<)\sin(\tau(y_> - s))}{\tau\cos(\tau s)} f_{1\alpha}(y')dy'$$

where

$$f_{1\alpha}(y') = -j\alpha E_x(y,0_+) + \frac{\partial}{\partial z}E_x(y,0_+) = -j\alpha E_x(y,0_+) - j\omega\mu_1 H_y(y,0_+)$$

The solution of the homogeneous wave equation vanishing on the boundary $y = s$ is given by

$$A(\alpha)\sin(\tau(y-s))$$

Consequently, we get

$$\tilde{E}_{x+}(y,\alpha) = \frac{\int_0^y \cos(\tau y')f_{1\alpha}(y')dy' \sin(\tau(y-s)) + \int_y^s \sin(\tau(y'-s))f_{1\alpha}(y')dy' \cos(\tau y)}{\tau\cos(\tau s)} + A(\alpha)\sin(\tau(y-s))$$

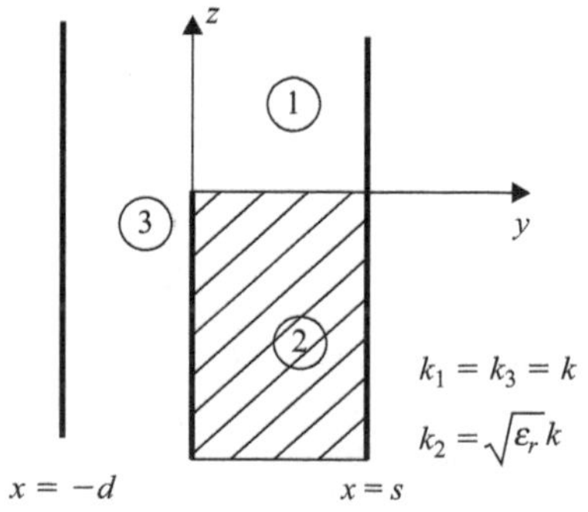

Fig. 10: Geometry of the problem

Setting $y = s$ we get the expected result $\tilde{E}_x(s, \alpha) = 0$. Conversely, putting $y = 0$ we get

$$\tilde{E}_{x+}(0, \alpha) = \frac{\int_0^s \sin(\tau(y' - s)) f_{1\alpha}(y') dy'}{\tau \cos(\tau s)} - A(\alpha) \sin(\tau s) = V_+(\alpha) \tag{12}$$

The field $\tilde{H}_{z+}(y, \alpha) = \frac{1}{j\omega\mu_1} \frac{\partial \tilde{E}_{x+}(y, \alpha)}{\partial y}$ presents a similar expression that here is reported only for $y = 0$. This expression is simple since the integral in it contains an integrand $\sin(\tau y_<)$ that vanishes for $y = 0$. We get

$$\tilde{H}_{z+}(0, \alpha) = \frac{\tau}{j\omega\mu_1} A(\alpha) \cos(\tau s) \tag{13}$$

Elimination of $A(\alpha)$ in (12) and (13) leads to

$$Y_s V_+(\alpha) = Y_s \frac{\int_0^s \sin(\tau(y' - s)) f_{1\alpha}(y') dy'}{\tau \cos(\tau s)} - \tilde{H}_{z+}(0, \alpha) \tag{14}$$

where

$$Y_s = -j \frac{\sqrt{k^2 - \alpha^2}}{k\, Z_o} \cot\left[\sqrt{k^2 - \alpha^2} s\right]$$

By repeating the same reasoning for region 2 and taking into account that the solution of the homogeneous equation vanishes, we get

$$\tilde{E}_{x-}(y, \alpha) = \int_{-\infty}^{0} E_x(y, z) e^{j\alpha z} dz, \quad \tilde{H}_{z-}(y, \alpha) = \int_{-\infty}^{0} H_z(y, z) e^{j\alpha z} dz$$

$$\tilde{H}_{z-}(0_+, \alpha) = \frac{\int_0^s \sin(\tau_2(y' - s)) f_{2\alpha}(y') dy'}{j\omega\mu_2 \sin(\tau_2 s)} \tag{15}$$

where

$$f_{2\alpha}(y') = j\alpha E_x(y, 0_-) - \frac{\partial}{\partial z} E_x(y, 0_-) = j\alpha E_x(y, 0_-) + j\omega\mu_2 H_y(y, 0_-)$$

In region 3 the transmission line equation yields

$$\tilde{H}_{z+}(0, \alpha) + \tilde{H}_{z-}(0_-, \alpha) = Y_d V_+(\alpha) \tag{16}$$

where

$$Y_d = -j \frac{\sqrt{k^2 - \alpha^2}}{k\, Z_o} \cot\left[\sqrt{k^2 - \alpha^2} d\right]$$

Equations (14) through (16) yield the incomplete W-H equation

$$(Y_s + Y_d)V_+(\alpha) = \frac{\int_0^s \sin(\tau(y' - s))f_{1\alpha}(y')dy'}{j\omega\mu_1 \sin(\tau s)} + \frac{\int_0^s \sin(\tau_2(y' - s))f_{2\alpha}(y')dy'}{j\omega\mu_2 \sin(\tau_2 s)} + A_-(\alpha) \quad (17)$$

where $A_-(\alpha) = \tilde{H}_{z-}(0_-, \alpha) - \tilde{H}_{z-}(0_+, \alpha)$.

To have a closed mathematical problem we must relate the integrals present in (17) to the W-H unknowns. To this end, we consider the modal representation of the field in waveguide 2:

$$E_x(y, 0_-) = \sum_{n=1}^{\infty} C_n \sin\left(\frac{n\pi}{s}y\right), \quad H_y(y, 0_-) = \sum_{n=1}^{\infty} Y_{2n} C_n \sin\left(\frac{n\pi}{s}y\right) \quad (18)$$

where

$$Y_{2n} = \frac{\chi_{2n}}{\omega\mu_2}, \quad \chi_{2n} = \sqrt{k_2^2 - \left(\frac{n\pi}{s}\right)^2}$$

For the sake of simplicity, in the following we consider $\mu_1 = \mu_2 = \mu$. Taking into account that

$$\int_0^s \sin(\tau(y' - s)) \sin\left(\frac{n\pi}{s}y'\right)dy' = \frac{n\pi}{s}\frac{\sin(\tau s)}{\chi_n^2 - \alpha^2}$$

where $\chi_n = \sqrt{k^2 - \left(\frac{n\pi}{s}\right)^2}$, and substituting expressions (18) in $f_{1\alpha}(y)$ of (14) yields

$$Y_s V_+(\alpha) = \frac{1}{\omega\mu}\sum_{n=1}^{\infty}\frac{n\pi}{s}\frac{\alpha - \chi_{2n}}{\alpha^2 - \chi_n^2}C_n - \tilde{H}_{z+}(0, \alpha)$$

The poles $\alpha = -\chi_n$ in the sum are present only in Y_s. Thus, by comparing the residues of both members at these poles we get

$$C_n = j\frac{V_+(-\chi_n)}{\chi_n + \chi_{2n}}$$

The previous equation completes the mathematical formulation of the problem. In fact, eq. (17) can be rewritten as

$$(Y_s + Y_d)V_+(\alpha) = \frac{1}{j\omega\mu}\sum_{n=1}^{\infty}\frac{n\pi}{s}\left[\frac{\alpha - \chi_{2n}}{\alpha^2 - \chi_n^2} - \frac{\alpha - \chi_{2n}}{\alpha^2 - \chi_{2n}^2}\right]\frac{1}{\chi_n + \chi_{2n}}V_+(-\chi_n) + A_-(\alpha) \quad (19)$$

Even though eq. (19) can be solved directly by the technique for incomplete W-H equations, it is instructive to reduce it to a transverse modified W-H equation. To this end we observe that

$$\frac{1}{j\omega\mu}\sum_{n=1}^{\infty}\frac{n\pi}{s}\left[-\frac{\alpha-\chi_{2n}}{\alpha^2-\chi_{2n}^2}\right]\frac{1}{\chi_n+\chi_{2n}}V_+(-\chi_n)=-\frac{1}{j\omega\mu}\sum_{n=1}^{\infty}\frac{n\pi}{s}\frac{1}{\alpha+\chi_{2n}}\frac{1}{\chi_n+\chi_{2n}}V_+(-\chi_n)$$

$$= \text{minus function}$$

$$\frac{1}{j\omega\mu}\sum_{n=1}^{\infty}\frac{n\pi}{s}\frac{\alpha-\chi_{2n}}{\alpha^2-\chi_n^2}\frac{1}{\chi_n+\chi_{2n}}V_+(-\chi_n)$$

$$=\frac{1}{j\omega\mu}\sum_{n=1}^{\infty}\frac{n\pi}{s}\frac{\alpha-\chi_{2n}}{\alpha^2-\chi_n^2}\frac{1}{\chi_n+\chi_{2n}}[V_+(-\chi_n)-V_+(-\alpha)]+H(\alpha)V_+(-\alpha)$$

where

$$H(\alpha)=\frac{1}{j\omega\mu}\sum_{n=1}^{\infty}\frac{n\pi}{s}\frac{\alpha-\chi_{2n}}{\alpha^2-\chi_n^2}\frac{1}{\chi_n+\chi_{2n}}$$

The function in the first term of the second member also is a minus function. Hence, we can rewrite eq. (19) as the modified W-H equation:

$$(Y_s+Y_d)V_+(\alpha)=H(\alpha)V_+(-\alpha)+X_-(\alpha)$$

where

$$X_-(\alpha)=A_-(\alpha)-\frac{1}{j\omega\mu}\sum_{n=1}^{\infty}\frac{n\pi}{s}\frac{1}{\alpha+\chi_{2n}}\frac{1}{\chi_n+\chi_{2n}}V_+(-\chi_n)$$
$$+\frac{1}{j\omega\mu}\sum_{n=1}^{\infty}\frac{n\pi}{s}\frac{\alpha-\chi_{2n}}{\alpha^2-\chi_n^2}\frac{1}{\chi_n+\chi_{2n}}[V_+(-\chi_n)-V_+(-\alpha)]$$

10.8 Some problems involving dielectric slabs

Many electromagnetic problems can be formulated with the technique illustrated in this section. However, we must remember that when the W-H formulation is very difficult to obtain there may be some alternative methods of yielding sufficiently accurate results in an easier way. In the following we consider some problems studied in the literature.

10.8.1 Semi-infinite dielectric guides

The truncated dielectric slab (Fig. 11) has been considered by Aoki and Uchida (1982). The W-H formulation of a truncated optical fiber has been studied by Daniele (2004a).

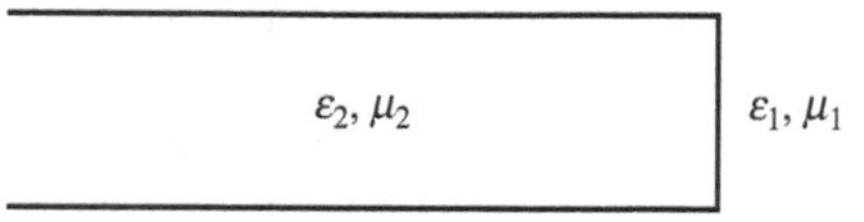

Fig. 11: Geometry of the problem

10.8.2 The junction of two semi-infinite dielectric slab guides

This problem has been studied by Ittipiboon and Hamid (1981):

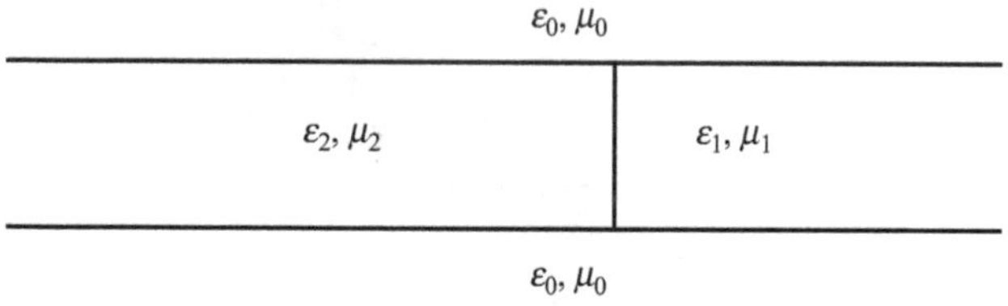

10.8.3 Some problems solved in the literature

Problem (a): Scattering of a plane wave by two parallel semi-infinite overlapping screens with dielectric loading (Büyükaksoy, Uzgoren & Birbir, 2001).

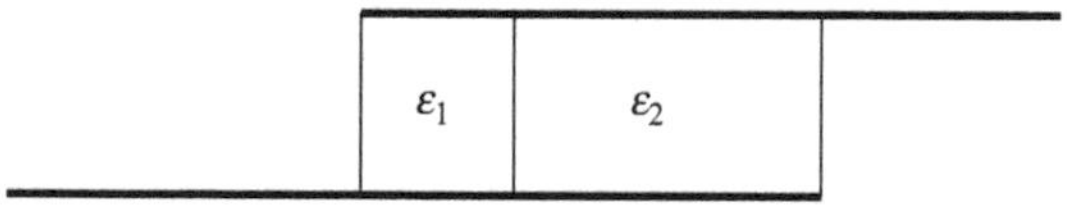

Problem (b): Analysis of the radar cross section of parallel-plate waveguide cavities (Kobayashi & Koshikawa, 1996).

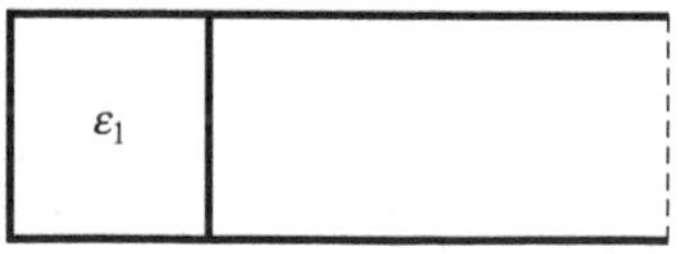

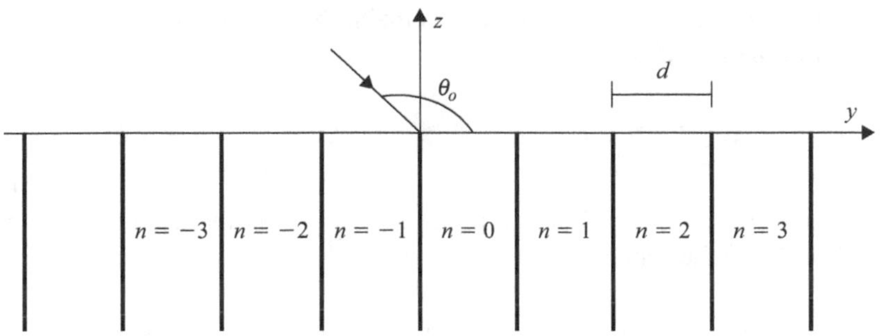

Fig. 12: Geometry of the problem

10.9 Some problems involving periodic structures

10.9.1 Diffraction by an infinite array of equally spaced half-planes immersed in free space

The problem illustrated in Fig. 12 consists of equally spaced infinite PEC half-planes immersed in free space. For the sake of simplicity we consider as source the E-polarized plane wave

$$E_x^i = E_o e^{-j\eta_o y + j\tau_o z} \quad (\eta_o = k \sin\theta_o, \ \ \tau_o = k\cos\theta_o)$$

The W-H equation can be obtained considering the circuit model shown in Fig. 13.

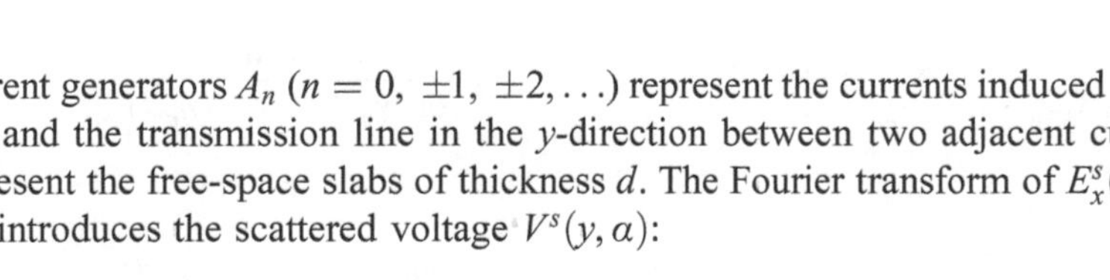

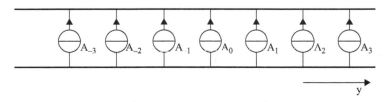

Fig. 13: Circuit model for the problem of Fig. 12

The current generators A_n ($n = 0, \pm 1, \pm 2, \ldots$) represent the currents induced in the PEC half-planes and the transmission line in the y-direction between two adjacent current generators represent the free-space slabs of thickness d. The Fourier transform of $E_x^s(y,z)$ in the z-direction introduces the scattered voltage $V^s(y,\alpha)$:

$$V^s(y,\alpha) = \int_{-\infty}^{\infty} E_x^s(y,z) e^{j\alpha z} dz, \quad E_x^s(y,z) = \frac{1}{2\pi}\int_{-\infty}^{\infty} V^s(y,\alpha) e^{-j\alpha z} d\alpha$$

It is obtained by considering, in the spectral domain, the total field radiated by all the currents induced in the half-planes:

$$V^s(y,\alpha) = \frac{1}{2} Z_a \sum_{n=-\infty}^{\infty} A_n(\alpha) e^{-j\chi_a|y-nd|}, \quad \left(Z_a = \frac{\omega\mu}{\chi_a}, \ \chi_a = \sqrt{k^2 - \alpha^2} \right) \tag{20}$$

Observe that the functions $A_n(\alpha)$ are all minus functions. On the apertures $y = md, \ z > 0$ $(m = 0, \pm 1, \pm 2, \ldots)$ the functions

$$V_+(md,\alpha) = V_+^s(md,\alpha) + V_+^i(md,\alpha), \quad V_+^i(md,\alpha) = -E_o \frac{j}{\alpha + \tau_o} e^{-j\eta_o md}$$

are plus functions.

Letting $y = md$ in eq. (20) yields

$$V_-^s(md,\alpha) = \frac{1}{2} Z_a \sum_{n=-\infty}^{\infty} A_{n+}(\alpha) e^{-j\chi_a|m-n|d} - E_o \frac{j}{\alpha + \tau_o} e^{-j\eta_o md} \tag{21}$$

Taking into account the Floquet theorem

$$E_x(y + md, z) = E_x(y,z) e^{-j\eta_o md}$$

we get

$$V_-^s(md,\alpha) = V_-^s(0,\alpha) e^{-j\eta_o md}, \quad A_{n+}(\alpha) = A_{o+}(\alpha) e^{-j\eta_o nd}$$

Substitution in (21) yields the W-H equation

$$V_-^s(0,\alpha) = Z_e(\alpha) A_{0+}(\alpha) - E_o \frac{j}{\alpha + \tau_o}$$

where the kernel is

$$Z_e(\alpha) = \frac{1}{2} Z_a \sum_{n=-\infty}^{\infty} e^{-j\chi_a|m-n|d} e^{-j\eta_o(m-n)d}$$

The sum in the previous equation is a geometric series that can be evaluated in closed form, yielding

$$\sum_{n=-\infty}^{\infty} e^{-j\chi_a|m-n|d} e^{-j\eta_o(m-n)d} = j \frac{\sin(\chi_a d)}{\cos(\chi_a d) - \cos(\eta_o d)}$$

The kernel

$$Z_e(\alpha) = \frac{1}{2} Z_a j \frac{\sin(\chi_a d)}{\cos(\chi_a d) - \cos(\eta_o d)} = \frac{\omega\mu}{2\chi_a} j \frac{\sin(\chi_a d)}{\cos(\chi_a d) - \cos(\eta_o d)}$$

is meromorphic and can be factorized by using the Weirstrass factorization discussed in chapter 3, section 3.2.4.

The problem of this section was considered for the first time by Carlson and Heins (1947). The extension to impedance half-planes has been studied by several authors, and in particular by Lüneburg (1993).

10.9.2 Other problems solved in the literature

Diffraction problems by structures that generalize the illustration in Fig. 12 are very important. In particular, many geometries are studied in Mittra and Lee (1971) by using the W-H technique or other methods related to the W-H technique. In the following we will indicated some important geometries.

Problem (a): Corrugated surfaces (Hurd, 1954; Weinstein, 1969, p. 240).

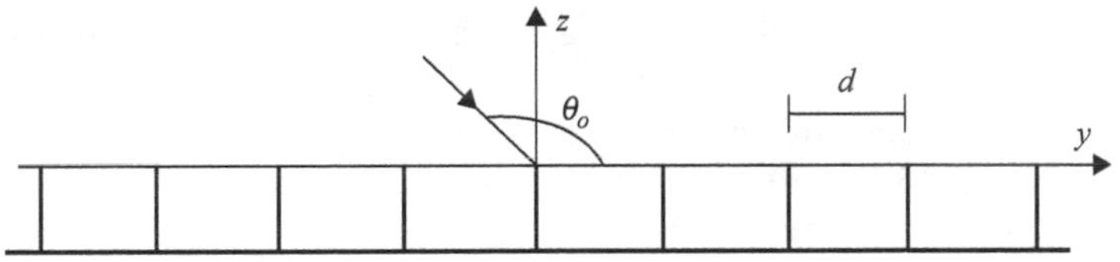

Problem (b): Parallel-plate media for microwave lenses (Heins, 1950b; Whitehead, 1951).

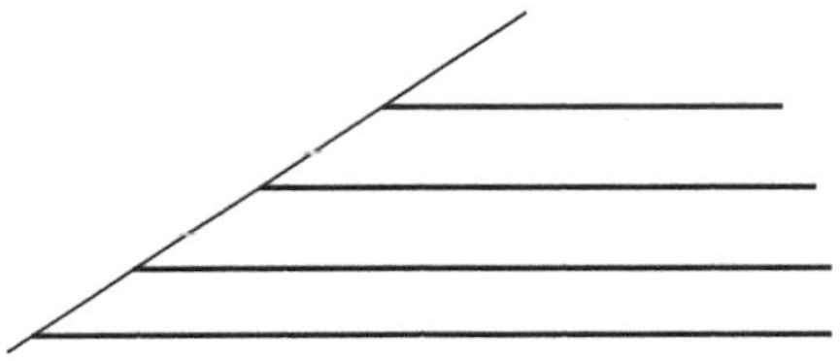

Problem (c): Modulated corrugated surface excited by a waveguide (Karjala & Mittra, 1966).

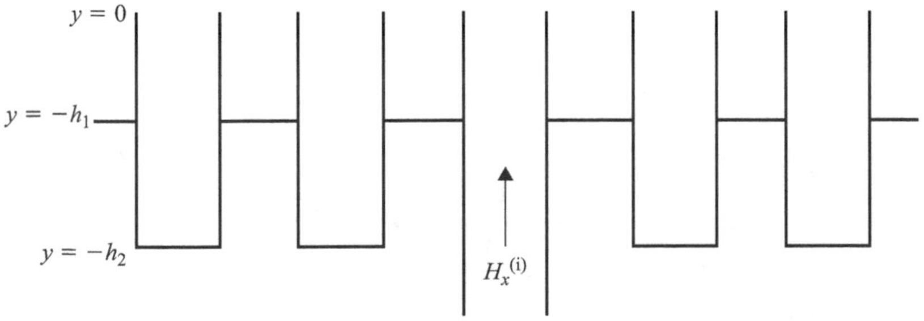

Problem (d): Radiation from infinite aperiodic array of parallel-plate waveguides (Lee, 1967).

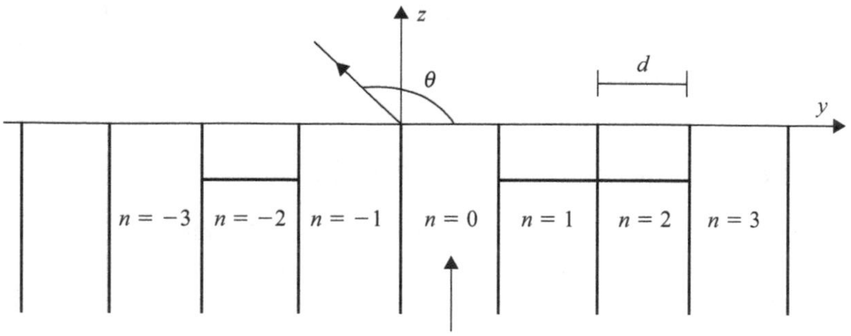

10.10 Diffraction by infinite strips

Figure 14 illustrates the original problem: a plane wave $E_y^i = e^{jkx}$ in the direction $-y$ is incident on a plane where infinite periodic PEC strips are present. Figure 14b illustrates the key problem where the apertures in the discontinuity plane are substituted by PMC strips.

According to Fig. 15 the original problem can be solved by summing the solution of the key problem with the solution of the very simple problem of reflection by a PEC plane.

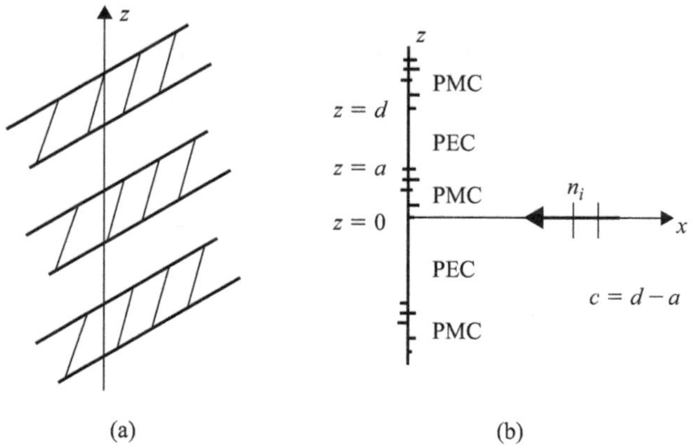

Fig. 14: (a) Original problem; (b) key problem

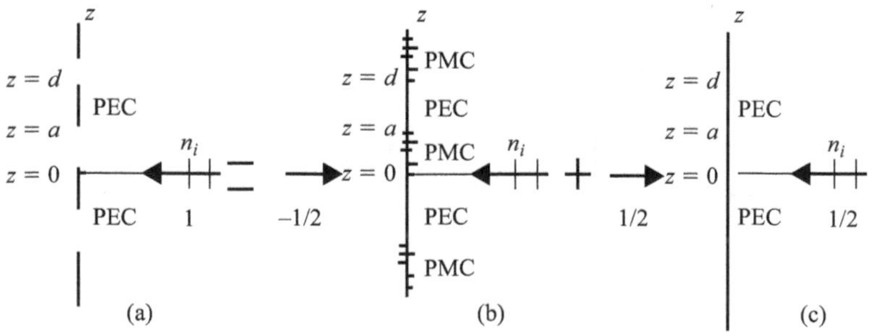

Fig. 15: Reduction of the strip problem to two simpler problems

In particular, by indicating with $\Phi(x,z)$ the solution of the key problem for $x>0$, the solution of the problem indicated in Fig. 15b is

$$\frac{1}{2}\Phi(x,z) \quad \text{for } x>0 \quad \text{and} \quad \frac{1}{2}\Phi(-x,z) \quad \text{for } x<0$$

The solution of the simple problem indicated in Fig. 15c is

$$\frac{1}{2}e^{jkx}-\frac{1}{2}e^{-jkx} \quad \text{for } x>0 \quad \text{and} \quad -\frac{1}{2}e^{-jkx}+\frac{1}{2}e^{jkx} \quad \text{for } x<0$$

Consequently, the solution of the original problem in Fig. 14a is

$$E_y(x,z)=\frac{1}{2}e^{jkx}-\frac{1}{2}e^{-jkx}+\frac{1}{2}\Phi(x,z) \quad \text{for } x>0$$
$$E_y(x,z)=-\frac{1}{2}e^{-jkx}+\frac{1}{2}e^{jkx}-\frac{1}{2}\Phi(-x,z) \quad \text{for } x<0$$

10.10.1 Solution of the key problem

The equation to be solved is

$$\left(\frac{\partial^2}{\partial x^2}+\frac{\partial^2}{\partial z^2}+k^2\right)\Phi(x,z)=0$$

with $\Phi(0,z)=0$ on the PEC strips and $\frac{\partial\Phi(0,z)}{\partial x}=0$ on the PMC strips. The W-H formulation of the problem can be obtained by considering the two waveguides shown in Fig. 16.

By putting $\Phi(x,z)=E_y(x,z)$, the guide in Fig. 16a involves even Fourier transforms and the guide in Fig. 16b odd Fourier transforms.

Guide of Fig.15a

$$E_y(x,z)=2\int_0^\infty V_e(\alpha,z)\cos(\alpha x)d\alpha$$

$$-H_x(x,z)=2\int_0^\infty I_e(\alpha,z)\cos(\alpha x)d\alpha$$

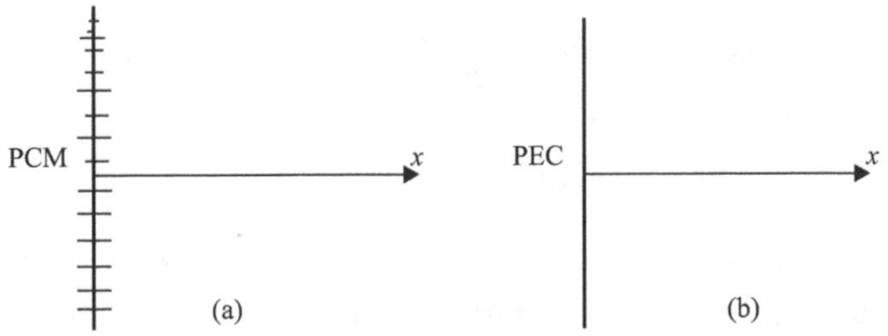

Fig. 16: The two waveguides involved in the key problem

Consideration of the transmission line in the slab between $z = 0_+$ and $z = a_-$ yields

$$\begin{aligned} V_e(\alpha, 0_+) &= \cos(\tau\, a) V_e(\alpha, a_-) + jZ_\alpha \sin(\tau\, a) I_e(\alpha, a_-) \\ I_e(\alpha, 0_+) &= jY_\alpha \sin(\tau\, a) V_e(\alpha, a_-) + \cos(\tau\, a) I_e(\alpha, a_-) \end{aligned} \tag{22}$$

with $\tau = \sqrt{k^2 - \alpha^2}$, $Z_\alpha = \frac{1}{Y_\alpha} = \frac{\omega\mu}{\tau}$.

Taking into account that the even Fourier transforms are related to the standard Fourier transforms through the operator Σ (section 9.1.2), we get

$$F_e(\alpha, z) = \frac{1}{\pi}\int_0^\infty f(x,z)\cos(\alpha x)dx = \frac{1}{2\pi}[F_+(\alpha, z) + F_+(-\alpha, z)] = \frac{1}{2\pi}\Sigma F_+(\alpha, z)$$

Consequently, eq. (22) can be rewritten as

$$\begin{aligned} \Sigma V_+(\alpha, 0_+) &= \cos(\tau\, a)\Sigma V_+(\alpha, a_-) + jZ_\alpha \sin(\tau\, a)\Sigma I_+(\alpha, a_-) \\ \Sigma I_+(\alpha, 0_+) &= jY_\alpha \sin(\tau\, a)\Sigma V_+(\alpha, a_-) + \cos(\tau\, a)\Sigma I_+(\alpha, a_-) \end{aligned} \tag{23}$$

Guide of Fig. 16b

$$E_y(x,z) = 2\int_0^\infty V_o(\alpha, z)\sin(\alpha x)d\alpha$$

$$-H_x(x,z) = 2\int_0^\infty I_o(\alpha, z)\sin(\alpha x)d\alpha$$

Consideration of the transmission line between $z = -c_+$ and $z = 0_-$ yields

$$\begin{aligned} V_o(\alpha, -c_+) &= \cos(\tau\, a) V_o(\alpha, 0_-) + jZ_\alpha \sin(\tau\, a) I_o(\alpha, 0_-) \\ I_o(\alpha, -c_+) &= jY_\alpha \sin(\tau\, a) V_o(\alpha, 0_-) + \cos(\tau\, a) I_o(\alpha, 0_-) \end{aligned} \tag{24}$$

with $\tau = \sqrt{k^2 - \alpha^2}$, $Z_\alpha = \frac{1}{Y_\alpha} = \frac{\omega\mu}{\tau}$.

Taking into account that the even Fourier transforms are related to the standard Fourier transforms through the operator Δ (section 9.1.2), we get

$$F_o(\alpha, z) = \frac{1}{\pi}\int_0^\infty f(x,z)\sin(\alpha x)dx = \frac{1}{2\pi}[F_+(\alpha, z) - F_+(-\alpha, z)] = \frac{1}{2\pi}\Delta F_+(\alpha, z)$$

Therefore, eq. (24) can be rewritten as

$$\begin{aligned} \Delta V_+(\alpha, -c_+) &= \cos(\tau\, a)\Delta V_+(\alpha, 0_-) + jZ_\alpha \sin(\tau\, a)\Delta I_+(\alpha, 0_-) \\ \Delta I_+(\alpha, -c_+) &= jY_\alpha \sin(\tau\, a)\Delta V_+(\alpha, 0_-) + \cos(\tau\, a)\Delta I_+(\alpha, 0_-) \end{aligned} \tag{25}$$

10.10.2 Boundary conditions

The continuity of the electromagnetic field at $z = 0$ forces the continuity of the Fourier transforms

$$V_+(\alpha, 0_+) = V_+(\alpha, 0_-), \quad I_+(\alpha, 0_+) = I_+(\alpha, 0_-) \tag{26}$$

Taking into account that the source field does not contain a propagation factor in the z-direction, the Floquet theorem states that the fields and their Fourier transforms must be periodic with periodicity d. For instance,

$$V_+(\alpha, -c_+) = V_+(\alpha, a_-), \quad I_+(\alpha, -c_+) = I_+(\alpha, a_-) \tag{27}$$

The previous equations can be rewritten in a compact form by introducing the matrices

$$\Psi_+(\alpha) = \begin{vmatrix} V_+(\alpha, 0_+) \\ I_+(\alpha, 0_+) \end{vmatrix} = \begin{vmatrix} V_+(\alpha, 0_-) \\ I_+(\alpha, 0_-) \end{vmatrix}$$

$$\Psi_{1+}(\alpha) = \begin{vmatrix} V_+(\alpha, -c_+) \\ I_+(\alpha, -c_+) \end{vmatrix} = \begin{vmatrix} V_+(\alpha, a_-) \\ I_+(\alpha, a_-) \end{vmatrix}$$

$$M_s = \exp\left[j\tau\, s \begin{vmatrix} 0 & Z_a \\ Y_a & 0 \end{vmatrix} \right] = \begin{vmatrix} \cos(\tau\, s) & jZ_a \sin(\tau\, s) \\ jY_a \sin(\tau\, s) & \cos(\tau\, s) \end{vmatrix}$$

With these notations and taking into account that M_a is even in α, eqs. (23) and (25) become

$$\Psi_+(\alpha) + \Psi_+(-\alpha) = M_a \Psi_{1+}(\alpha) + M_a \Psi_{1+}(-\alpha)$$
$$M_c \Psi_+(\alpha) - M_c \Psi_+(-\alpha) = \Psi_{1+}(\alpha) - \Psi_{1+}(-\alpha)$$

or, in matrix form,

$$G(\alpha)\theta_+(\alpha) = \theta_+(-\alpha) \tag{28}$$

where

$$G(\alpha) = \begin{vmatrix} -\mathbf{1} & M_a \\ M_c & -\mathbf{1} \end{vmatrix}^{-1} \cdot \begin{vmatrix} \mathbf{1} & -M_a \\ M_c & -\mathbf{1} \end{vmatrix}$$

$$\theta_+(\alpha) = \left| V_+(\alpha, 0) \quad I_+(\alpha, 0) \quad V_+(\alpha, a) \quad I_+(\alpha, a) \right|^t$$

Taking into account that M_a, M_c, and M_d commute, we get

$$G(\alpha) = (M_d - \mathbf{1})^{-1} \cdot \begin{vmatrix} \mathbf{1} + M_d & -2M_a \\ 2M_c & -\mathbf{1} - M_d \end{vmatrix}$$

10.10.3 Solution of the W-H equation

Canonical form of the W-H equation

Equation (28) is a homogeneous W-H equation. The source term can be obtained taking into account that both unknowns are nonstandard for the presence of the pole $\alpha_o = -k$ in $\theta_+(\alpha)$

and $\alpha_o = k$ in $\theta_+(-\alpha)$. The characteristic part $\theta^i_+(\alpha)$ relevant to the nonstandard pole can be accomplished by observing that

$$\theta^i_+(\alpha) = \left| V^i_+(\alpha,0) \quad I^i_+(\alpha,0) \quad V^i_+(\alpha,a) \quad I^i_+(\alpha,a) \right|^t = j\frac{1}{\alpha+k}\mathbf{q}_o$$

where $\mathbf{q}_o = |1 \quad Y_o \quad 1 \quad Y_o|^t$. By putting $\theta_+(\alpha) = \theta^s_+(\alpha) + \theta^i_+(\alpha)$, we can rewrite (28) in the inhomogeneous form

$$G(\alpha)\theta^s_+(\alpha) = \theta^s_+(-\alpha) + n(\alpha) \tag{29}$$

where

$$n(\alpha) = j\frac{\mathbf{q}_o}{-\alpha+k} - G(\alpha)\cdot j\frac{\mathbf{q}_o}{\alpha+k} = |n_1(\alpha) \quad n_2(\alpha)|^t$$

Factorization of $G(\alpha)$

The exact factorization of $G(\alpha) = G_-(\alpha)G_+(\alpha)$ can be obtained only when $a = c = \frac{d}{2}$.
This is the only case that we will study in the following. Putting

$$M_a = M_c = M, \quad M_d = M_{a+c} = M_a M_c = M^2$$

eq. (29) is equivalent to

$$\frac{\mathbf{1}+M^2}{-\mathbf{1}+M^2}\Psi^s_+(\alpha) - \frac{2M}{-\mathbf{1}+M^2}\Psi^s_{1+}(\alpha) = \Psi^s_+(-\alpha) + n_1(\alpha)$$
$$\frac{2M}{-\mathbf{1}+M^2}\Psi^s_+(\alpha) - \frac{\mathbf{1}+M^2}{-\mathbf{1}+M^2}\Psi^s_{1+}(\alpha) = \Psi^s_{1+}(-\alpha) + n_2(\alpha)$$

By summing these equations we get

$$Q(\alpha)X_+(\alpha) = X_-(\alpha) + n_p(\alpha) + n_1(\alpha) + n_2(\alpha) \tag{30}$$

Where

$$Q(\alpha) = \frac{(\mathbf{1}+M)^2}{-\mathbf{1}+M^2} = -\frac{\mathbf{1}+M}{\mathbf{1}-M} = -j\cot\frac{\tau a}{2}\begin{vmatrix} 0 & \frac{\omega\,\mu}{\tau} \\ \frac{\tau}{\omega\,\mu} & 0 \end{vmatrix}$$

$$n_p(\alpha) = n_1(\alpha) + n_2(\alpha) = \frac{2j}{-\alpha+k}\begin{vmatrix} 1 \\ Y_o \end{vmatrix}$$

$$X_+(\alpha) = \Psi^s_+(\alpha) - \Psi^s_{1+}(\alpha), \quad X_-(\alpha) = \Psi^s_+(-\alpha) + \Psi^s_{1+}(-\alpha)$$

The factorizaction of $Q(\alpha) = Q_-(\alpha)Q_+(\alpha)$ is immediate since it is skew diagonal. We get

$$Q_-(\alpha) = -j\frac{2}{a}g_-(\alpha)\begin{vmatrix} 0 & \frac{1}{k+\alpha} \\ 1 & 0 \end{vmatrix}, \quad Q_+(\alpha) = g_+(\alpha)\begin{vmatrix} \frac{1}{\omega\mu} & 0 \\ 0 & \frac{\omega\mu}{k-\alpha} \end{vmatrix}$$

where the factorization of the meromorphic function $g(\alpha) = -j\tau \cot\frac{\tau\alpha}{2}$ can be accomplished with the method indicated in section 3.2.4. Summing the solution $X_+(\alpha)$ and $X_-(-\alpha)$ of (30), we get

$$\Psi_+(\alpha) = \begin{vmatrix} V_+(\alpha,0) \\ I_+(\alpha,0) \end{vmatrix} = \left[j\frac{Q_-(-\alpha)}{\alpha+k} - j\frac{Q_+^{-1}(-\alpha)}{\alpha-k} \right] Q_-^{-1}(k) \begin{vmatrix} 1 \\ Y_o \end{vmatrix}$$

Of course the spectrum is discrete and the poles are defined by $\pm\alpha_m$, where

$$\alpha_m = \sqrt{k^2 - \left(\frac{2\pi m}{d}\right)^2}, \quad m = 0, \pm 1, \pm 2, \ldots$$

By assuming the inverse Fourier transform we get

$$\Phi(x,0) = e^{jkx} + \sum_{m=-\infty}^{\infty} \Phi_m e^{-j\sqrt{k^2-\left(\frac{2\pi m}{d}\right)^2}x}$$

where Φ_m is related to the residues at the poles α_m.

The solution of the key problem $\Phi(x,z)$ can be obtained by observing that the Floquet theorem provides the expression

$$\Phi(x,z) = e^{jkx} + \sum_{m=-\infty}^{\infty} F_m e^{-j\sqrt{k^2-\left(\frac{2\pi m}{d}\right)^2}x} e^{-j\frac{2\pi m}{d}z}$$

By putting $z = 0$, we find that $F_m = \Phi_m$.

The problem considered in this section is a very important one that has been studied by many researchers. Wiener-Hopf solutions of this problem have been considered in Weinstein (1969), Lukyanov (1980), and Daniele, Gilli, and Viterbo (1990). A comparison of the W-H technique with other techniques is presented in Lüneburg (1993).

10.11 Presence of an inductive iris in rectangular waveguides

Let us assume an incident TE_{10} mode in a rectangular waveguide where an iris is located in the section $x = 0$ (Fig. 17):

$$E_y^i = e^{j\chi_{10}x} \sin\left[\frac{\pi}{A}(z+c)\right], \quad A = c + a$$

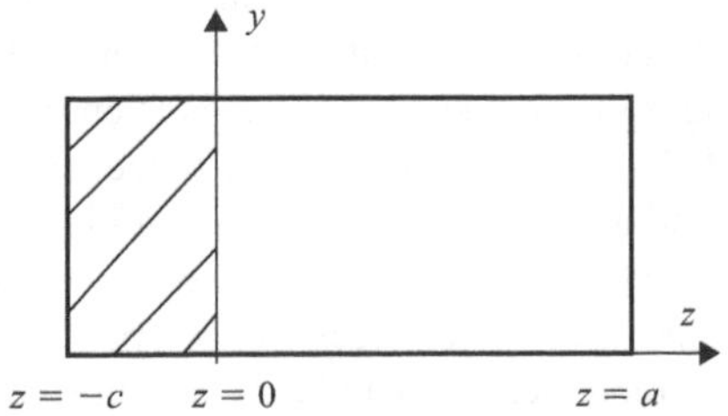

Fig. 17: Inductive iris in a rectangular waveguide

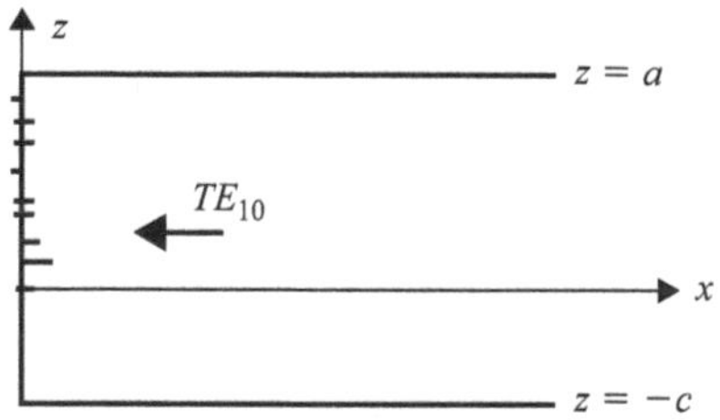

Fig. 18: The key problem for the inductive iris in a rectangular waveguide

For the presence of the iris, the incident TE_{10} excites only modes TE_{m0} (the modes TM_{m0} are vanishing). The problem reduces to the solution of the key problem shown in Fig. 18.

By comparing Fig. 18 with Fig. 14b, the only difference is constituted by the new boundary conditions indicated in eq. (31) that substitute the periodicity conditions (27). Hence, the new boundary conditions are

$$\begin{aligned} &V_+(\alpha,0_+) = V_+(\alpha,0_-), \quad I_+(\alpha,0_+) = I_+(\alpha,0_-) \\ &V_+(\alpha,\ -c_+) = V_+(\alpha,a_-) = 0 \end{aligned} \tag{31}$$

Taking into account again the transmission lines relevant to even and odd Fourier transforms, we get

$$\begin{aligned} V_+(\alpha,0) + V_+(-\alpha,0) &= jZ_a \tan(\tau a)[I_+(\alpha,0) + I_+(-\alpha,0)] \\ V_+(\alpha,0) - V_+(-\alpha,0) &= jZ_a \tan(\tau c)[I_+(\alpha,0) - I_+(-\alpha,0)] \end{aligned}$$

This constitutes a system of two W-H equations. Again, closed-form factorization can be accomplished if $c = a$.

10.12 Presence of a capacitive iris in rectangular waveguides

The geometry considered in section 9.4, includes this problem as a particular case.

10.13 Problems involving semi-infinite periodic structures

A periodic roughness in the form of corrugations is present in many waveguide walls. The radiation from, and the excitation of, these corrugated waveguides give rise to W-H problems in which semi-infinite periodic structures are present. Modeling the corrugated surfaces with appropriate anisotropic impedance surfaces (Senior and Volakis, 1995) reduces these problems to standard W-H geometries. For instance, the junction of two corrugated circular waveguides or the radiation by a circular waveguide have been studied in Daniele, Montrosset, and Zich (1979, 1981). In general, periodic planar structures can be modeled by surface impedances. However, these impedances can be spatially dispersive. For example, at low frequency a wire mesh screen located in a plane (x, y) presents the following impedance:

$$Z_s = Z_a(\hat{x}\hat{x} + \hat{y}\hat{y}) + Z_b \nabla_t \nabla_t$$

where Z_a and Z_b are scalars that depend on the geometry of wire mesh, and ∇_t is the transverse del operator defined by

$$\nabla_t = \hat{x}\frac{\partial}{\partial x} + \hat{y}\frac{\partial}{\partial y}$$

In these cases the W-H technique can be applied successfully (Gilli & Daniele, 1995). It is interesting to observe that particular semi-infinite periodic structures have been studied in an exact way without resorting to the approximation of the impedance surface. In these cases, system of infinite W-H equations are involved (Hills & Karp, 1965).

10.14 Problems involving impedance surfaces

The approximate boundary conditions obtained with impedance surfaces (Senior & Volakis, 1995) introduce important problems that are very suitable to be formulated and solved by W-H technique. In this way the number of the electromagnetic canonical problems that we are able to solve has considerable increased. Many important works that assume surface impedances instead of PEC or PMC boundaries for the geometries considered in this chapter are reported in the references at the back of this book. Some problems are outlined in the following.

Problem (a): Scattering at the junction formed by a PEC half-plane and an half-plane with anisotropic conductivity (Sendag & Serbest, 2001).

Problem (b): Plane wave diffraction by two parallel overlapped thick semi-infinite impedance plates (Birbir & Büyükaksoy, 1999).

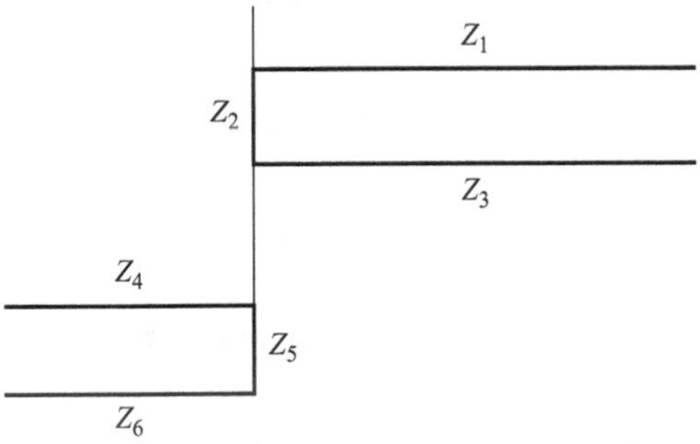

Problem (c): Plane wave diffraction by two oppositely placed, parallel two-part planes (Tayyar & Büyükaksoy, 2003).

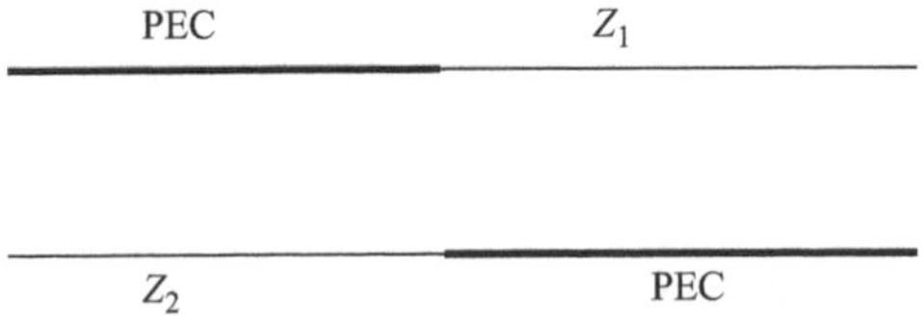

Problem (d): Diffraction of a normally incident plane wave by three parallel half-planes with different face impedances (Cinar & Büyükaksoy, 2004).

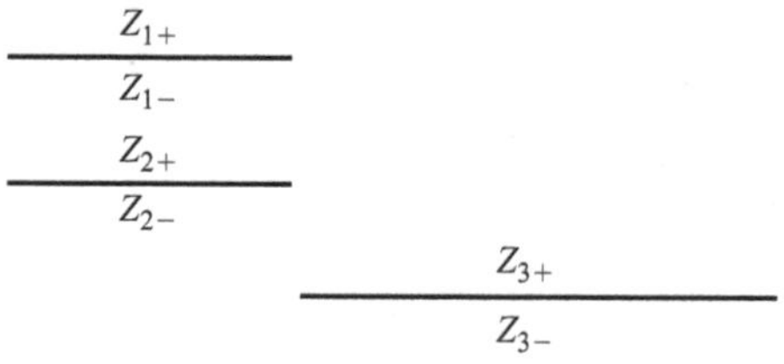

Problem (e): Scattering of electromagnetic waves by a rectangular impedance cylinder (Topsakal, Büyükaksoy & Idemen, 2000).

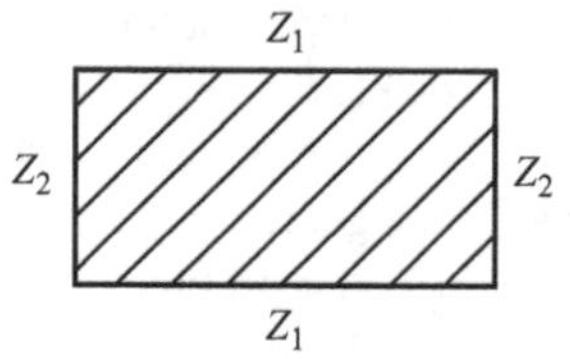

10.15 Some problems involving cones

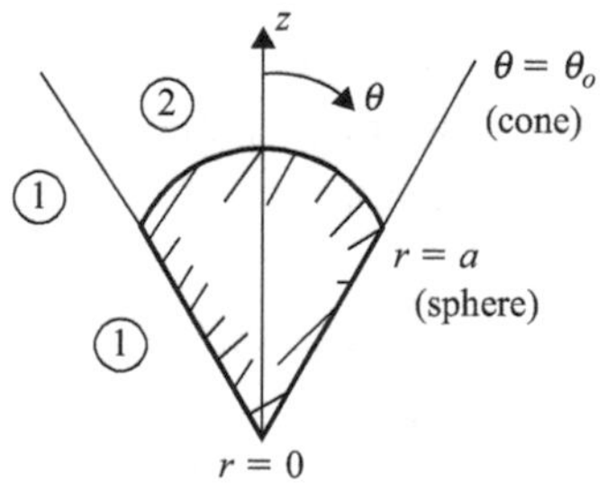

Fig. 19: The ice cone problem

Let us consider an axially symmetric $\left(\frac{\partial}{\partial\varphi} = 0\right)$ distribution of electrical charges located on the conducting truncated cone shown in Fig. 19.[3] The problem consists in finding an electrostatic potential $\psi(r, \theta)$ that satisfies the following conditions:

$$\begin{aligned} &\nabla^2\psi(r,\theta) = 0 \\ &\psi(r,\theta) = \psi_o, \text{ constant on truncated cone} \\ &\psi(r,\theta) \to 0 \quad \text{as } \frac{1}{r} \text{ when } r \to \infty, \text{ uniformly in } \theta \end{aligned}$$

[3] Note that the case $\theta_o = \pi/2$ yields a hemisphere.

We divide the space around the cone into two regions, separated by the boundary $\theta = \theta_o$:

$$\text{region 1: } \theta_o \leq \theta \leq \pi, \quad r \geq 0$$
$$\text{region 2: } 0 \leq \theta \leq \theta_o, \quad r \geq a$$

Consider Laplace's equation in these two regions using spherical coordinates:

$$\frac{\partial^2 \psi}{\partial r^2} + \frac{2}{r}\frac{\partial \psi}{\partial r} + \frac{1}{r^2 \sin\theta}\frac{\partial}{\partial \theta}\left(\sin\theta \frac{\partial \psi}{\partial \theta}\right) = 0$$

and adopt the change of variable

$$r = a\,e^x \begin{cases} 0 \leq r \leq a \to -\infty < x \leq 0 \\ a \leq r < \infty \to 0 \leq x \leq \infty \end{cases}$$

whence

$$\frac{\partial^2 g}{\partial x^2} + \frac{\partial g}{\partial x} + \frac{1}{\sin\theta}\frac{\partial}{\partial \theta}\left(\sin\theta \frac{\partial g}{\partial \theta}\right) = 0 \tag{32}$$

where

$$g(x,\theta) = \psi(r,\theta)$$
$$(g \to 0 \text{ as } e^{-x} \quad \text{when } x \to \infty)$$

Region 1

Introduce the Laplace transform

$$G(\alpha,\theta) = \int_{-\infty}^{\infty} g(x,\theta)e^{-px}dx = \int_{-\infty}^{\infty} \frac{\psi(r,\theta)}{r}\left(\frac{r}{a}\right)^{-p} dr, \quad p = -\frac{1}{2} - j\alpha$$

whence

$$\frac{1}{\sin\theta}\frac{d}{d\theta}\left(\sin\theta \frac{dG}{d\theta}\right) - \left(\alpha^2 + \frac{1}{4}\right)G = 0$$

The solution of this equation regular in $\theta = \pi$ is

$$G(\alpha,\theta) = A(\alpha)P_{j\alpha-1/2}(-\cos\theta) \tag{33}$$

where $P_\nu(u)$ is the Legendre function. At $\theta = \theta_o$

$$G(\alpha,\theta_o) = \int_{-\infty}^{0} \psi_o e^{-px}dx + \int_{0}^{\infty} g(x,\theta_o)e^{-px}dx = G_-^i(\alpha) + G_+(\alpha)$$

where $G_-^i(\alpha) = \frac{\psi_o}{j\alpha + 1/2}$. Substituting in $G(\alpha,\theta)$ yields

$$A(\alpha) = \frac{G_+(\alpha) + G_-^i(\alpha)}{P_{j\alpha-1/2}(-\cos\theta_o)} \tag{34}$$

Take $\left(\frac{\partial}{\partial\theta}\right)_{\theta=\theta_o}$ in (33):

$$M(\alpha) = M_-(\alpha) + M_+(\alpha) = A(\alpha)P'_{j\alpha-1/2}(-\cos\theta_o)\sin\theta_o$$

where

$$M_+(\alpha) = \int_0^\infty \frac{\partial g(x,\theta_o)}{\partial\theta_o} e^{-px}dx, \quad M_-(\alpha) = \int_{-\infty}^0 \frac{\partial g(x,\theta_o)}{\partial\theta_o} e^{-px}dx$$

Substituting (34) in the previous equation yields

$$M_-(\alpha) + M_+(\alpha) = \overleftarrow{P}(\alpha)\sin\theta_o[G_+(\alpha) + G^i_-(\alpha)] \tag{35}$$

where

$$\overleftarrow{P}(\alpha) = \frac{P'_{j\alpha-1/2}(-\cos\theta_o)}{P_{j\alpha-1/2}(-\cos\theta_o)}$$

Note that $\overleftarrow{P}(\alpha)$ is even because $P_\nu(u) = P_{-\nu-1}(u)$.

Region 2
Introduce the one-sided Laplace transform[4]

$$G_+(\alpha,\theta) = \int_0^\infty g(x,\theta)e^{-px}dx, \quad p = -\frac{1}{2} - j\alpha$$

From the Laplace transform properties we have

$$\int_0^\infty \frac{\partial g(x,\theta)}{\partial x}e^{-px}dx = pG_+(\alpha,\theta) - g(0,\theta) = pG_+(\alpha,\theta) - \psi_o$$

$$\int_0^\infty \frac{\partial^2 g(x,\theta)}{\partial x^2}e^{-px}dx = p^2G_+(\alpha,\theta) - p\,\psi_o - \left.\frac{\partial g}{\partial x}\right|_{x=0}$$

Applying the previous equation in $\frac{\partial^2 g}{\partial x^2} + \frac{\partial g}{\partial x} + \frac{1}{\sin\theta}\frac{\partial}{\partial\theta}\left(\sin\theta\frac{\partial g}{\partial\theta}\right) = 0$ yields

$$\frac{1}{\sin\theta}\frac{d}{d\theta}\left(\sin\theta\frac{dG_+}{d\theta}\right) - \left(\alpha^2 + \frac{1}{4}\right)G_+ = p\,\psi_o + \left.\frac{\partial g}{\partial x}\right|_{x=0}$$

Rewrite the previous equation with $\alpha \to -\alpha$, and then subtract the two equations:

$$\left[\frac{1}{\sin\theta}\frac{d}{d\theta}\left(\sin\theta\frac{d}{d\theta}\right) - \left(\alpha^2 + \frac{1}{4}\right)\right]\Delta G_+ = -j2\alpha\psi_o$$

where

$$\Delta G_+(\alpha,\theta) = G_+(\alpha,\theta) - G_+(-\alpha,\theta)$$

---

[4] Note that $G_+(\alpha,\theta_o) = G_+(\alpha)$.

The solution of this equation that is regular at $\theta = 0$ is[5]

$$\Delta G_+(\alpha, \theta) = B(\alpha)P_{j\alpha-1/2}(\cos\theta) + \frac{j2\alpha}{\alpha^2 + 1/4}\psi_o$$

It follows that

$$B(\alpha) = \frac{1}{P_{j\alpha-1/2}(-\cos\theta)}[\Delta G_+(\alpha) + \Delta G^i_-(\alpha)]$$

Taking the derivate of $\Delta G_+(\alpha, \theta) = B(\alpha)P_{j\alpha-1/2}(\cos\theta) + \frac{j2\alpha}{\alpha^2+1/4}\psi_o$, now we have

$$\begin{aligned}\Delta M_+(\alpha) &= \Delta\frac{\partial}{\partial\theta_o}G_+(\alpha, \theta) = \frac{\partial}{\partial\theta_o}\Delta G_+(\alpha, \theta) = B(\alpha)P'_{j\alpha-1/2}(\cos\theta_o)(-\sin\theta_o) \\ &= -\sin\theta_o \vec{P}(\alpha)[\Delta G_+(\alpha) + \Delta G^i_-(\alpha)] \qquad (36)\end{aligned}$$

where $\vec{P}(\alpha) = \frac{P'_{j\alpha-1/2}(\cos\theta_o)}{P_{j\alpha-1/2}(\cos\theta_o)} = \vec{P}(-\alpha)$.

Apply Δ to eq. (35), and then subtract from (36):

$$\Delta M_-(\alpha) = \sin\theta_o \overleftrightarrow{P}(\alpha)[\Delta G_+(\alpha) + \Delta G^i_-(\alpha)] \qquad (37)$$

where $\overleftrightarrow{P}(\alpha) = \vec{P}(\alpha) + \overleftarrow{P}(\alpha)$.

Setting $\alpha \to -\alpha$ in (35) and adding to (37) yields the transverse modified W-H equation

$$\overleftrightarrow{P}(\alpha)G_+(\alpha) - \vec{P}(\alpha)G_+(-\alpha) - X_-(\alpha) = \vec{P}(\alpha)G^i_-(-\alpha) - \overleftrightarrow{P}(\alpha)G^i_-(\alpha)$$

where $X_-(\alpha) = \frac{M_-(\alpha)+M_+(-\alpha)}{\sin\theta_o}$.

A similar treatment for the hollow cone of Fig. 20 yields the W-H equation

$$\overleftrightarrow{P}(\alpha)G_+(\alpha) - X_-(\alpha) = -\overleftrightarrow{P}(\alpha)G^i_-(\alpha)$$

The problem of Fig. 20 in static form has been considered by Karp (1950). The dynamic case requires the introduction of the Kontorovich-Lebedev transform (besides the related Fourier-Laplace-Mellin trio). However, the W-H technique may also be applied to this geometry (Noble, 1958, pp. 215–219).

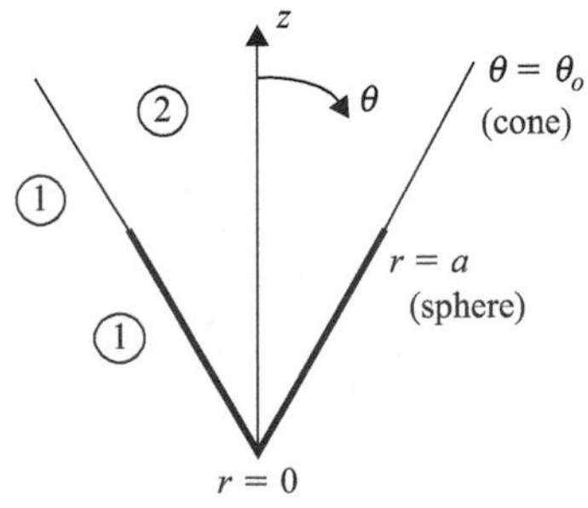

Fig. 20: The truncated hollow-cone problem

[5] Note that $\Delta G_+(\alpha, \theta_o) = \Delta G_+(\alpha)$, $\Delta G^i_-(\alpha) = -\frac{j2\alpha}{\alpha^2+1/4}\psi_o$.

10.16 Diffraction by a PEC wedge by an incident plane wave at skew incidence

We introduce the Laplace transforms *($e^{-j\alpha_o z}$ is omitted)*

$$V_{z+}(\eta,\varphi) = \int_0^\infty E_z(\rho,\varphi)e^{j\eta\rho}d\rho, \quad I_{z+}(\eta,\varphi) = \int_0^\infty H_z(\rho,\varphi)e^{j\eta\rho}d\rho$$
$$V_{\rho+}(\eta,\varphi) = \int_0^\infty E_\rho(\rho,\varphi)e^{j\eta\rho}d\rho, \quad I_{\rho+}(\eta,\phi) = \int_0^\infty H_\rho(\rho,\varphi)e^{j\eta\rho}d\rho \tag{38}$$

The following functional equations hold (Daniele, 2001, 2003):

$$\xi\; V_{z+}(\eta,0) - \frac{\tau_o^2}{\omega\,\varepsilon}I_{\rho+}(\eta,0) - \frac{\alpha_o\,\eta}{\omega\,\varepsilon}I_{z+}(\eta,0) = -n\,V_{z+}(-m,\Phi) - \frac{\tau_o^2}{\omega\,\varepsilon}I_{\rho+}(-m,\Phi) + \frac{\alpha_o\,m}{\omega\,\varepsilon}I_{z+}(-m,\Phi)$$
$$\xi\; I_{z+}(\eta,0) + \frac{\tau_o^2}{\omega\,\mu}V_{\rho+}(\eta,0) + \frac{\alpha_o\,\eta}{\omega\,\mu}V_{z+}(\eta,0) = -n\,I_{z+}(-m,\Phi) + \frac{\tau_o^2}{\omega\,\mu}V_{\rho+}(-m,\Phi) - \frac{\alpha\,m}{\omega\,\mu}V_{z+}(-m,\Phi)$$
$$\xi\; V_{z+}(\eta,0) + \frac{\tau_o^2}{\omega\,\varepsilon}I_{\rho+}(\eta,0) + \frac{\alpha_o\,\eta}{\omega\,\varepsilon}I_{z+}(\eta,0) = -n\,V_{z+}(-m,\,-\Phi) + \frac{\tau_o^2}{\omega\,\varepsilon}I_{\rho+}(-m,\,-\Phi) - \frac{\alpha_o\,m}{\omega\,\varepsilon}I_{z+}(-m,\,-\Phi)$$
$$\xi\; I_{z+}(\eta,0) - \frac{\tau_o^2}{\omega\,\mu}V_{\rho+}(\eta,0) - \frac{\alpha_o\,\eta}{\omega\,\mu}V_{z+}(\eta,0) = -n\,I_{z+}(-m,\,-\Phi) - \frac{\tau_o^2}{\omega\,\mu}V_{\rho+}(-m,\,-\Phi) + \frac{\alpha_o\,m}{\omega\,\mu}V_{z+}(-m,\,-\Phi)$$

where $\tau_o = k\sin\beta$, $\alpha_o = k\cos\beta$

$$\xi = \xi(\eta) = \sqrt{\tau_o^2 - \eta^2}, \quad m = m(\eta) = -\eta\cos\Phi + \xi\sin\Phi, \quad n = n(\eta)$$
$$= -\xi\cos\Phi - \eta\sin\Phi.$$

If the wedge is perfectly conducting we obtain the following GWHE:

$$\xi\; V_{z+}(\eta,0) - \frac{\tau_o^2}{\omega\varepsilon}I_{\rho+}(\eta,0) - \frac{\alpha_o\,\eta}{\omega\varepsilon}I_{z+}(\eta,0) = -\frac{\tau_o^2}{\omega\varepsilon}I_{\rho+}(-m,\Phi) + \frac{\alpha_o m}{\omega\,\varepsilon}I_{z+}(-m,\Phi)$$
$$\xi\; I_{z+}(\eta,0) + \frac{\tau_o^2}{\omega\mu}V_{\rho+}(\eta,0) + \frac{\alpha_o\,\eta}{\omega\mu}V_{z+}(\eta,0) = -n\,I_{z+}(-m,\Phi)$$
$$\xi\; V_{z+}(\eta,0) + \frac{\tau_o^2}{\omega\varepsilon}I_{\rho+}(\eta,0) + \frac{\alpha_o\,\eta}{\omega\varepsilon}I_{z+}(\eta,0) = +\frac{\tau_o^2}{\omega\varepsilon}I_{\rho+}(-m,\,-\Phi) - \frac{\alpha_o m}{\omega\varepsilon}I_{z+}(-m,\,-\Phi)$$
$$\xi\; I_{z+}(\eta,0) - \frac{\tau_o^2}{\omega\mu}V_{\rho+}(\eta,0) - \frac{\alpha_o\,\eta}{\omega\mu}V_{z+}(\eta,0) = -n\,I_{z+}(-m,\,-\Phi)$$

Summing the first and the third equations yields

$$2\xi\; V_{z+}(\eta,0) = -\frac{\tau_o^2}{\omega\varepsilon}[I_{\rho+}(-m,\Phi) - I_{\rho+}(-m,\,-\Phi)] + \frac{\alpha_o\,m}{\omega\,\varepsilon}[I_{z+}(-m,\Phi) - I_{z+}(-m,\,-\Phi)] \tag{39}$$

Subtracting the first equation from the third,

$$2\left[\frac{\tau_o^2}{\omega\varepsilon} I_{\rho+}(\eta, 0) + \frac{\alpha_o \eta}{\omega\varepsilon} I_{z+}(\eta, 0)\right] = \frac{\tau_o^2}{\omega\varepsilon}[I_{\rho+}(-m, \Phi) + I_{\rho+}(-m, -\Phi)]$$

$$-\frac{\alpha_o\, m}{\omega\varepsilon}[I_{z+}(-m, \Phi) + I_{z+}(-m, -\Phi)] \qquad (40)$$

Summing the second and the four equations yields

$$2\,\frac{\xi}{n}\; I_{z+}(\eta, 0) = -[I_{z+}(-m, \Phi) + I_{z+}(-m, -\Phi)] \qquad (41)$$

Subtracting the four equation from the second,

$$2\frac{1}{n}\left[\frac{\tau_o^2}{\omega\mu} V_{\rho+}(\eta, 0) + \frac{\alpha_o\, \eta}{\omega\mu} V_{z+}(\eta, 0)\right] = -[I_{z+}(-m, \Phi) - I_{z+}(-m, -\Phi)] \qquad (42)$$

Equations (39) through (42) are four decoupled GWHE. In the $\overline{\eta}$ – plane defined by

$$\eta = -\tau_o \cos\left[\frac{\Phi}{\pi}\arccos\left[-\frac{\overline{\eta}}{\tau_o}\right]\right]$$

these equations become classical W-H equations (chapter 1, section 1.5.4.1). Their solution is based on the classical factorization in the $\overline{\eta}$ – plane of

$$n = \sqrt{\tau_o^2 - m^2}, \quad \text{and} \quad \xi = \sqrt{\tau_o^2 - \eta^2}$$

For instance, the factorization of the scalar $\xi = \sqrt{\tau_o^2 - \eta^2} = \sqrt{\tau_o^2 - \left(-\tau_o \cos\left[\frac{\Phi}{\pi}\arccos\left[-\frac{\overline{\eta}}{\tau_o}\right]\right]\right)^2}$ in the $\overline{\eta}$ – plane was accomplished in the example of section 3.3.2. By introducing the w – plane defined by

$$\eta = -\tau_o \cos w, \quad \text{or} \quad \overline{\eta} = -\tau_o \cos\left(\frac{\pi}{\Phi} w\right)$$

we get

$$\xi_- = \sqrt{\frac{\tau_o + \overline{\eta}}{2}} = \sqrt{\tau_o}\,\sin\frac{\pi\, w}{2\,\Phi}, \quad \xi_+ = -\sqrt{\tau_o}\,\frac{\sin w}{\sin\frac{\pi w}{2\Phi}}$$

Similarly, we can obtain

$$n_+ = \sqrt{\frac{\tau_o - \overline{\eta}}{2}} = \sqrt{\tau_o}\,\cos\frac{\pi w}{2\Phi}, \quad n_- = \sqrt{\tau_o}\,\frac{\sin(w + \Phi)}{\cos\frac{\pi w}{2\Phi}}$$

The solution of the homogeneous equation requires the source, as discussed in section 2.4.2.

Let us consider (Fig. 5 of chapter 1)

$$\varphi_o < \Phi - \frac{\pi}{2}$$

In this case in the direction $\varphi = 0$ geometrical optic contains only the incident field. We get

$$\begin{aligned} V_{z+}(\eta,0) &= V^i_{z+}(\eta,0) + V^s_{z+}(\eta,0), \quad V_{\rho+}(\eta,0) = V^i_{\rho+}(\eta,0) + V^s_{\rho+}(\eta,0) \\ I_{z+}(\eta,0) &= I^i_{z+}(\eta,0) + I^s_{z+}(\eta,0), \quad I_{\rho+}(\eta,0) = I^i_{\rho+}(\eta,0) + I^s_{\rho+}(\eta,0) \end{aligned} \tag{43}$$

where the geometrical optics contribution is obtained by the Laplace transform of the incident plane wave:

$$V^i_{z+}(\eta,0) = \frac{jE_o}{\eta + \tau_o \cos\varphi_o}, \quad V^i_{\rho+}(\eta,0) = j\frac{\alpha_o \cos\varphi_o E_o + kZ_o \sin\varphi_o H_o}{\tau_o} \frac{1}{\eta + \tau_o \cos\varphi_o}$$

$$I^i_{z+}(\eta,0) = \frac{jH_o}{\eta + \tau_o \cos\varphi_o}, \quad I^i_{\rho+}(\eta,0) = j\frac{\alpha_o Z_o \cos\varphi_o H_o - k \sin\varphi_o E_o}{\tau_o Z_o} \frac{1}{\eta + \tau_o \cos\varphi_o}$$

Passing to the $\overline{\eta}$-plane, the pole $\eta = -\tau_o \cos\varphi_o$ becomes

$$\overline{\eta}_o = -\tau_o \cos\frac{\pi}{\Phi}\varphi_o$$

Taking into account that $\varphi_o < \Phi - \frac{\pi}{2} < \frac{\Phi}{2}$, this pole is nonconventional for the plus functions of the $\overline{\eta}$-plane. Therefore, provided that $\varphi_o < \Phi - \frac{\pi}{2}$, in the $\overline{\eta}$-plane the minus functions are conventional, and the nonconventional parts of the plus functions are expressed by

$$\overline{V}^i_{z+}(\overline{\eta},0) = \frac{jE_o}{\overline{\eta} - \overline{\eta}_o} \frac{d\overline{\eta}}{d\eta}\bigg|_{\overline{\eta}=\overline{\eta}_o} = \frac{jE_o}{\overline{\eta} - \overline{\eta}_o} \frac{\pi}{\Phi} \frac{\sin\frac{\pi}{\Phi}\varphi_o}{\sin\varphi_o}$$

$$\overline{V}^i_{\rho+}(\overline{\eta},0) = j\frac{\alpha_o \cos\varphi_o E_o + kZ_o \sin\varphi_o H_o}{\tau_o(\overline{\eta} - \overline{\eta}_o)} \frac{d\overline{\eta}}{d\eta}\bigg|_{\overline{\eta}=\overline{\eta}_o}$$

$$= j\frac{\alpha_o \cos\varphi_o E_o + kZ_o \sin\varphi_o H_o}{\tau_o(\overline{\eta} - \overline{\eta}_o)} \frac{\pi}{\Phi} \frac{\sin\frac{\pi}{\Phi}\varphi_o}{\sin\varphi_o}$$

$$\overline{I}^i_{z+}(\overline{\eta},0) = \frac{jH_o}{\overline{\eta} - \overline{\eta}_o} \frac{d\overline{\eta}}{d\eta}\bigg|_{\overline{\eta}=\overline{\eta}_o} = \frac{jH_o}{\overline{\eta} - \overline{\eta}_o} \frac{\pi}{\Phi} \frac{\sin\frac{\pi}{\Phi}\varphi_o}{\sin\varphi_o}$$

$$\overline{I}^i_{\rho+}(\overline{\eta},0) = j\frac{\alpha_o Z_o \cos\varphi_o H_o - k \sin\varphi_o E_o}{\tau_o Z_o(\overline{\eta} - \overline{\eta}_o)} \frac{d\overline{\eta}}{d\eta}\bigg|_{\overline{\eta}=\overline{\eta}_o}$$

$$= j\frac{\alpha_o Z_o \cos\varphi_o H_o - k \sin\varphi_o E_o}{\tau_o Z_o(\overline{\eta} - \overline{\eta}_o)} \frac{\pi}{\Phi} \frac{\sin\frac{\pi}{\Phi}\varphi_o}{\sin\varphi_o}$$

The W-H technique applied to the four uncoupled equations yields the solution

$$V_{z+}(-\tau_o \cos w, 0) = 2j\pi \frac{E_o \cos\left(\frac{\pi}{2\Phi}\varphi_o\right) \sin\left(\frac{\pi}{2\Phi} w\right)}{\tau_o \Phi \sin w\left[-\cos\left(\frac{\pi}{\Phi} w\right) + \cos\left(\frac{\pi}{\Phi}\varphi_o\right)\right]}$$

$$V_{\rho+}(-\tau_o \cos w, 0) = j2\pi \frac{E_o \cos\left(\frac{\pi}{2\Phi}\varphi_o\right) \cot w \cot\beta \sin\left(\frac{\pi}{2\Phi} w\right) + Z_o H_o \cos\left(\frac{\pi}{2\Phi} w\right) \csc\beta \sin\left(\frac{\pi}{2\Phi}\varphi_o\right)}{\tau_o \Phi\left[-\cos\left(\frac{\pi}{\Phi} w\right) + \cos\left(\frac{\pi}{\Phi}\varphi_o\right)\right]}$$

$$I_{z+}(-\tau_o \cos w, 0) = j\pi \frac{H_o \sin\left(\frac{\pi}{\Phi} w\right)}{\tau_o \Phi \sin w\left[-\cos\left(\frac{\pi}{\Phi} w\right) + \cos\left(\frac{\pi}{\Phi}\varphi_o\right)\right]}$$

$$I_{\rho+}(-\tau_o \cos w, 0) = j\pi \frac{-Z_o H_o \cot w \cot\beta \sin\left(\frac{\pi}{\Phi} w\right) + E_o \csc\beta \sin\left(\frac{\pi}{\Phi}\varphi_o\right)}{Z_o \tau_o \Phi\left[\cos\left(\frac{\pi}{\Phi} w\right) - \cos\left(\frac{\pi}{\Phi}\varphi_o\right)\right]}$$

The Sommerfeld functions are related to the W-H plus functions through the following equations (Daniele, 2003b):

$$s_E(w) = \frac{j}{2}\left[-\tau_o \sin w V_{z+}(-\tau_o \cos w, 0) + \frac{\tau_o^2}{\omega\varepsilon} I_{\rho+}(-\tau_o \cos w, 0) - \frac{\alpha_o \tau_o \cos w}{\omega\varepsilon} I_{z+}(-\tau_o \cos w, 0)\right]$$

$$s_H(w) = \frac{j}{2}\left[-\tau_o \sin w I_{z+}(-\tau_o \cos w, 0) - \frac{\tau_o^2}{\omega\mu} V_{\rho+}(-\tau_o \cos w, 0) + \frac{\alpha_o \tau_o \cos w}{\omega\mu} V_{z+}(-\tau_o \cos w, 0)\right]$$

Substituting we get

$$s_E(w) = \pi \frac{E_o \cos\left(\frac{\pi}{2\Phi}\varphi_o\right)}{2\Phi\left[\sin\left(\frac{\pi}{2\Phi} w\right) - \sin\left(\frac{\pi}{2\Phi}\varphi_o\right)\right]}$$

$$s_H(w) = \pi \frac{H_o \cos\left(\frac{\pi}{2\Phi} w\right)}{2\Phi\left[\sin\left(\frac{\pi}{2\Phi} w\right) - \sin\left(\frac{\pi}{2\Phi}\varphi_o\right)\right]}$$

Hence, the following representation of the longitudinal components valid for every value of φ follows:

$$E_z(\rho, \varphi) = \frac{1}{2\pi j}\left[\int_\gamma s_E[w + \varphi] e^{+j\tau_o \cos[w]\rho} dw\right]$$

$$H_z(\rho, \varphi) = \frac{1}{2\pi j}\left[\int_\gamma s_H[w + \varphi] e^{+j\tau_o \cos[w]\rho} dw\right]$$

where $\gamma = C_1 \cup C_2$ is the Sommerfeld contour (see chapter 2, Fig. 4).

Use of the saddle point method on the previous equations (Senior & Volakis, 1995) yields the far field evaluation

$$E_z(\rho, \varphi) = E_z^g(\rho, \varphi) + E_z^d(\rho, \varphi)$$
$$H_z(\rho, \varphi) = H_z^g(\rho, \varphi) + H_z^d(\rho, \varphi)$$

where E_z^g, H_z^g represent the geometrical optics contribution, and E_z^d, H_z^d is the diffracted field. We have that

$$E_z^g = e^{-j\alpha_o z}[E_o u(\pi - |\varphi - \varphi_o|)e^{j\tau_o \rho \cos(\varphi - \varphi_o)} - u(\pi - |\varphi + \varphi_o - 2\Phi|)e^{j\tau_o \rho \cos(\varphi + \varphi_o - 2\Phi)}$$
$$+ -u(\pi - |\varphi + \varphi_o + 2\Phi|)e^{j\tau_o \rho \cos(\varphi + \varphi_o + 2\Phi)}]$$

$$H_z^g = e^{-j\alpha_o z}[H_o u(\pi - |\varphi - \varphi_o|)e^{j\tau_o \rho \cos(\varphi - \varphi_o)} + u(\pi - |\varphi + \varphi_o - 2\Phi|)e^{j\tau_o \rho \cos(\varphi + \varphi_o - 2\Phi)}$$
$$+ u(\pi - |\varphi + \varphi_o + 2\Phi|)e^{j\tau_o \rho \cos(\varphi + \varphi_o + 2\Phi)}] \tag{44}$$

where $u(x)$ is the unit step function. The diffracted fields arise from the saddle points at $w = \pm\pi$ and have the form

$$E_z^d(\rho, \varphi, z) = e^{-j\alpha_o z}\frac{e^{-j\left(\tau_o \rho + \frac{\pi}{4}\right)}}{\sqrt{2\pi\,\tau_o\,\rho}}[s_E(\varphi - \pi) - s_E(\varphi + \pi)]$$
$$H_z^d(\rho, \varphi, z) = e^{-j\alpha_o z}\frac{e^{-j\left(\tau_o \rho + \frac{\pi}{4}\right)}}{\sqrt{2\pi\,\tau_o\,\rho}}[s_H(\varphi - \pi) - s_H(\varphi + \pi)] \tag{45}$$

A diffracted ray constitutes a generatrix of Keller's cone and is defined by the angular spherical coordinates β and φ. Taking into account that the incident field has angular spherical coordinates β and φ_o, it is convenient to relate the transversal component E_β^d, E_φ^d of the diffracted ray to the transversal component E_β^i, $E_{\varphi_o}^i$ of the incident ray. By geometrical consideration we have

$$E_\beta^d = -\frac{1}{\sin\beta}E_z^d, \quad E_\varphi^d = -\frac{1}{\sin\beta}Z_o H_z^d$$
$$E_z^i = \sin\beta\, E_\beta^i, \quad H_z^i = \frac{1}{Z_o}\sin\beta\, E_{\varphi_o}^i \tag{46}$$

Equations (46) cannot be used when the observation point approaches the shadow boundaries of the incident and the reflected waves. Uniform expressions are reported in Daniele and Lombardi (2006).

10.17 Diffraction by a right PEC wedge immersed in a stratified medium

For the sake of simplicity we consider only the case of E-polarization:

$$\mathbf{E} = \hat{z}\,E_z, \quad \left(\frac{\partial}{\partial z} = 0\right)$$

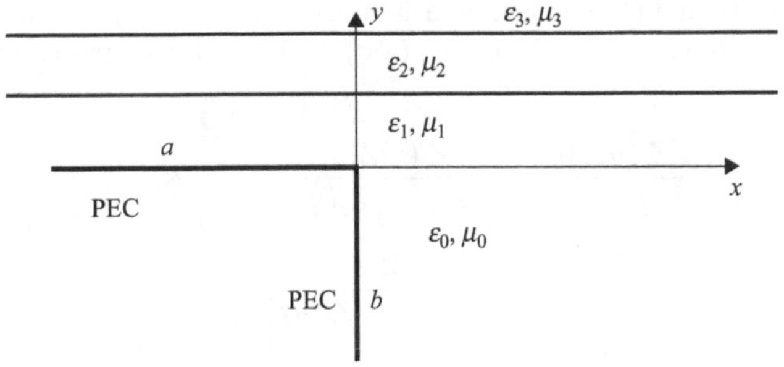

Fig. 21: Right wedge immersed in a stratified medium

By modeling the space $y > 0$ with circuits as indicated in chapter 7 and taking into account that on the face aE_z is zero, we obtain (Fig. 21)

$$V_+(\eta) = Z(\eta)[I_+(\eta) + I_{a-}(\eta)] \tag{47}$$

where

$$V_+(\eta) = \int_0^\infty E_z(x,0)e^{j\eta x}dx, \quad I_+(\eta) = \int_0^\infty H_x(x,0)e^{j\eta x}dx, \quad I_{a-}(\eta) = \int_{-\infty}^0 H_x(x,0)e^{j\eta x}dx,$$

and $Z(\eta)$ is the impedance of the stratified medium. For instance, if we have free space for $y > 0$

$$Z(\eta) = Z_c(\eta) = \frac{\omega\mu_o}{\sqrt{k_o^2 - \eta^2}} \tag{48}$$

Taking into account that on the face $b\,E_z$ is zero and using the equations that relate the Laplace transforms on the boundary of an angular region (Daniele, 2003b, 2004b) we get

$$V_+(\eta) + Z_c(\eta)I_+(\eta) = Z_c(\eta)I_{b+}\left(-\sqrt{k_o^2 - \eta^2}\right) \tag{49}$$

where $Z_c(\eta) = \frac{\omega\mu_o}{\sqrt{k_o^2-\eta^2}}$ and, using polar coordinates, $I_{b+}(\eta) = \int_0^\infty H_\rho(\rho, -\frac{\pi}{2})e^{j\eta\rho}d\rho$.

We can eliminate $I_{b+}(-\sqrt{k_o^2 - \eta^2})$ by letting $\eta \to -\eta$ in eq. (49). We get

$$V_+(\eta) + Z_c(\eta)I_+(\eta) = V_+(-\eta) + Z_c(\eta)I_+(-\eta) \tag{50}$$

Setting $\eta \to -\eta$ in eq. (47) we get

$$V_+(-\eta) = Z(-\eta)[I_+(-\eta) + I_{a-}(-\eta)] \tag{51}$$

Equations (47), (50), and (51) constitute a homogeneous W-H system having as unknown the three plus functions $V_+(\eta)$, $I_+(\eta)$, $I_{a-}(-\eta)$ and the three minus functions $V_+(-\eta)$, $I_+(-\eta)$, $I_{a-}(\eta)$:

$$\begin{vmatrix} \dfrac{Z(\eta)}{Z(\eta)+Z_c(\eta)} & \dfrac{Z_c(\eta)Z(\eta)}{Z(\eta)+Z_c(\eta)} & \dfrac{Z_c(\eta)Z(\eta)}{Z(\eta)+Z_c(\eta)} \\ \dfrac{1}{Z(\eta)+Z_c(\eta)} & \dfrac{Z_c(\eta)}{Z(\eta)+Z_c(\eta)} & -\dfrac{Z(\eta)}{Z(\eta)+Z_c(\eta)} \\ \dfrac{1}{Z(\eta)} & -1 & 0 \end{vmatrix} \cdot \begin{vmatrix} V_+(\eta) \\ I_+(\eta) \\ I_{a-}(-\eta) \end{vmatrix} = \begin{vmatrix} V_+(-\eta) \\ I_+(-\eta) \\ I_{a-}(\eta) \end{vmatrix} \tag{52}$$

This formula is consistent, since the matrix kernel $g_e(\eta)$ satisfies

$$g_e(\eta) = \begin{vmatrix} \dfrac{Z(\eta)}{Z(\eta)+Z_c(\eta)} & \dfrac{Z_c(\eta)Z(\eta)}{Z(\eta)+Z_c(\eta)} & \dfrac{Z_c(\eta)Z(\eta)}{Z(\eta)+Z_c(\eta)} \\ \dfrac{1}{Z(\eta)+Z_c(\eta)} & \dfrac{Z_c(\eta)}{Z(\eta)+Z_c(\eta)} & -\dfrac{Z(\eta)}{Z(\eta)+Z_c(\eta)} \\ \dfrac{1}{Z(\eta)} & -1 & 0 \end{vmatrix}$$

$$g_e(-\eta) \cdot g_e(\eta) = \mathbf{1}$$

The difficulty in factorizing the matrix $g_e(\eta)$ exactly depends on the impedance $Z(\eta)$.

For instance, if the wedge is located in free space, eq. (48) holds and $g_e(\eta)$ assumes the form

$$g_e(\eta) = \begin{vmatrix} \dfrac{1}{2} & 0 & 0 \\ 0 & \dfrac{1}{2} & -\dfrac{1}{2} \\ 0 & -1 & 0 \end{vmatrix} [\mathbf{1} + Z_c(\eta)\mathbf{P}(\eta)]$$

where the polynomial matrix $\mathbf{P}(\eta)$ is given by

$$\mathbf{P}(\eta) = \begin{vmatrix} 0 & 1 & 1 \\ -\dfrac{1}{Z_c^2(\eta)} & 0 & 0 \\ -\dfrac{2}{Z_c^2(\eta)} & 0 & 0 \end{vmatrix} = \begin{vmatrix} 0 & 1 & 1 \\ -\dfrac{k_o^2-\eta^2}{(\omega\,\mu)^2} & 0 & 0 \\ -2\dfrac{k_o^2-\eta^2}{(\omega\,\mu)^2} & 0 & 0 \end{vmatrix}$$

The matrix $m(\eta) = \mathbf{1} + Z_c(\eta)\mathbf{P}(\eta)$ commutes with $\mathbf{P}(\eta)$ and can be factorized by using the methods of chapter 4, section 4.9. For example,

$$\log[m(\eta)] = \psi_o(\eta)\mathbf{1} + \psi_1(\eta)P(\eta) + \psi_2(\eta)P^2(\eta)$$

Since

$$P^2(\eta) = \frac{1}{Z_c^2}\begin{vmatrix} -3 & 0 & 0 \\ 0 & -1 & -1 \\ 0 & -2 & -2 \end{vmatrix}$$

taking into account the $P^2(\eta)$ commutes with $P(\eta)$ we get

$$m_\pm(\eta) = \exp[\psi_{o\pm}\mathbf{1}]\cdot\exp[\psi_{1\pm}P(\eta)]\cdot\exp\left[\left\{\frac{\psi_2}{Z_c^2}\right\}_\pm \begin{vmatrix} -3 & 0 & 0 \\ 0 & -1 & -1 \\ 0 & -2 & -2 \end{vmatrix}\right]$$

Alternatively, since the eigenvalues of $m(\eta)$ are constants, we could reduce the order of the matrix. In effect, we have

$$m(\eta) = \begin{vmatrix} 1 & Z_c(\eta) & Z_c(\eta) \\ -\frac{1}{Z_c(\eta)} & 1 & 0 \\ -\frac{2}{Z_c(\eta)} & 0 & 1 \end{vmatrix}$$

In scalar form the W-H equations are

$$\begin{aligned} F_{1+} + Z_c(F_{2+} + F_{3+}) &= F_{1-} \\ -\frac{1}{Z_c(\eta)}F_{1+} + F_{2+} &= F_{2-} \\ -\frac{2}{Z_c(\eta)}F_{1+} + F_{3+} &= F_{3-} \end{aligned}$$

Summing the second and third equations yields the second-order system

$$\begin{aligned} F_{1+} + Z_c(F_{2+} + F_{3+}) &= F_{1-} \\ -\frac{3}{Z_c(\eta)}F_{1+} + (F_{2+} + F_{3+}) &= F_{2-} + F_{3-} \end{aligned}$$

that requires the factorization of the matrix of order two:

$$\begin{vmatrix} 1 & Z_c \\ -\frac{3}{Z_c} & 1 \end{vmatrix}$$

This matrix has the Daniele form $\begin{vmatrix} 1 & a \\ b & 1 \end{vmatrix}$, where $b/a = -3\frac{k_o^2-\eta^2}{(\omega\mu)^2}$.

10.18 Diffraction by a right isorefractive wedge

Figure 22 shows the geometry that we will study. Region 2 is constituted by a right wedge with permettivity ε_2 and permeability μ_2 that is isorefractive (or diaphanous) to the

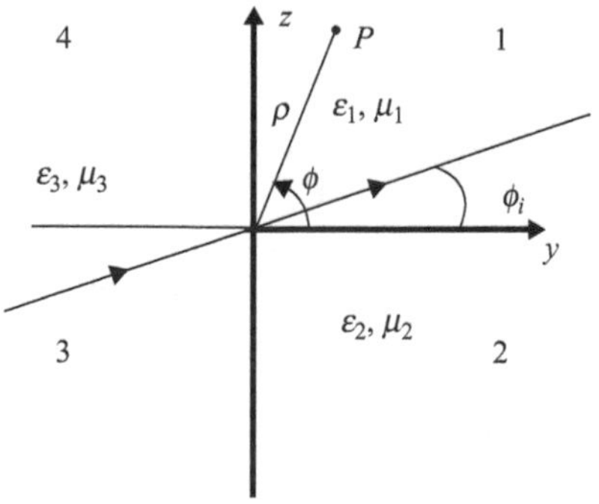

Fig. 22: Geometry of the problem

remaining media characterized by the permettivity ε_1 and the permeability μ_1 in region 2 and the permettivity ε_3 and the permeability μ_3 in regions 3 and 4. The isorefractivity of all the media implies that $\varepsilon_1\mu_1 = \varepsilon_2\mu_2 = \varepsilon_3\mu_3$. The problem of diffraction by an isorefractive wedge is constituted by a wedge (say, region 2) immersed in a homogeneous isorefractive medium filling the space complementary to 2. Of course this problem is a particular case ($\varepsilon_1 = \varepsilon_3$ and $\mu_1 = \mu_3$) of the problem shown in Fig. 22. The source of the electromagnetic field is an E-polarized plane wave with direction $\varphi = \varphi_i$ as indicated in figure. The problem is two-dimensional since there are not field variations in the x-direction.

By indicating with E_x the only nonvanishing component of the electric field, we must solve the following wave equation in every region of space:

$$(\nabla^2 + k^2)E_x = 0 \tag{53}$$

where k is the propagation constant ($k^2 = \omega^2\varepsilon_1\mu_1 = \omega^2\varepsilon_2\mu_2 = \omega^2\varepsilon_3\mu_3$), and ω is the angular frequency.

In region 1 defined by $y \geq 0,\ z \geq 0$, we introduce the Laplace transform (or unilateral Fourier transform) in the z-direction:

$$\tilde{E}_x(y,\alpha) = \int_0^\infty E_x(y,z)e^{j\alpha z}dz = \mathbf{F}_+[E_x]$$

To avoid singularities on the real axis of α, we assume that the imaginary part of k is negative and vanishingly small. From the theory of the Laplace transform we have

$$\mathbf{F}_+\left[\frac{\partial}{\partial z}E_x\right] = -j\alpha\,\tilde{E}_x(y,\alpha) - E_x(y,0_+)$$

$$\mathbf{F}_+\left[\frac{\partial^2}{\partial z^2}E_x\right] = -\alpha^2\,\tilde{E}_x(y,\alpha) + j\alpha\,E_x(y,0_+) - \frac{\partial}{\partial z}E_x(y,0_+)$$

and by substituting in eq. (53)

$$\frac{\partial^2}{\partial y^2}\tilde{E}_x + \tau^2\tilde{E}_x = -j\alpha\,E_x(y,0_+) + \frac{\partial}{\partial z}E_x(y,0_+) = f_{1\alpha} \tag{54}$$

where

$$\tau = \sqrt{k^2 - \alpha^2} \tag{55}$$

and the branches are chosen such that $\mathrm{Im}[\tau] \leq 0$.

A general solution of (54) is given by

$$\tilde{E}_x(y,\alpha) = \int_0^\infty g(y,y')f_{1\alpha}(y')dy' + A(\alpha)e^{-j\tau y} \tag{56}$$

where

$$g(y,y') = -\frac{e^{-j\tau|y-y'|}}{2j\tau} \tag{57}$$

and $A(\alpha)$ is for the time being an arbitrary function of α.

The Laplace transform $\tilde{H}_z(y,\alpha) = F_+[H_z]$ of H_z can be obtained by the Maxwell equation

$$\begin{aligned}\tilde{H}_z(y,\alpha) &= \frac{1}{j\omega\mu_1}\frac{\partial}{\partial y}\tilde{E}_x(y,\alpha)\\ &= \frac{1}{j\omega\mu_1}\left(\int_0^\infty \frac{e^{-j\tau|y-y'|}}{2} Sign(y-y')f_{1\alpha}(y')dy' - j\tau A(\alpha)e^{-j\tau y}\right)\end{aligned} \tag{58}$$

and by putting

$$y = 0_+, \quad \tilde{H}_z(0_+,\alpha) = -I_+(\alpha), \quad \tilde{E}_x(0_+,\alpha) = V_+(\alpha) \tag{58}$$

from (56) through (58) we obtain the functional equation

$$j\omega\mu_1 I_+(\alpha) - j\tau\, V_+(\alpha) = -j\alpha f_1(\alpha) - j\omega\mu_1 g_1(\alpha) \tag{59}$$

$$A(\alpha) = \frac{\omega\mu_1 I_+(\alpha) + \tau\, V_+(\alpha)}{2\tau} \tag{60}$$

where

$$f_1(\alpha) = \int_0^\infty e^{-j\tau y'}E_x(y',0_+)dy', \quad g_1(\alpha) = \int_0^\infty e^{-j\tau_1 y'}H_y(y',0_+)dy' \tag{61}$$

From eqs. (61) we observe that $f_1(\alpha)$ and $g_1(\alpha)$ are even function of α. By setting $\alpha \to -\alpha$ in (59), we can eliminate these two functions and obtain

$$\begin{aligned}f_1(\alpha) &= -\frac{-\omega\mu_1 I_+(-\alpha) + \omega\mu_1 I_+(\alpha) + \tau V_+(-\alpha) - \tau V_+(\alpha)}{2\alpha}\\ g_1(\alpha) &= -\frac{\omega\mu_1 I_+(-\alpha) + \omega\mu_1 I_+(\alpha) - \tau V_+(-\alpha) - \tau V_+(\alpha)}{2\omega\mu_1}\end{aligned} \tag{62}$$

By repeating the same reasoning in region 2 ($y \geq 0,\ z \leq 0$), we get

$$f_2(\alpha) = \frac{-\omega\mu_2 I_-(-\alpha) + \omega\mu_2 I_-(\alpha) + \tau V_-(-\alpha) - \tau V_-(\alpha)}{2\alpha}$$
$$g_2(\alpha) = \frac{\omega\mu_2 I_-(-\alpha) + \omega\mu_2 I_-(\alpha) - \tau V_-(-\alpha) - \tau V_-(\alpha)}{2\omega\mu_2} \tag{63}$$

which give

$$V_-(\alpha) = \int_{-\infty}^{0} e^{j\alpha z} E_x(0,z)dz$$
$$I_-(\alpha) = -\int_{-\infty}^{0} e^{j\alpha z} H_z(0,z)dz \tag{64}$$
$$f_2(\alpha) = \int_0^{\infty} e^{-j\tau y'} E_x(y',0_-)dy', \quad g_2(\alpha) = \int_0^{\infty} e^{-j\tau y'} H_y(y',0_-)dy'$$

It is important to observe that the continuity of the tangential electric and magnetic fields on the half-plane ($y \geq 0,\ z = 0$) implies that $f_1(\alpha) = f_2(\alpha)$ and $g_1(\alpha) = g_2(\alpha)$. This allows us to eliminate these functions from eqs. (63) and (62) and to obtain the following two functional equations:

$$+\omega\mu_1 I_+(-\alpha) - \omega\mu_1 I_+(\alpha) - \tau V_+(-\alpha) + \tau V_+(\alpha)$$
$$= -\omega\mu_2 I_-(-\alpha) + \omega\mu_2 I_-(\alpha) + \tau V_-(-\alpha) - \tau V_-(\alpha) \tag{65}$$

$$-\frac{\omega\mu_1 I_+(-\alpha) + \omega\mu_1 I_+(\alpha) - \tau V_+(-\alpha) - \tau V_+(\alpha)}{\omega\mu_1}$$
$$= \frac{\omega\mu_2 I_-(-\alpha) + \omega\mu_2 I_-(\alpha) - \tau V_-(-\alpha) - \tau V_-(\alpha)}{\omega\mu_2} \tag{66}$$

We can apply the same procedure to regions 3 ($y \leq 0,\ z \leq 0$) and 4 ($y \geq 0,\ z \geq 0$). Taking into account that these regions are homogeneous, we get

$$+\omega\mu_3 I_+(-\alpha) - \omega\mu_3 I_+(\alpha) + \tau V_+(-\alpha) - \tau V_+(\alpha)$$
$$= -\omega\mu_3 I_-(-\alpha) + \omega\mu_3 I_-(\alpha) - \tau V_-(-\alpha) + \tau V_-(\alpha) \tag{67}$$

$$-\omega\mu_3 I_+(-\alpha) - \omega\mu_3 I_+(\alpha) - \tau V_+(-\alpha) - \tau V_+(\alpha)$$
$$= \omega\mu_3 I_-(-\alpha) + \omega\mu_3 I_-(\alpha) + \tau V_-(-\alpha) + \tau V_-(\alpha) \tag{68}$$

Note that $V_+(\alpha)$ and $I_+(\alpha)$, $V_-(-\alpha)$ and $I_-(-\alpha)$ are plus functions. Conversely, $V_+(-\alpha)$ and $I_+(-\alpha)$, $V_-(\alpha)$ and $I_-(\alpha)$ are minus functions.

Equations (65) through (68) constitute a system of four Wiener-Hopf equations in the four plus functions $V_+(\alpha)$, $I_+(\alpha)$, $V_-(-\alpha)$ and $I_-(-\alpha)$ and in the four minus functions $V_+(-\alpha)$, $I_+(-\alpha)$, $V_-(\alpha)$ and $I_-(\alpha)$. It is convenient to put this system in the matrix form

$$g_e(\alpha)x_+(\alpha) + x_-(\alpha) = 0 \tag{69}$$

where

$$g_e(\alpha) = \begin{vmatrix} -\frac{(\mu_1-\mu_2)\mu_3}{(\mu_1+\mu_2)(\mu_1+\mu_3)} & \frac{\omega\mu_1(\mu_1-\mu_2)\mu_3}{\tau(\mu_1+\mu_2)(\mu_1+\mu_3)} & \frac{\mu_1(\mu_1+\mu_2+2\mu_3)}{(\mu_1+\mu_2)(\mu_1+\mu_3)} & \frac{\omega\mu_1(\mu_1-\mu_2)\mu_3}{\tau(\mu_1+\mu_2)(\mu_1+\mu_3)} \\ \frac{\tau(\mu_1-\mu_2)}{\omega(\mu_1+\mu_2)(\mu_1+\mu_3)} & -\frac{(\mu_1-\mu_2)\mu_1}{(\mu_1+\mu_2)(\mu_1+\mu_3)} & \frac{\tau(-\mu_1+\mu_2)}{\omega(\mu_1+\mu_2)(\mu_1+\mu_3)} & 1-\frac{(\mu_1-\mu_2)\mu_1}{(\mu_1+\mu_2)(\mu_1+\mu_3)} \\ \frac{\mu_2(\mu_1+\mu_2+2\mu_3)}{(\mu_1+\mu_2)(\mu_2+\mu_3)} & -\frac{\omega\mu_2(\mu_1-\mu_2)\mu_3}{\tau(\mu_1+\mu_2)(\mu_2+\mu_3)} & \frac{(\mu_1-\mu_2)\mu_3}{(\mu_1+\mu_2)(\mu_2+\mu_3)} & -\frac{\omega\mu_2(\mu_1-\mu_2)\mu_3}{\tau(\mu_1+\mu_2)(\mu_2+\mu_3)} \\ \frac{\tau(\mu_1-\mu_2)}{\omega(\mu_1+\mu_2)(\mu_2+\mu_3)} & 1+\frac{(\mu_1-\mu_2)\mu_2}{(\mu_1+\mu_2)(\mu_2+\mu_3)} & -\frac{\tau(\mu_1-\mu_2)}{\omega(\mu_1+\mu_2)(\mu_2+\mu_3)} & \frac{(\mu_1-\mu_2)\mu_2}{(\mu_1+\mu_2)(\mu_2+\mu_3)} \end{vmatrix} \tag{70}$$

$$x_+(\alpha) = \begin{vmatrix} V_+(\alpha) \\ I_+(\alpha) \\ V_-(-\alpha) \\ I_-(-\alpha) \end{vmatrix}, \quad x_-(\alpha) = \begin{vmatrix} V_+(-\alpha) \\ I_+(-\alpha) \\ V_-(\alpha) \\ I_-(\alpha) \end{vmatrix} = x_+(-\alpha) \tag{71}$$

Note that $g_e(\alpha)$ is an even function of α and has the property

$$g_e(\alpha) \cdot g_e(\alpha) = \mathbf{1} \tag{72}$$

Next we take into account the presence of the source: an incident E-polarized plane wave defined by (Fig. 22)

$$E_x^i(y,z) = E_0 e^{-jk\sin\varphi_i\, z} e^{-jk\cos\varphi_i\, y} \tag{73}$$

and write

$$x_-(\alpha) = x_-^s(\alpha) + x_-^g(\alpha) \tag{74}$$

where $x_-^s(\alpha)$ and $x_-^g(\alpha)$ represent the scattered and the geometrical optics fields, respectively.

The geometrical optic field $x_-^g(\alpha)$ being dominating for $z \to \pm\infty$, can be evaluated directly by considering simple problems of reflection by infinite planes. The solution of these problems leads to

$$x_-^g(\alpha) = -2j \begin{vmatrix} \frac{E_0\mu_1}{(\alpha+\alpha_i)(\mu_1+\mu_3)} \\ \frac{H_0\mu_3}{(\alpha+\alpha_i)(\mu_1+\mu_3)} \\ \frac{E_0\mu_2}{(\alpha-\alpha_i)(\mu_2+\mu_3)} \\ \frac{H_0\mu_3}{(\alpha-\alpha_i)(\mu_2+\mu_3)} \end{vmatrix} \tag{75}$$

where

$$\alpha_i = k\sin\varphi_i, \quad H_o = E_o k \frac{\cos\varphi_i}{\omega\mu_3} \tag{76}$$

Thus, the W-H eq. (69) assumes the inhomogeneous form

$$g_e(\alpha)x_+(\alpha) + x_-^s(\alpha) = -x_-^g(\alpha) \tag{77}$$

10.18.1 Solution of the W-H equations

The standard method for solving the W-H system (77) is based on the matrix factorization of $g_e(\alpha)$:

$$g_e(\alpha) = g_{e-}(\alpha)g_{e+}(\alpha) \tag{78}$$

The factorized matrices $g_{e-}(\alpha)$ and $g_{e+}(\alpha)$ (with their inverses) must be regular in the half-planes $\mathrm{Im}[\alpha] \leq 0$ and $\mathrm{Im}[\alpha] \geq 0$, respectively, and present an algebraic behavior for $\alpha \to \infty$. Taking into account that $g_e(\alpha)$ contains only the singular function $\tau = \sqrt{k^2 - \alpha^2}$, the singularities of $g_{e-}(\alpha)$ and $g_{e+}(\alpha)$ (and their inverses) are a branch point at $\alpha = -k$ and $\alpha = k$, respectively.

The process of factorization of the matrix $g_e(\alpha)$ and the subsequent elaborations are very cumbersome and require a computer manipulator. For this purpose we used MATHEMATICA.

After the matrix factorization is obtained, we apply the W-H technique to (77), leading to

$$g_{e+}(\alpha)x_+(\alpha) - K(\alpha) = -(g_{e-}(\alpha))^{-1}x_-^s(\alpha) - (g_{e-}(\alpha))^{-1}x_-^g(\alpha) - K(\alpha) = w(\alpha) \tag{79}$$

where $K(\alpha)$ is the characteristic part of the known term $-(g_{e-}(\alpha))^{-1}x_-^g(\alpha)$. From (69), this function is also the characteristic part of $(g_{e+}(\alpha))x_+(\alpha)$, and it has only first order poles at $\alpha = \pm\alpha_i$ that correspond to the geometrical optics field. $K(\alpha)$ is given by

$$K(\alpha) = \begin{vmatrix} \dfrac{A_1}{\alpha - \alpha_i} + \dfrac{B_1}{\alpha + \alpha_i} \\ \dfrac{A_2}{\alpha - \alpha_i} + \dfrac{B_2}{\alpha + \alpha_i} \\ \dfrac{A_3}{\alpha - \alpha_i} + \dfrac{B_3}{\alpha + \alpha_i} \\ \dfrac{A_4}{\alpha - \alpha_i} + \dfrac{B_4}{\alpha + \alpha_i} \end{vmatrix} \tag{80}$$

where A_i, B_i ($i = 1, 2, 3, 4$) are constants that depend only on the permeabilities μ_1, μ_2, and μ_3. We calculated them with MATHEMATICA. For the sake of brevity we report only the expression of A_1:

$$A_1 = -2jE_o\left[\frac{2jn_1\sqrt{(n_1 - n_2)(p_1 + p_2)}\sin\left[c(\frac{\pi}{2} - \varphi_1)\right](\mu_1 + \mu_2)}{(n_1 - n_2)(p_1 + p_2)\lambda\sqrt{1 - \lambda^2}(\mu_1^2 + 6\mu_1\mu_2 + \mu_2^2)}\right.$$

$$+ \frac{2n_1(p_1 + p_2)\lambda\cos\left[c(\frac{\pi}{2} - \varphi_1)\right]\mu_2(\mu_1 + \mu_2)}{(n_1 - n_2)(p_1 + p_2)\lambda\sqrt{1 - \lambda^2}(\mu_1^2 + 6\mu_1\mu_2 + \mu_2^2)}$$

$$\left. + + \frac{(p_1 + p_2)\lambda\sqrt{1 - \lambda^2}\mu_2[(-n_1 + 3n_2)\mu_1 + (n_1 + n_2)\mu_2])}{(n_1 - n_2)(p_1 + p_2)\lambda\sqrt{1 - \lambda^2}(\mu_1^2 + 6\mu_1\mu_2 + \mu_2^2)}\right]$$

where

$$c = \frac{\sqrt{n_2 - n_1}\sqrt{p_2 + p_1}}{\pi}$$

$$\lambda = \sqrt{-\frac{(\mu_1 - \mu_2)^2 \mu_3 (\mu_1 + \mu_2 + 2\mu_3)(2\mu_1\mu_2 + \mu_3(\mu_1 + \mu_2)}{(\mu_1\mu_2(\mu_1 + \mu_2) + 4\mu_1\mu_2\mu_3 + \mu_3{}^2(\mu_1 + \mu_2))^2}}$$

$$n_1 = \frac{(\mathrm{Log}[1 - \lambda] - Log[1 + \lambda])\mu_1(\mu_1 - \mu_2)\mu_3(\mu_2 + \mu_3)}{2(\mu_1\mu_2(\mu_1 + \mu_2) + 4\mu_1\mu_2\mu_3 + \mu_3{}^2(\mu_1 + \mu_2))}$$

$$n_2 = -\frac{(\mathrm{Log}[1 - \lambda] - Log[1 + \lambda])\mu_2(\mu_1 - \mu_2)\mu_3(\mu_1 + \mu_3)}{2(\mu_1\mu_2(\mu_1 + \mu_2) + 4\mu_1\mu_2\mu_3 + \mu_3{}^2(\mu_1 + \mu_2))}$$

$$p_1 = \frac{(\mathrm{Log}[1 - \lambda] - Log[1 + \lambda])(\mu_2 + \mu_3)((\mu_1 + \mu_2)\mu_3{}^2 + 4\mu_1\mu_2\mu_3 + \mu_1\mu_2(\mu_1 + \mu_2))}{2\mu_3(\mu_1 - \mu_2)(\mu_1 + \mu_2 + 2\mu_3)(2\mu_1\mu_2 + \mu_3(\mu_1 + \mu_2))}$$

$$p_2 = \frac{(\mathrm{Log}[1 - \lambda] - Log[1 + \lambda])(\mu_1 + \mu_3)((\mu_1 + \mu_2)\mu_3{}^2 + 4\mu_1\mu_2\mu_3 + \mu_1\mu_2(\mu_1 + \mu_2))}{2\mu_3(\mu_1 - \mu_2)(\mu_1 + \mu_2 + 2\mu_3)(2\mu_1\mu_2 + \mu_3(\mu_1 + \mu_2))}$$

An examination of the singularities of the first and second members of (79) shows that the third member $w(\alpha)$ must be an entire vector. By indicating with $w_i(\alpha)$ ($i = 1, 2, 3, 4$), the four entire functions that define the components of $w(\alpha)$, from eqs. (79) and (80) we get

$$\begin{vmatrix} V_+(\alpha) \\ I_+(\alpha) \\ V_-(-\alpha) \\ I_-(-\alpha) \end{vmatrix} = g_{e+}^{-1}(\alpha) \begin{vmatrix} \dfrac{A_1}{\alpha - \alpha_i} + \dfrac{B_1}{\alpha + \alpha_i} \\ \dfrac{A_2}{\alpha - \alpha_i} + \dfrac{B_2}{\alpha + \alpha_i} \\ \dfrac{A_3}{\alpha - \alpha_i} + \dfrac{B_3}{\alpha + \alpha_i} \\ \dfrac{A_4}{\alpha - \alpha_i} + \dfrac{B_4}{\alpha + \alpha_i} \end{vmatrix} + g_{e+}^{-1}(\alpha) \begin{vmatrix} w_1(\alpha) \\ w_2(\alpha) \\ w_3(\alpha) \\ w_4(\alpha) \end{vmatrix} \qquad (81)$$

Taking into account that the first member must be a plus Fourier transform (i.e., it must vanish for $\alpha \to \infty$) we can evaluate the entire functions $w_i(\alpha)$ by considering an asymptotic evaluation for $\alpha \to \infty$ of the second member. Following this procedure, after algebraic manipulation of the second member of (81) it can be shown that the entire vector $w(\alpha)$ is a constant vector w given by

$$w = \begin{vmatrix} 0 \\ \dfrac{(A_1 - A_3 + B_1 - B_3)p_1\lambda_1}{\sqrt{n_2 - n_1}\sqrt{p_2 + p_1}\omega} \\ 0 \\ \dfrac{(A_1 - A_3 + B_1 - B_3)p_2\lambda_1}{\sqrt{n_2 - n_1}\sqrt{p_2 + p_1}\omega} \end{vmatrix}$$

We have been able to show, using many numerical simulations, that the coefficients A_i, B_i ($i = 1, 2, 3, 4$) satisfy

$$A_1 - A_3 + B_1 - B_3 = 0$$

Consequently, the entire vector $w(\alpha)$ is null, thus yielding

$$x_+(\alpha) = g_{e+}^{-1}(\alpha) \cdot K(\alpha)$$

Substituting we obtain

$$x_+(\alpha) = \begin{vmatrix} V_+(\alpha) \\ I_+(\alpha) \\ V_-(-\alpha) \\ I_-(-\alpha) \end{vmatrix} = r_0(\alpha) + \frac{1}{2}\left[\left[\frac{j\sqrt{k^2-\alpha^2}-k-\alpha}{j\sqrt{k^2-\alpha^2}+k+\alpha}\right]^c + \left[\frac{j\sqrt{k^2-\alpha^2}-k-\alpha}{j\sqrt{k^2-\alpha^2}+k+\alpha}\right]^{-c}\right] r_1(\alpha)$$
$$+\frac{1}{2}\sqrt{k^2-\alpha^2}\left[\left[\frac{j\sqrt{k^2-\alpha^2}-k-\alpha}{j\sqrt{k^2-\alpha^2}+k+\alpha}\right]^c - \left[\frac{j\sqrt{k^2-\alpha^2}-k-\alpha}{j\sqrt{k^2-\alpha^2}+k+\alpha}\right]^{-c}\right] r_2(\alpha) \tag{82}$$

where $r_i(\alpha)$ $(i = 0, 1, 2)$ are rational vectors of α and $j\omega$ given by

$$r_0(\alpha) = \begin{vmatrix} \dfrac{-n_2A_1+n_1A_3}{(n_1-n_2)(\alpha-\alpha_i)} - \dfrac{n_2B_1-n_1B_3}{(n_1-n_2)(\alpha+\alpha_i)} \\ \dfrac{p_2A_2-p_1A_4}{(p_1+p_2)(\alpha-\alpha_i)} + \dfrac{p_2B_2-p_1B_4}{(p_1+p_2)(\alpha+\alpha_i)} \\ \dfrac{-n_2A_1+n_1A_3}{(n_1-n_2)(\alpha-\alpha_i)} - \dfrac{n_2B_1-n_1B_3}{(n_1-n_2)(\alpha+\alpha_i)} \\ \dfrac{-p_2A_2+p_1A_4}{(p_1+p_2)(\alpha-\alpha_i)} - \dfrac{p_2B_2-p_1B_4}{(p_1+p_2)(\alpha+\alpha_i)} \end{vmatrix}$$

$$r_1(\alpha) = \begin{vmatrix} \dfrac{2n_1(A_1-A_3)\alpha_i}{(n_1-n_2)(\alpha-\alpha_i)(\alpha+\alpha_i)} \\ \dfrac{p_1[(B_2+B_4)(\alpha-\alpha_i)+(A_2+A_4)(\alpha+\alpha_i)]}{(p_1+p_2)(\alpha-\alpha_i)(\alpha+\alpha_i)} \\ \dfrac{2n_2(A_1-A_3)\alpha_i}{(n_1-n_2)(\alpha-\alpha_i)(\alpha+\alpha_i)} \\ \dfrac{p_2[(B_2+B_4)(\alpha-\alpha_i)+(A_2+A_4)(\alpha+\alpha_i)]}{(p_1+p_2)(\alpha-\alpha_i)(\alpha+\alpha_i)} \end{vmatrix} \tag{83}$$

$$r_2(\alpha) = \begin{vmatrix} \dfrac{jn_1\omega[(B_2+B_4)(\alpha-\alpha_i)+(A_2+A_4)(\alpha+\alpha_i)]}{\lambda\sqrt{(-n_1+n_2)(p_1+p_2)}(k^2-\alpha^2)(\alpha^2-\alpha_i^2)} \\ \dfrac{2jp_1\lambda(A_1-A_3)\alpha_i}{\omega\sqrt{(-n_1+n_2)(p_1+p_2)}(\alpha^2-\alpha_i^2)} \\ \dfrac{jn_2\omega[(B_2+B_4)(\alpha-\alpha_i)+(A_2+A_4)(\alpha+\alpha_i)]}{\lambda\sqrt{(-n_1+n_2)(p_1+p_2)}(k^2-\alpha^2)(\alpha^2-\alpha_i^2)} \\ \dfrac{2jp_2\lambda(A_1-A_3)\alpha_i}{\omega\sqrt{(-n_1+n_2)(p_1+p_2)}(\alpha^2-\alpha_i^2)} \end{vmatrix}$$

10.18.2 Matrix factorization of $g_e(\alpha)$

The matrix $g_e(\alpha)$ defined by (70) can be put in the form

$$g_e(\alpha) = u_o + q(\alpha) \tag{84}$$

where the constant matrix u_o and the matrix $q(\alpha)$ depending on α through the function

$$\tau = \sqrt{k^2 - \alpha^2}$$

are given by

$$u_o = \begin{vmatrix} -\dfrac{(\mu_1-\mu_2)\mu_3}{(\mu_1+\mu_2)(\mu_1+\mu_3)} & 0 & \dfrac{\mu_1(\mu_1+\mu_2+2\mu_3)}{(\mu_1+\mu_2)(\mu_1+\mu_3)} & 0 \\ 0 & -\dfrac{(\mu_1-\mu_2)\mu_1}{(\mu_1+\mu_2)(\mu_1+\mu_3)} & 0 & 1-\dfrac{(\mu_1-\mu_2)\mu_1}{(\mu_1+\mu_2)(\mu_1+\mu_3)} \\ \dfrac{\mu_2(\mu_1+\mu_2+2\mu_3)}{(\mu_1+\mu_2)(\mu_2+\mu_3)} & 0 & \dfrac{(\mu_1-\mu_2)\mu_3}{(\mu_1+\mu_2)(\mu_2+\mu_3)} & 0 \\ 0 & 1+\dfrac{(\mu_1-\mu_2)\mu_2}{(\mu_1+\mu_2)(\mu_2+\mu_3)} & 0 & \dfrac{(\mu_1-\mu_2)\mu_2}{(\mu_1+\mu_2)(\mu_2+\mu_3)} \end{vmatrix}$$

$$q(\alpha) = \begin{vmatrix} 0 & \dfrac{\omega\mu_1(\mu_1-\mu_2)\mu_3}{\tau(\mu_1+\mu_2)(\mu_1+\mu_3)} & 0 & \dfrac{\omega\mu_1(\mu_1-\mu_2)\mu_3}{\tau(\mu_1+\mu_2)(\mu_1+\mu_3)} \\ \dfrac{\tau(\mu_1-\mu_2)}{\omega(\mu_1+\mu_2)(\mu_1+\mu_3)} & 0 & \dfrac{\tau(-\mu_1+\mu_2)}{\omega(\mu_1+\mu_2)(\mu_1+\mu_3)} & 0 \\ 0 & -\dfrac{\omega\mu_2(\mu_1-\mu_2)\mu_3}{\tau(\mu_1+\mu_2)(\mu_2+\mu_3)} & 0 & \dfrac{\omega\mu_2(\mu_2-\mu_1)\mu_3}{\tau(\mu_1+\mu_2)(\mu_2+\mu_3)} \\ \dfrac{\tau(\mu_1-\mu_2)}{\omega(\mu_1+\mu_2)(\mu_2+\mu_3)} & 0 & \dfrac{\tau(-\mu_1+\mu_2)}{\omega(\mu_1+\mu_2)(\mu_2+\mu_3)} & 0 \end{vmatrix}$$

Equation (84) can be rewritten in the form

$$g_e(\alpha) = u_o \cdot (\mathbf{1} + u_o^{-1} \cdot q(\alpha)) = u_o \cdot (\mathbf{1} + q_1(\alpha)) = u_o \cdot g(\alpha)$$

where

$$q_1(\alpha) = u_o^{-1} \cdot q(\alpha) = \frac{1}{\tau} q_e(\alpha)$$

and

$$q_e(\alpha) = \begin{vmatrix} 0 & a & 0 & a \\ \tau^2 b & 0 & -\tau^2 b & 0 \\ 0 & e & 0 & e \\ \tau^2 d & 0 & -\tau^2 d & 0 \end{vmatrix}$$

is a polynomial matrix, where

$$a = -\frac{\omega\mu_1(\mu_1 - \mu_2)\mu_3(\mu_2 + \mu_3)}{\mu_1\mu_2(\mu_1 + \mu_2) + 4\mu_1\mu_2\mu_3 + (\mu_1 + \mu_2)\mu_3^2}$$

$$b = \frac{(\mu_1 - \mu_2)(\mu_2 + \mu_3)}{\omega\left[\mu_1\mu_2(\mu_1 + \mu_2) + 4\mu_1\mu_2\mu_3 + (\mu_1 + \mu_2)\mu_3^2\right]}$$

$$e = \frac{\omega\mu_2(\mu_1 - \mu_2)\mu_3(\mu_2 + \mu_3)}{\mu_1\mu_2(\mu_1 + \mu_2) + 4\mu_1\mu_2\mu_3 + (\mu_1 + \mu_2)\mu_3^2}$$

$$d = \frac{(\mu_1 - \mu_2)(\mu_1 + \mu_3)}{\omega\left[\mu_1\mu_2(\mu_1 + \mu_2) + 4\mu_1\mu_2\mu_3 + (\mu_1 + \mu_2)\mu_3^2\right]}$$

We observe that $g(\alpha) = \mathbf{1} + q_1(\alpha)$ does commute with the polynomial matrix $q_e(\alpha)$. It follows that we can accomplish the matrix factorization using a logarithmic decomposition of the matrix

$$f(\alpha) = \log[g(\alpha)]$$

By using the computer manipulator MATHEMATICA we obtain

$$f(\alpha) = f_0 + \frac{1}{\tau}p(\alpha)$$

where f_o and the polynomial matrix $p(\alpha)$ do commute and are given by

$$f_0 = \begin{vmatrix} a_1 & 0 & -a_1 & 0 \\ 0 & b_1 & 0 & b_1 \\ -a_2 & 0 & a_2 & 0 \\ 0 & b_2 & 0 & b_2 \end{vmatrix}, \quad p(\alpha) = \begin{vmatrix} 0 & \dfrac{n_1\omega}{\lambda} & 0 & \dfrac{n_1\omega}{\lambda} \\ \dfrac{p_1\lambda\tau^2}{\omega} & 0 & -\dfrac{p_1\lambda\tau^2}{\omega} & 0 \\ 0 & \dfrac{n_2\omega}{\lambda} & 0 & \dfrac{n_2\omega}{\lambda} \\ \dfrac{p_2\lambda\tau^2}{\omega} & 0 & -\dfrac{p_{21}\lambda\tau^2}{\omega} & 0 \end{vmatrix}$$

In the previous equations a_1, a_2, b_1, b_2 are given by

$$a_1 = \frac{(\mathrm{Log}[1-\lambda] + Log[1+\lambda])\mu_1(\mu_2 + \mu_3)}{4\mu_1\mu_2 + 2\mu_3(\mu_1 + \mu_2)}$$

$$a_2 = \frac{(\mathrm{Log}[1-\lambda] + Log[1+\lambda])\mu_2(\mu_2 + \mu_3)}{4\mu_1\mu_2 + 2\mu_3(\mu_1 + \mu_2)}$$

$$b_1 = \frac{(\mathrm{Log}[1-\lambda] + Log[1+\lambda])(\mu_2 + \mu_3)}{2(\mu_1 + \mu_2 + 2\mu_3)}$$

$$b_2 = \frac{(\mathrm{Log}[1-\lambda] + Log[1+\lambda])(\mu_1 + \mu_3)}{2(\mu_1 + \mu_2 + 2\mu_3)}$$

Taking into account that

$$g_e(\alpha) = u_o \exp[\log[g(\alpha)] = u_o \exp[f_0]\exp\left[\frac{1}{\tau}p(\alpha)\right]$$

the factorization of $g_e(\alpha)$ is reduced to the factorization of the matrix $\exp\left[\frac{1}{\tau}p(\alpha)\right]$. Since $p(\alpha)$ is a polynomial matrix, this factorization is easily performed by a standard decomposition into plus and minus parts (chapter 3, section 3.1.1):

$$\frac{1}{\tau} = \frac{1}{j\pi\tau}\theta(\alpha) + \frac{1}{j\pi\tau}\theta(-\alpha)$$

where

$$\theta(\alpha) = \log\left[-\frac{k+\alpha-j\tau}{k+\alpha+j\tau}\right]$$

and the plus and minus terms are the first and second term of the second member, respectively. Hence, $g_{e+}(\alpha)$ and its inverse $[g_{e+}(\alpha)]^{-1}$ can be defined by

$$g_{e+}(\alpha) = \exp\left[\frac{\theta(\alpha)}{j\pi\tau}p(\alpha)\right]$$
$$= \begin{vmatrix} \frac{n_2 - n_1\mathrm{Cosh}[c\theta]}{n_2 - n_1} & -j\frac{n_1\omega\mathrm{Sinh}[c\theta]}{\sqrt{n_2-n_1}\sqrt{p_2+p_1}\lambda\tau} & \frac{n_1(-1+\mathrm{Cosh}[c\theta])}{n_2-n_1} & -j\frac{n_1\omega\mathrm{Sinh}[c\theta]}{\sqrt{n_2-n_1}\sqrt{p_2+p_1}\lambda\tau} \\ -j\frac{p_1\mathrm{Sinh}[c\theta]\lambda\tau}{\sqrt{n_2-n_1}\sqrt{p_2+p_1}\omega} & \frac{p_2+p_1\mathrm{Cosh}[c\theta]}{p_2+p_1} & j\frac{p_1\mathrm{Sinh}[c\theta]\lambda\tau}{\sqrt{n_2-n_1}\sqrt{p_2+p_1}\omega} & \frac{p_1(-1+\mathrm{Cosh}[c\theta])}{p_1+p_2} \\ -\frac{n_2(-1+\mathrm{Cosh}[c\theta])}{n_2-n_1} & -j\frac{n_2\omega\mathrm{Sinh}[c\theta]}{\sqrt{n_2-n_1}\sqrt{p_2+p_1}\lambda\tau} & -\frac{n_1-n_2\mathrm{Cosh}[c\theta]}{n_2-n_1} & -j\frac{n_2\omega\mathrm{Sinh}[c\theta]}{\sqrt{n_2-n_1}\sqrt{p_2+p_1}\lambda\tau} \\ -j\frac{p_2\mathrm{Sinh}[c\theta]\lambda\tau}{\sqrt{n_2-n_1}\sqrt{p_2+p_1}\omega} & \frac{p_2(-1+\mathrm{Cosh}[c\theta])}{p_1+p_2} & j\frac{p_2\mathrm{Sinh}[c\theta]\lambda\tau}{\sqrt{n_2-n_1}\sqrt{p_2+p_1}\omega} & \frac{p_1+p_2\mathrm{Cosh}[c\theta]}{p_2+p_1} \end{vmatrix} \tag{85}$$

We observe that the inverse of $g_{e+}(\alpha)$ can be obtained from $g_{e+}(\alpha)$ by changing the sign of τ. By using (85), we may also write explicitly the minus factorized matrices $g_{e-}(\alpha)$ and $[g_{e-}(\alpha)]^{-1}$. In fact, taking into account (78), we have the following formulas that complete the factorization process:

$$g_{e-}(\alpha) = g_e(\alpha)[g_{e+}(\alpha)]^{-1}, \quad [g_{e-}(\alpha)]^{-1} = g_{e+}(\alpha)g_e(\alpha)$$

10.18.3 Near field behavior

In this section we study the near field ($\rho \approx 0$); for the sake of simplicity we assume $\mu_1 = \mu_3$. We begin our analysis by examining the field on the plane $y = 0$. It is possible to investigate the behavior for $z \approx 0$ by using Watson's lemma (Jones, 1964, p. 438), without the necessity to perform the inverse transform of $x_+(\alpha)$. To this end, we must know only the asymptotic behavior of $x_+(\alpha)$ for $\alpha \to \infty$. Analytical manipulations done using MATHEMATICA allow us to accomplish this task starting from the exact solution eq. (82), leading to

$$x_+(\alpha) \approx r_0(\alpha) + u(\alpha)\left(-\frac{\alpha}{k}\right)^c \quad \alpha \to \infty \tag{86}$$

where $r_0(\alpha)$ has been defined by the first of (83) and

$$u(\alpha) = \begin{vmatrix} \dfrac{2^{c-1}\left[-\omega(A_2+A_4+B_2+B_4)n_1\sqrt{n_2-n_1}+\lambda(A_1-A_3-B_1+B_3)n_1\sqrt{p_1+p_2}\alpha_i\right]}{\alpha^2\lambda(n_1-n_2)\sqrt{p_1+p_2}} \\ \dfrac{2^{c-1}p_1\left[\dfrac{A_2+A_4+B_2+B_4}{p_1+p_2}-\dfrac{\lambda(A_1-A_3-B_1+B_3)\alpha_i}{\omega\sqrt{n_2-n_1}\sqrt{p_1+p_2}}\right]}{\alpha} \\ \dfrac{2^{c-1}\left[-\omega(A_2+A_4+B_2+B_4)n_2\sqrt{n_2-n_1}+\lambda(A_1-A_3-B_1+B_3)n_2\sqrt{p_1+p_2}\alpha_i\right]}{\alpha^2\lambda(n_1-n_2)\sqrt{p_1+p_2}} \\ \dfrac{2^{c-1}p_2\left[\dfrac{A_2+A_4+B_2+B_4}{p_1+p_2}-\dfrac{\lambda(A_1-A_3-B_1+B_3)\alpha_i}{\omega\sqrt{n_2-n_1}\sqrt{p_1+p_2}}\right]}{\alpha} \end{vmatrix}$$

Remembering that $\alpha_i = k\sin\varphi_i$, from the previous equations it is evident that the dominant dependence on α and ω is given by

$$r_0(\alpha) \propto \begin{vmatrix} \dfrac{1}{\alpha} \\ \dfrac{1}{\alpha} \\ \dfrac{1}{\alpha} \\ \dfrac{1}{\alpha} \end{vmatrix}, \quad u(\alpha) \propto \begin{vmatrix} \dfrac{\omega}{\alpha^2} \\ \dfrac{\omega^0}{\alpha} \\ \dfrac{\omega}{\alpha^2} \\ \dfrac{\omega^0}{\alpha} \end{vmatrix}$$

Hence, applying Watson's lemma (or initial-value theorem) to $x_+(\alpha)$, we obtain

$$\begin{aligned} E_x(0,z) &= a_0 + a_1(kz)^{1-c} \quad (z \geq 0) \\ H_z(0,z) &= b_1(kz)^{-c} \quad (z \geq 0) \\ E_x(0,z) &= c_0 + c_1(kz)^{1-c} \quad (z \leq 0) \\ H_z(0,z) &= d_1(kz)^{-c} \quad (z \leq 0) \end{aligned}$$

where the coefficients a_i, b_i, c_i, d_i can be obtained from the expressions of $r_0(\alpha)$ and $u(\alpha)$.

The near-field behavior in the plane $y = 0$ can be extended to every points of space. First we evaluate the electric field everywhere using the Fourier inversion of (56). Taking into account the equation that expresses the Fourier transform of the Hankel function,[6] after analytical manipulations we obtain $E_x(\rho,\varphi)$ and $H_\rho(\rho,\varphi)$ through a Green representation in

[6] We have the identity (Felsen & Marcuvitz, 1973, p. 478, section 1.1.1): $\mathbb{F}\left[H_0^{(2)}(|x|), x, \alpha\right] = \frac{2}{\sqrt{k^2-\alpha^2}}$.

the spatial domain, whence we get the near field:

$$\begin{aligned} E_x(\rho,\varphi) &= a_0(\varphi) + a_1(\varphi)(k\rho)^{1-c} \quad (\rho \to 0) \\ H_\rho(\rho,\varphi) &= b_1(\varphi)(k\rho)^{-c} \quad (\rho \to 0) \end{aligned} \tag{87}$$

We remark that the coefficient c satisfies the condition $0 < c < 1$ and does not depend either on the spatial frequency α or on the temporal frequency ω.

Looking at (87), we observe that the radial component of the magnetic field is singular as $\rho \to 0$ for the presence of the term $b_1(\varphi)(k\rho)^{-c}$.

To make a comparison between the near-field behavior of static and dynamic fields, we consider the magnetostatic problem of a right wedge made of magnetic material of permeability μ_2 immersed in a homogeneous magnetic material of permeability μ_1. The exact solution of this problem can be accomplished by applying the duality principle to the well-studied static dielectric wedge. In particular, starting from the exact solution, we can obtain the following static near field behavior (van Bladel, 1996, p. 153):

$$H_\rho(\rho,\varphi) = b_{2s}(\varphi)\rho^{-c} \quad (\rho \to 0) \tag{88}$$

where we can show that the parameter c is the same parameter considered previously for the isorefractive wedge.

By comparing (87) and (88), it is evident that the static behavior is the same as that present in the dynamic isorefractive wedge.

It is possible to investigate the edge near-field behavior for a non-isorefractive wedge starting from the results obtained for the isorefractive wedge. In particular, the dominant near-edge behavior is the same for both problems (Daniele & Uslenghi, 2000).

10.19 Diffraction by an arbitrary dielectric wedge

The diffraction by a dielectric wedge has constituted a very important and challenging problem during the past century. The GWHE of this problem were obtained and solved in Daniele (2010, 2011).

References

Abrahams, I.D. (1998) On the noncommutative factorization of Wiener-Hopf kernels of Khrapkov type, *Proc. Roy. Soc. London, A*, 454, pp. 1719–1743.

Abrahams, I.D. (2000) The application of Pade approximants to Wiener-Hopf factorization, *IMA J. Appl. Math.*, 65, pp. 257–281.

Abrahams I.D. and G.R. Wickham (1988) On the scattering of sound by two semi-infinite parallel staggered plates-I. Explicit matrix Wiener-Hopf factorization, *Proc. Roy. Soc. London, A*, 420, pp. 131–156.

Abrahams, I.D. and G.R. Wickham (1990) General Wiener-Hopf factorization of matrix kernel with exponential phase factors, *SIAM J. Appl. Math.*, 50, 3, pp. 819–838.

Abrahams, I.D. and G.R. Wickham (1991) The scattering of water wave by two semi-infinite opposed vertical walls, *Wave Motion*, 14, pp. 145–168.

Abramovitz, M. and I. Stegun (1964) *Handbook of Mathematical Functions*, Dover Publications, New York.

Achenbach, J.D. and A.K. Gautesen (1977) Geometrical theory of diffraction for 3D elastodynamics, *J. Acoust. Soc. Am.*, 61, pp. 413–421.

Albani, M. (2007) On Radlow's quarter-plane diffraction solution, *Radio Science*, 42, p. 10.

Albertsen, N.C. (1997) Diffraction by a quarterplane of the field from a halfwave dipole. *Inst. Elect. Eng. Proc. Microwave Antennas Propagat.*, 144, pp. 191–196.

Antipov, Y.A. and V.V. Silvestrov (2002) Factorization on a Riemann surface in scattering theory, *Quart. J. Mech. Appl. Math.*, 55, 607–654.

Antipov, Y.A. and V.V. Silvestrov (2004a) Vector functional-difference equation in electromagnetic scattering, *IMA J. Appl. Math.*, 69, 1, pp. 27–69.

Antipov, Y.A. and V.V. Silvestrov (2004b) Second-order functional-difference equations. II: Scattering from a right-angled conductive wedge for E-polarization. *Quart. J. Mech. Appl. Math.*, 57, 2, pp. 267–313.

Antipov, Y.A. and V.V. Silvestrov (2006) Electromagnetic scattering from an anisotropic half-plane at oblique incidence: the exact solution, *Quart. J. Mech. Appl. Math.*, 59, pp. 211–251.

Aoki, K. and K. Uchida (1978) Scattering of plane electromagnetic waves by a conducting rectangular cylinder-E polarized wave-, *Mem. Fac. Eng. Kyushu Univ.*, 38, pp. 153–175.

Aoki, K., A. Ishizu, and K. Uchida (1982) Scattering of surface waves in semi-infinite slab dielectric guide, *Mem. Fac. Eng. Kyushu Univ.*, 42, pp. 197–215.

Baker, H.F. (2006) *Abelian Functions*, Cambridge Mathematical Library, New York.

Ball, J.A. and A.C.M. Ran (1986) Left versus right canonical Wiener-Hopf factorization, in *Constructive Methods of Wiener-Hopf Factorization*, vol. 21, *Operator Theory: Advances and Applications*, edited by I. Gohberg and M.A. Kaashoek, Birkhäuser Verlag Basel.

Bart, H., I. Goheberg, and M.A. Kaashoek (1979) *Minimal Factorization of Matrix and Operator Functions,* Birkhäuser Werlag, Basel.

Bazer, J. and S.N. Karp (1962) Propagation of electromagnetic waves past a shore line, *Radio Sci., J.Res. NBS*, 66D, 3, pp. 319–333.

Birbir, F. and A. Büyükaksoy (1999) Plane wave diffraction by two parallel overlapped thick semi-infinite impedance plates, *Can. J. Phys.*, 77, pp. 873–891.

Bliss, G.A. (1966) *Algebraic Functions*, Dover Phoenix Editions, New York.

Bouwkamp, C.J. (1954) Diffraction theory, *Rept. Progr. Phys.*, 17, pp. 35–100.

Bowman, J.J., T.B.A. Senior, and P.L.E. Uslenghi (1969) *Electromagnetic and Acoustic Scattering by Simple Shapes*, North-Holland Publishing Company, Amsterdam.

Brekhovskikh, L.M. (1960) *Waves in Layered Media*, Academic Press, New York.

Bresler, A.D. and N. Marcuvitz (1956) Operator methods in electromagnetic field theory, Report R-495,56, PIB-425, MRI Polytechnic Institute of Brooklyn, pp. 34–36.

Bucci, O. and G. Franceschetti (1976) Electromagnetic scattering by an half plane with two face impedances, *Radio Science*, 11, pp. 49–59.

Buldyrev, V.S. and M.A. Lyalinov (2001) *Mathematical Methods in Modern Diffraction Theory*, Science House Inc., Tokyo.

Büyükaksoy, A. and F. Birbir (1999) Plane wave diffraction by two parallel overlapped thick semi-infinite impedance planes, *Can. J. Phys.*, 77, pp. 873–891.

Büyükaksoy, A. and A.H. Serbest (1993) Matrix Wiener-Hopf factorization methods and applications to some diffraction problems, in *Analytical and Numerical Methods in Electromagnetic Wave Theory*, edited by M. Hashimoto, M. Idemen, and O.A. Tretyakov, Science House Co. Ltd., Tokyo.

Büyükaksoy, A., G. Uzgoren, and F. Birbir (2001) The scattering of a plane wave by two parallel semi-infinite overlapping screens with dielectric loading, *Wave Motion*, 34, pp. 375–389.

Camara, M.C., A.F. dos Santos, and N. Manojlovich (2001) Generalized factorization for $N \times N$ Daniele-Khrapkov matrix functions, *Math. Meth. Appl. Sci.*, 24, pp. 993–1020.

Canavero, F., V. Daniele, and R. Graglia (1988) Spectral theory of transmission lines in presence of external electromagnetic sources, *Electromagnetics* (Special Issue on Electromagnetic Coupling to Transmission Lines), 8, 2–4, pp. 125–157.

Carleman, T. (1922) Sur la resolution de certain equation integrals, *Arkiv Mat. Astr. Och. Phys. Bd.*, 16, pp. 1–19.

Carlson, J.F. and A.E. Heins (1947) The reflection of an electromagnetic plane wave by an infinite set of plates I, *Quart. Appl. Math*, 4, pp. 313–329.

Chang, D.C. (1969) Propagation along a mixed path in the curved earth-ionosphere waveguide, *Radio Science*, 4, 4, pp. 335–345.

Chebotarev, G.N. (1956) On the closed form solution of the Riemann boundary value problem for a system on n pairs of functions, *Scientific Trans of the Kazan State University*, 116, 4, pp. 31–58.

Cinar, G. and A. Büyükaksoy (2004) Diffraction on a normally incident plane wave by three parallel half-planes with different face impedances, *IEEE Trans. Antennas Propag.*, 52, pp. 478–486.

Clemmow, P.C. (1953) Radio propagation over a flat earth across a boundary separating two different media, *Phil. Trans. Roy. Soc. London A*, 246, 905, pp. 1–55.

Copson, E.T. (1945) On an integral equation arising in the theory of the diffraction, *Quart. J. Math.*, 17, 2, pp. 19–34.

Daniele, V. (1971) Wave propagation in stratified multifluid plasma, *Alta Frequenza*, 40, pp. 904–914.

Daniele, V. (1978) On the factorization of W-H Matrices in problems solvable with Hurd's methods, *IEEE Trans. on Antennas and Propag.*, AP-26, pp. 614–616.

Daniele, V., I. Montrosset, and R. Zich (1979) Wiener-Hopf solution for the junction between a smooth and corrugated cylindrical waveguide, *Radio Science*, 14, pp. 943–956.

Daniele, V. and R. Zich (1980) Classe di discontinuita' in guide d'onda analizzabili con sistemi di equazioni Wiener-Hopf, *Terza Riunione Nazionale di Elettromagnetismo applicato*, Bari 2-4 giugno, pp. 103–106.

Daniele, V., I. Montrosset, and R. Zich (1981) Wiener-Hopf solution for the radiation from a cylindrical truncated waveguide with anisotropic boundary conditions on the walls, *Radio Science*, 16, pp. 1115–1118.

Daniele, V. and P.L.E. Uslenghi (1981b) Exact scattering by an imperfectly conducting thick half-plane, IEEE/AP International Symposium, Los Angeles, CA, June, pp. 657–661.

Daniele, V. and P.L.E. Uslenghi (1982) Exact scattering from a thick half-plane with different face impedances, IEEE/AP International Symposium, Albuquerque, NM, May, pp. 9–12.

Daniele, V. (1983) Wiener-Hopf Methods in scattering problems (invited paper), URSI International Electromagnetic Symposium, Santiago de Compostela, Spain, August, pp. 101–105.

Daniele, V., R. Tascone, and R. Zich (1983b) Scattering from structures involving steps, URSI International Electromagnetic Symposium, Santiago de Compostela, Spain, August, pp. 107–110.

Daniele, V. (1984a) On the solution of two coupled Wiener-Hopf equations, *SIAM Journal of Applied Mathematics*, 44, pp. 667–680.

Daniele, V.G. (1984b) On the solution of vector Wiener-Hopf equations occurring in scattering problems, *Radio Science*, 19, pp. 1173–1178.

Daniele, V. (1986) Factorization of meromorphic matrices occurring in scattering problems, 1986-URSI International Symposium on Electromagnetic theory, Budapest, Ungheria, August, pp. 84–86.

Daniele, V., M. Gilli, and E. Viterbo (1990) Diffraction of a plane wave by a strip grating, *Electromagnetics*, 10, 3, pp. 245–269.

Daniele, V., M. Gilli, and S. Pignari (1996) Spectral theory of a semi-infinite transmission line over a ground plane, *IEEE Trans. Electromagn. Compat.*, 38, 3, pp. 230–236.

Daniele, V. and M. Gilli (1997) A spectral technique for the steady-state analysis of switched distributed networks, 1997 IEEE International Symposium on Circuits and Systems, Hong Kong, June, pp. 992–996.

Daniele, V.G. (2000) Generalized Wiener-Hopf technique for wedge shaped regions of arbitrary angles, 2000 International Conference on Mathematical Methods in electromagnetic theory (MMET 2000), Kharkov, Ukraine, September 12–15, pp. 432–434.

Daniele, V.G. and P.L.E. Uslenghi (2000) Wiener-Hopf solution for right isorefractive wedges, Rapporto ELT-2000-2, Politecnico di Torino, September, http://personal.delen.polito.it/vito.daniele/

Daniele, V.G. (2001) New analytical methods for wedge problems, 2001 International Conference on Electromagnetics in Advanced Applications (ICEAA01), Torino, Italy, September 10–14, pp. 385–393.

Daniele, V. and R. Zich (2002) Approximate factorizations for the kernels involved in the scattering by a wedge at skew incidence (invited paper), 2002 International Conference on Mathematical Methods in Electromagnetic Theory (MMET 2002), Kiev, Ukraine, September 10–13, pp. 130–135.

Daniele, V.G. (2003a) Rotating waves in the Laplace domain for angular regions, *Electromagnetics*, 23, 3, pp. 223–236.

Daniele, V. (2003b) The Wiener-Hopf technique for impenetrable wedges having arbitrary aperture angle, *SIAM J. Appl. Math.*, 63, 4, pp. 1442–1460.

Daniele, V.G. and G. Lombardi (2003c) Generalized Wiener-Hopf technique for wedges problems involving arbitrary linear media, International Conference on Electromagnetics in Advanced Applications (ICEAA03), Torino, Italy, September 8–12, pp. 761–765.

Daniele, V. (2004a) An introduction to the Wiener-Hopf Technique for the solution of electromagnetic problems, Rapporto ELT-2004-1, Politecnico di Torino, September, http://personal.delen.polito.it/vito.daniele/

Daniele, V. (2004b) The Wiener-Hopf technique for wedge problems, Rapporto ELT-2004-2, Politecnico di Torino, October, http://personal.delen.polito.it/vito.daniele/

Daniele, V.G. (2005) The Wiener-Hopf technique for penetrable wedge problems, URSI General Assembly 2005, New Delhi, India, October 23–29.

Daniele, V. and G. Lombardi (2006) Wiener-Hopf solution for impenetrable wedges at skew incidence, *IEEE Trans. Antennas Propag.*, 54, pp. 2472–2485.

Daniele, V. and R. Graglia (2007) Diffraction by an imperfect half plane in a bianisotropic medium, *Radio Science*, 42, pp. RS6–S05.

Daniele, V. and G. Lombardi (2007) Fredholm Factorization of Wiener-Hopf scalar and matrix kernels, *Radio Science*, 42, pp. RS6–S01.

Daniele, V.G. (2010) The Wiener-Hopf formulation of the penetrable wedge problem: part I, *Electromagnetics*, 30, 8, pp. 625–643.

Daniele, V.G. (2011) The Wiener-Hopf formulation of the penetrable wedge problem: part II, *Electromagnetics*, 31, 8, pp. 1–17.

Felsen, L.B and N. Marcuvitz (1973) *Radiation and Scattering of Waves*, Prentice-Hall, Englewood Cliffs, NJ.

Gakhov, F.D. (1966) *Boundary Value Problems*, Dover Publications, New York.

Gilli, M. and V. Daniele (1995) Low frequency diffraction by a planar junction of a metallic and a wire-mesh halfplane, *IEEE Trans. Electromagn. Compat.*, 37, 3, pp. 343–357.

Gohberg, I.C. and M.G. Krein (1960) Systems of integral equation on a half line with kernels depending on the difference of arguments, *Amer. Math. Soc. Transl.* 13, 2, pp. 185–264.

Gohberg, I., P. Lancaster, and L. Rodman (1982) *Matrix Polynomials*, Academic Press, New York.

Gohberg, I.P. and N. Krupnik (1992) *One Dimensional Linear Singular Integral Equations*, Birkhauser, Basel.

Gradshteyn I.S. and I.M. Ryzhik (1965) *Table of Integrals, Series and Products*, Academic Press, New York.

Hallen, E. (1962) *Electromagnetic Theory*, Chapman & Hall.

Hashimoto, M., M. Idemen, and O.A. Tretyakov (eds.) (1993) *Analytical and Numerical Methods in Electromagnetic Wave Theory*, Science House Co. Ltd., Tokyo.

Heins, A.E. (1950a) *System of Wiener-Hopf Equations, Proc. of Symposia in Applied Math. II*, MacGraw-Hill, pp. 76–81.

Heins, A.E. and J.F. Carlson (1947) Tee reflection of an electromagnetic plane wave by an infinite set of plates II, *Quart. Appl. Math*, 5, pp. 82–88.

Heins, A.E. (1950b) The reflection of an electromagnetic plane wave by an infinite set of plates III, *Quart. Appl. Math*, 8, pp. 281–291.

Hills, N.L. and A.N. Karp (1965) Semi-infinite Diffraction Gratings-I, *Communications on Pure and Applied Mathematics*, 38, pp. 203–233.

Hurd, R.A. and H. Gruenberg (1954) H-plane bifurcation of rectangular waveguides, *Can. J. Phys.*, 32, pp. 727–734.

Hurd, R.A. and S. Przezdziecki (1981) Diffraction by a half-plane perpendicular to the distinguished axis of gyrotropic medium (oblique incidence), *Can. J. Phys.*, 59, pp. 403–424.

Hurd, R.A. and S. Przezdziecki (1985) Half-plane diffraction in a gyrotropic medium, *IEEE Trans. Antennas Propag.* AP-33, 8, pp. 813–822.

Hurd, R.A. and E. Luneburg (1985) Diffraction by an anisotropic impedance half plane, *Can. J. Phys.*, 63, 9, pp. 1135–1140.

Hurd, R.A. (1954) The propagation of an electromagnetic wave along an infinite corrugated surface, *Can. J. Physics*, 32, pp. 727–734.

Hurd, R.A. (1976) The Wiener-Hopf–Hilbert method for diffraction problems, *Can. J. Phys.* 54, 7, pp. 775–780.

Hurd, R.A. (1987) The explicit factorization of 2×2 Wiener-Hopf matrices, Technische Hochschule Darmstadt, *Preprint-Nr.* 1040, Marz.

Idemen, M. (1979) A new method to obtain exact solutions of vector Wiener-Hopf equations, *Z. Angew. Math. Mekh. (ZAMM)*, 59, pp. 656–568.

Idemen, M. and L.B. Felsen (1981) Diffraction of a whispering gallery mode by the edge of a thin concave cylindrically curved surface, *IEEE Trans. Antennas Propag.*, AP-29, 4, pp. 571–579.

Idemen, M. (1977) On the exact solutions of vector Wiener-Hopf equation, Technical Report, TÜBİTAK Marmara Research Institute, Türkiye, N.39.

Ittipiboon, A. and M. Hamid (1981) Application of the Wiener-Hopf technique to dielectric slab waveguides discontinuities, *IEE Proc.*, 128, pp. 188–196.

Ivanov, V.V. (1976) *The Theory of Approximate Methods*, Noordhoff International Publishing, Leyden.

Jones, D.S. (1952a) A simplifying technique in the solution of a class of diffraction problems, *Quart. J. Math.*, 3, 2, pp. 189–196.

Jones, D.S. (1952b) Diffraction by a waveguide of finite length, *Proc. Camb. Phil, Soc.*, 48, pp. 118–134.

Jones, D.S. (1953) Diffraction by a thick semi-infinite plane, *Proc. Roy. Soc. A*, 217, pp. 153–175.

Jones, D.S. (1955) The scattering of a scalar wave by a semi-infinite rod of circular cross section, *Phil. Trans. Roy. Soc. A*, 247, pp. 499–528.

Jones D.S. (1964) *The Theory of Electromagnetism*, Pergamon Press, Oxford.

Jones D.S. (1973) Double knife-edge diffraction and ray theory, *Quart. J. Mech. Appl. Math.*, 26, pp. 1–18.

Jones, D.S. (1984) Commutative Wiener-Hopf factorization of a matrix, *Proc. Roy. Soc. London, A*, 393, pp. 185–192.

Jones, D.S. (1986) Diffraction by three semi-infinite half planes, *Proc. Roy. Soc. London, A*, 404, pp. 299–321.

Jones, D.S. (1991) Wiener-Hopf splitting of a 2×2 matrix, *Proc. R. Soc. London, A*, 434, pp. 419–433.

Jones, D.S. (1994) *Methods in Electromagnetic Wave Propagation*, Clarendon Press, Oxford.

Jull, E.V. (1964) Diffraction by a conducting half-plane in an anisotropic plasma, *Can. J. Phys.*, 42, pp. 1455–1468.

Kantorovich, L.V. and V.I. Krylov (1967) *Approximate methods of higher analysis*, Interscience, New York, Noordhoff, Gronigen

Karjala, D.S. and R. Mittra (1966) Radiation from a modulated corrugated surface excited by a waveguide, *Proc. IEE (London)*, 113, pp. 1143–1150.

Karlovich, Y. and I.M. Spitkovsky (1995) Factorization of almost periodic matrix functions, *J. Math. Anal. Appl.*, 193, pp. 209–232.

Karp, S.N. (1950) Wiener-Hopf techniques and mixed boundary value problems, *Commun. Pure Appl. Math.*, 3, pp. 411–426.

Keller, J.B. (1962) Geometrical theory of diffraction, *J. Opt. Soc. Am.*, 52, pp. 116–130.

Keller, J.B. (1979) Progress and prospects in the theory of linear wave propagation, *SIAM Review*, 21, 2, pp. 229–245.

Khrapkov, A.A. (1971) Certain cases of the elastic equilibrium of infinite wedge with a non symmetric notch at the vertex, subjected to concentrated forces, *Prikl. Math. Mekh.*, 35, pp. 625–637.

Kobayashi, K. (1982a) Diffraction of a plane electromagnetic waves by a rectangular conducting rod (I), *Bull. Facult. Sci. & Eng., Chuo University*, 25, pp. 229–261.

Kobayashi, K. (1982b) Diffraction of a plane electromagnetic wave by a rectangular conducting rod (II), *Bull. Facult. Sci. & Eng., Chuo University*, 25, pp. 263–282.

Kobayashi, K. (1984) On the Factorization of certain kernels arising in functional equations of the Wiener-Hopf type, *J. Phys. Soc. Japan*, 53, pp. 2885–2898.

Kobayashi, K. (1993) Some diffraction problems involving modified Wiener-Hopf geometries, in *Analytical and Numerical Methods in Electromagnetic Wave Theory*, edited by M. Hashimoto, M. Idemen, and O.A. Tretyakov, Science House Co. Ltd., Tokyo, pp. 147–228.

Kobayashi, K. and S. Koshikawa (1996) Wiener-Hopf analysis of the radar cross section of parallel-plate waveguides cavities, Department of Electrical and Electronics Engineering, Technical Report No. KK96-3-8, 1-13-27 Kasuga, Bunkyo-ku, Tokyo, Japan, March, 112.

Kobayashi, K. and S. Sapmaz (1997) Wiener-Hopf analysis of the diffraction by strip and slit geometries in the proximity of medium discontinuities, Department of Electrical and Electronics Engineering, Technical Report No. KK97-3-21, 1-13-27 Kasuga, Bunkyo-ku, Tokyo, Japan, March, 112.

Kong, J.A. (1975) *Theory of Electromagnetic Waves*, John Wiley and Sons, New York.

Kraut, E.A. and G.W. Lehaman (1969) Diffraction of electromagnetic waves by a right-angled dielectric wedge, *J. Math. Phys*., 10, pp. 1340–1348.

Lee, S.W. (1967) Radiation from the infinite aperiodic array of parallel-plate waveguides, *IEEE Trans*., AP-15, pp. 598–606.

Levine, H. and J. Schwinger (1948) On the radiation of sound from an unflanged circular pipes, *Phys. Rev*., 73, pp. 383–340.

Levine, H. and J. Schwinger (1949) On the theory of electromagnetic wave diffraction by an aperture in an infinite plane screen II, *Phys. Rev*., 75, pp. 1423–1432.

Lüneburg E. (1993) Diffraction by an infinite set of parallel half-planes and by infinite strip grating: Comparison of different methods, in *Analytical and Numerical Methods in Electromagnetic Wave Theory*, edited by M. Hashimoto, M. Idemen, and O.A. Tretyakov, Science House Co. Ltd., Tokyo, pp. 317–372.

Lüneburg, E. and W. Westpfahl (1971) Diffraction of plane waves by an infinite strip grating, *Ann. der Physik*, 27, pp. 257–288.

Lüneburg, E. and R.A. Hurd (1984) On the diffraction of a half-plane with different face impedances, *Can. J. Phys.*, 62, pp. 853–860.

Lüneburg, E. and A.H. Serbest (2000) Diffraction of a obliquely incident plane wave by a two-face impedance half plane: Wiener-Hopf approach, *Radio Science*, 35, pp. 1361–1374.

Lukyanov, V.D. (1980) Exact solution of the problem of diffraction of an obliquely incident wave at a grating, *Sov. Phys. Dokl*., 25, 11, pp. 905–906.

MacDonald, H.M. (1915) A class of diffraction problems, *Proc. London Math. Soc.,* 14, pp. 419–427.

Maliuzhinets, G.D. (1958a) Relation between the inversion formula for the Sommerfeld integral and the formulas of Kontorovich-Lebedev, *Soviet Phys. Dokl*., 3, pp. 266–268.

Maliuzhinets, G.D. (1958b) Excitation, reflection and emission of surface waves from a wedge with given face impedance, *Soviet Phys. Dokl*., 3, pp. 752–755.

Manara, G., P. Nepa, G. Pelosi, and A. Vallecchi (2004) Skew incidence diffraction by an anisotropic impedance half plane with a PEC face and arbitrarily oriented anisotropy axes, *IEEE Trans. Antennas Propag.*, 52, 2, pp. 487–496.

Meister, E. and F.O. Speck (1979) Some multidimensional Wiener-Hopf equations with applications, in *Trends in Applic. Pure Math. Mech.*, vol. 2, edited by H. Zorski, Pitman, London, pp. 217–262.

Meister, E. and F.O. Speck (1989) Wiener-Hopf factorization of certain non-rational matrix functions in mathematical physics, in *The Gohberg Anniversary Collection*, vol. 2, Birkhauser, Base, pp. 385–394.

Mittra, R. and S.W. Lee (1971) *Analytical Techniques in the Theory of Guided Waves*, MacMillan Company, New York.

Moiseyev, N.G. (1989) Factorization of matrix function of special form, *Dokl. AN SSSR*, 305, pp. 44–47.

Montgomery, J.P. (1979) Electromagnetic boundary-value problems based upon a modification of residue calculus and function theoretic techniques, NBS Monography 164, National Bureau of Standards, Boulder, CO.

Morse and Feshbach (1953) *Methods of Theoretical Physics*, McGraw-Hill, New York.

Munk, B.A. (2000) *Selective Surfaces: Theory and Design*, Wiley Interscience, New York.

Muskhelishvili, N.I. (1953) *Singular Integral Equation*, P. Noordhoff N.V. Groningen-Holland.

Newcomb, R.W. (1966) *Linear Multiport Synthesis*, McGraw-Hill, New York.

Noble, B. (1958) *The Wiener-Hopf Technique*, Pergamon Press, London.

Noble, B. (1988) *The Wiener-Hopf Technique*, 2d ed., Chelsea Pub. Co.

Nosich, A.I. (1993) *Green's function-dual* series approach in wave scattering by combined resonant scatterers, in *Analytical and Numerical Methods in Electromagnetic Wave Theory*, edited by M. Hashimoto, M. Idemen, and O.A. Tretyakov, Science House Co. Ltd., Tokyo, pp. 419–469.

Osipov, A.V. and A.N. Norris (1999) The Malyuzhinets theory for scattering from wedge boundaries: a review, *Wave Motion*, 29, pp. 313–340.

Paul, C.R. (1975) Useful matrix chain parameter identities for the analysis of multiconductor transmission lines (short papers), *IEEE Transactions on Microwave Theory and Techniques* (Short Papers), 23, 9, pp. 756–760.

Paul, C.R. (1992) *Introduction to Electromagnetic Compatibility*, John Wiley & Sons, New York.

Papadopoulos, V.M. (1957) The scattering effect of a junction between two circular waveguides, *Quart. J. Mech. Appl. Math.,* 10, pp. 191–209.

Pease, M.C. (1965) *Methods of Matrix Algebra*, Academic Press, New York.

Peters, A.S. (1952) Water waves over sloping beaches and the solution of a mixed boundary value problem for $\Delta^2\phi - k^2\phi = 0$ in a sector, *Commun. Pure Appl. Math.*, 5, pp. 97–108.

Poincare, H. (1892) Sur la polarization par diffraction, *Acta Math.*, 16, pp. 297–339.

Przezdziecki, S. (2000) Half-plane diffraction in a chiral medium, *Wave Motion*, pp. 157–200.

Radlow, J. (1961) Diffraction by a quarter plane, *Arch. Rat. Mech. Anal.*, 8, pp. 139–158.

Radlow, J. (1964) Diffraction by a right-angled dielectric wedge, *Intern. J. Enging. Sci.*, 2, pp. 275–290.

Radlow, J. (1965) Note on the diffraction at a corner, *Arch. Rat. Mech. Anal.*, 19, pp. 62–70.

Rawlins, A.D. (1975) The solution of mixed boundary-value problem in the theory of diffraction by a semi-infinite plane, *Proc. Roy. Soc. London, A*, 346, pp. 469–484.

Rawlins, A.D. (1980) A note on the factorization of matrices occurring in Wiener-Hopf problems, *IEEE Trans. Antennas Propag.*, AP-28, 6, pp. 933–934.

Sendag, R. and H. Serbest (2001) Scattering at the Junction formed by a PEC half-plane and an half-plane with anisotropic conductivity, *Electromagnetics*, 21, pp. 415–434.

Senior, T.B.A. (1978) Some problems involving imperfect half planes, in *Electromagnetic Scattering*, edited by P.L.E. Uslenghi, Academic Press, New York, pp. 185–219.

Senior, T.B.A. and J.L. Volakis (1995) *Approximate Boundary Conditions in Electromagnetics*, Institution of Electrical Engineers, London.

Senior, T.B.A. (1959) Diffraction by an imperfectly conducting wedge, *Comm. Pure Appl. Math.*, 12, pp. 337–372.

Serbest, A.H. and A. Büyükaksoy (1993) Some approximate methods related to the diffraction by strips and slits, in *Analytical and Numerical Methods in Electromagnetic Wave Theory*, edited by A. Hashimoto, M. Idemen, and O.A. Tretyakov, Science House Co. Ltd., Tokyo, pp. 229–256.

Seshadri, S.R. and A.K. Rajagopal (1963) Diffraction by a perfectly conducting semi-infinite screen in an anisotropic plasma, *IEEE Trans. Antennas Propag.*, 11, pp. 497–502.

Shestopalov, V.P. (1971) *The Riemann Method in the Theory of Diffraction and Propagation of Electromagnetic Waves*, Isdatelstvo Kharkov University, Karkov (in Russian).

Smythe, William R. (1989) *Static and Dynamic Electricity*, Hemisphere, New York.

Sommerfeld, A. (1896) Mathematische theorie der diffraktion, *Math. Ann.*, 47, pp. 317–341.

Spiegel, M.R. (1964) *Complex Variables*, Schaum, New York.

Springer, Link Encyclopaedia of Mathematics, http://eom.springer.de/

Springer, G. (1957) *Introduction to Riemann Surfaces*, Chelsea Publishing Company, New York.

Tayyar, I.H. and A. Büyükaksoy (2003) Plane wave diffraction by two oppositely placed, parallel two part planes, *IEE Proc. Sci. Meas. Technol.*, 153, pp. 169–176.

Thompson, J.R. (1962) Radio propagation over a sectionally homogeneous surface, *Proc. Roy. Soc., A*, 267, 1329, pp. 183–196.

Topsakal, E., A. Büyükaksoy, and M. Idemen (2000) Scattering of electromagnetic waves by a rectangular impedance cylinder, *Wave Motion*, 31, pp. 273–296.

Van Bladel, J. (2007) *Electromagnetic Fields*, John Wiley & Sons, New York.

Van Bladel, J. (1996) *Singular Electromagnetic Fields and Sources*, Clanderon Press, Oxford.

Vekua, N.P. (1967) *Systems of Singular Integral Equations*, P. Noordhoff, Gronigen.

Volakis, J.L. and T.B.A. Senior (1987) Diffraction by a thin dielectric half plane, *IEEE Trans. Antennas Propag.*, 35, pp. 1483–1487.

Wait, J.R. (1970) Factorization method applied to electromagnetic wave propagation in a curved waveguide with non uniform walls, *Radio Science*, 5, pp. 1059–1068.

Weinstein, L.A. (1969) *The Theory of Diffraction and the Factorization Method*, Golem Press, Boulder, CO.

Whitehead, E.A.N. (1951) The theory of parallel-plate media for microwave lenses, *Proc. IEE (London)*, 98, 3, pp. 133–140.

Wiener, N. and E. Hopf (1931) Über eine Klasse singulärer Integralgleichungen, *S.B. Preuss. Akad. Wiss.*, pp. 696–706.

Wiener, N. (1949) *Extrapolation, Interpolation and Smoothing of Stationary Time Series*, John Wiley, New York.

Wiener-Masani (1958) The prediction theory of multivariate stochastic processes, Part II., The linear predictor, *Acta Math.*, 99, pp. 93–137.

Williams, W.E. (1954) Diffraction by two parallel planes of finite length, *Proc. Camb. Phil. Soc.*, 50, pp. 309–318.

Williams, W.E. (1956) Diffraction by a cylinder of finite length, *Proc. Camb. Phil. Soc.*, 52, pp. 322–335.

Williams, W.E. (1959) Diffraction of an E-polarized plane wave by an imperfectly conducting wedge, *Proc. R. Soc. Lond., A*, 252, pp. 376–393.

Zverovich, E.I. (1971) Boundary value problems in the theory of analytic functions in Holder classes on Riemann surfaces, *Russian Math. Surveys*, 26, pp. 117–192.

Index